Astronomy and Power: How Worlds Are Structured

Proceedings of the SEAC 2010 Conference

Edited by

Michael A. Rappenglück
Barbara Rappenglück
Nicholas Campion
Fabio Silva

BAR International Series 2794

2016

First Published in 2016 by
British Archaeological Reports Ltd
United Kingdom

BAR International Series 2794

Astronomy and Power: How Worlds Are Structured

ISBN: 978 1 4073 1441 9

Cover Image:

Astrological divination board made of ivory, probably used for magical and medical purposes, from Grand (Dép. Alsace-Lorraine, France), 2nd c. AD.
Musée d'Archéologie Nationale, St. Germain-en-Laye (Yvelines), France, N# 83675.
© Michael A. Rappenglück 2009.

Printed in England

All BAR titles are available from:

British Archaeological Reports Ltd
Oxford
United Kingdom
Phone +44 (0)1865 310431
Fax +44 (0)1865 316916
Email: info@barpublishing.com
www.barpublishing.com

TABLE OF CONTENTS

PRIFYSGOL CYMRU
Y Drindod Dewi Sant
UNIVERSITY OF WALES
Trinity Saint David

This volume has been made possible with the aid of a generous grant from

The Sophia Centre for the Study of Cosmology in Culture,

School of Archaeology, History and Anthropology,

University of Wales Trinity Saint David

www.uwtsd.ac.uk/sophia

and

The Sophia Centre Press

www.sophiacentrepress.com

PREFACE

Throughout the course of history, from early prehistory to the Space Age, power structures have existed which have been more or less derived from, or correlated to, astronomical phenomena or certain cosmologies and cosmovisions. These have significantly affected and formed the economic, social, political, artistic and religious life of people across different cultures. Cosmographies, time reckoning and calendar systems, celestial navigation techniques, landscape and architectural models of cosmic potency, celestial divination and astrological ideas, cosmic clothing, and other related concepts have been used successfully by interest groups to establish, maintain, and expand psychological, social, religious and political power. Furthermore, the celestial sphere and its inhabitants have also been closely connected and partially interwoven with the concept of the manifestation of cosmic order and power both in nature and in culture.

Power can be defined as the capacity to effect, affect, and control entities and environments. Sometimes it is possible to measure its strength and effects. It is a kind of enforcement of its own status within a dynamic structure of other rival entities. There are several theories of power, rooted in philosophy, psychology, sociology, political science, economic science, ethnology, ethology, game theory and other disciplines. They deal with the sources, types, bases, balances and effects of power.

Places have certain powers related to their position within a certain spatial structure. Spheres of spatial influence are established by the surveying, mapping, planning and erection of dwellings, towns, socio-political and religious architecture, as well as by structuring the surrounding landscape. The positional power of authority, ancestry, sovereignty, advising, spiritual leadership and sacred kingship often found its legitimation by a heavenly mandate. The techniques of orientation and navigation are also a result of the need for spatiotemporal positioning.

To organize the world, it was essential to set and respect physical, psychological and social properties and boundaries. These allowed activities to be structured and directed, physical and psychological power to be focused, and therefore the establishment and protection of a steady human life. Such structured worlds—having hierarchically ordered power strata, as well as places of transition and exchange between them—served to establish and maintain individual and social power. They were based on properties of the wider environment, topographical conditions and celestial essentials.

To obtain and to exert power meant that the right person had to do the right thing at the right place and the right time. This established the roots and potency of astrology and alchemy, as well as several divination techniques. People also thought that heavenly prototypes of humans, animals, plants and inanimate objects, represented by celestial (including today's meteorological) entities, empowered and disempowered earthly beings and vice-versa. This idea might be at the origins of the concept of astral magic and was used to justify the authority of people who were believed to manage such cosmic powers. People respected and at the same time feared those spiritual and worldly rulers who had superior abilities to handle time reckoning and the needs of spatial orientation, whether of a quantitative or a qualitative nature. Timekeepers and chronologists based their special power on knowing how clocks and calendars worked, and how to align individual, natural and social life to them. This allowed for the functionalization of time as a basis for biological and socio-political domination. Moreover, the synchronization of the different natural and social cycles allowed for networking among different social groups, motivated the exchange of material artefacts and knowledge, and strengthened the power of individuals within and outwith the community. People thought it would be possible to gain mastery over time, especially mastery over their own future and fate. Another way of obtaining cosmic power was, supposedly, a more passive one: sky symbols or other peculiar celestial events frequently presaged and/or legitimized the election of specific people for a particular task: the chosen ones.

The power of transformation was related to concepts of fertility, production, destruction and renewal. Such concepts not only feature in the medieval alchemical writings, but also in the rites of passage and renewal of several different cultures. Another materialization of the power of transformation is in the transfiguration of mundane life into cosmic immortality, through the aid of the heavens above, such as appears in rituals of foundation and dedication, where cosmic powers are invoked to vivify dwellings, to establish order, and to animate the living space.

While reward power occasionally appears connected to astronomical phenomena and concepts, the astronomically related power of reference frequently showed up in architectural representations, costuming, regalia, coins and rituals.

Today, the natural sciences have rationalized the world and, with it, many of these notions seem to have disappeared. However, remnants of astronomical forms of power are still present in architecture, on flags, on advertising media, as military or police insignia, and in form of common newspaper horoscopes. The power of modern astronomy is also expressed in the achievements of science and engineering, especially aerospatial technology. The role of the space race during the Cold War can be understood as a powerplay for control of the heavens above and, with it, legitimacy to be the dominating world superpower. Also in the 20th and 21st centuries AD, certain interest-groups have exploited, and continue to exploit, archaeoastronomy and ethnoastronomy for their own interests. It might be better then to think of such astronomical forms of power has having not disappeared in the modern world but simply having found new ways to materialize themselves.

A total of sixty-four researchers from around the world attended the conference. During the meeting, fifty-three talks and posters were presented, with six invited lectures. Two public lectures were given: Clive Ruggles presented 'Heavenly Power in Worldly Hands: Ancient Sky Perceptions and Social' and Michael A. Rappenglück spoke on 'Archaische Kosmovisionen im Schamanismus (Archaic Cosmovisions in Shamanism)'. In addition, a round-table discussion on methodological issues within cultural astronomy, chaired by Stanislaw Iwaniszewski and Emilia Pasztor, and a public exhibition on the Antikythera Mechanism, curated by Xenophon Moussas, were offered.

A special objective of the conference was the discussion of cultural astronomy, presenting the interaction of different disciplines and methodologies. Thus the programme of scientific sessions and published papers covered numerous aspects of thisbroad topic. Starting with General Ideas and following a chronological sequence we have: Neolithic Cultures; Megalithic Cultures; Power of Calendar and Clocks; Chalcolithic, Bronze Age and Iron Age Cultures; Egypt and Minoan Culture; Cultures in Europe, Asia, Oceania and Africa; Cultures of North America, Mesoamerica and South America; Antiquity; Medieval Time in Orient and Occident; and finally Astronomy and Politics in the Modern Age. With few exceptions the forty-three papers succeeded in passing an in-depth peer review process, including revisons, to ensure a high-quality scientific standard for this volume.

We hope that the proceedings of SEAC 2010 will be a fruitful source for further scholarly work in the field of cultural astronomy.

The editors:

Michael A. Rappenglück
Barbara Rappenglück
Nicholas Campion
Fabio Silva

ACKNOWLEDGEMENTS

The conference was hosted and organized by the Adult Education Center and Observatory vhs Gilching. We thank the managing committee of the vhs (Fritz Wauthier, Oliver Kübrich, Hans-Dieter Moehring, Erika Spude). Moreover we are very much indebted to Karin Zauchner, principal of the James-Kruess elementary school, Gilching, close to the Adult Education Centre (vhs), who provided the conference room and the Aula for our use.

We gratefully acknowledge the logistic work and practice of the Local Organization Committee (LOC): Michael A. Rappenglück (vhs Gilching / INFIS, Germany), Barbara Rappenglück (INFIS, Germany), Eva-Maria Hackstein and Rainer Hackstein (vhs Gilching), Iris Janisch (vhs Gilching) Uschi and Oliver Kübrich (vhs Gilching), Karin Thiel (vhs Gilching), Thomas Rose (Astrogilde), Kirsten Weinfurtner (vhs Gilching), Michaela Frey und Julia Huber. The amateur astronomical group 'Astrogilde Gilching & Fürstenfeldbruck'provided a lot of assistence in organizing the conference: Herbert Birzele, Monika und Walter Gallinat, Hans and Karin Gnädig, Philip Hedrich, Max Kiermayer, Uschi Kösters, Thomas Rose and Günter Wagner.

The Scientific Organizing Team (SOC) actively supported the LOC. Members of the SOC were: Michael A. Rappenglück (SEAC secretary, vhs Gilching, INFIS, Germany)—Chairperson; Barbara Rappenglück (INFIS, Germany)—Chairperson; Gudrun Wolfschmidt (University of Hamburg, Germany)—Chairperson, Juan Belmonte Aviles (IAC & President of SEAC, Spain); Lis Brack-Bernsen (University of Regensburg, Germany); Nicholas Campion (University of Wales Lampeter, UK); Stanislaw Iwaniszewski (National Institute of Anthropology and History, Mexico); David Pankenier (University of Lehigh, USA); Emilia Pasztor (Hungary); Clive Ruggles (Leicester University, UK); Lionel Sims (University of East London, UK); and Ivan Sprajc (Scientific Research Centre of Slovenian Academy of Sciences and Arts, Slovenia).

We are much obliged to Uschi Kübrich for establishing and maintaining the SEAC 2010 website. We would particularly like to thank Xenophon Moussas, University of Athens, for presenting an excellent replica of the Antikythera Mechanism and an associated poster exhibition (translated into German for the first time by Michael A. Rappenglück). We are indebted for all those who provided private accommodation, which helped to minimize the costs for colleagues lacking enough funding to make their participation in the the conference possible. This was a great act of hospitality. The opening dinner was accompanied by the folk group 'Guichinger Brauchtum', who performed beautiful Bavarian traditional dances. Christian Schindler and Christine Bochtook took care of the catering that evening. Günter Wagner and his jazz band accompanied the closing festivities.

Without funding by sponsoring institutions the conference would not havebeen possible: the Adult Education Centre (vhs) Gilching (Germany); the Sophia Centre for the Study of Cosmology in Culture, School of Archaeology, History and Anthropology, University of Wales Trinity Saint David (UK); Baader Planetarium, Mammendorf (Germany); Sparkasse München-Starnberg (now München-Starnberg-Ebersberg, Germany); PC Markt Gilching (Germany); and the Mayor of the Municipality of Gilching (Germany).

We appreciate the peer reviewers for their advice and suggestions, which helped to ensure a very high scientific standard of the publication.

Finally we are deeply indebted to Fabio Silva, Nicholas Campion and their co-workers at the Sophia Centre for the Study of Cosmology in Culture, School of Archaeology, History and Anthropology, University of Wales Trinity Saint David (UK) and the Sophia Centre Press (UK) for their editing work, reviewing English grammar and style, converting the papers into a publishable product, and managing communications with BAR. Special thanks are due to Kate Namous for copy-editing skills. Their support was absolutely essential for achieving the publication of the SEAC 2010 proceedings at a time when the first editor Michael Rappenglück was burdened with health problems of himself and his family.

Michael A. Rappenglück
Barbara Rappenglück

LOCAL ORGANIZING COMMITTEE

Michael A. Rappenglück – VHS Gilching, INFIS, Germany
Barbara Rappenglück – INFIS, Germany
Eva-Maria Hackstein – VHS Gilching
Uschi Kuebrich – VHS Gilching
Thomas Rose –Astrogilde
Rainer Hackstein – VHS Gilching

SCIENTIFIC ORGANIZING COMMITTEE

Michael A. Rappenglück – VHS Gilching, INFIS, Germany
Barbara Rappenglück – INFIS, Germany
Gudrun Wolfschmidt – University of Hamburg, Germany
Juan Antonio Belmonte– IAC, Spain
Lis Brack-Bernsen – University of Regensburg, Germany
Nick Campion – University of Wales Trinity Saint David, UK
Stanislaw Iwaniszewski – National Institute of Anthropology and History, Mexico
David Pankenier – University of Lehigh, USA
Emilia Pasztor – Hungary
Clive Ruggles – Leicester University, UK
Lionel Sims – University of East London, UK
Ivan Šprajc – Scientific Research Centre of Slovenian Academy of Sciences and Arts, Slovenia

List of Participants

Adams, William Ben (USA)
Antonello, Elio (Italy)
Artelaris, Gerasimos
Belmonte, Juan Antonio (Spain)
Blomberg, Mary (Sweden)
Blomberg, Peter (Sweden)
Bulloni, Marco (Italy)
Campion, Nick (UK)
Chapman-Rietschi, Peter (Switzerland)
Ciancia, Guiseppe (Italy)
Chukwuezi, Barth K. (Nigeria)
Dmitrieva, Nataliya (Russia)
Draxler, Sonja (Austria)
Engfer, Roland (Switzerland)
Fisher, David (UK)
Frank, Roslyn (USA)
Galindo Trejo, Jesús
García, A. César Gonzáles (Spain)
Gómez, Cecilia Paula (Argentina)
Gropp, Harald (Germany)
Hadžibegović, Zalkida (Bosnia-Herzegovina)
Henriksson, Göran (Sweden)
Henty, Liz (UK)
Holbrook, Jarita C. (USA)
Incerti, Manuela (Italy)
Iwaniszewski, Stanislaw (Poland, Mexico)
Kerek, Franz (Germany)
Kõiva, Mare (Estonia)
Koleva, Vesselina Petrova (Bulgaria)
Krauss, Rolf (Germany)
Kuperjanov, Andres (Estonia)
Leihmlehner, Astrid (Austria)
Lippitsch, Max E. (Austria)
Liritzis, Ioannis (Greece)
López, AlejandroMartín (Argentina)
Martin, Jean-Pierre (France)
Masse, W. Bruce (USA)
Mosenkis, Iurii Leonidovich (Ukraine)
Mosoia, Cătălin (Romania)
Moussas, Xenophon (Greece)
Mozaffari, Seyyed Mohammad (Iran)
Pankenier, David (USA)
Pasztor, Emilia (Hungary)
Pérez-Gutiérrez, Manuel (Spain)
Pimenta, Fernando (Spain)
Pereverziev, Dmytro (Ukraine)
Polcaro, Vito Francesco (Italy)
Ranieri, Marcello (Italy)
Rappenglück, Barbara (Germany)
Rappenglück, Michael A. (Germany)
Ridderstad, Marianna (Finland)
Rothwangl, Sepp (Austria)
Ruggles, Clive (UK)
Saletta, Morgan (Australia)
Schmidl, Petra G. (Germany)
Silva, Fabio (UK)
Šprajc, Ivan (Slovenia)
Stanescu, Florin (Romania)
Szücs-Csillik, Iharka (Romania)
Tauber, Heidi (Germany)
Vafea, Flora (Greece)
Vickers, Doris (Austria)
Vodolazhskaya, Larisa N. (Russia)
Wolfschmidt, Gudrun (Germany)
Zito, Richard R. (USA)
Zotti, Georg (Austria)

IMPRESSIONS OF THE CONFERENCE

Group photo: Particpants of SEAC 2010 in Gilching. © vhs Gilching, 2010.

Arrival: Welcome of participants in the lobby of the adult education center Gilching. ©vhs Gilching, 2010.

Opening evening: The folk dance group 'Guichinger Brauchtum', internationally well-known, perfoms traditional Bavarian dances. © vhs Gilching, 2010.

In the yard: Relaxation combined with vivid discussions during the breaks. © vhs Gilching, 2010.

The public exhibition of a replica of the Antikythera mechanism (3rd / 2nd c. BC), presented by Professor Xenophon Moussas got a lot of interest. © vhs Gilching, 2010.

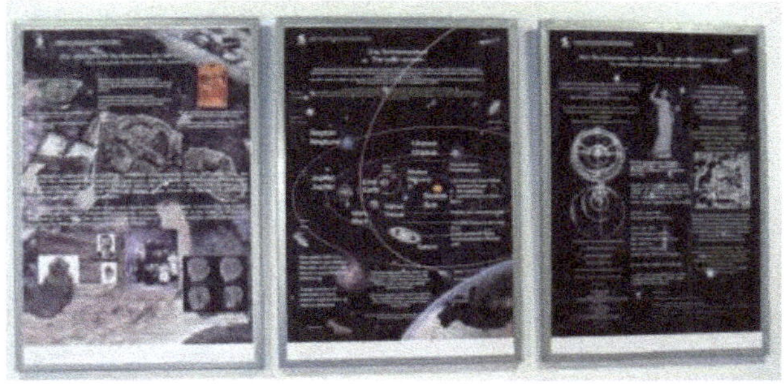

The exhibition accompanying the presentation of the Antikythera Mechanism. © vhs Gilching, 2010

In front of the Hohle Fels cave, Swabian Alps, Schelklingen, Baden Wurttemberg, Germany, yealding famous Palaeolithic artwork (40000-35000 years ago). © vhs Gilching, 2010.

Half day excursion to Munich. The group in front of the famous Munich Residenz, the royal palace of the Bavarian monarchs of the House of Wittelsbach.
© vhs Gilching, 2010.

At the entrance to the Hohle Fels cave.
© vhs Gilching, 2010.

St. Afra chapel, Schelklingen, Baden-Württemberg, Germany, housing famous frescoes from the Romanesque period (14th c.). © vhs Gilching, 2010.

The figurine of a Paleolithic pin-up girl from the Aurignacian period (40000-35000 years ago) was found in the Hohle Fels cave.
© vhs Gilching, 2010.

The famous 'Blautopf' in Blaubeuren, a very deep carst source linked to a giant cave system of the Blau and archaic Danubian river. © vhs Gilching, 2010.

GENERAL IDEAS

KEEPERS OF TIME AND GUARDIANS OF SPACE – SOME BASIC CONCEPTS OF ASTRONOMY AND POWER

MICHAEL A. RAPPENGLÜCK

Abstract: Starting with its formation in early prehistory, astronomy was closely connected and partially interwoven with the concept of the manifestation of cosmic order and power in nature and culture. The reduction of instincts and the appearance of an advanced cognitive ability of self-reflection and of knowledge of the world (the so-called subject-object division) caused human needs to establish and maintain order, rhythm, communication and power in personal and social life, with respect to given ecosystems. Therefore man was forced to think about his own and the world's whys and wherefores. Cultural systems superseded certain biological guidelines and helped to organize the world into a meaningful system of related parts, integrating and orientating man within the changes of nature. As a part of this, concepts of cosmovisions and power, correlated to each other, were developed. The interdisciplinary approach summarizes and categorizes the main types of power concepts existing in different cultures across the world, throughout the epochs and corresponding to astronomical phenomena as well as archaic cosmovisions.
Keywords: Power, Astronomy, Power of Space, Power of Time, Anthropology

Definition of Power

Science describes and analyzes power as the capacity to effect, affect, and control entities and environments (Foucault, 1980). Sometimes it is possible to measure its strength and effects (Hinkin and Schriesheim, 1989). It is a kind of enforcement of its own status within a dynamic structure of other rival entities. There exist several theories of power, rooted in philosophy, psychology, sociology, political science, economic science, ethnology, ethology, game theory, and other disciplines (French, 1956; Dahl, 1957; Hosking and Morley, 1991; Steven, 2005). They deal with the sources, types, bases, balances and effects of power (French and Raven, 1959; Morgan, 1986; Raven, 1992; Mallory, Segal-Horn and Lovitt, 2002; Raven, 2004).

The World as a Competition of Powers

In ancient times, people conceived the world as a spatiotemporal domain of interacting powers mostly thought to be individual and collective beings. Cultures of indigenous people worldwide and across time understood their world as animistic and totemic (Quaritch Wales, 1959; Elkin, 1969; Kuper, 1973; Kolig, 1988; Silverman, 1996; Rappenglück, 1999; Glowczewski, 1999; Harrison-Buck, 2012). However, sometimes people also thought that impersonal principles, e.g., Ṛta/Dharma (ancient India), Maat (ancient Egypt), or logos (ancient Greece) triggered the behaviour of nature (Broadie and Macdonald, 1978; Premnath, 1994; Horsch and Whitaker, 2004). The growing influence of science coined by Greek natural philosophers promoted this concept (Vlastos, 1947; Guthrie, 1952).

Among the interacting powers of nature, the bodies and certain appearances of the celestial sphere played an outstanding and important role: people thought that they set up, animated and influenced permanently the spatiotemporal world, in particular living beings and man (Rappenglück, 2009). Hence, ancient people felt compelled to watch very carefully the phenomena between heaven and Earth, because they wanted to know what place and time was best suited for organizing human individual, biological, economic, socio-political and religious life (Stanner, 1963; Tannenbaum, 1984; Rappenglück, 2009, 2013). Knowledge and skills to discover the natural order, to structure and rate the world by measuring time, space and matter, and to associate with spiritual entities, gave rise to the expert power of the sage, magician, shaman, medicine man and later to the scientist (Rappenglück, 1999, 2009; Ryan, 1999). First man tried qualifying spatiotemporal events. To get and to exert power meant to do the right thing at the right place and the right time by the right person. This established the roots and potency of astrology and alchemy, as well as several divination techniques. The idea of terrestrial phenomena, here the functions of the human body, being related to heavenly ones is expressed, for example, by the mantic discipline of iatromathematics (Barton 1994, 179-180; Lancellotti 2001, 438-443). People also thought that the heavenly prototypes of humans, animals, plants and inanimate objects, represented by celestial (including today's meteorological) entities, empowered and disempowered earthly beings (Rappenglück, 1999, 2009). Conversely, they took the view that an influence of earthly creatures onto the heavenly archetypes was possible, too. This founded the concept of astral magic and justified the authority of people who seemed to manage cosmic powers.

Keepers of Time: Empowered by Time–Positional Power of Time

Already in the parietal and mobile art from earlier prehistory different phenologically designed and astronomical related natural calendars existed which used a system of images and signs to denote certain periods of biological time together with astronomical time (Rappenglück, 1999, 2010, 2014b, 2014c). People watched the Moon's phases (synodic month), its celestial orbit related to stars (sidereal month), its local change on the horizon, and its course over several lunar years (Rappenglück, 2010, 2014b, 2014c).

Apart from lunar time reckoning, people used solar and lunisolar time-factored notations, too. Among mobile objects from the Aurignacian to the Azilian, several tally sticks and tally pebbles (Azilian) illustrate different kinds of time reckoning (Rappenglück, 2010, 2014c). Often, combinations of the astronomical periods with biological rhythms of certain animals and/or the human female (menstruation, pregnancy) exist. This kind of representation could be designated best of all with the term 'Palaeolithic Almanac' (Rappenglück, 2008, 2010, 2014b, 2014c), which combines calendar dates, a calendar and important information of the respective ecosystems. Furthermore, in some very complicated cases, especially if longer periods are noted, a clearly structured counting–a so-called 'arithmetic' notation–of astronomically significant time units has been used (Rappenglück, 2010, 2014c). There is evidence for the knowledge of certain asterisms (today's Northern Crown, Pleiades, Hyades, Cygnus and others) during the Upper Palaeolithic, which were important for time reckoning or well suited for orientation purposes (Rappenglück, 2010, 2014b). Moreover, the use of simple measuring instruments for astronomical purposes (protractors, rods and ropes for measurement, plumb bobs and gnomons) is evident (Rappenglück, 2014a). Already in that bygone time, the first attempts to quantify times and to determine the most exact date of an event occurred (Rappenglück, 2010, 2014b, 2014c). People respected and at the same time feared those spiritual and worldly rulers who had superior abilities to handle qualitative and quantitative time reckoning and needs of spatial orientation (Rappenglück, 2009).

Timekeepers and chronologists based their special power on many abilities (Krupp, 1983, 1997; Aveni, 2000; Kelley and Milone, 2011; Stern, 2012; Hill Boone, 2013): They established time reckoning by natural and artificial clocks. They kept and archived time by ancestral traditions, almanacs and calendars, structuring time, fixing critical dates and evaluating unusual events (eclipses, appearance of comets, meteor showers, supernovae and novae). They glimpsed at the future by fortune telling and divination with respect to the regular movements of celestial bodies. People supposed that there is a correlation of astronomical, biological and sociological time. Knowing how clocks and calendars worked, and how to align individual and social life to them, vested a few persons with power over the others. People thought it would be possible to gain mastery about time, especially about the future and fate. However, frequently they were also afraid of the undeterminable and opaque effects of destiny. Masters of magic and the mantic are welcome by people at all times. Today some of the ancient ideas are continued or refreshed, for example in today's astrology, sometimes in organic farming, e.g., the anthroposophical based biodynamic agriculture (Haverkort, Hooft and Hiemstra, 2003; Smith, 2009), and in the approaches of chronobiology (Dunlap, Loros and Decoursey, 2004; Zürcher and Schlaepfer, 2014).

The attempt to control time as accurately as possible led to the development of certain tools for time reckoning: certain simple clocks, complex time machines and calendars (Krupp, 1983, 1997; Aveni, 2000; Kelley and Milone, 2011; Stern, 2012). They allowed step-by-step to functionalize time as a basis for biological and socio-political domination. Moreover, the synchronization of different time cycles in nature and culture networked societies, pushed forward the exchange of material and knowledge, and strengthened the power of individuals and the community.

Finally, people were very interested in arranging themselves in the ancestral line referring to their cosmic origin (Eliade, 1971). This allowed them to re-connect their own time with cosmic time and especially with the primordial point of creation.

Guardians of Space: Empowered by Space–Positional Power of Space

Positional power establishes spheres of spatial influence by the surveying, mapping, planning and foundation of dwellings, towns and socio-political and religious architecture, and by structuring the landscape and architecture (Rappenglück, 2005, 2009, 2013). Moreover, it describes and evaluates positions and relations between locations in space. Ancient cultures evaluated positions in space according to the idea of quality, quite similar to the concept of different qualities of times. Places have certain powers related to their position within a certain spatial structure (Rappenglück, 2005, 2009, 2013).

Cosmographical concepts often legitimated positional power (Rappenglück, 2005, 2009, 2013). These are, for example, the zenithal and polar centricity of the world axis; cardinality; the hierarchy of cosmic strata, including ideas of ascent and descent; the linking functionality of the Milky Way (Rappenglück, 2000; Rappenglück, 2009; Rappenglück 2014d, 298-299) and the rainbow; and the orbits of the sky and the celestial bodies. The positional power of authority, ancestry, sovereignty, advising, spiritual leadership or sacred kingship often got its legitimation by the mandate of the heavens (Pankenier, 1995, 1998; Krupp, 1997; Allan, 2007).

Finally, the techniques of orientation and navigation are a result of the need for spatiotemporal positioning. They also allowed the expanding, structuring and controlling of orbits (Rappenglück, 2009, 2015).

Spatiotemporal Centricity

Since the Lower Palaeolithic epoch, the producing and keeping of fire has been a very important human activity (Rappenglück, 2009). Sitting around a fireplace helped to protect against the attacks of wild animals. It was necessary to keep the right distance from the fire to profit from its power and at the same time to avoid being burned. This led to a natural domain, in which activities are concentrated and distributed around a fireplace. There exist three distances which define the human intercourse with fire: looking from the centre to the periphery there is a first inner domain where one can put things into the fire for transformation, such as cooking, burning clay and destruction purposes. A second sphere of activity has its borderline where one is not able anymore to feel the heat dissipation. Finally, the area where one is not able to see the light emission defines the outer domain around a fireplace. Setting up a fireplace leads to the experience of centricity, centripedality and centrifugality and causes the distribution of activities around the fire according to an ordinary scale. Examples handed down by people worldwide show that producing, taming and using fire means to have great power and to control acts of production, transformation and destruction (Rappenglück, 2009). The fireplace and the hearth, according to the traditions of ancient cultures and archaeological records, symbolize the act of creation, which inherently contains birth and death, and is related to the origin of both, to individuals and the community, but also to the whole world. Often the hearth symbolizes the sexual union of man and woman or the male and the female powers, which set up and move the whole world. People thought that making fire by drilling a stick into a wooden plate imitates the creative coition and the origin of both the world and man. That explains the importance of fire-rituals across different cultures. The fire-drilling procedure, however, was recognized also in looking at the rotation of the Sun and the whole sky around the polar axis, modelled by the shadow-stick as a phallus upon the plane of the ground, which resembles a vulva (Rappenglück, 2009). According to ancient conceptions, the drilling world-axis and the rotating sun, both symbols of heaven, exerted their power upon the susceptible earth, producing and preserving the life in the world. Thus, it is understandable why people thought that the fireplace represents a creative womb, which they compared and related to the Sun as symbol for the celestial source of power in the world. The close relation of the female principal to the fireplace/hearth is reflected in the performance of special fire rituals carried out by women within the domestic space (Rappenglück, 2009). The ancient Greek ideas relating the goddess Hestia to fire, the celestial pole, the Earth and the hearth, belong to the same topic (Merkelbach, 1980, 81, 89). A derivative of such ideas is the identification of the fireplace or hearth with both the navel of the human body and the centre of the world, which denote the origin of transformational physical and spiritual power. Thus according to ancient people the place at which a fire is kindled offers the possibility to get in contact with the primordial origin of the cosmos and to communicate with the ancestors, who are present there (Rappenglück, 2009). The fireplace and hearth anchored a family or a clan at a place, giving life, protection, stability and unity, and ensuring linkage and communication with the ancestral domain or even the powerful centre of the world itself. The position of fire-altars in the dwellings, close to the centre and the main support, if one exists, follows a similar symbolism (Rappenglück, 2009). A late echo of the idea of the central fire appears in the Pythagorean worldview, allowing both a geocentric and a heliocentric interpretation of that concept (Richardson, 1926; Wiersma, 1941). The heliocentric system illustrates another version of a fire, which places the Sun in the centre of the universe (Bosworth Burch, 1954). The Sun empowering the ruler was an important concept in some cultures (Hombert, 1945; El-Khachab, 1961, 1971; Loeb and Klein, 1962; Hoffmann, 1963; Kirkland, 1997; Gury 2000, 580, 582; Fischer, 2002; Marlowe, 2006; Cizek, 2006). The centred Sun within the planetary system was so attractive that it served as an image of the wise ruler, who is named 'sol' in the 'The City of the Sun' (*La città Del Sole*, 1602 / *Civitas solis*, 1623) (Ernst and Firpo, 1997), a Utopia written by Fra' Tommaso Campanella (1568-1639). Ludwig XVI of France (1638-1715), who was named the 'Solar King' (*Le Roi-Soleil*), gives another example for solar political ideology (Reichardt and Cohen 1998, 97-101).

Ancient people thought that the centre is storage of creation and pure dynamic intensity, a point at which opposing forces and domains, such as life and death, female and male, the material and the spiritual world, are meeting and intersecting with each other, coexisting and balanced in harmony (Rappenglück, 2005, 2009). People considered such places, where one can encounter the original cosmic power, sacred (Rappenglück, 2005, 2009). The scope of the vigour defines a boundary, where the sacred area changes into a profane. People considered these sacred places charged with energy, which made contact dangerous (Rappenglück, 2005, 2009). Therefore, they respected the particularities. Sensitive persons, for example shamans and rulers, tried to reside at the middle of the world, to access cosmic power and knowledge for their purposes. Ordinary people also hoped to participate in the forces of the centre and therefore settled closely nearby. Only at the centre of the world is it possible to get in contact with creative power and knowledge (Nibley, 1951; Rappenglück, 2005, 2009; Mabbett 1983, 77, 83). Therefore, it is quite understandable why some ancient cultures compared the original point of space and time with the 'navel' of the world (*Omphalos motif*; Roscher, 1913, 1915, 1918; Gaerte, 1914; Herrmann, 1959; Terrien, 1970). Natural objects, like a mountain, a cave, a well, a big stone, a tree and also artificial ones, megaliths or buildings for example, all having special properties, served to mark the centre within a human sphere (Rappenglück, 2009).

Organizing the World around a Centre

Human self awareness defines a topocentric system of observation and order (Müller, 1978; Rappenglück, 2009). Thus, the addiction to centre the world is an anthropological *a priori*. It is quite understandable why ancient cosmovisions tend to be preferably geocentric. Cosmogonical myths often illustrate how, starting from the point of the origin, the cosmic realms, directions and strata are unfolded and arranged around it.

Membranes, Structured Worlds, and Passages

To organize the world, it was essential to set and respect physic, psychic and social boundaries. These allowed activities to be structured and directed, physical and mental power to be concentrated, and the establishment and protection of a steady state of human life (Rappenglück, 2009, 2013). In cosmogonies, creation frequently started with a first separation of sky and earth from a pre-existing singularity, which people often imagined as an egg, as a primordial chaotic substance–mostly water–or as a kind of living being. In any case, the original entity is somehow divided into two parts, which sets up a first polarity. Their interaction generates the power figures of the world in an iterative process of binary divisions, creating interplay of further dichotomies. The first splitting-off, which started creation, is compared to a potent primordial sacrifice of a giant cosmic living entity, e.g., a turtle, a shell, a cow, a human, and others. Ancient people tried to repeat the process of primeval creation ritually at every founding of habitations and sacred areas (Rappenglück, 2009, 2013).

This was intended to involve them in the cosmogonical strength of creation.

The concept of a semi-permeable membrane represented by rock faces in caves, walls in artificial buildings and non-domestic structures, such as stone circles, offered humans spatiotemporal enclosures (Rappenglück, 2009, 2013). Setting an enclosure founded a first partition, which separates an outer from an inner living space (exosphere/endosphere), the wild and the domestic sphere, proximity and distance, kinship and foreign parts, the sacred and the profane, the male and female area, and other dichotomies. In the case of a cave the two realms are defined by a dichotomy and polarity of experiences: there exists a world of light (day) and another one of darkness (night), an area of high and low sensual perception, an ordinary and a miraculous space-time, and also a profane and sacral sphere, related to each other (Rappenglück, 2009, 2013). The openings in the membrane and the partition within the enclosure, a cave or a dwelling for example, follow sets of binaries (Rappenglück, 2009, 2013). They divide the spatiotemporal landscape and the whole range of objects, living beings, events, properties, social relationships, myths and rituals and spiritual powers of the world into paired opposites (Rappenglück, 2009, 2013).

Places of transition and of exchange are given by openings in the membrane, such as the entrance to a cave, a door, or a window in a structure (Rappenglück, 2009, 2013). According to very old concepts, a cave enables access to an interior landscape behind the ordinary world (Rappenglück, 2009, 2013). Concerning some artificial buildings, the smoke outlet of a tent or a house, for example, opened the membrane of the roof to align the dwelling to the highest point in the sky and at the same time to make the structure in principle susceptible for celestial powers, such as light, wind or rain (Rappenglück, 2009, 2013). There the spatiotemporal organization of the building gets transparent and other strata of the world, equipped with their own vigour, are accessible.

There is archaeological evidence of order and orientation patterns concerning the design of monuments, such as dwellings and tombs, or the location of special 'sacred' places (e.g., rocks, hills, mountains, sources, lakes, rivers, coastline and caves), the layout of early villages or cities and the existence of sacred geographies, in which monuments are embedded and related to each other (Rappenglück, 2009). Examples come from cultures all over the world and across all epochs. Frequently, order, distribution of power and orientation is given by particular 'alignments' with focus on one-dimensional sequencing and directionality. In addition one can find domains in which ordering and intensification of certain objects within a special flat space follows a net-like reference system, or it is done around a centre, often according to an ordinary scale of weighting the arrangement with reference to the middle (Rappenglück, 2009). Another set of spatial order and orientation is indicated by divisions of a domain along a binary iterated scheme (Rappenglück, 2009). These concepts allowed a grading and distribution of power in societies (Rappenglück, 2009).

In the case of buildings, often a post or pillar completely or partially shows a quadrangular shape, which symbolizes not only vertical, but also horizontal, axiality (Rappenglück, 2009, 2013). It indicates the cardinal directions to the important spatiotemporal domains, containing grades of powers, objects, different qualities, and ancestors in the world, and aligns the parts of the enclosed human habitat to them (Rappenglück, 2009, 2013). A similar symbolism is associated with certain objects, such as remains of plants, animals, humans, food, masks and other things, which, charged with mythological meaning, are located at special places close to the main support or fixed upon its surface, if they follow the principle of cardinality (Rappenglück, 2009, 2013). Additional supports, occasionally equipped with special masks or depictions and put at the corners of a structure, hold up the world at the cardinal points (Rappenglück, 2009, 2013). Frequently they are associated with the main powers, world quarters, colours, plants, animals, gods, the ancestors. These powerful supports carry the world and act as guardians, are related to seasonality, ensure the cycle of life and death, and protect the descendants in the living community (Rappenglück, 2009, 2013).

The setting-out of creation from the centre into the space implies a first world-axis, which is given by the human sense for gravity and fixed as the vertical line between zenith, location and nadir (Rappenglück, 2005). A second, but much more important world axis, is given by aligning the location to the celestial pole, because this was considered to embody the energetic origin of movements in the world (Rappenglück, 2005). Both world axes fundamentally define a tripartite cosmos, consisting of the Earth in the centre, and the polarity of the lower and the upper worlds (Rappenglück, 2005). This way the idea of a spectrum of increasing or decreasing cosmic power, flowing up and down along the world-axis, and the movements and interactions between a spiritual and a material realm, was illustrated.

Hierarchical Cosmic Power Strata

The recognition and ordinary scaling of the third dimension was an important factor of order and orientation. There are examples concerning the importance of vertically-arranged strata related to caves, hills, mountains and artificial structures (Rappenglück, 2009, 2013). The partitions of the vertical and horizontal strata followed social, psychical, religious and natural properties. The latter were founded in the ecosystem and topographical and astronomical conditions. The spatiotemporal courses of the Sun, Moon and the stars especially helped to perceive directionality and rhythm of life in cultures worldwide. Thus it is understandable why astronomy can be found reflected in the human organization of landscape and buildings.

Considering architecture, the main, strong support of a structure, a dwelling or a temple, for instance, frequently was identified with the axis mundi (Rappenglück, 2009,

2013). Occasionally structures have been set upon one single post to symbolize the world turning around the polar world-axis (Rappenglück, 2009, 2013). In addition, ancient people thought the column of smoke, embodying spiritual power and rising up from the fire-place/altar, to be a symbol of verticality, joining and relating to each other the cosmic strata between heaven and earth (Rappenglück, 2009, 2013). The world-axis holds up and connects, but also pierces the hierarchically layered cosmos. These were considered to be 'other worlds' and often are equated with different physical and psychical levels (Rappenglück, 2009, 2013). Representations of certain cosmic landscapes and the plants, animals, humans or gods living there signify the respective strata (Rappenglück, 2009, 2013).

A tree with its branches, a mountain having different levels, a terraced artificial building or the central post in a dwelling, shaped as a ladder with its steps, served as other models of the hierarchical stratified cosmos (Rappenglück, 2009, 2013). However, while the latter only shows the layering, the former by their conical shape pinpoint a highest possible level in the world. Sometimes ancient people mirrored the above mentioned objects vertically to illustrate the structure of the underworld (Rappenglück, 2009, 2013). Thus, they occasionally created the motif of the reversed world tree or the terraced world building to pinpoint the deepest possible level, too. Since Palaeolithic times different levels in caves, blind ends of rearmost galleries or deep shafts also have been attributed to the layered underworld.

Access to Cosmic Power Levels–World Tree, Ladder, Mountain

In any case, ancient people took the world axis as a way to reach and to travel through the hierarchical layered but coexisting realms of the universe (Rappenglück, 1999, 2005, 2009, 2012). At locations which were recognized as the centre and the place of the cosmic axis, sensitive persons, for example shamans, fell in ecstasy, rulers were enthroned, jurisdiction took place, sacrifices were carried out, and meeting places, houses, shrines, temples, cities, or states were founded (Eisler 1910, 35-36; Herzfeld, 1920; Heine-Geldern, 1942; Rappenglück, 2005, 2009, 2013; Palmer 1991, 84). A nice example for these ideas comes from ancient China: 'The House of Man', the forerunner of the later 'House of Calendar', was created as an underground cavern, in which a central sacred pole, a substitute for the world axis and the king's highway, was set up vertically (Rappenglück 1999, 276-277). A prince who wanted to mount the throne of the emperor had to undergo a special examination. He had to climb the mast to reach the ceiling of the grotto, which represents the sky's roof and was named 'The Bell of the Heaven'. There he was allowed to suckle the nipples of the heavens, the stalactites of the cave roof, which are the breasts of the cosmic mother. Thus, he proved his heavenly lineage and his ability to connect the sky and Earth to each other. From then on his body's size served to determine the length of a standard gnomon and also of an important musical instrument, the whistle, constructed according a standard measure. In addition, thrones were built directly above the centre or navel of the universe, set upon the world axis or made of the world tree's wood (Eisler 1910, 36, 76; Weitzmann, 1973, 1974). Moreover, sometimes people thought that the secular lords owned the power of the world axis, symbolized by the polar star or at least an asterism close to it (Rappenglück 1999, 98-100, 154, 241, 255, 276-277; Kanas 2009, 20; Baumann, 2013). In 1813, even Napoleon Bonaparte was regarded to be a 'polar star' (Stephens and George, 1870; BM Satires 12112; 'Napoleon le Grande'). The king's subjects should be related to the ruler and revolving around the centre of power like the circumpolar stars (Allan, 2007). Social class systems ordered around a monarchic ruler mirrored the idea of a hierarchical ordered natural world around a central axis (Nibley, 1951; Meyer, 1996/1997; Baumann, 2013). Finally, in several cases people considered the shaman's or ruler's power stick or sceptre, indicating sovereignty and frequently crowned by the bird of the sky, as a handy shaped world axis (Rappenglück 1999, 241; Rappenglück, 2005). This idea seems to be rooted in the Upper Palaeolithic (Magdalenian; 17-12 ka BP) epoch (Rappenglück, 1999).

Getting Cosmic Power

From what has been said it is quite understandable why people tried getting, embodying, and distributing cosmic power. According to archaic concepts, there are two kinds of participating in universal energies: if one knows the regularities, one can force the sky phenomena to act according to his own wishes. This idea gave rise to different versions of astral magic (Gundel and Gundel, 1933; Gundel, 1981; Reiner, 1995; Schwartz, 2003). Interestingly the development of natural sciences is rooted in the principle idea of magic, recognizing the world as a network of interactions which man can manipulate in some way (Thorndike, 1923-1958). The idea that the universe has chosen one for a particular task is the base of the other possibility for getting cosmic power. Symbols of the sky and peculiar celestial events frequently presaged and legitimated the specific election (Krupp 1997, 153-243).

In addition, the power of transformation describes another kind of strength dedicated to influencing and managing mind and matter for empowering and vivifying the individual human being, the society and the natural living space (Rappenglück, 2009, 2013). It is related to concepts of fertilizing, production, change, destruction and renewal, which manifest themselves in cosmic elements like earth, water, air, fire and light and in the life cycles of plants, animals and humans.

Rituals of founding, renewal and dedication served to vivify the habitation area by cosmic powers, to establish order and to animate the living space (Rappenglück, 2009, 2013). The landscape or the building is considered a living being, which in some cases is designed as an enormous animal (turtle, cow, shell and others) or a giant (Rappenglück, 2009, 2013). People often identified the latter with the world mother or father. The peculiarities of the territory or the construction elements of the building correspond to the body parts of the cosmic creature. The terrestrial and the subterranean landscape are loaded with signs of female and male sexuality indicating fertilizing but also destructive power. Ancient people especially equated the structure of the female genitals with the topography of a cavern. Both the cave portal and the vulva permit the passage between two worlds, an outer and an inner one, and vice versa. Moreover, at another level of understanding, the entry into a cavern and the passage through the inner space-time of the world was interpreted as a voyage to the generative interior of an enormous female, representing the Great Mother of the World. The cosmos and the cave, the outer and the inner structure of the world, run parallel to the physical body and the psychonoetic essence of a human being. The cave was thought to be a kind of cosmic vessel, later substituted by sacred temples or alchemistic furnaces, in which the primordial elements being in a state of chaos are collected, mixed and transformed into objects and creatures, so that a cosmos is formed. It acted as a giant placenta, which conceived and fed the new embryonic beings (Rappenglück, 2009, 2013). These are the celestial bodies (Sun, Moon, and stars), plants, animals, humans and sometimes gods. The transmutation consisted of the two main processes which configure the diversity of the world: creation and destruction, or at the level of living beings, birth and death.

These conceptions motivated people to tame, to treat with care, to purify periodically, and to feed the territory or the building as a whole or in its parts (Rappenglück, 2009, 2013). Ritual processions related to the landscape and the structures, time-factored by the biotope or astronomical phenomena, served to animate and empower the living space and to renew cosmogony. Moreover, ancient people thought that features of the landscape and the framework of the building were closely related to the ancestors, who founded, guided, and protected the clan through time and space (Müller, 1978; Rappenglück, 2009, 2013). Going to special sacred places at sacred times was to get in contact with the original power and to chain oneself to the archetypical pattern, displayed by the ancestors, who influence and dominate the world.

In some rare cases, reward power is connected to astronomy. Meteorites falling down from heaven to Earth give one example: the owner of the object was thought to have an invaluable part of the sky, which vested him with special power (Rappenglück, 'Astral high fashion clothing', this volume).

A special motif of the power of transformation was to transfigure mundane life into cosmic immortality by the heavens' support. The concept of transformational power links alchemistical concepts with astronomical ones (Heinze, 1981; Rappenglück 2007, 66; Eliade 2009, 233-244).

Power of Reference

The personification of celestial power or the participation in astral might illustrates the power of reference. It appears, for example, in architectural representations, costuming, regalia, coins and certain rituals (Eisler, 1910; Krupp, 1997; Joost-Gaugier, 1998; Strong, 1916; Gariboldi, 2004; Hannah, Magli and Palmieri, 2013; Rappenglück, 'Astral high fashion clothing', this volume). Secular and spiritual rulers (like gods) exhibited and legitimated their ownership of cosmic power, or at least their claim for it, by special cosmic clothing and power instruments like swords, shields, drums and other things (Eisler, 1910; Schafer, 1977; Hardie, 1985; Chevalier and Gheerbrant 1996, 874-876, 900-901, 959-960; Krupp, 1997; Rappenglück, 'Astral high fashion clothing', this volume; Taragan, 2003; Maeder, 2007; Bechtold, 2011). The rulers themselves or their devotees sometimes set exceptional asterisms, which today are mostly obsolete, in the sky (Kanas 2009, 128-130). In any case, interested parties intended to show heavens' mandate quite plainly.

Power and Astronomy–Some Elements Have Survived

Though many of the old and mostly all of the archaic concepts connecting astronomy and power have vanished because of the rationalizing work of modern natural sciences, some of them–weakened and hidden–are still present. The power of modern astronomy is expressed in the enlargement and enhancement of instruments, including the big machines of particle physics, e.g., the LHC. The extension of research fields into the planetary system by space flight is another variant of modern astronomical power. In future, it may be connected anew with claims of political and economic rulers. Remnants of astronomical power, however, are still alive in architecture, on flags, on advertising media, as military or police insignia, and in form of common newspaper horoscopes.

However, globally growing light pollution causes the sky view to be deprived of power. Most notably for the nonprofessional, especially children, is the loss of being impressed by a beautiful star-spangled sky with a broad, shining Milky Way. Nevertheless, professionals also are setting out in retreat in outlands to seek a dark night sky for their powerful terrestrial instruments. There are only a few possibilities left to get inspired by the heavens' power and to feel connected with natural strengths. This should engage us to preserve this original power of astronomy. It is also a world heritage.

References

Allan, S. 2007. On the Identity of Shang Di 上帝 and the Origin of the Concept of a Celestial Mandate (Tiang Ming 天命). Early China 31, 1-46.

Allen, N. J. 1991. Some gods of Pre-Islamic Nuristan. Revue de l'histoire des religions 208, no. 2, 141-168.

Aveni, A. 2000. Empires of Time. Calendars, Clocks, and Cultures. London and New York, Tauris Park Paperbacks.

Barton, T. S. 1994. Power and Knowledge. Astrology, Physiognomies, and Medicine under the Roman Empire. Ann Arbor, University of Michigan Press.

Baumann, B. 2013. By the Power of Eternal Heaven: The Meaning of Tenggeri to the Government of the Pre-Buddhist Mongols. Extrême-Orient Extrême-Occident 35, 233-284.

Bechtold, C. 2011. Gott und Gestirn als Präsenzformen des toten Kaisers: Apotheose und Katasterismos in der politischen Kommunikation der römischen Kaiserzeit und ihre Anknüpfungspunkte im Hellenismus. Göttingen, V & R Unipress.

Bosworth Burch, G. 1954. The Counter-Earth. Osiris 11, 267-294.

Broadie, A., and Macdonald, J. 1978. The concept of cosmic order in ancient Egypt in Dynastic and Roman Times. L'Antiquité Classique 47, no. 1, 106-128.

Chevalier, Jean, and Gheerbrant, Alain. 1996. Dictionary of Symbols, trans. John Buchanan-Brown. London, The Penguin.

Cizek, E. 2006. L'esprit militant des stoïciens et le premier État communiste de l'histoire. Latomus 65, no. 1, 49-61.

Dahl, R. 1957. The Concept of Power. Behavioral Science 2 , no. 3, 201-215.

Dunlap, J. C., Loros, J. J., and DeCoursey, P. J. 2004. Chronobiology: Biological Timekeeping. Sunderland MA, Sinauer Associates, Inc.

Eisler, R. 1910. Weltenmantel und Himmelszelt. Religionswissenschaftliche Untersuchungen zur Urgeschichte des antiken Weltbildes. 2 Vol. München, C. H. Beck.

Eliade, M. 1971. The Myth of the Eternal Return: Cosmos and History. Princeton, Princeton University Press.

Eliade, M. 2009. Yoga: Immortality and Freedom. Princeton, Princeton University Press.

El-Khachab, A. M. 1961. 'Ο "ΚΑΡΑΚΑΛΛΟΣ" ΚΟΣΜΟΚΡΑΤωΡ. The Journal of Egyptian Archaeology 47, 119-113.

El-Khachab, A. M. 1971. Some Gem-Amulets Depicting Harpocrates Seated on a Lotus Flower. The Journal of Egyptian Archaeology 57, 132-145.

Elkin, A. P. 1969. Elements of Australian Aboriginal Philosophy. Oceania 40, no. 2, 85-98.

Ernst, G., and Firpo, S. L. (eds). 1997. La città Del Sole, new edition. Rome and Bari, Laterza.

Fischer, C. 2002. Twilight of the Sun God. Iraq 64, 125-134.

Foucault, M. 1980. Power/Knowledge: Selected interviews and other writings 1972-1977. New York, Pantheon Books.

French, J. R. P. 1956. A Formal Theory of Social Power. Psychological Review 63, no. 3, 181-194.

French, J. R. P., and Raven, B. H. 1959. The bases of social power. In D. Cartwright (ed.), Studies in Social Power, 150-167. Ann Arbor MI, University of Michigan Press.

Gaerte, W. 1914. Kosmische Vorstellungen im Bilde prähistorischer Zeit: Erdberg, Himmelsberg, Erdnabel und Weltenströme. Anthropos 9, nos. 5/6, 956-979.

Gariboldi, A. 2004. Astral Symbology on Iranian Coinage. East and West 54, no. 1/4, 31-53.

Glowczewski, B. 1999. Dynamic Cosmologies and Aboriginal Heritage. Anthropology Today 15, no. 1, 3-9.

Gundel, W. 1981. Sterne und Sternbilder im Glauben des Altertums und der Neuzeit. Hildesheim, Olms.

Gundel, W., and Gundel, H.-G. 1933. Sternglaube, Sternreligion und Sternorakel. Aus der Geschichte der Astrologie. Leipzig, Quelle & Meyer.

Gury, F. 2000. L'idéologie impériale et la lune: Caligula. Latomus 59, no. 3, 564-595.

Guthrie, W. K. C. 1952. The Presocratic World-Picture. The Harvard Theological Review 45, no. 2, 87-104.

Handy, C. 1976. Understanding Organizations. London, Penguin Books.

Hannah, R., Magli, G., and Palmieri, A. 2013. Nero's "solar" kingship and the architecture of Domus Aurea. arXiv:1312.7583v2 [physics.hist-ph].

Hardie, P. R. 1985. Imago Mundi: Cosmological and Ideological Aspects of the Shield of Achilles. The Journal of Hellenic Studies 105, 11-31.

Harrison-Buck, E. 2012. Architecture as Animate Landscape: Circular Shrines in the Ancient Maya Lowlands. American Anthropologist 114, no. 1, 64-80.

Haverkort, B., Hooft, K. van, and Hiemstra, W. (eds). 2003. Ancient Roots, New Shoots. Endogenous development in practice. Leusen and London, ETC/Compas in association with Zed Books Ltd.

Heine-Geldern, R. 1942. Conceptions of State and Kingship in Southeast Asia. Far Eastern Quarterly 2, 15-30.

Heinze, R.-I. 1981. The Nine Imperial Gods in Singapore. Asian Folklore Studies 40, no. 1, 151-165.

Herrmann, H.-V. 1959. Omphalos. Orbis Antiquus 13. Münster, Aschendorff.

Herzfeld, E. 1920. Der Thron des Khosrô. Quellenkritische und ikonographische Studien über Grenzgebiete der Kunstgeschichte des Morgen- und Abendlandes. Jahrbuch der Preuszischen Kunstsammlungen 41, 1-24.
Hill Boone, E. 2013. Cycles of Time and Meaning in the Mexican Books of Fate. Austin, University of Texas Press.

Hinkin, T. R., and Schriesheim, C. A. 1989. Development and application of new scales to measure the French and Raven (1959) bases of social power. Journal of Applied Psychology 74, 561-567.

Hoffmann, H. 1963. Helios. Journal of the American Research Center in Egypt 2, 117-124.

Hombert, P. 1945. SARAPIS ΚΟΣΜΟΚΡΑΤΩΡ ET ISIS ΚΟΣΜΟΚΡΑΤΕΙΡΑ: A propos de quelques terres cuites inédites. L'Antiquité Classique 14, no. 2, 319-329.

Horsch, P., and Whitaker, J. L. 2004. From Creation Myth to World Law: The Early History of "Dharma". Journal of Indian Philosophy 32, no. 5/6, 423-448.

Hosking, D.-M. and Morley, I. E. 1991. A social psychology of organizing: people, processes and contexts. London, Harvester Wheatsheaf.

Joost-Gaugier, C. L. 1998. The Iconography of Sacred Space: A Suggested Reading of the Meaning of the Roman Pantheon. Artibus et Historiae 19, no. 38, 21-42.

Kanas, N. 2009. Star Maps. History, Artistry, and Cartography, 2nd edition. New York, Heidelberg, Dordrecht and London, Springer.

Kelley, D. H., and Milone, E. F. 2011. Exploring Ancient Skies: A Survey of Ancient and Cultural Astronomy, 2nd edition. New York, Dordrecht, Heidelberg and London, Springer.

Kirkland, R. 1997. The Sun and the Throne. The Origins of the Royal Descent Myth in Ancient Japan. Numen 44, no. 2, 109-152.

Kolig, E. 1988. Australian Aboriginal Totemic Systems: Structures of Power. Oceania 58, no. 3, 212-230.

Krupp, E. C. 1983. Echoes of the Ancient Skies: The Astronomy of Lost Civilizations. New York, Harper Collins.

Krupp, E. C. 1997. Skywatchers, Shamans & Kings: Astronomy and the Archaeology of Power. New York, John Wiley & Sons, Inc.

Kuper, H. 1973. Costume and Cosmology: The Animal Symbolism of the Ncwala Man. New Series 8, no. 4, 613-630.

Lambek, M. (ed.). 2007. A Reader in the Anthropology of Religion. Malden, Blackwell Publishing Ltd.

Lancellotti, M. G. 2001. Médecine et religion dans les gemmes magiques. Revue de l'histoire des religions 218, no. 4, 427-456.

Loeb, E. M., and Klein, H. 1962. Staatsfeuer und Vestalinnen. Paideuma 8, no. 1, 1-24.

Mabbett, I. W. 1983. The Symbolism of Mount Meru. History of Religions 23, no. 1, 64-83.

Maeder, S. 2007. Pilze, Paranüsse, Teewärmer und eine Schattenseite des Mondes. Schwertknäufe als Quellen zur Geschichte der Himmelsbeobachtung im Mittelalter. Available at: http://www.archaeologie-online.de/magazin/thema/archaeoastronomie/pilze-paranuesse-teewaermer/seite-1/ (Acessed 12 December 2010).

Mallory, G., Segal-Horn, S., and Lovitt, M. 2002. Organisational Capabilities: Culture and Power. Milton Keynes, The Open University.

Marlowe, E. 2006. Framing the Sun: The Arch of Constantine and the Roman Cityscape. The Art Bulletin 88, no. 2, 223-242.

Merkelbach, R. 1980. Der Kult der Hestia im Prytaneion der griechischen Städte. Zeitschrift für Papyrologie und Epigraphik 37, 77-92.

Meyer, J. F. 1996/1997. The Eagle and the Dragon: Comparing the Designs of Washington and Beijing. Washington History 8, no. 2, 4-21.

Morgan, G. 1986. Images of Organizations. London, Sage Publications, Inc.

Müller, K. E. 1987. Das magische Universum der Identität. Elementarformen sozialen Verhaltens. Ein ethnologischer Grundriss. Frankfurt am Main and New York, Campus.

Nibley, H. 1951. The Hierocentric State. The Western Political Quarterly 4, no. 2, 226-253.

Palmer, E. 1991. Land of the rising sun. The predominant east-west axis among the Early Japanese. Monumenta Nipponica 46, no. 1, 69-90.

Pankenier, D. W. 1995. The cosmo-political background of the heaven's mandate. Early China 20, 121-176.

Pankenier, D. W. 1998. The Mandate of Heaven. Archaeology 51 (2), 26-34.

Premnath, D. N. 1994. The Concepts of Ṛta and Maat: A Study in Comparison. Biblical Interpretation: A Journal of Contemporary Approaches 2, no. 3, 325-339.

Quaritch Wales, H. G. 1959. The Cosmological Aspect of Indonesian Religion. Journal of the Royal Asiatic Society of Great Britain and Ireland 91, no. 3/4, 100-139.

Rappenglück, M. 1999. Eine Himmelskarte aus der Eiszeit? Ein Beitrag zur Urgeschichte der Himmelskunde und zur paläoastronomischen Methodik. Frankfurt a. Main, Peter Lang.

Rappenglück, M. 2000. The Milky Way: Its Concept, Function and Meaning in Ancient Cultures. In T. M. Potyomkina and V. N. Obridko (eds), Astronomy of Ancient Sciences, 270-279. Moscow, Nauka.

Rappenglück, M. 2005. The pivot of the cosmos. The Concept of the world axis across cultures. In M. Kõiva, I. Pustylnik and L. Vesik (eds), Cosmic catastrophes. A collection of articles, 157-165. Tartu.

Rappenglück, M. 2007. Copying the cosmos: The archaic concepts of the sacred cave across cultures. Iin H. Jung and M. Rappenglück (eds), Signaturen des Lebens: Bilder und Zeichen von Kosmos und Bios und Symbole des Alltags - Alltag der Symbole, Symbolon NF. 16, 63-84. Frankfurt am Main, Peter Lang.

Rappenglück, M. 2008. Tracing the Celestial Deer – An Ancient Motif and its Astronomical Interpretation Across Cultures, in J. Vaiškūnas (ed.), Astronomy and Cosmology in Folk Traditions and Cultural Heritage (Archaeologia Baltica 10), 62-65, Klaipeda, Klaipeda University Press.

Rappenglück, M. 2009. Constructing Worlds: Cosmovisions as Integral Parts of Human Ecosystems. In J. A. Rubifio-Martin, J. A. Belmonte, F. Prada and A. Alberdi (eds), Cosmology across Cultures (ASP Conference Series 409), 107-115. San Francisco.

Rappenglück, M. 2010. Earlier prehistory. In C. Ruggles, and M. Cotte (eds), Heritage Sites of Astronomy and Archaeoastronomy in the context of the UNESCO World Heritage Convention. A Thematic Study, 13-27, International Council on Monuments and Sites, ICOMOS/IAU.

Rappenglück, M. 2013. The Housing of the World: The Significance of Cosmographic Concepts for Habitation. Nexus Network Journal 15, no. 3, 387-422.

Rappenglück, M. 2014a. Stone Age People Controlling Time and Space: Evidences for Measuring Instruments and Methods. In G. Katsiampoura (ed.), Scientific Cosmopolitism and Local Cultures: Religions, Ideologies, Societies, Proceedings, 5th International Conference of the European Society for the History of Science, Athens, 1-3 November 2012, 466-474.

Rappenglück, M. 2014b. Possible astronomical depictions in Franco-Cantabrian Palaeolithic rock art. In C. L. N. Ruggles (ed.), Handbook of Archaeoastronomy and Ethnoastronomy, 1205-1212. New York, Springer Science+Business Media.

Rappenglück, M. 2014c. Possible calendrical inscriptions on Palaeolithic artifacts. In C. L. N. Ruggles (ed.), Handbook of Archaeoastronomy and Ethnoastronomy, 1197-1204. New York, Springer Science+Business Media.

Rappenglück, M. 2014d. The cosmic deep blue: The significance of the celestial water world sphere across cultures. Mediterranean Archaeology and Archaeometry 14, no. 3, 293-305.

Rappenglück, M. 2015. Voyages guided by the skies: Ancient concepts of exploring and domesticating time and space across cultures. In F. Pimenta, N. Ribeiro, F. Silva, N. Campion, A. Joaquinitio and L. Tirapicos (eds), Stars and Stones. BAR International Series (in print).

Raven, B. H. 1992. A power/interaction model on interpersonal influence: French and Raven thirty years later. Journal of Social Behavior and Personality 7, no. 2, 217-244.

Raven, B. H. 2004. Power, Six Bases of. In the Encyclopedia of Leadership, 1242-1249. Thousand Oaks CA, SAGE.

Reichardt, R., and Cohen, D. L. 1998. Light against Darkness: The Visual Representations of a Central Enlightenment Concept. Representations 61, 95-148.

Reiner, E. 1995. Astral Magic in Babylonia. Transactions of the American Philosophical Society, New Series 85, no. 4, i-xiii, 1-150.

Richardson, H. 1926. The Myth of Er (Plato, Republic, 616b). The Classical Quarterly 20, no. 3/4, 113-133.

Roberts, A. F. 1982. Comets importing change of times and states: Ephemerae and process among the Tabwa of Zaire. American Ethnologist 9, 4, 712-729.

Roscher, W. H. 1913. Omphalos. Eine philologisch-archäologisch-volkskundliche Abhandlung über die Vorstellungen der Griechen und anderer Völker vom „Nabel der Erde". Abhandlungen der philologisch-historischen Klasse der königlichen sächsischen Gesellschaft der Wissenschaften 29, no. 9. Leipzig, B. G. Teubner.

Roscher, W. H. 1915. Neue Omphalosstudien. Ein archäologischer Beitrag zur vergleichenden Religionswissenschaft. Abhandlungen der philologisch-historischen Klasse der königlichen sächsischen Gesellschaft der Wissenschaften Kl. 31. Leipzig, B. G. Teubner.

Roscher, W. H. 1918. Der Omphalos-gedanke bei verschiedenen Völker bes. den semitischen. Abhandlungen der philologisch-historischen Klasse der königlichen sächsischen Gesellschaft der Wissenschaften. Kl. 70, 2. Leipzig, B. G. Teubner.

Ryan, R.E. 1999. The Strong Eye of Shamanism: A Journey into the Caves of Consciousness. Rochester VT, Inner Traditions.

Sammells, C. A. 2011. The Aymara year count: Calendrical translations in Tiwanaku, Bolivia. Ethnology 50, no. 3, 245-258.

Schafer, E. H. 1977. Pacing the void: T'ang Approaches to the Stars. Berkeley and Los Angeles CA, University of California Press.

Schwartz, D. 2003. Conceptions of Astral Magic within Jewish Rationalism in the Byzantine Empire. Aleph 3, 165-211.

Silverman, E. K. 1996. The Gender of the Cosmos: Totemism, Society and Embodiment in the Sepik River Oceania 67, no. 1, 30-49.

Smith, R. T. 2009. Cosmos, Earth and Nutrition. The Biodynamic Approach to Agriculture. London, Rudolph Steiner Press.

Stanner, W. E. H. 1963. On Aboriginal Religion VI. Cosmos and Society Made Correlative. Oceania 33, no. 4, 239-273.

Stein, R. A. 1987. Le Monde en Petit. Paris, Flammarion.

Stephens, F. G., and George, M. D. 1870. Catalogue of Political and Personal Satires in the Department of Prints and Drawings in the British Museum, 11 vols. London, BMP.

Stern, S. 2012. Calendars in Antiquity: Empires, States, and Societies. Oxford, University Press.

Steven, L. 2005. Power. A Radical View. Palgrave, London.

Strong, S. A. 1916. A Bronze Bust of a Iulio-Claudian Prince (? Caligula) in the Museum of Colchester; With a Note on the Symbolism of the Globe in Imperial Portraiture. The Journal of Roman Studies 6, 27-46.

Tannenbaum, N. 1984. Shan Calendrics and the Nature of Shan Religion. Anthropos 79, no. 4/6, 505-515.

Taragan, H. 2003. Atlas Transformed-Interpreting the 'Supporting' Figures in the Umayyad Palace at Khirbat al Mafjar. East and West 53, no. 1/4, 9-29.

Terrien, S. 1970. The Omphalos Myth and Hebrew Religion. Vetus Testamentum 20, no. 3, 315-338.

Thorndike, L. 1923-1958. A History of Magic and Experimental Science, 8 vols. New York, Macmillan.

Trebsche, P., Müller-Scheeßel, N., and Reinhold, S. 2010. Der gebaute Raum: Bausteine einer Architektursoziologie vormoderner Gesellschaften. Münster, Waxmann.

Verellen, F. 1995. The Beyond Within: Grotto-heavens (dongtian) in Taoist ritual and cosmology. Cahiers d'Extrême-Asie 8, 265-290.

Vlastos, G. 1947. Equality and Justice in Early Greek Cosmologies. Classical Philology 42, no. 3, 156-178.

Weitzmann, K. 1973. The Heracles Plaques of St. Peter's Cathedra. The Art Bulletin 55, no. 1, 1-37.

Weitzmann, K. 1974. An Addendum to "The Heracles Plaques of St. Peter's Cathedra". The Art Bulletin 56, no. 2) 248-25.

Wheatley, P. 1971. The Pivot of the Four Quarters. Edinburgh, Edinburg University Press.

Wiersma, W. 1941. Die Fragmente des Philolaos und das sogenannte philolaische Weltsystem. Mnemosyne, Third Series 10, no. 1, 23-32.

Zürcher, E., and Schlaepfer, R. 2014. Lunar Rhythmicities in the Biology of Trees, Especially in the Germination of European Spruce (Picea abies Karst.): A New Statistical Analysis of Previously Published Data. Journal of Plant Studies 3 no. 1, 103-113.

THE SOCIAL LIFE OF CELESTIAL BODIES: THE SKY IN CULTURAL PERSPECTIVE

STANISLAW IWANISZEWSKI

Abstract: The paper is divided into two major parts. The first section consists of a discussion that examines the double role of astronomical objects in cultural astronomy: they are either treated as physical or material objects belonging to the domain of astronomers, or as social objects belonging to the domain of social scientists. The second part deals with a series of theoretical frameworks capable of transcending the traditional subject-object dichotomy imposed by modern astronomy.
Keywords: Cultural astronomy, agency, hybrids, ontology, epistemology

May the great gods of the night:
Shining Fire-star,
Heroic Irra,
Bow-star, Yoke-star,
Orion, Dragon-star,
Wagon, Goat-star,
Bison-star, Serpent-star
Stand by and
Put a propitious sign
In the lamb I am blessing now
For the haruspicy I will perform (at dawn)
(Prayer to the Gods of the Night, Reiner 1995, 1)

1. Introduction

Traditionally, when we describe the relationship between the sky and society, we tend to treat them like opposed and autonomous categories, each of them standing apart from the other. In examining the relationships between these two categories, we usually examine the ways in which the sky may shape various political, ideological, religious, symbolic and economic aspects of the social life, or we investigate how societies may impose specific cultural meanings on the sky. However, if we observe that the distinction made between the sky on the one hand and human society on the other is not a universal but a relative one, we are in danger of casting models of classical Western duality (subject-object, mind-nature, culture-nature) onto non-Western cultural patterns. Certainly, there are distinct differences between human life and the 'life' of the stars in the sky. Nevertheless, the ontological separation between the sky and society which is imposed by the ethnocentric concepts of modern science does not necessarily refer to the ways in which non-western societies conceptualize their skies and their relation to it (Descola and Pálsson, 1996). It is therefore important to redesign cultural-astronomical (archaeo- and ethno-astronomical) analysis of the relationships between man and the sky to be able to capture the nuances in those diverse perceptions of the sky. For the purposes of this paper, there cannot be any *a priori* existing sky-society dichotomy and both categories should be viewed as intertwined and forming the same cosmos.

Having said this I want to discuss the following problem. On certain occasions several celestial bodies observed in the starry night sky may appear to behave as if they were living persons or animate objects. In other words, astronomical objects perceived in the sky may be conceived of as entities capable of signalling to humans: the Pleiades often announce the arrival of the agricultural or rain season, the Moon regulates the moments of sowing and harvesting, the planet Venus announces the moments appropriate for hunting or military assault, etc. If astronomical objects can be perceived as entities communicating with humans or displaying human-like behaviour, then the objects perceived in the sky by modern astronomers do not refer to the objects perceived by the humans who 'dwell-in-the-world'. The things belonging to the human lifeworld are mediating between peoples and are invested with social meanings, that is, they stay between the human world and the realm of physics. It is therefore important to distinguish between the features of the common-sense world and the objects belonging to the material world of astronomers.

The matter is, however, not so simple. By including the 'sky' in the cultural discourse, we remove it from the category of nature and from the domain of modern astrophysics. A star in the sky ignored by culture is just a physical object located within the world of astronomical events, called the 'natural world'. This star belongs to the realm of astronomy and evidences, through different kinds of radiation, its place in the evolution of the universe. However, when somebody picks out the star for any cultural purpose or activity, it becomes immersed in human culture. Now, regarding our modern distinctions made between nature and society, between the sky and humankind, it appears that many non-western societies made no such division (Latour, 1993). Drawing upon Bruno Latour's (1993) description of modernity, conceived as the process in which objects are purified (i.e., separated) from their social and cultural uses, non-humans are separated from humans, material objects and artefacts are alienated from society, etc. It is easy to deduce that the star's 'naturalness' is lost when it is 'culturalized' or 'humanized'. However, according to Latour, our modern distinction, carefully made to separate nature from society, has never been successfully completed, leading to the creation of systems of mixed things rather than in the establishment of purified objects. For that reason, in the human world there are many things that are not purified: they simultaneously are natural and cultural, belonging to the class of objects called 'hybrids' or 'quasi-objects' by Latour. Taking this argument, it follows that 'stars' constitute a hybrid category: originally it has no cultural value and belongs to the material ('natural') world; later however it is evaluated according to its uses and meanings with regard to human culture. As a natural object each (any) star functions according to the physical laws of the universe, but as a social object it behaves like all other living organisms. Following Latour (1993) one may conclude that these dichotomous concepts (nature/culture, mind/body, sky/earth) were necessary to order Western thought, providing the ground for the foundation of modern epistemology. Today however, the dyadic model implied by those dichotomies is often taken as axiomatic by social scientists, who *a priori* consider that astronomical objects, such as the luminaries, are not animate entities.

In recent debates in archaeology and anthropology, the categories of 'person', 'agency', 'power' and 'gender' are increasingly discussed in relation to things. The categories mentioned here are relative; they do not possess physical attributes. This means they cannot be simply 'found' in material data, they can only be inferred from them. Orientation patterns and mythological narratives clearly indicate that some type of agency can be attributed to diverse celestial bodies. If people emphasized their connections to specific objects perceived in the sky, they performed activities in a patterned way, so we can interpret their activity patterns in terms of some dependency on what they perceived in the sky.

In this context I want to raise a question regarding the agency of celestial bodies. If certain celestial bodies 'act' as if they were agent-like persons, or if the actions of celestial bodies are perceived as possessing some capacity to act or a capability to impose their own will upon human beings, then we can ask how human capacities of agency can be attributed to celestial bodies.

2. Celestial bodies as meanings

Non-western societies imagined the sky in many forms, and as possessing different powers over diverse spheres of everyday life. The study of Mesopotamian divination texts allows us to infer about the nature of the relationships between the objects and events perceived in the sky and humans. Celestial divination has been only a part of the whole Mesopotamian divinatory tradition, which encompassed the interpretation of many terrestrial signs. Techniques of divination ranged from provoked omens (thus 'giving the deity the opportunity of directly affecting an object activated by the diviner', Oppenheim 1964, 208) to unprovoked ones (see Koch 1995, 9-12). Diverse divinatory techniques such as extispicy, lecanomancy, libanomancy and the casting of lots belonged to the former category; divination based on the behaviour of birds and stars, birth and dream *omina* belonged to the latter. The Gods of the Night, or *ilī mušīti*, invoked in the prayer mentioned in the beginning of this paper, were of course the stars and constellations perceived in the night sky (Reiner 1995, 1-3). The *bāru* diviner looked at the sky to see the favourable moment in time to interpret the liver of a killed lamb. The diviner, whose name meant 'one who inspects', carefully examined the exta, but in other cases he might have observed the configurations of the oil in a basin of water, the smoke from an incense burner, or other phenomena provoked by him 'for the purpose of soliciting a response from the gods' (Rochberg 2004, 84). As the stars belonged to the category of unprovoked omens, they were regarded as 'signs' (*ittu*) sent by gods.
The stars scintillating in the night sky were defined as a heavenly script (the 'heavenly writing', *šiṭir šamê,* Reiner 1995, 9). They were considered as 'signs' and Mesopotamian diviners attempted to decode the messages conveyed by astronomical phenomena. On many occasions two-way communication, query-and-answer, was available. Another text says: 'Yoke star standing at the right, Yoke star standing at the left, the god sends you to man, and man to the god...' (Reiner 1995, 15).

On the other hand, the stars and planets were addressed either under their astral names or the deities they represented. The planets, stars and individual constellations were the celestial manifestations of gods, but sometimes they acted as if they were gods in their own right (Koch 1995, 115-132). It may be deduced from the following statement: 'The lower heaven, of jasper, is of the stars. He (Bel) drew the constellations of the gods on it' (Koch 1995, 115). They acted to exert their influence in the omens activated by diviners. For example, an Old Babylonian prayer says 'O my lord Ninsianna (Morning Star), accept this offering, be present in my offering, and place in it a portent of well-being and life for your servant Ur-Utu' (Reiner 1985, 591). It seems therefore that in ancient Mesopotamia there were two distinct levels on which the stars acted. The stars acted either directly, through astral irradiation as divine signs, or indirectly as mediators between gods and humans.

On other occasions, however, the relation between celestial signs and their interpretations obeyed the same rules as other fields of inquiry and depended on the interconnections made between the terrestrial and celestial omens This may be exemplified by a text called 'A Babylonian diviner's manual' (tablets K.2848 and K.2847, lines 22-24, 38-40) which says 'sky and earth both produce portents though appearing separately, they are not separate (because) sky and earth are related' (Oppenheim 1974, 204). Both Ulla Koch (1995, 97-99) and Francesca Rochberg (2004, 260) provide lists of diverse schemes (or schemata) based on binary oppositions and standardized qualifications (colours, directions, locations in the sky and time intervals) that served to classify all celestial events used in celestial divination. Celestial events interesting for a Mesopotamian diviner included 'heliacal and acronical risings and settings, stationary points, conjunctions and other positions in relation to a particular celestial body, eclipses, colours and other optical phenomena' (Koch 1995, 97). Schematic manipulation of binary oppositions combined with the use of qualifiers served to correlate celestial and terrestrial phenomena (compare Rochberg-Halton, 1984). What I want to emphasize is the fact that Mesopotamian methods for prognostication on the basis of celestial omens were the same as those used for the interpretation of terrestrial phenomena. No specific technical terms were invented to describe the heavenly motions.

The hybrid (*sensu* Latour, 1993) nature of Mesopotamian heavenly bodies is clear. On the one hand, the Enuma-Anu-Enlil scribes regularly observed the appearances and disappearances of stars and planets to predict (establish) patterns in the recurrence of their movements; on the other hand, the Enuma-Anu-Enlil diviners observed stellar motions in the sky to interpret them as presaging certain events and to classify them according to their prognostication concerns (Oppenheim 1978, 643-644).

3. The hybrid nature of the objects perceived in the sky

Historians of astronomy have long demonstrated that the sky, as an integral part of human environment, was essential to human existence. It can be argued that celestial events entered human life as part of man's natural environment, but as its significant components they started to assume specific meanings in relation to him. It was in this way, therefore, that the regularities perceived in the motions of celestial objects provided the necessary context upon which specific cultural patterns were created to regulate human activities. From the celestial vault, and from naked-eye observations, ancient societies gained practical knowledge of their environment and the correlations they made between terrestrial and celestial events and processes served to understand how to conduct the human life on earth. Later, skywatching became associated with calendar making, time-reckoning, and freeing humankind from the regime of irregular and unpredictable fluctuations of diverse environmental cycles.

However, in light of the abovementioned discussion, this is only one part of a whole story. The sky provides meanings through connections to origin myths, lifeworld categories and rituals. As people define themselves in part through their physical environment, they feel a type of attachment to their surroundings; it then seems natural that those conceptual elements that relate the human body and its functions to the house and its components are the same that relate them to celestial bodies and the calendrical cycles related to their movements. In phenomenology, the Lifeworld (*Lebenswelt*) is conceived as the practical and often pre-theoretical 'field' of and for human understanding, including scientific, or rational, understanding (Husserl, date). Human understanding functions by interpretation and in order to interpret people have to ascribe meanings to the things they interpret. It is a collective enterprise. Lifeworld refers to the common products of human understanding. As both Schütz (Schütz and Luckmann 2003, 25-29) and Habermas (2005) propose, the structures of the Lifeworld belong to the public 'domain' where people share the products of human understanding by common habits of practice and by the use of language and other language like-media (communication through visual arts, music, dress codes, etc.) which are semiotic in nature. The philosophical notion of the Lifeworld has been applied to archaeology through Schütz and Heidegger and for the purposes of the present paper may roughly be defined as not only as *the familiar world of everyday life*, but also as *the patterned ways in which a (physical, social, psychodynamical) environment is functionally meaningful within some activity*.

The order perceived in the sky or imposed by the rotating heavens gave structure to the ways with which the people perceived their realm as a structurally ordered entity, so they were able to conceptually organize their world in the form of cosmographies, worldviews and cosmologies. They conceived their Lifeworld as an extension of their own bodies and the patterns perceived in the surrounding world as being analogous to the patterns of their life cycle. So when they observed the sky, they perceived patterns which were explained in terms of metaphors based on the universal human sensory-motor experience. Of course, what I mean by observing the skies has nothing to do with the appreciation of the beauty of the skies or with the contemplation of the rotating firmament by a neutral or culturally disinterested skywatcher. In accordance with what Ingold (2000, 40-60) proposed, acting in the Lifeworld is the practitioners' way of knowing. In other words, the knowledge of the world is inseparably linked to acting in the world, or to the process of dwelling-in-the-world.

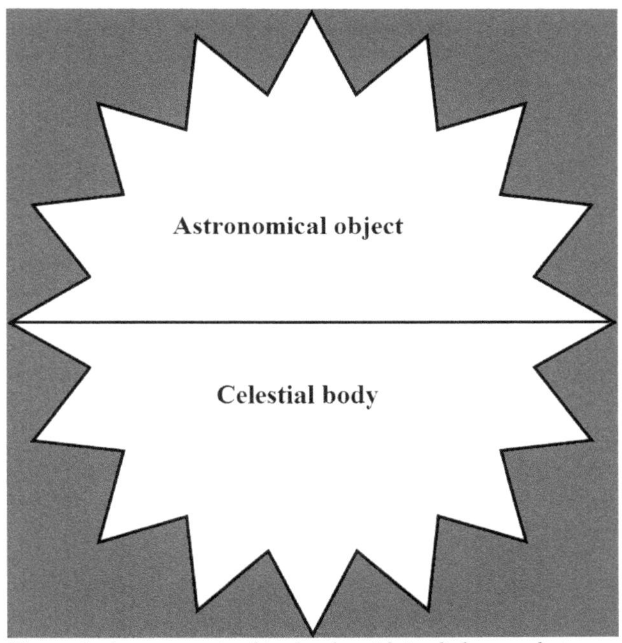

Figure 1. A star perceived in the sky with its dichotomy between physical substance (astronomical object) and conceptual form (celestial body).

As pointed out, the observation of heavenly bodies takes place on two distinct levels. The stars in the sky belong to physical reality and to the social world. Astronomical objects are parts of physical reality, but they also manifest some features which are associated with the domain of social and cultural activities (Figure 1). Thus for astronomers, astronomical objects belong to the world of physics; for people they belong to the Lifeworld (or the common-sense world). Celestial bodies stand for people, they embody people's sense of attachment to the world, of 'being-in-the-world'; they embody social relationships, identities and practices. The Lifeworld is a world of things which are used for various practical (utilitarian, functional) purposes, things which can exist always *in situ*, that is, in an environment of other things.

Paradoxically, cultural astronomy may be described as a hybrid (*sensu* Latour, 1993) discipline simultaneously dealing with two categories of objects.

4. What celestial objects?

In my view cultural astronomy is, in the first instance, a discipline that studies people's understanding of the universe, not because all human groups live under the skies, but because as social beings they all construct their identities through beliefs about how the world operates. In many non-modern societies, the definition of self has been quite different from modern concepts that usually define self as a monolithic, homogeneous and bounded individual identity that lives independently of its social and cultural context. Identities of ancient and non-modern societies consisted of correspondences, analogies and identities between celestial bodies, landscape features, and particular animals and ancestors on the one hand and everyday relationships and events on the other. For instance, Duncan Earle's (1986) descriptions of a Maya Quiché family shows how the activities of women and men at different ages correspond to the cycles of the Sun and Moon on their daily, monthly and yearly courses. In the home, each day the woman rises with the Morning Star to bring the fire back to life in the hearth. Then she works continuously all day, always close to the earth, cooking over the three-stones hearth on the ground or weaving while sitting on the ground, etc. Men get up with the Sun, warm themselves by the fire, then set out for agricultural fields, and as the Sun sets they return home. While women's activities are basically constant throughout the day and the year, men's activities vary seasonally: before the rainy season and during the first month of the season they work intensively in the field, during the dry season they rest or emigrate to work outside. The Maya see the genders as operating on different temporal cycles (Earle 1986, 160). The male cycle of 365 days is the solar year; the female cycle approximates 9 lunations (the human gestation period) and the 260-day cycle.

In the same vein Pierre Bourdieu (1977) reports on the gender-specific everyday activities of women and men living in a Kabyle village synchronized with shifts in seasonal subsistence and economic activities. When men set out to work on their fields, women start to weave on their looms outside their houses; however, when the agricultural cycle is over, men spend more time in the houses or in villages, so women start to weave inside their houses. Bourdieu (1977, 154) argues that all spheres of the human Lifeworld are permeated by the same organizing principles: 'the whole of human existence that, being the product of the same system of schemes, is organized in a manner homologous to that of the agrarian year and the other great "series"'. For example, in Kabylia there is a system of homologous correspondence between human gestation, the underground life of the grain, night, winter and north; another homology exists between spring, childhood, morning, inaugural periods; and still another homology is found between old age, the west direction, the setting sun, night and death (Bourdieu 1977, 154).

Additionally, in both descriptions of the organizing symbolic principles submitted by Earle and Bourdieu, the parts of the house are gendered, with a male roof and female hearth which correspond to the concepts of the male sky and female earth. Since men rise and work with the Sun, and come back home as the Sun sets, and women's activities are pretty constant throughout the day, month and the year, like the Moon, which is visibly in both the day and night, both luminaries are perceived as gendered: the Sun is male and the Moon is female. Furthermore, they behave like human beings, and often represent types of behaviour paradigmatic of both sexes.

It is evident that the celestial object and events do not produce meanings in themselves unless they are perceived as meaningful by the humans. They may be perceived as meaningful if it is possible to convert them into the categories understandable within a certain cultural context. This context stems from the schemes of perception developed through active engagement with the environment, in the process called 'dwelling-in-the-world' (Iwaniszewski 2010, 127). According to anthropological descriptions of symbolic classifications, the schemes of perception and cognition that produce knowledge of the sky can only become objective if they stem from the domain of traditional practices founded in the doxic mode, that is, in the world of values which either are taken for granted or experienced as natural or undisputed. The legitimacy of the dominant organizing principles that allow for the adscription of a celestial object or event into a meaningful category originates in the ontology of things immersed in the human Lifeworld rather than from epistemological speculation.

5. Celestial bodies as synthetic super objects

The engagement with the heavenly bodies as they are, and not their contemplation from the perspective of a detached observer, allowed humans to experience them in themselves. Thus the order perceived or imposed by in the rotating heavens was not used to inscribe a cultural order on the lifeworld, but the sky was the way that revealed how an order might have operated. Once this order was fixed in the sky vault, it could have objectified the order of that world.

Certainly the sky provides vast imaginative potential to carry diverse cultural meanings. Many of the issues surrounding the sky happen again in different cultural contexts, enabling productive ethnographic comparisons. The cross-cultural commonality in the meanings encoded in the sky raises important questions about universalities of human experience. This is particularly relevant in considering sensory and cognitive processes and their influence on cultural concepts practices and concepts (Bloch, 1998).

We see therefore that heavenly bodies were part of the engendered universe, constituting part of the culturally constructed archetypes within which ancient or non-western women and men negotiated their daily lives. The key cultural elements commonly attributed or symbolized by the Sun, the Moon, the Morning Star and the Evening Star in many societies should be seen as gender-specific and paradigmatic ones. Many typical masculine activities such as hunting and war were sometimes synchronized with the movements of Venus, so the planet was perceived as masculine. Sometimes the planet was associated with the feminine principle, and the differences between both sexes were symbolized through the Evening Star–Morning Star dichotomy. Paradoxically, the study of gendered celestial bodies may offer more important glimpses at people's concepts of self than traditional studies of material culture. The social life of the stars mirrors that of the humans.

6. Celestial objects as social objects

Observing that there is some difference between astronomical objects seen in their own materialities and celestial bodies representing human relations, I want to introduce the analytic category of a social object. Elsewhere (Iwaniszewski, 2007) I used this term to conceptually separate objects and things possessing the ability to act from those that lack this capacity. Behaving like human persons, social objects act within the dimension of the common-sense world. They exist within the Lifeworld structures, in an environment of other things, and participate (functionally, practically) in human life. As they fall outside the narrowly understood domain of physics, they have their own lives.

Introducing the study of social objects into the research field of cultural astronomy means that we move from entirely epistemological explanations of facts to the domain of ontological enquiries (see Figure 2). What is observed in the sky is perceived in correlation with the skywatcher: his individual position in the world, his ongoing engagement with the environment and within the context of his life. Stars in the sky interact with the human world through sets of rules, so they are seen as exercising power. This affects our notions of agency, because stars in the sky are not always seen as passive objects and humans as active.

7. The agency of celestial bodies

Recent developments in archaeology emphasized the agency of things. Many new theoretical directions examined western ontology, regarding its nature/culture dichotomy as an ethnocentric and misguiding methodology incapable of describing non-western relational epistemologies. There is no room for discussing them in this paper. I will mention only a few: 'naturalism' (Descola, 1996), 'relational epistemology' (Bird-David, 1999), 'dwelling ontology' (Ingold, 2000; 2006), 'Amerindian perspectivism' (De Castro, 1998), and 'ecosemiotics' (Hornborg 1996; 2001), and 'actor-network theory' (Latour, 2008). Following the papers collected by Appadurai (1986) archaeologists produced diverse books discussing the agency of things: Rival (1998), van Binsbergen and Geschiere (2005), Henare et al. (2007), Santos-Graneo (2009).

Astronomical objects (physical objects)	**Celestial bodies (social objects)**
Physical, material world	Social world
Not related to humans	Connected in complex relationships to humans
Separated from the domain of human everyday activities or from human conceptions of the world	Constitute a context for social and cultural activities
Maintain nature-culture dichotomy	Constitute one world
Obey mechanistic laws	Display human-like behaviour, or exhibit *rites-de-passage* symbolic-classificatory logic

Figure 2. Physical versus social objects. To emphasize differences between them, physical objects are here represented by astronomical objects, social objects are rendered as celestial bodies.

As the relationship between human beings and things has attracted much theoretical interest in current archaeology and anthropology, it is important to see these developments in the research field of cultural astronomy. The relational constitution of the world as perceived by many non-western societies is potentially a new topic which suggests a reinterpretation of orientation patterns. The Sun, the Moon and the brilliant planets and stars may be seen as agents interacting with pre-modern societies either through already built alignments or as entities (social objects) capable of influencing human societies in their decisions of how to align their houses and temples or where to orient the bodies and graves of their ancestors. In my opinion, today we are able to overcome the various disciplinary intellectual deficiencies preventing us from a better understanding of the skylore of the past. In western science, the culture/nature division is used as a central analytic device for studying the properties of the matter. In cultural astronomy, however, we are studying astronomical objects as natural, common phenomenal objects only in order to understand how they were transformed into social objects. This means we should substitute traditional epistemologies with ontologies derived from the common-sense world.

References

Appadurai, A. (ed.) 1986. The social life of things: Commodities in cultural perspective. Cambridge, Cambridge University Press.

Bird-David, N. 1999. Animism revisited: personhood, environment, and relational epistemology. Current Anthropology 40, Suppl. 167-191.

Bloch, M. 1998. Why trees, too, are good to think with: Towards an anthropology of the meaning of life. In L. Rival (ed.), The Social Life of Trees: Anthropological Perspectives on Tree Symbolism, 39-55. Oxford, Berg.

Bourdieu, P. 1977. Outline of a theory of Practice. Cambridge, Cambridge University Press.

De Castro, E. Viveiros. 1998. Cosmological Deixis and Amerindian Perspectivism. The Journal of the Royal Anthropological Institute 4, no. 3, 469-488.

Descola, P., and Pálsson, G. 1996. Introduction. In P. Descola and G. Pálsson (eds.), Nature and Society. Anthropological Perspectives, 1-21. London and New York, Routledge.

Descola, P. 1996. Constructing natures: Symbolic ecology and social practice. In P. Descola and G. Pálsson (eds.), Nature and Society. Anthropological Perspectives, 82-102. London and New York, Routledge.

Earl, D. M. 1986. The Metaphor of the Day in Quiché: Notes of the Nature of Everyday Life. In G. H. Gosen (ed.), Symbol and Meaning Beyond the Closed Community: Essays in Mesoamerican Ideas, 155-172. Albany NY, The University of Albany, State University of New York.

Habermas, J. 2005. Teoría de la acción comunicativa, II. Madrid, Editorial Taurus.

Henare, A., Holbraad, M., and Wastell, S. (eds.) 2007. Thinking Through Things: Theorising Artefacts Ethnographically. London and New York, Routledge.

Hornborg, A. 1996. Ecology as semiotics. Outlines of a contextualist paradigm for human ecology. In P. Descola and G. Pálsson (eds.), Nature and Society. Anthropological Perspectives, 45-62. London and New York, Routledge.

Hornborg, A. 2001.Vital signs: An ecosemiotic perspective on the human ecology of Amazonia. Sign Systems Studies 29, no. 1, 121-147.

Ingold, T. 2000. The perception of the environment. Essays in livelihood, dwelling and skill. London and New York, Routledge.

Ingold, T. 2006. Rethinking the animate, re-animating thought. Ethnos 71, no. 1, 9-20.

Iwaniszewski, S. 2007. La arqueología de alta montaña frente al paisaje montañés en México central – problemas, interpretaciones, perspectivas epistemológica. In M. Loera Chávez y Peniche, S. Iwaniszewski and R. Cabrera (eds.), Páginas en la Nieve. Estudios sobre la Montaña en México, 8-31. México, D.F., Consejo Nacional para la Cultura y el Arte-Instituto Nacional de Antropología e Historia-Escuela Nacional de Antropología e Historia.

Iwaniszewski, S. 2010. Cultural impacts of astronomy. In A.T. Tymieniecka and A. Granpierre (eds.), Astronomy and Civilization in the New Enlightenment. Passions of the Skies. Analecta Husserliana 107, 123-128. Dordrecht, London and New York, Springer.

Koch, U. S. 1995. Mesopotamian Astrology: An Introduction to Babylonian and Assyrian Celestial Divination. Copenhagen, The Carsten Niebuhr Institute of Near Eastern Studies, Museum Tusculanum Press.

Latour, B. 1993. We have never been modern. Cambridge MA, Harvard University Press.

Latour, B. 2008. Reensamblar lo social: una introducción a la teoría del actor-red. Buenos Aires, Ediciones Manantial.

Oppenheim, A. L. 1964. Ancient Mesopotamia. Portrait of a Dead Civilization. Chicago and London, The University of Chicago Press.

Oppenheim, A. L. 1974. A Babylonian Diviner's Manual. Journal of New Eastern Studies 33, no. 2, 197-219.

Oppenheim, A. L. 1978. Man and Nature in Mesopotamian Civilization. In C. C. Gillespie (ed.), Dictionary of Scientific Biography, Vol. 15, Supplement I, 634-666. New York, Scribner's.

Reiner, E. 1985. The Uses of Astrology. Journal of the American Oriental Society 105, no. 4, 589-595.

Reiner, E. 1995. Astral Magic in Babylonia. Transactions of the American Philosophical Society 85, Part 4. Philadelphia, The American Philosophical Society.

Rival, L. (ed.) 1998. The Social Life of Trees: Anthropological Perspectives on Tree Symbolism. Oxford, Berg.

Rochberg-Halton, F. 1984. New Evidence for the History of Astrology. Journal of New Eastern Studies 43, 2, 115-140.

Rochberg, F. 2004. The Heavenly Writing: Divination, Horoscopy, and Astronomy in Mesopotamian Culture. Cambridge, Cambridge University Press.

Santos-Granero, Fernando (ed.) 2009. The Occult Life of Things: Native Amazonian Theories of Materiality and Personhood. Tucson, The University of Arizona Press.

Schutz, A., and Luckmann, T. 2003. Las estructuras del mundo de la vida. Buenos Aires, Amorrortu.

van Binsbergen, W. M. J., and Geschiere, P. L. (eds.) 2005. Commodification: Things, Agency, and Identities (The Social Life of Things revisited). Münster, Lit Verlag.

ASTRAL HIGH-FASHION CLOTHING: RELATIONS BETWEEN COSTUMES AND ASTRONOMY

MICHAEL A. RAPPENGLÜCK

Abstract: Across cultures and through time people used costuming for presenting essential parts of their worldviews. They often linked dress, appliqués and accessories to certain concepts of cosmography and cosmogony, including the interplay of forces in nature. Costumes showed the intermediary and changeable position of man in nature and his relationship with other cosmic entities. Besides tattoos and body painting, people associated a wide range of garments with astronomical phenomena or cosmovisions: astral crowns and sky coats, special designed garments for passages through cosmic realms, astral necklaces, girdles or veils symbolizing the zodiac or the Milky Way, head coverings and dresses decorated with astral symbols (Sun, Moon, constellations, selected fixed and wandering stars), celestial footwear, feathered and cornuted crowns etc. The sky itself was considered to be a star-spangled skin of an animal (later a coat or a textile). The study presents an overview about the topic.
Keywords: Costumes, ethnoastronomy, cosmovisions, power, astral symbolism

Since the time humans started to dress themselves, clothing was important, and not only for the needs of biological viability in special environmental conditions. Moreover, costuming served as a special medium of self-awareness, image cultivation, masquerade and magical shape shifting (Weiner and Schneider, 1989; Payne, 1997; Arthur, 2000; Rieff Anawalt, 2007). Later, garments and accessories were used for defining, controlling, and representing social order and power (Weiner and Schneider, 1989; Arthur, 2000; Rieff Anawalt, 2007). Finally, particular kinds of costuming allowed and supported symbolic interaction and communication with the entities of the world and beyond (Heiler 1979, 118-122; Weiner and Schneider, 1989; Arthur, 2000; Rieff Anawalt, 2007).

Celestial garments

The astrologers' and also the magicians' 'overall' is well-known to everybody: at work he wears a high conical cap, holds a sphere in one hand and a magic wand in the other, has a long white beard and is clothed in a cloak covered with the Sun, Moon and stars (Eisler 1910, 1). Though that stereotype is entirely a modern assembly, the picture of the celestial robe and the headdress showing astral figures and the magic ward can be traced far back in time.

In the Cinderella fairy tale it is said that the poor young woman wore a dress of the Sun, woven of sunbeams, shining like the Sun or with the Sun ornamented on it. In some variants of the garment the Moon replaced the Sun. According to other tales, the dress, coloured in a saturated blue of a noontide sky, shines like the stars in a golden light or has stars appliquéd as diamond or pearls. Frequently the robe is a kind of metal clothing on which is embroidered the heavens with the Sun, Moon, planets, stars and other celestial phenomena (Cox, 1967; Winton 2000, 14-15). Additionally, it is said that the clothing shows the Sun at the front and the Moon at the back or the Sun by day and the Moon by night. 'Zistel im Körbel', an Austrian fairy tale (Zingerle and Zingerle 1911, 3-14) and 'Allerleirauh' (Grimm and Grimm 1812/1815, KHM 65), a German fairy tale, which both present variants of the Cinderella motif, elaborate on the idea that there are three robes: the Sun dress, the Moon dress and the star dress. These correspond to the main three celestial timers and orbits, which define certain heavenly realms.

The concept of celestial or, in a broader sense cosmic, garments can be found in different cultures all over the world (Mannhardt 1875, 80, 284-285, 296 and fn. 1, 316; Eisler, 1910; Weidner, 1931-1932; Cammann, 1947, 1948, 1951; Oppenheim, 1949; Zerries, 1977, 1979; Schafer 1981, 402-403 fn. 98; Parpola, 1985a, 1985b; Cohen, 1987; Baumgärtel-Fleischmann, 1990; Miller, 1991; Leopold, 1995; Huth, 1996; Podella, 1996).

The sky's skin and the cosmic garment

The notion of the sky as a kind of garment is rooted in the very archaic idea of the sky as an animal's skin (Eisler 1910, 79-83; Freedman, 1972; MacDonald 1987, 23-25; Miller, 1991; Oppenheim 1949, 182, figure 2, 187, fn. 25; Rist 1967, 188-198; Sauren, 1984; Widengren 1945, 50-55, 76-83; Muhammad 2011, fn. 22). The Babylonians speak of the hide of the great bull as an emblem of An(u), the god of heavens and the lord of constellations, who dwells in the highest region of the sky (Muhammad 2011, fn. 32). Some examples from ancient Egypt show a leopard skin with metal stars appliquéd (Oppenheim 1949, 187, fn. 25). In ancient Iranian tradition (Zan-Ākāsīh: Iranian or Greater Bundahišn 189, 8) the cosmic body is said to have a skin like the sky (Muhammad 2011, fn. 22). In Zurvanism the anthropomorphic body of the deity of time, Zurvan, appears as the blue firmament, which is thought to be a blue garment, too (Zaehner 1955, 11-12, 122). The Manicheans gave us the idea that the Mother of Life rolls out the sky, which is the skin of the Sons of Darkness (Muhammad 2011, fn. 22). The Jewish tradition is aware of the giant skin, which as the firmament spreads out and like a coat covers all things below (Muhammad 2011, fn. 32). As in the case of the celestial hide, which is decorated by stars, constellations and planets, including the Sun and Moon, the skin of the human body is spangled with hidden figures which are similar to the heavenly ones. The statue of a sacrificer from Susa shows stars incised as tattoos upon the bare chest and the upper arms (Oppenheim 1949, 187, fn. 25). According to other concepts the stars are celestial light beams, which emanate from the hair follicles of the cosmic skin (Muhammad 2011, fn. 32).

Thus it seems that people originally recognized the sky, and even the complete world, shaped as a kind of a primeval giant being's body, an animal or anthropoid, which shows the parts and phenomena of the cosmos

(Baumann 1995, 277-293, and map 4; Rappenglück, 1999; Muhammad 2011, fn. 26-30). A reminiscence of that idea is given in the Maya king's body metaphor, modelling the macrocosmic structure of the world into an anthropoid shape (Baudez, 2000): the structure of the cosmos (strata, cardinality and axis) together with symbols of the Sun and the Milky Way are shown.

Later people reduced the image to the skin, with which as hunter-gatherers they were acquainted. After having invented weaving, they finally replaced the hide by fabric (Rappenglück 2007, 166-169).

The 'cloak of stars' (Baumgärtel-Fleischmann 1990, 105-125) was often worn by gods and worldly and spiritual rulers. The cosmic or celestial cloaks originally belonged to certain divine beings. The replica of a sculpture shows Caelus (Coelus), the Roman sky-god, wrapping himself up in a cloak embroidered with stars (Abguss-Sammlung, Berlin, FU, VIII 696, Inv.-Nr. SH 340). Another sculpture, dating from about the 1st century AD, presents the moon goddess Selene dressed in a sky coat (Neugebauer 1921, 111-112). A similar belief is handed down by the Latvian people, who think that the Moon is wearing a star cloak, which is the night sky with the stars embroidered (Mannhardt 1875, 316). Zeus Heliopolitanus (2nd / 3rd century AD) wears a celestial robe, decorated with the planets Mercury, Venus, Mars, Jupiter, and the Sun and Moon (Eisler 1910, 72). The same is known from representations of Zeus Sosipolis (Eisler 1910, 72). According to Pherekydes of Syros (6th century BC) Zeus wove a magnificent nuptial garment for his bride Chthonie, which shows depictions of Gaia, the Okeanos and the houses of the heavens (Rappenglück 2007, 166). The cosmic nuptial garment is also depicted on a Tyrrhenian amphora painting of Timiades (565 BC-550 BC) exhibited in the G.A. Sanna Archaeological Museum in Sassari, Sardinia. In Greek (Orphic) tradition, Chronos, the god of space and time, weaves the cosmic primordial thread in twists and turns into the garment of the world (Eisler 1910, 391; Rappenglück 2007, 166-167). Japanese myths tell about the 'Holy Hall of Weaving', which is located in the centre of the cosmos. (Rappenglück 2007, 167). In this hall a goddess of the Sun, Amaterasu (Wakahirume) weaves a divine celestial garment.

According to the Homeric Hymn 6 to Aphrodite (7th to 5th century BC), the Greek goddess was clothed with heavenly garments and wore a fine, well-wrought crown of gold on her head (Evelyn-White 1914, Hymn 6 to Aphrodite).

The Babylonian god Marduk and the Mesopotamian god Adad sometimes appear clothed in a sky cloak, too (Eisler 1910, 60-61). The vestment of Nâbu is decorated with sewn-on stars (Oppenheim 1949, 180-181, 187). Moreover the clouds, containing the fertilizing rain and drifting along the sky, are depicted on the celestial robes. They are thought to be pieces of a fleece, a concept, which calls to mind the ancient Greek motif of the 'Golden Fleece'. The Mesopotamian tradition of 'golden sky garments' goes back to the 4th millennium BC (Parpola 1985, 35).

The ancient Indian gods Varuna, Mitra and Soma are also equipped with the celestial cloak and related to the power of cosmic fertilizing water, present in rainclouds, lightning and thunder (Eisler 1910, 101-102; Parpola 1985a, 37-341). Varuna wears the garment of the night sky adorned with celestial jewels: stars, constellations and planets (Parpola 1985, 41-42). Sometimes Síva wears the celestial garment, showing the Moon on his chest. In addition the heavenly *Ganga* (Milky Way) is clipped in his hair (Parpola 1985a, 41). The *tarpya* garment of the Vedic ritual is also associated with the sky, with cosmic waters and souls seen as stars and fireplaces (Parpola 1985a, 205; Parpola, 1985b). It is used during royal consecrations and in rituals performed by the priest. In particular this celestial robe was used to enable ascending to the heavens.

The god Mithras has a star-spangled cloak (Eisler 1910, 62-63; Beck 2006, 106-107). The Zoroastrian god Ahura Mazda owns a woven sky coat, embroidered with stars (Eisler 1910, 94; Parpola 1985a, 140-142).

The 'Queen of the Heaven', having various names in different cultures (Tanit, Astarte, Aphrodite, Artemis, Venus, Regina Coeli, Fortuna Coelestis, later the Saint Mary) is vested with a star cloak, too (Eisler 1910, 66-85). Isis wears the cosmic *pallium*, a narrow band wrapped around her body and decorated with multiple crescents and stars (Eisler 1910, 69-70, 86, 102-103, 135 fn. 7).) The shirt of a Sioux woman presents the ocean, the sky with stars and clouds, and the cosmic turtle (Müller 1970, 330-334, and figure 56).

The Baltic sun goddess Saule has a beautiful cloth with embroidered solar symbols (McCrickard 1990, 75-77). The colours (red, white, silver and gold), used for dying the costume, symbolize the power of solar light. A robe of the Blackfeet (USA), with depictions of butterflies and the Morning Star, illustrates the cosmic power of the Sun (Taylor 1993, 85, and figure 21).

The mantle of Alkisthenes of Sybaris (probably 5th century BC), is said to have shown weavings of zodiacal constellations (Eisler 1910, 33-34). Alexander the Great (356-323 BC), Nero (37-68 AD), Julius Caesar (100-44 BC), Augustus (63 BC-14 AD), Domitian (51-96 AD), Trajan (53-117 AD) and other emperors are said to have been garbed in cosmic cloaks, showing some celestial phenomena like the stars (Eisler 1910, 39-45; Freyer-Schauenburg, 2007). Augustus of Primaporta (after 14 AD) with respect to the *traditio legis*, shows the god Caelus with the sky blanket (Berger, Beinert, Wetzel and Kehl et al. 2006, 148). The emperors Otto III (980-1002) of the Ottonian Dynasty and Heinrich III (1017-1056) of the Salian Dynasty of Holy Roman Emperors were clothed in such a celestial or even cosmic robe (Eisler 1910, 5-26; Zinner, 1939). The vestment of the pope, a Byzantine work dated to the 12th century AD (and since the 17th century titled as 'Dalmatika Leonis III' or 'Dalmatic of Emperor Charles the Great'), shows stars on the blue coloured front side and the cosmogram of the nine choirs of angels on the back side (Eisler 1910, 19-21).

Later this idea was assigned to Jesus Christ (Berger, Beinert, Wetzel and Kehl et al. 2006, 148-149). Even in the Christian Middle Ages, Alanus ab Insulis (c. 1120-1202), in *De Planctu Naturae (*1168 und 1176) described nature as a beautiful woman (*natura*), clothed in a cosmic garment illustrating the realms and creatures of the world and wearing a crown, which shows the movement of the celestial bodies (Knowlton 1920, 242-243).

The costume of the Taoists' 'Heavenly Thearch of Prime Initiation', described in scripts of the 8th century AD, is associated to solar energy, the aurora and stars (Kochab [β UMi, 2.07 mag] and Alcor [80 UMa, 3.99 mag]) close to the northern sky pole (Schaffer 1981, 402-403, fn. 98). The female Taoist initiate dressed herself in a rosy robe and donned a star crown on her head (Schafer 1978, 43-45).

The ancient Chinese emperors dressed up themselves in the 'dragon robe', which showed the emblematic figures of the Sun, the Moon, the stars, the dragon, the Twelve Symbols, the World Mountain, the 'cloud collar motif', and other things and beings in the cosmos. The imperial clothing was worn at sacrificial ceremonies and served to signify the universal dominion of the ruler, illustrating his divinity (Cammann 1947, 7-17; Cammann, 1948, 1951; Smith 1919, 39, 95, 97; Vollmer, 2000). It was thought that the emperor's body clothed in such a robe corresponded to the world axis stretching the canopy of heavens (Cammann 1951, 4). Tibetan and Mongolian traditions, too, considered the decorated emperor's robe as a symbol of the world, cosmic structure and power (Cammann 1951, 5). The Skidi Pawnee (North America) Sun Chief is wrapped in a sky blanket (Chamberlain 1982, 20, figure 3; Taylor 1995, 168). The poncho of a chief of the Shipibo-Conibo (Peru), exhibited at the Linden-Museum in Stuttgart, Germany, shows the Southern Cross, which in today's context is interpreted as the Christian cross.

Why cosmic clothing is needed?

In any case the exceptional celestial or more extended cosmic clothing was thought to enable the socio-political and the medical-spiritual specialist to get the necessary and specific power and knowledge from cosmic spheres to carry out his intentions. Weaving textiles for clothing corresponds to the process of world creation (Prechtel and Carlsen, 1988; Rappenglück, 2007).

By changing the costume his nature is transformed so that he can be adapted to another realm (Rappenglück 1999, 252-254; Hoppál 2002, 94-115). Most notably this is the case during shamanistic trance (ecstasy). According to ancient views the passage through the cosmic strata to the upper or lower worlds requires certain 'overalls' and sometimes protective clothing depending on the riskiness of contacts with powerful entities, among them most notably the celestial bodies (Rappenglück 1999, 252-254; Hoppál 2002, 94-11). Thus 'cosmic' clothes and accessories support and certify the success of the earthly specialists, serving at the same time as apotropaic protections against any calamities. This is why specially designed cosmic clothing was thought to have magical influence, to make a person able for divination, and to be a powerful support for rulers, who felt called upon to move and dominate processes in the world or even the whole universe. Cosmic clothing also allowed shape shifting into certain other beings–plants, animals, other humans and supernatural creatures, which can operate benevolently or malevolently (white or black magic). People wanted to adopt the characteristics of these beings, in particular their vital force and their knowledge. At least they hoped to participate in the qualities of those beings. In addition people were interested to be protected, shepherded and blessed by certain astral beings. The background for such an idea is given by an archaic totemistic-shamanistic perception of the world: the celestial beings, Sun, Moon, stars, constellations etc. are considered to be a kind of primordial and lasting kinship localized in the different realms of the cosmos. The astral totems guaranteed the order, classification, relations, adjustment, balance and stability of the vital and inanimate beings in nature. They were thought to be responsible for creating and preserving individual and social life. They existed in the shape of fixed and wandering stars, the Pole Star (if there are any in the epoch) or circumpolar stars, asterisms, the Milky Way and others. In addition, meteorological phenomena, like rainclouds, lighting, thunder etc., were included. Emblems of certain astral totems, tutelary spirits, mothers and lords of the animals put on dresses as an appliqué or woven in the textile were intended to help to avoid misfortune. Thus the phrase 'to live under a good star' is quite understandable. Nevertheless astral power is ambivalent and, as everything in human hands, could be used for bad activities, too. Following the path of black magic, celestial emblems on clothing were also applied for destructive purposes.

Both the socio-political and the spiritual leaders dressed themselves in cosmic garments to show everyone quite plainly that they really were the embodiment of the primeval and lasting strength of the universe. The celestial robes served to illustrate the claim to power, authority and superior expert knowledge. A nice example is given by the dress of the Pitahawirata Pawnee chief Jim (Chamberlain 1982, 109, figure 26): he wore a hat with an embroidered four-rayed star emblem, symbolizing the chief star (Polaris). The members of the tribe should be related to the chieftain like the stars to the polar point, both presenting powerful beings. The concept of identification of the celestial pole with the earthly political ruler or spiritual leader occurred often in various cultures worldwide. A photograph of 'Sun Chief', a Pawnee, from about 1870, shows the native leader clothed in a robe decorated by five-rayed stars related to the sky and the Morning Star (Taylor 1995, 168-169). The Omaha robe presents the night time and Morning Star designs (Miller 1997, 233, figure 11.6).

Often special designs express the cosmic meaning of the clothes: the Hmong people of Southeastern Asia applied a cross symbol on their textiles, including clothes, which illustrates the structure of the cosmos and at the same time served for protective magic purposes (Cohen, 1987). A Nganasan shaman's costume is painted half in red (right side), illustrating day and spring, and half in black (left side), indicating the time of night and winter (Hoppál

2002, 104). On a Darchat's shaman costume, above a depiction of the vertebral column ('tail') appears the swastika as a Sun symbol (Hoppál 2002, 104). On an Evenk shaman's costume, the world tree and the upper, middle and lower world are displayed (Okladnikova 1998, 335, figure 8.5). In general the Siberian shaman's costume represents cosmic structures (Hoppál, 2004).

Besides this, on a lower level of meaning, certain cosmic attributes shown on the clothing helped to classify members of social groups and to integrate them in the society. Elements of cosmic symbolism are often used as emblems to document the status related to kinship and to ethnic groups. These concepts are well known from several ancient and recent cultures worldwide. They have survived until now: imperial, military and constabulary insignia, emblems of leagues, and others.

In addition, accessories like certain musical instruments, insignia, sceptres, weapons and shields were also often considered to be a part of the cosmic robe and charged with supernatural power. They played an important role in the rituals of investiture (Latin: *investire* 'to clothe in a robe') of spiritual and/or political rulers, from shamans to emperors. There is however no space left to interpret these fascinating accessories as well.

Finally, people, e.g., in Siberia, northwestern North America and in East Greenland (Chaussonet 1988, 210; Buijs and Petersen 2004, 84-85), paid attention to wear carefully and beautifully made, well preserved clothing at special places and times to please the spirit of the plants, animals, humans and gods they were related to and upon whom they depended. From different cultures all over the world and through the epochs it is known that the best suited clothing, the newest one, was worn at seasonal high times (solstices, equinoxes) during the year and the calendric fixed festivals (Bekbassar 2005, 112). This tradition–without the cosmological and spiritual background–is preserved in the custom to appear at feasts in one's Sunday dress.

The ideas of astral and cosmic clothing expressed in symbols, myths and rituals are closely connected to archaic shamanistic-totemistic cosmovisions.

The cosmic costumes based on ancient shamanistic-totemistic worldviews

A shaman is specialized in mastering altered states of consciousness. Special garments, including masks, crowns, jewellery applications, music instruments and shields enabled a well-suited, well-trained and sensitive person for shape-shifting and spiritual transformation. This allowed him to travel through the cosmic strata (Ridington and Ridington 1970, 50, 61). A shaman's main mission is to visit other realms of the universe and to get in contact with supernatural beings. These provide him with the right power and knowledge to take care of his group's survival and especially to heal people.

Shamans consider their costumes to be a model of the tripartite cosmos, which they create new by making the dress (Hoppál 2002, 94, 104; Hoppál, 2004). The upper world, the sky, is represented by the crown; the middle world, the Earth, is shown by the body clothing (skeleton, metal rings, disks, figurines etc.); and the lower world, the underworld, is visualized by the trousers and footwear (feet of animals). Masks, belts, bells and drums are additional elements of the costume.

Frequently shamans, and rulers, too, dressed themselves as birds to adopt their powers to fly, to walk, and to dive (Ellis and Hammack 1968, 27-28, 35, 37, 39-40; Rappenglück 1999, 247-254). Birds equipped with wings have the unique ability and mobility to dominate verticality through all cosmic strata, allowing them to develop and control three-dimensional space. They can fly up to the highest levels of the air and access the realm of the powerful Sun, the Moon and the stars. Thus birds mostly are considered to be the powerful and dominating animal lords or gods controlling aerial and celestial places, regions and bodies. From their high viewpoint in the sky they are able to consider the events on Earth and to collect knowledge of all that is going on beneath. In addition, aquatic birds are able to swim and to dive; to manage the domain of water. The multicolour plumage of some birds symbolized the vital power of the Sun, air, sky, fire and light, and the play of colours at dawn or dusk. Birds carried the shaman through holes in the cosmic strata from one realm to another right up to the celestial pole, the world's point of origin (Rappenglück 1999, 248-252). They believed that getting there would put them in contact with the power which caused and maintained the world.

The crown as a heavenly headdress

The feather headdress in particular symbolized the fiery mental power during ecstasy, which people thought to be solar and to originate at the centre of the cosmos. This is well-known from different cultures worldwide and through the epochs (Zerries, 1977; Rappenglück, 1999). The traditional Latvian costume of women, who are thought to represent the Sun's daughters, has a golden crown with beads and long ribbons, visualizing the Sun's rays (McCrickard 1990, 75-77). Other decorations, e.g., a belt, a necklace, or a golden goblet, are Sun symbols, too. The Slavic traditions associate the woman's voluminous headdress of the folkloristic costume, looking like a peacock's tail or a bunch of flowers, embroidered with glass balls or beads, with the Sun (McCrickard 1990, 89-91). A silver or golden crown as part of the traditional woman's costume, used during ceremonial dances of the Khasi people in India, represents the Sun, too (McCrickard 1990, 175).

South American Amerindians linked the feather crowns with the soul, the Sun and solar power, but also with the Moon (Nimuendajú 1919-1920, 1010; Reichel-Dolmatoff 1974, 24, 72, 98, 116; Zerries 1977, 284, 294-295, 300-301). Feather crowns played an important role for enabling the ecstatic journey through the spheres of the cosmos: the hat of a Peyoteros shows feathers of the turkey as a symbol of the sun god and other elements related to the theft of fire (Deimel 2000, 167).

A good example for the deeper symbolism is given by the ceremonial bonnet of the Crow Natives (North America): It consists of 28 Golden Eagle feathers, which correspond to the 28 ribs of the sacred bison and the 28 days of a sidereal month. The Golden Eagle was thought to have led man from the realms of the stars down to an earthly life. Twelve pinions of the Golden Eagle were included in the bonnet: they represented the corona of sunbeams and the number of months in a year. The bonnet allowed direct contact to the high god Wakan Tanka (Owusu 1997, 154-155). According to the ancient Egyptians the double feather crown, worn by some gods, was so high that it reached the heavens and pierced through the starry sky. This piercing symbolizes the claim to power over the sky, over light and darkness, over the whole cosmos and the ability to predict the future (Budde, 2002). A little sun disk, with or without the serpent Uraeus, signifies the solar power. There also exists the idea that the horns of a crown touch and open the sky (Budde 2002, 72). This point of view is not casual. Cornuted crowns, like feather bonnets, indicate the shaman's transformation during altered states of consciousness (Rappenglück 2008, 64; Rappenglück 2009, 147-148). The Indo-European word **ker- (*er-")* meaning 'horn' and 'head' is the root of Latin cerebrum 'brain' (Pokorny 1959, 574-577). The Hebraic *qaran* means to emit rays, to shine, and horned, that is to have and to present power. Ancient and recent cultures of the northern and some of the southern hemisphere compared the aureole of sun beams with the widely ramified deer antlers (Rappenglück 2007, 64). They linked the sacral cosmic deer with a shamans' trance: the 'fiery', 'hot' and 'solar' nature of hallucinogenic plants, which they used as an aid among others to get into an ecstatic state, produced a subtle emanation through the brain (cerebrum) and the skull into an aura around the head (horns, antlers). Cornuted animals (deer, bovine etc.) symbolized the climax of astral power (solar, lunar, stars) and the highest states of trance (Hoppál 2002, 111). From this they derived the idea of the sacred cornuted animal, which acts as spirit helper and safely guides a shaman (or also a dead person) on his way through alternative states of consciousness, imagined as different spatiotemporal, transcendent worlds. Equipped with crescent horns they were related to lunar power. In any case, as a being of the primeval time of creation they were responsible for fertilizing the world by water, light and fire, for the cycles of living beings, and last not least for providing hallucinogens. Animals having antlers were associated with solar power and the tree at the centre of the cosmos.

This topic appears again embodied in Korean shaman and royal crowns (Zolla 1985, 101). They were thought to have originated in the northern quadrant: they show trees having either three or four branches, placed one upon the other. The first represented spiritual power, which allowed connecting of the three cosmic strata. The second symbolized secular power, which permitted the domination of the Earth spreading out towards the cardinal points (Ogg 1993, 300).

Obviously the shamanistic concept of a cosmic (celestial) crown, cap, or hat appears to be continued in the ruler's crowns, well-known in various cultures to today. The cap of the Phrygian and Greek god Attis, covered with stars, the Moon and Sun, and the hat of the Germanic god Odin (Wōdan), associated with the wind, the clouds and the sky, are forerunners of the magician's hat (Eisler 1910, 64-65). The Sassanian crown shows star(s); the crescent Moon and the Sun, symbolizing the cosmic and the mundane (Erdmann, 1951). The Sun has a crown (of rays) in Serbian and Latvian sun myths (Mannhardt 1875, 79-80, 97, 296 fn.1). In a Latvian fairy tale the mythical princess Karalune, a beautiful young woman, wears the Sun as a crown and a star cloak (Mannhardt 1875, 296 fn. 1). Finally even the Great Star (Venus) as a headdress played an important role in Hopi rituals (Geertz 1987, pl. XXII). In addition the Zuni headdress worn during the Corn Maiden ceremony shows stars, the Sun and the Moon (Miller 1997, 182, fig. 10.6).

Celestial masquerade and face paintings

Masks give the shaman another face and support him on his journey through the cosmos. Therefore they sometimes show the Sun and Moon, day and night, as the principal female/male polarity of powers in the universe (Jones and Molyneaux 2002, 28, 51, 54, 194, 201; Krupp 1987, 5; Krupp 1991, 13). They also served for incorporating the cosmic power of the Sun, Moon, Venus, or certain asterisms. The Tsimshian (Pacific Northwest Coast) deliver a myth telling that one of the two sons of the sky god wears a flaming sun mask made out of pine wood. Asleep he emits sparks out of his mouth, which are the stars (Boas 1908, 777). Paintings of the Inca Manco Capac and his wife, Mama Occlo, show him holding up the sun mask, while she presents the mask of the Moon (Jones and Molyneaux 2002, 201, 202).

There are many examples of sun and moon masks all over the world (Krupp 1990, 7-9; Willis and Walter 1994, 216, 256). Especially interesting is an Inuit mask of the Full Moon ghost (Willis and Walter 1994, 216): The wood around the Moon's face signifies the air around the Moon. The concentric rings mark the different cosmic strata (oceanic waters, the Earth, and the sky. According to the Inuit, the souls of the deceased first ascend into heaven to get spiritual power. Then, together with the Moon, they descend to Earth and assume the shape of animals or humans.

The Hopi Mastop Kachina mask, which was worn during winter solstice ceremony, shows the cluster of the Pleiades over each eye and the Big Dipper on the cheeks (Krupp 1991, 13).

The belt as Milky Way and Zodiac

The belt signifies the border between the upper and lower realms of the cosmos, a male and female, a light and a dark, a spiritual and a material sphere. Such a girdle is depicted on a Koryak shaman's celestial costume (Okladnikova 1998, 333-334, 336, figure 8.8, and plate 14). While the leathern disks signify the stars, asterisms of summer and winter, the embroidered satin belt marks the Milky Way. The Zoroastrianism idea of a very ancient belt or girdle, embroidered with stars and circular wrapped around the sky, the Milky Way (Parpola 1985, 141-142) may be rooted in such shamanistic conceptions. It is divided into three cords with 24 threads each (= 72). The Luiseño natives (North America) had a headband in their medicine bundles which was made of white eagle down feathers and signified the primordial Milky Way (Krupp 1993, 7). The significance of the Milky Way is understandable, because people considered the Milky Way to be their path between the cosmic strata (Rappenglück 1999, 133-134).

Later variants dating to the Roman Principate (27 BC-284 AD) put the Zodiac instead of the Milky Way, as is shown for example by the *baltei*, a textile band of the zodiac worn across a shoulder down to the hip. The torso of Helios (?), the Artemis of Ephesos (probably 1st century BC), or the female figure (Virtus?), are well-kwon examples (Eisler 1910, 96; Gundel 1992, 101, 213, 216, 214 no. 29, 215 nos. 30 and 31).

Celestial footwear

Usually footwear is symbolical related to the Earth or even to the netherworld (Hoppál 2002, 114), but there are examples which show a celestial meaning. The legging of an Arapaho girl (Miller 1997, 260, figure 11.13) shows two triangles (mountains) near the ankle, and between them the Morning Star is situated. Along the back above the heel a diamond signifies the rising Morning Star. Its position high in the sky is marked by two crosses. The leather strip running up the front of the legging signifies the Milky Way along its outer edges. The six diamonds ringing the foot display an asterism, probably Camp Circle (CrB). In addition the Skidi-Pawnee put the symbol of the Morning Star on moccasins (Taylor 1995, 76.)

Body paintings and tattoos

The body painting of certain asterisms like the Pleiades (Krupp 1991, 13, 16) or the Big Dipper (Krupp 1987, 5) belongs to the make-up of shamans, too, and indicated the direction of their cosmic journey or acted as tutelary spirits. Bull Child, a shaman of the Blackfeet (USA) is dressed with symbols of the heavens (Krupp 1987, 5). The pendent made of a shell is thought to be a gift of the creator sun. It is used for purposes of weather magic to cause sunshine or rain. The Moon is displayed by symbols on the chest and the cheeks. Blue points on the yellow painted body symbolize the starry sky. The pattern of points on the left cheek indicates the asterism of the Big Dipper, the husband of the creator. The injured leg (Moon) had born these seven star children. In autumn the Chumash (USA) celebrate the ceremony of Hutash, their earth goddess. During the festivity the shaman paints seven white points, the Pleiades, on one of his cheeks (Krupp 1991, 13, 16). Apache Mountain spirit dancers in Arizona (USA) have a four-pointed star painted on their chests, indicating cosmic cardinality (Krupp 1983, 287-288)

A Tabwa woman (Zaire) shows a peculiar tattoo (Basset 1998, 30-13, and figures 3.7 and 3.8): the V-shaped sign and vertical line separates the right side (maleness, birth, vitality, the good, east and light) from the left (femaleness, death, dependency, the bad, west and darkness). The vertical line represents the north-south axis, the Milky Way, drainage divides and migration paths of mythic heroes. The isosceles triangle marks the new Moon, related to cyclical renewal, and the ambivalent power of the chief. During the New Year Sun Dance ceremony of the Mandan (USA), body painted dancers are representing 'Night' and 'Day' (Müller 1970, 305-315, 313, figure 52). The Omaha (USA) decorate girls during special ceremonies with cosmic tattoos (Ridington 1988, 1998).

Astral appliqués

Beside pure decorative purposes, astral appliqués mostly served for getting cosmic power, as support during initiation, dreams and trance, or for showing the horoscope at birth or an important stage of life. Some examples, coming from different cultures and times, illustrate this.

The Northern Asian shaman's dress is regularly equipped with appliqués of cosmic power animals or tutelary spirits, bones, depicted skeletons, fringes, bells and metal pieces (Rappenglück, 1999; Hoppál 2002, 53, 71, 101, and 104; Storm 2000, 214). The costume of an Altai shaman from the 19th century AD, preserved at the Museum of Anthropology and Ethnology, Peter the Great, St. Petersburg, Russia, e.g., has celestial pendants fixed on the back: a half-disk symbolizing the Moon, a full disk signifying the Sun, and a ring as representation of the rainbow (Storm 2000, 214).

Among the metal appliqués full disks and crescents symbolize the polarity of cosmic powers given by the Sun and the Moon. They also served as a kind of mirror to look into the future and as guides on the shaman's cosmic journey. Sometimes pendants are decorated with representations of sketchy cosmological representations, as in the case of a Siberian shaman (Hoppál 2002, 53, 101). Pearl shells, representing Venus, are worn as pendants by the Ngalea people in Australia (Tindale 1996, 371, figure 6). The celestial symbolism of other objects like pendants, aprons, or blankets is associated with concepts and rituals of fertility. Other appliqués indicate that the respective holder is equipped with cosmic power: Mostly the basic powers of the Sun and the Moon are shown, but stars and asterisms, thunder and lightning, and in rare cases meteorites, are shown.

The Huichol people of Mexico wear certain discs showing cosmological motives on the back or chest of costumes during their ceremonies (Deimel 2000, 156-157). A disk together with a feathered stick was offered to the sun god Tayaupá. At the centre the Sun is displayed and 13 temples are set around it. Symbols of the gods in the sky, of healers, the double-headed eagle, the jaguar, the fire, the sky bird, the stag, water and the four cardinal directions etc. are shown.

The headdresses of the Crow (USA), from about 1880, occasionally show the symbol of the Morning Star (Taylor 1995, 77). Star motives as appliqués are often used by the Crow, Cheyenne and Apache natives (Taylor 1995, 77).

A pendant worn by a shaman of the Nerchinsk Evenk Tyngirin (Siberia) shows a dotted sky (the stars), the Sun in the centre, the division into the four parts of the world,

Venus, and other planets (Okladnikova 1998, 333, 335 Fig. 8.7). A pendant of the chiefs and kings of the Ashanti (Ghana) shows the crescent combined with the horns of a ram, related to fertility (moon, menstruation, rain (Owusu 1998, 202-203).

An apron of a Thompson boy (British Columbia) with two Moons and six stars was worn during his puberty ceremony (Miller 1997, 112, figure 7.2).

Meteorite has been used as appliqués on headdresses, e.g., in the Hopewell Culture, 400 BC to 400 AD (Prufer 1961, 342, 348).

The headdress of an Apache (USA) had a meteorite fixed upon it (Jones and Molyneaux 2002, 17). A falling star, seen during a dream, is considered to indicate the meeting with a powerful protective spirit. The badge of a dead shaman of the Dolgans shows the symbols of the Sun and Moon (Hoppál 2002, 71).

A ring from Tartus in Syria has the names of zodiacal signs (Taurus, Leo, Scorpio and Capricorn) and planets on it. It is a birth horoscope and dates from 327 AD (Gundel 1992, 300, no. 344). Another polygonal ring showing the images of the twelve zodiacal signs comes from a place in the Eastern Mediterranean area (Gundel 1992, 232, no. 79). A silver chain of the Roman Principate (27 BC-284 AD) consists of 12 pendants: four platelets, six circular and two oval discs, showing on each the image of one zodiacal sign (Gundel 1992, 264, no. 204). The ideas behind astral appliqués still existed in more or less obvious ways in later times: even Queen Elizabeth I (1533-1603), wearing a medallion with the depiction of an armillary sphere on it (Wilson 2006, 154, 4a), liked to display herself thereby as the Muse Urania, representing astronomy and astrology, as well as a lot of other connotations (Wilson 2006).

A few examples of today's cosmic fashion

Even today a few examples exist which illustrate the vestiges of the original ideas, though often barely understood: jewellery, badges, flags costumes, promotional signs of anything and everything, and much more, are decorated or shaped according to celestial objects. Frequently they serve as badges, which gives social status or they are used as a nice decoration. They however still function as cosmic amulets for those who believe in another transcendent sphere, providing ‘power’, ‘protection’ and ‘guiding’.

References

Arthur, L. B. (ed). 2000. Undressing Religion: Commitment and Conversion from a Cross-Cultural Perspective. Dress, Body, Culture Series. Oxford and New York, Berg Publishing.

Basset, T. J. 1998. Indigenous Mapmaking in Intertropical Africa. In D. Woodward and G. M. Lewis (eds), The History of Cartography, Vol. 2, Book 3, 24-48. Chicago and London, The University of Chicago Press.

Baudez, C.-F. 2000. The Maya King's Body, Mirror of the Universe. Anthropology and Aesthetics 38, 134-143.

Baumann, H. 1995. Das doppelte Geschlecht. Studien zur Bisexualität in Ritus und Mythos. Berlin, Dietrich Reimer.

Baumgärtel-Fleischmann, R. 1990. Der Sternenmantel Kaiser Heinrichs II. und seine Inschriften. In W. Koch, Epigraphik 1988, Fachtagung für mittelalterliche und neuzeitliche Epigraphik. Graz, Wien Österreichische Akademie der Wissenschaften, philosophisch-historische Klasse, Denkschriften 213, 105-125, Wien.

Beck, R. 2006. The Religion of the Mithras Cult in the Roman Empire: Mysteries of the Unconquered Sun. Oxford, Oxford University Press.

Bekbassar, N. M. 2005. Astronomical Practices and Ritual Calendar of Euro-Asian Nomad. Folklore 31, 101-120.

Berger, H., Beinert, W., Wetzel, C., and Kehl, M. et al. 2006. Bilder des Himmels. Die Geschichte des Jenseits von der Bibel bis zur Gegenwart. Freiburg, Basel and Wien, Herder.

Bezold, C. 1926. Babylonisch-assyrisches Glossar. Heidelberg, C. Winter.

Boas, F. 1908. Eine Sonnensage der Tsimschian. Zeitschrift für Ethnologie 40, no. 5, 776-797.

Budde, D. 2002. Die den Himmel durchsticht und sich mit den Sternen vereint. Zur Bedeutung und Funktion der Doppelfederkrone in der Götterikonographie. Studien zur Altägyptischen Kultur 30, 57-102.

Buijs, C. and Petersen, M. 2004. Festive Clothing and National Costumes in 20th Century East Greenland. Études/Inuit/Studies 28, no. 1, 83-10.

Caelus / Coelus, Collection of Castings, Berlin, FU, VIII 696. Accession number SH 340.

Cammann, S. 1947. A Robe of the Ch'ien-lung Emperor. Journal of the Walters Gallery 9, 7-17.

Cammann, S. 1948. Cosmic Symbolism on the Dragon Robes of the Ch'ing Dynasty. Art and Thought, London, 116-128.

Cammann, S. 1951. The Symbolism of the Cloud Collar Motif. The Art Bulletin 33, no. 1, 1-9.

Chamberlain, V. D. 1982. When Stars Came Down to Earth: Cosmology of the Skidi Pawnee Indians of North America. Los Altos, Ballena Press.

Chaussonet, V. 1988. Needles and Animals: Women's Magic, in W. W. Fitzhugh and A. Crowell (eds), Crossroads of continents. Washington D.C., Smithsonian.

Cohen, E. 1987. The Hmong Cross: A Cosmic Symbol in Hmong (Meo) Textile Designs. Anthropology and Aesthetics 14, 27-45.

Cox, M. R. (1967, 2nd revised edition). Cinderella: Three Hundred and Forty-Five Variants of Cinderella, Catskin, and Cap o' Rushes, Abstracted and Tabulated, With a Discussion of Mediaeval Analogues, and Notes. Nendeln, Liechtenstein, Kraus Reprint Limited.

Deimel, C. 2000. Aus Symbolen eine Welt. In C. Deimel and E. Ruhnau (eds), Jaguar und Schlange. Der Kosmos der Indianer in Mittel- und Südamerika, 151-168. Berlin, Dietrich Reimer.

Eisler, R. 1910. Weltenmantel und Himmelszelt. Eine religionsgeschichtliche Untersuchung zur Urgeschichte des Antiken Weltbildes, 2 Vol. München, C.H. Beck'sche Verlagsbuchhandlung.

Elliot Smith, G. 1919. The Evolution of the Dragon. Manchester, University Press.

Ellis, F. H., and Hammack, L. 1968. The Inner Sanctum of Feather Cave, A Mogollon Sun and Earth Shrine Linking Mexico and the Southwest. American Antiquity 33 no.1, 25-43.

Erdmann, K. 1951. Die Entwicklung der sāsānidischen Krone. Ars Islamica 15/16, 87-123.

Evelyn-White, H. G. 1914. The Homeric Hymns and Homerica with an English Translation by Homeric Hymns. Cambridge MA, Harvard University Press; London, William Heinemann.

Freedman, D. 1972. "şubāt bāšti": A Robe of Splendor. Journal of the Ancient Near Eastern Society 4, no. 2, 91-95.

Freyer-Schauenburg, B. 2007. Die Sternenmantel des Kaisers Trajan. In V. Brinkmann and W. Wünsche (eds), Bunte Götter. Die Farbigkeit antiker Skulptur. Hamburg, Museum für Kunst und Gewerbe.

Geertz, A. 1987. Hopi Altar Iconography. Iconography of Religions 10, no. 5. Leiden, E. J. Brill.

Greater Bundahišn (=Zand-Ākāsīh: Iranian or Greater Bundahišn). Behramgore Tahmuras Anklesaria and Dastur Framroze Ardeshir Bode, eds. Bombay, Published for the Rahnumae Mazdayasnan Sabha by its Honorary Secretary Dastur Framroze A. Bode, 1956.

Hoppál, M. 2002. Das Buch der Schamanen. Europa und Asien. München, Econ Ulstein List.

Hoppál, M. 2004. Cosmic Symbolism in Siberian Shamanhood. In E. P. Firnhaber and A. Leete (eds), A. Shamanism in the Cultural Context, 178-190. Boca Raton, BrownWalker Press.

Grimm, J., and Grimm W. 1812/1815. Kinder- und Hausmärchen, vol. 1. Berlin, Realschulbuchhandlung.

Gundel, H. G. 1992. Zodiakos. Tierkreisbilder im Altertum (= Kulturgeschichte der antiken Welt 54). Mainz, Philipp von Zabern.

Heiler, F. 1979 (2nd edition). Erscheinungsformen und Wesen der Religion. Stuttgart, W. Kohlhammer.

Huth, O. 1996. Sonne-, Mond- und Sternenkleid. In W. Laiblin, and V. Kast (eds), Märchenforschung und Tiefenpsychologie, 151-160. Darmstadt, Primus.

Jones, D. M. and Molyneaux, B. L. 2002. Mythologie der neuen Welt. Die Enzyklopädie der Mythen in Nord-, Meso- und Südamerika. Reichelsheim, Edition XXL.

Knowlton, E. C. 1920. The Goddess Nature in Early Periods. The Journal of English and Germanic Philology 19, no. 2, 224-253.

Krupp, E. C. 1987. The Dipper in Disguise/Ursa-Major. Griffith Observer 51, no. 12, 3-18.

Krupp, E. C. 1990. Facing the Sun. Griffith Observer 54, no. 1, 2-13.

Krupp, E. C. 1991. Seven Sisters. Griffith Observer 55, no. 1, 2-16.

Krupp, E. C.1983. Echoes of the Ancient Skies. The Astronomy of lost Civilizations. New York and Oxford, Oxford University Press.

Lee, S. E. 1993 (5th edition). A History of Far Eastern Art. New York, Prentice Hall.

Leopold, J. 1995. Elementare kosmische Symbolik in Federkronen der nördlichen Plains. KULT-UR-NOTIZEN 18, 17-21.

MacDonald, D. R. 1987. There is no Male and Female: The Fate of a Dominical Saying in Paul and Gnosticism Philadelphia, Fortress Press.

Mannhardt, W. 1875. Die lettischen Sonnenmythen. Zeitschrift für Ethnologie 7, 73-330.

McCrickard, J. 1990. Eclipse of the Sun. An Investigation into Sun and Moon Myths. Glastonbury, Gothic Image Publications.

Miller, A. 1991. The Garments of the Gods in Japanese Ritual. Journal of Ritual Studies 5, 33-55.

Miller, D. S. 1997. Stars of the First People. Boulder CO, Pruett.

Müller, W. 1970. Glauben und Denken der Sioux. Zur Gestalt archaischer Weltbilder. Berlin, Diedrich Reimer.

Muhammad, W. 2011. The Shadow of God: Incarnation and the Divine Body in Second Temple Priestly Tradition. http://drwesleywilliams.com/yahoo_site_admin/assets/docs/Sapphiric_Part_I.187113017.pdf

Muhammad, W. (Williams, W.) 2015. Sapphiric God: Esoteric Speculation on the Divine Body in Post-Biblical Jewish Tradition. Harvard Theological Review, in print.

Neugebauer, K. A. 1921. Antike Bronzestatuetten. Berlin, Schoetz und Parrhysius.

Nimuendajú, C. 1919/1920. Bruchstücke aus Religion und Überlieferung der Šipáia-Indianer. Beiträge zur Kenntnis der Indianerstämme des Xingú-Gebietes, Zentralbrasilien. Anthropos 14/15 (4/6), 1002-1039.

Ogg, L. 1993. Korean Mythology, in Y. Bonnefoy and W. Doniger (eds), Asian Mythologies, 296-300. Chicago and London, The University of Chicago Press.

Okladnikova, E. 1998. Traditional Cartography in Arctic and Subarctic Eurasia. In D. Woodward and G. M. Lewis (eds.), The History of Cartography, Vol. 2, Book 3, 329-349. Chicago and London, The University of Chicago Press.

Oppenheim, A. L. 1949. The Golden Garments of the Gods. Journal of Near Eastern Studies 8, no. 3, 172-193.

Owusu, H. 1997. Symbole Afrikas. Darmstadt, Schirner.

Parpola, A. 1985a. The Sky-Garment. A Study of the Harappan Religion and Its Relation to the Mesopotamian and Later Indian

Religions. Studia Orientalia 57, Helsinki, Societas Orientalis Fennica.

Parpola, A. 1985b. The Harappan Priest-King's Robe and the Vedic Tarpya Garment: Their Interrelation and Symbolism (astral and procreative). In J. Schotsmans and M. Taddei (eds), South Asian Archaeology 1983 (Istituto Universitario Orientale, Dipartimento di Studi Asiatici, Series Minor, 23), Vol. I, 385-403. Naples, Istituto Universitario Orientale, Dipartimento di Studi Asiatici.

Payne, B. 1997 (2nd edition). The History of Costume: From Ancient Mesopotamia through the Twentieth Century. Upper Saddle River, New Jersey, Prentice Hall.

Podella, T. 1996. Das Lichtkleid JHWHs: Untersuchungen zur Gestalthaftigkeit Gottes im Alten Testament und seiner altorientalischen Umwelt. Forschungen zum Alten Testament 15. Tübingen, Mohr Siebeck.

Pokorny, J. 1959. Indogermanisches etymologisches Wörterbuch, Vol. 2. Bern, München, Francke.

Prechtel, M., and Carlsen, R. S. 1988. Weaving and Cosmos amongst the Tzutujil Maya of Guatemala. Anthropology and Aesthetics 15, 122-132.

Prufer, O. H. 1961. Prehistoric Hopewell Meteorite Collecting: Context and Implications. The Ohio Journal of Science 61, no. 6, 341-352.

Rapp. A. 1872. Die Mänade im griechischen Cultus, in der Kunst und Poesie. II. (Schluss). Rheinisches Museum für Philologie, Neue Folge 27, 562-611.

Rappenglück, M. 1999. Eine Himmelskarte aus dem Eiszeitalter? Frankfurt am Main, Peter Lang.

Rappenglück, M. 2007. Cosmic Spinning and Weaving: Making the Texture of the World. BAR International Series 1647, 161-171.

Rappenglück, M. 2008. Tracing the Celestial Deer – An Ancient Motif and its Astronomical Interpretation Across Cultures, in J. Vaiškūnas (ed.), Astronomy and Cosmology in Folk Traditions and Cultural Heritage (=Archaeologia Baltica 10), 62-65, Klaipeda, Klaipeda University Press.

Rappenglück, M. 2009. Heavenly Messengers: The Role of Birds in the Cosmographies and the Cosmovisions of Ancient Cultures. In J. A. Rubiño-Martín, J. A. Belmonte, F. Prada and A. Alberdi (eds), Cosmology across cultures (ASP Conference Series 409), 145-150. Orem UT, Astronomical Society of the Pacific (ASP).

Reichel-Dolmatoff, G. 1974 (2nd edition). Amazonian Cosmos: The Sexual and Religious Symbolism of the Tukano Indians. Chicago and London, The University of Chicago Press.

Ridgway, B. S. 1990. Birds, "Meniskoi", and Head Attributes in Archaic Greece. American Journal of Archaeology 94, no. 4, 583-612.

Ridington, R. 1988. Images of Cosmic Union: Omaha Ceremonies of Renewal. History of Religion 28, no. 2, 135-150.

Ridington, R. 1998. The Cry of the Living Creatures: An Omaha Performance of Blessing. Anthropologica 40, no. 2, 183-196.

Ridington, R. and Ridington, K. 1970. The Inner Eye of Shamanism and Totemism. History of Religion 10, no. 1, 49-61.

Rieff Anawalt, P. 2007. The Worldwide History of Dress. London, Thames and Hudson.

Rist, J. M. 1967. A Common Metaphor. In J.M. Rist (ed.) Plotinus: The Road to Reality, 188-198. London, Cambridge University Press.

Sauren, H. 1984. Die Kleidung der Götter. Visible Religion 2, 95-117.

Schafer, E. H. 1978. The Capeline Cantos: Verses on the Divine Loves of Taoist Priestesses. Asiatische Studien 32, 5-65.

Schafer, E. H. 1981. Wu Yün's "Cantos on Pacing The Void". Harvard Journal of Asiatic Studies 41, no. 2, 377-415.

Smith, G. E. 1919. The Evolution of the Dragon. Manchester, University Press.

Storm, R. 2000. Die Enzyklopädie der östlichen Mythologie. Reichelsheim, Edition XXL

Taylor, C. 1993. Saam: The Symbolic Content of Early Northern Plains Ceremonial Regalia. Wyk (Foehr), Verlag für Amerikanistik.

Taylor, C. F. 1995. Myths of the North American Indians. London, Laurence King Publishing.

Tedlock, B. 1983. Zuni Sacred Theater. American Indian Quarterly 7, no. 3, 93-110.

Tindale, N. B. 1996. Celestial Lore of Some Australian Tribes. In V. D. Chamberlain, J. B. Carlson and M. J. Young (eds), Songs from the Sky. Indigenous Astronomical and Cosmological Traditions of the World, 358-379. Bognor Regis, Ocarina Books.

Vollmer, J. E. 2000. Clothed to Rule the Universe. Art Institute of Chicago Museum Studies 26, no. 2, 12-51, 99-105.

Weidner, E. F. 1931-1932. Das Himmelskleid. Archiv für Orientforschung 7, 115-116.

Weiner, A. B., and Schneider, J. (eds). 1989. Cloth and Human Experience. Washington, Smithsonian Books.

Werness, H. B. 2000. The Continuum Encyclopaedia of Native Art. Worldview, Symbolism and Culture in Africa, Oceania and Native North America. New York and London, Continuum.

Wessing, R. 1986. Wearing the Cosmos: Symbolism in Batik Design. An Interdisciplinary Journal of Southeast Asian Studies 2, no. 3, 40-82.

Widengren, G. 1945. The Great Vohu Manah and the Apostle of God: Studies in Iranian and Manichaean Religion (Uppsala Universitets Åarsskrift 1945.5). Uppsala, A.-B. Lundequistska Bokhandeln.

Willis, R., and Walter, R. 1994. Bertelsmann Handbuch Mythologie. Gütersloh/München, Bertelsmann Lexikon Verlag.

Wilson, J. 2006. Queen Elizabeth I as Urania. Journal of the Warburg and Courtauld Institutes 69, 151-173.

Winton, A. F. 2000. Creative Clothing Design Inspired by Descriptions of Cinderella's Clothing Recorded in 322

Cinderella Story Variants. MA Thesis. Madison WI, The Graduate College, University of Wisconsin-Stout.

Wolff, M., and Opitz, D. 1935-1936. Jagd zu Pferde in der altorientalischen und klassischen Kunst. Archiv für Orientforschung 10, 317-359.

Zaehner, R. C. 1955. Zurvan: A Zoroastrian Dilemma. Oxford, Clarendon Press.

Zerries, O. 1977. Die Bedeutung des Federschmuckes des südamerikanischen Schamanen und dessen Beziehung zur Vogelwelt. Paideuma 23, 277-324.

Zerries, O. 1979. Das Federdiadem: Ein Sonnensymbol bei südamerikanischen Indianern. In R. Hartmann and U. Oberem, (eds), Estudios Americanistas II, 314-321. Anthropos 21. St. Augustin, Anthropos Institute.

Zingerle, I. V., and Zingerle, J. 1911. Kinder- und Hausmärchen aus Tirol. Innsbruck, Schwick.

Zinner, Ernst. 1939. Der Sternenmantel Kaiser Heinrichs. Himmelskunde und Rechenkunst im alten Bamberg. Bamberg, Kleine Bamberger Bücher 8.

Zolla, E. 1985. Korean Shamanism. Anthropology and Aesthetics 9, 101-113.

NEOLITHIC CULTURES

ASTRONOMY, LANDSCAPE AND POWER IN EASTERN ANATOLIA

JUAN ANTONIO BELMONTE AND A. CÉSAR GONZÁLEZ GARCÍA

Abstract: Since the earlier hunters-gatherers who built the impressive sanctuaries at Göbekli Tepe ten thousand years ago to the Sabeans of Harran, every single culture in eastern Anatolia looked to the sky in search of inspiration, guidance and political control. In a recent campaign, the Göbekli Tepe monuments were studied in a search for traces of astronomical skill in the builders of such impressive structures, the oldest stone monuments ever erected by humankind. The ancient ruins of the Kingdom of Commagene (first century BC), including the hierothesion of Antiochos I at the summit of Nemrud Dag and that of his wife at Karakush, were inspected in depth with the aim of challenging previous ideas and testing new hypotheses in relation to such evocative monuments. These were indeed impressive manifestations of power where astronomy presumably played a substantial role. Finally, we researched the ancient Sabean ruins at Sogmatar Harabesi, in northwestern Mesopotamia, where a set of buildings has been identified as a cluster of temples devoted to the seven planetary deities. Our new data, obtained on site, will be useful to test this speculative exercise.

Keywords: constellations, alignments, planetary divinities, Göbekli Tepe, Commagene, Karakush, Sogmatar, Sabeans

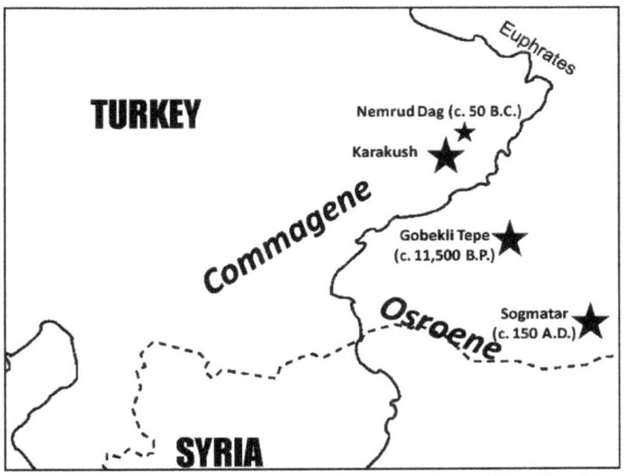

Figure 1. Map of the Euphrates Valley region where the three sites discussed in the text are located. The figure also illustrates the approximate chronology.

1. Introduction

Although our field campaign in Turkey during the summer of 2009 had the Hittites as the main objective (see González García and Belmonte, 2011), we were indeed interested in other aspects of Anatolian cultures where astronomy could be manifested within a certain cultural context. Notably, we concentrated in three sites in southeastern Anatolia where manifestations of power would have been reflected in astronomical terms. In subsequent sections, we will analyse each of them independently (see Figure 1).

2. Göbekli Tepe: the temples of 'ten thousand' years

Until very recent times, the megalithic monuments in Europe were credited as the earliest stone monuments showing hints of potential astronomical knowledge. However, a recent discovery in the steppes of southeast Anatolia has disrupted our previous ideas. There, on a barren isolated hill called Göbekli Tepe, Hill of the Navel, a German-Turkish team of archaeologists (Schmidt, 2006) are excavating a cluster of remarkable stone monuments constructed of large T-shaped pillars and dry-stone walls (see Figure 2). They were built by a completely unknown hunter-gatherer society more than 11,000 years ago, and there have been certain speculative claims on possible solstitial alignments in these structures.

Figure 2. One of the impressive T pillars in the centre of Göbekli Tepe temple D which might represent a very stylized human figure. The side of the pillar shows an enigmatic engraving, including what could be the earliest representation of the crescent and the disk, later symbol of the lunar divinity of Harran, a most important deity of the region for centuries. Photographs courtesy of M. Sanz de Lara and adapted from Schmidt (2006).

These monuments are mostly ellipsoidal in form and a favourite orientation could be established for the monuments' entrances. However, our inspection on site showed no apparent selected pattern and calls for solstitial alignments should be taken with scepticism. However, among these monuments one presents a nearly rectangular shape with sides almost perfectly aligned according to the cardinal directions (see Figure 3). This circumstance suggests that we are faced with a society that looked at the sky and used it as a guide to find appropriate ways of orientation in space and, almost certainly, also in time.

Figure 3. The walls of the rectangular structure built c. 8500 BC in the upper sector of Göbekli Tepe are perhaps the first orientated by humankind in the cardinal directions. One of the pillars (see Figure 4) was decorated with a lion. Either by chance or design, the equinoctial sun was easterly rising in conjunction with our Leo in that epoch. Adapted from a photograph by courtesy of M. Sanz de Lara.

Analysing the profuse decoration of the T-pillars we find motifs that may be interpreted as astronomical representations such as the crescent and the star, which are so common in later cultures of the Middle East and beyond. We even find totemic representations of animals. Allowing a little speculation, some of the animals remind us of constellations such as Leo and Scorpio (see Figure 4) that we can recognize in the skies of other cultures in the region several centuries later. A pictographic and statistical analysis of these animal figurative representations would be desirable to be able to be on firmer ground with these ideas.

However, when analyzing the data of Göbekli Tepe, we are facing two unmistakable problems: their singularity and the complete absence of texts with which the symbolism and the archaeological data could be confronted.

2. Thinking Commagene: aiming to the stars?

The Kingdom of Commagene was a small country between the upper course of the river Euphrates and the mountains of Anti-Taurus in southeast Anatolia. Commagene played, despite her tiny size, a relevant role

Figure 4. Two of the decorated T-pillars of temple D and the rectangular structure (see Figure 3), respectively. Several of these pillars show animal depictions—perhaps totemic ones—later to be identified with constellations of the people occupying the same area millennia later. Diagram adapted from a photograph by M. Sanz de Lara and adapted from Schmidt (2006).

in the history of the Middle East during the late Hellenistic and early Roman periods as a buffer state between the powerful Seleucid (later Roman) and Parthian Empires. Antiochos I Theos (c. 69-36 BC) arguably was the most important of her kings, governing for more than 30 years in one of the most challenging periods of the region's history.

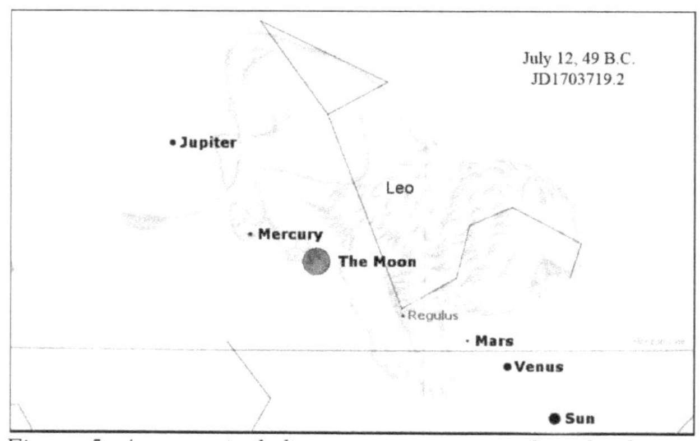

Figure 5. Astronomical diagram as represented in the famous lion slab of Nemrud Dag, corresponding to Leo's setting after sunset for 12 July 49 BC in the Gregorian proleptic calendar. This date has potential connections with those yielded by the orientation of the main architectural elements of the hierothesion (see Figure 4). Diagram created using StarryNight Pro 6.0 software.

The world heritage site of the *hierothesion* of Antiochos I at Mount Nemrud certainly constitutes one of the most fascinating historical enigmas in human culture worldwide. The monument includes the famous lion 'horoscope'. On the slab, a lion with stars on his body, likely the constellation of Leo, is represented together with a crescent moon on his chest and three planets, identified in Greek as Pyroeis of Heracles, Stilbon of Apollon and Phaeton of Zeus, standing for Mars, Mercury and Jupiter. The possibility that the slab depicts a real or schematic astronomical scene or an astrological image introduced the idea of dating the monument and interpreting its nature since the earliest archaeological studies of the site. The most accepted conclusion so far has the support of Neugebauer and Van Hoesen (1959) who argued that the scene might represent a sort of horoscope for the date 7 July 62 BC at the beginning of the reign of Antiochos I.

Figure 6. Antiochos I of Commagene facing sunset at Audnayos 16 in 49 BC, a date close to winter solstice, commemorating his birthday in the western terrace of his hierothesion at Nemrud Dag. His alter ego of the eastern terrace was facing sunrise at Loios 11 of the same year, commemorating his ascent to the throne. These facts suggest a planning of the monuments in the year 49 BC. Photograph courtesy of M. Sanz de Lara.

After a visit to the monument, with on site observations, in the summer solstice of 2009, we have proposed an alternative and more substantiated explanation which deals not only with the lion slab (see Figure 5) but also, and most importantly, with the orientation of the eastern and western terraces of the *hierothesion* and the religious tradition of the country as described in the monument by Antiochos himself (see Figure 6). Our conclusion is that Antiochos' monument reflects the situation of the skies at specific moments of the year 49 BC, as confirmed by the local monumental inscription or *nomos* (Belmonte and González García, 2010).

His son Mithradates II (or perhaps Antiochos himself) built a much smaller *hierothesion* for his mother Isias and other female members of the royal family at Karakush (see Figure 7). According to Beck (1998), certain elements of this smaller but still magnificent monument have suggested an astronomical interpretation. The main elements of the monuments consist of three groups of columns associated with a standing eagle, a recumbent bull and a possible pair (now vanished) of sitting lions related to a *dexiosis* scene. These structures were located at the south, northeast and northwest sectors of the tumulus (see Figure 7) and, according to Beck, might represent stellar alignments to the respective constellations (Aquila, Taurus and Leo) in June of 26 BC, related to a planetary conjunction in the same year.

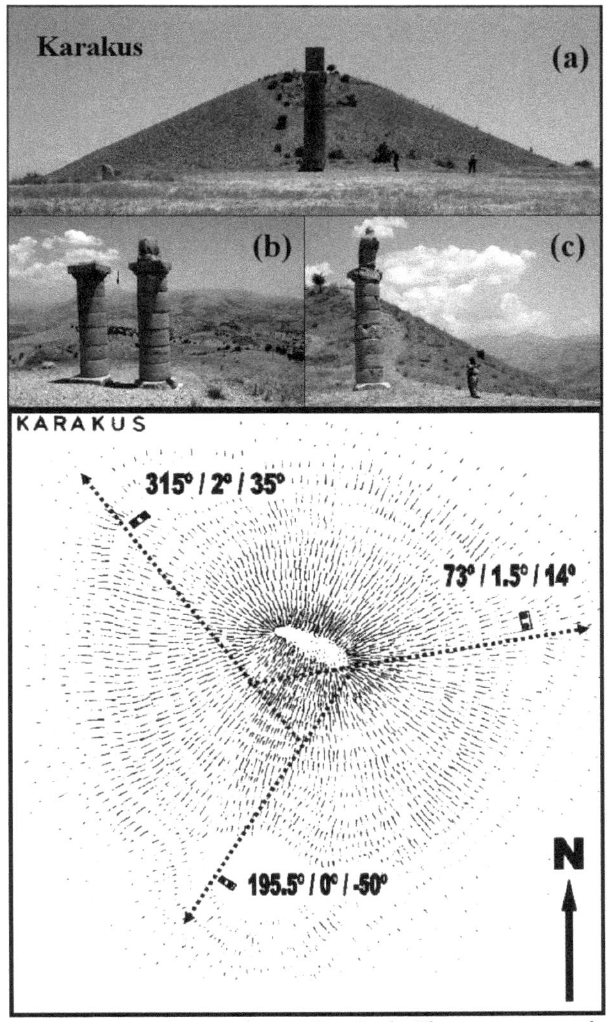

Figure 7. Top: the tumulus at Karakush. The most outstanding characteristic of the site are the three groups of columns related to a dexiosis scene plus two seated lions (a, badly preserved), a recumbent bull (b) and a standing eagle (c). Bottom: orientation data of the three groups of columns in the tumulus of Karakush. The numbers represent measured azimuths, angular heights and the computed declinations. They are not symmetrically arranged with the centre of the structure where the burial chamber, now vanished, was located. See the text for further discussions. Photographs courtesy of M. Sanz de Lara (c) and J. A. Belmonte (a and b).

Our measurements (see Figure 7) do not seem to support this hypothesis. The 'alignment' to Aquila, being a meridian one, is difficult to test. The Taurus relationship is possible. However, the actual orientation of the group related to Leo is far from ecliptic coordinates ($\delta \sim 35°$). So we can neither not support nor fully reject Beck's ideas.

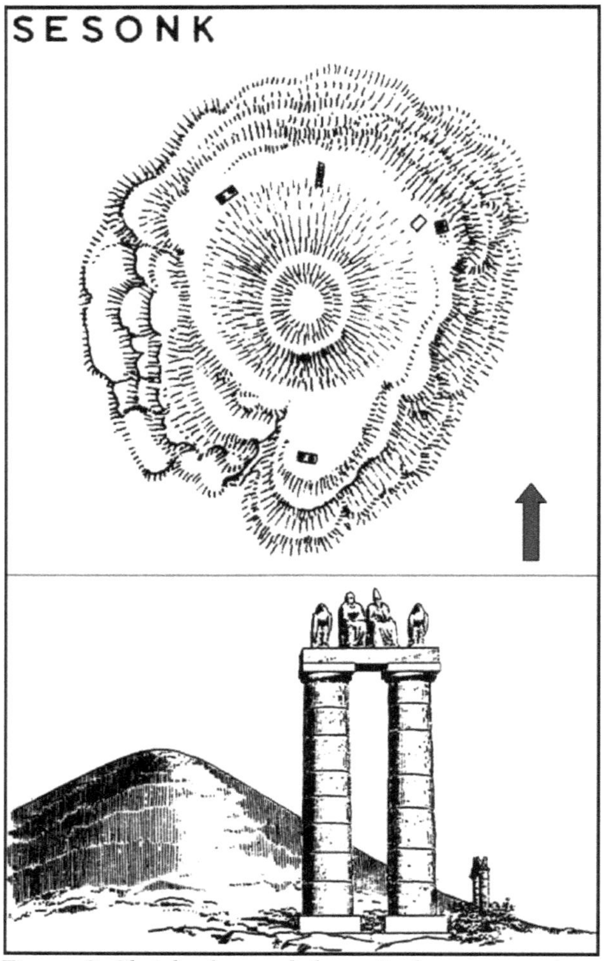

Figure 8. Sketch plan and theoretical reconstruction of what is suspected to be the hierothesion of Mithradates II at Sesonk, to the southwest of Commagene. Notice the general layout, strikingly similar to the one in Karakush. Adapted from Beck (1998).

The certainly curious planning of Karakush is repeated further southwest in the not-as–well-preserved monument of Sesonk (see Figure 8). This has also been attributed to Mithradates II and has a strikingly similar layout to Karakush, although the state of preservation of the column groupings do not permit further conclusions. An archaeo-astronomical study of the site would indeed prove certainly interesting.

3. Sogmatar Harabesi: temples to the planetary divinities?

The Sabeans of the Harran region, descendents of the pagans of the Kingdom of Osrhoene, with its capital at Edessa (present- day Sanliurfa), created in the first millennium of our era a selective and peculiar astral religion, and they were highly regarded as sky-watchers and astrologers (Segal, 1988). This religion was latter affected by the Christian conversion of the Edessa sovereigns, the Arab invasion in the 7th century AD, which converted most of their temples into Muslim sacred precincts, and the Mongol invasion of Mesopotamia in the 1248 AD Consequently, very little has been preserved.

We have been lucky enough to research the ancient Sabean ruins at Sogmatar Harabesi (see Figure 9) in northwestern Mesopotamia. A set of buildings on this site have been identified as a cluster of temples devoted to the seven planetary deities (Segal, 1953) and dated to the second half of the 2nd century (the date of 165 AD has been collected on site).

Figure 9. (a) General panorama of the currently barren and isolated landscape of Sogmatar Harabesi, showing structure I in the foreground. (b) A close-up of structure V (T5), showing the tower-like building, formed by a circular structure lapped on a square one, the underground excavated chamber and the dromos approaching it. Photographs by J. A. Belmonte.

Our idea was to test Segal's hypotheses by establishing a more precise map of the site, measuring the orientation of buildings where possible, and analysing the alignments between different structures. Our results are presented in Figure 10. The orientation of the underground chamber *dromoi* clearly illustrates the importance of what Segal called the 'central shrine' (T0, X) but in contrast to his proposal, not all the *dromoi* are orientated to this direction. Actually, the one belonging to structure T3, supposedly devote to the Moon, is orientated either by chance or by design to the winter solstice sunrise.

The alignments between different structures do not show any preferred pattern. The probability to have an astronomical target by chance for a certain alignment is not low, given the number of structures. Curiously, we have identified a possible Venus alignment between structures V (T6, assigned to Venus by Segal) and III (T2) but it is far from being statistically significant. The 'Sun' structure (II, T7) offers a possible lunar connection to structure I (T1) in contrast to the 'Moon'-assigned structure IV (T3) which, on the contrary, shows a summer solstice sunset alignment. Hence, our preliminary inspection of the site does not seem to support Segal's planetary temple affiliations. Our impression is that these peculiar structures are tombs rather than temples, the superstructures possibly being commemorative monuments. Indeed, a state of the art archaeological inspection (with techniques not available to Segal in the 1950s) and a deeper archaeoastronomical study will be necessary before Sogmatar will unveil all its secrets.

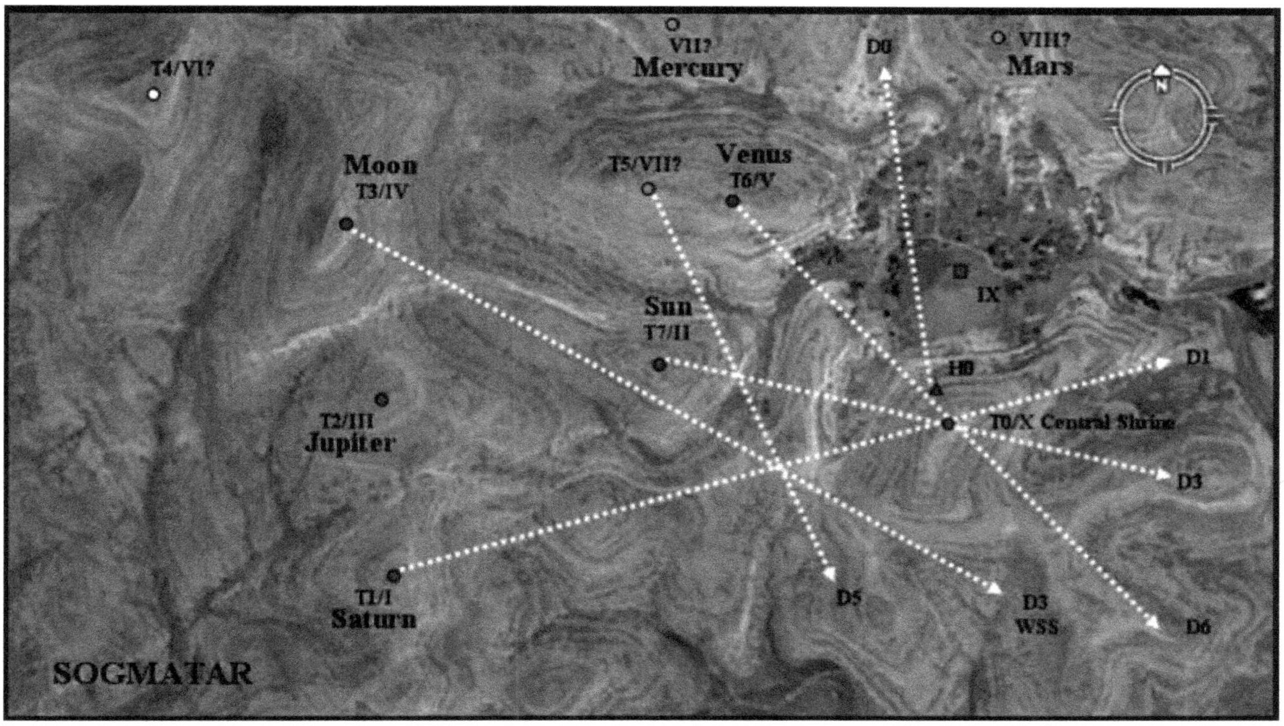

Figure 10. General map of Sogmatar showing the location of the different structures, identified by our own (T) and Segal's (1953) Roman numbering. Segal's hypothesis on 'temple' planetary adscription is also shown. Grey circles stand for super-structures which are apparently no longer visible. Dotted lines stand for the orientation of the different dromoi (D) of underground chambers related to other structures that we could identify on site. See the text for further discussions. Adapted from an image courtesy of Google Earth.

4. Final remarks

We have performed an astronomical analysis of three extremely interesting archaeological sites of Eastern Anatolia built in three *different* historical moments, namely Göbekli Tepe, Karakush and Sogmatar. Our results present hints of the possible existence of some interesting astronomical relations in these sites. This calls for a dedicated research in the region, probably including other nearby sites, before producing a definitively and conclusive cultural astronomy analysis of this fascinating area.

Acknowledgements:

We would like to express our gratitude to Margarita Sanz de Lara for her excellent images. This work is partially financed under the framework of the projects P310793 'Arqueoastronomía' of the IAC, and AYA2007-60213 'Orientatio ad Sidera II' of the Spanish MICINN.

References

Beck, R. 1998. The Astronomical Design of Karakush, A Royal Burial Site in Ancient Commagene: a Hypothesis. Culture and Cosmos 3, 10-31.

Belmonte, J. A. and González García, A. C. 2010). Antiochos' hierothesion at Nemrud Dag re-visited: adjusting the date in the light of new astronomical evidence. Journal for the History of Astronomy 41, 469-81.

González García, A. C. and Belmonte, J. A. 2011. Thinking Hattusha: astronomy and landscape in the Hittite lands. Journal for the History of Astronomy 42,461-94.

Neugebauer, O. and Van Hoesen, H. B. 1959. Greek Horoscopes, American Philosophical Society 48, 14-6.

Schmidt, K. 2006. Sie bauten die Ersten Temple. Munich, Verlag C.H. Beck.

Segal, J. B. 1953. Pagan Syriac monuments in the vilayet of Urfa. Anatolian Studies 3, 97-119.

Segal, J. B. 1988. Los Misteriosos Sabeos: el Culto a los Planetas en la Antigua Harran, in E. Bacon (ed.), Historia de las Civilizaciones II. Madrid, Alianza Editorial 1314, 265-302.

PREHISTORIC SANCTUARIES IN DAUNIA

ELIO ANTONELLO, VITO F. POLCARO, ANNA M. TUNZI AND MARIANGELA LO ZUPONE

Abstract: Beginning in the 5th millennium BC, the farmers living on the wide plain of Daunia dug *hypogea* and holes in sites which could be considered sanctuaries. The most impressive phenomenon is the presence of rows of hundreds of holes, with characteristics that suggest a ritual use and exclude agricultural and post hole applications. The analysis of the site of Trinitapoli indicated a few specific astronomical orientations for the rows. In 2009 a new sanctuary was discovered in Ordona. Just two small areas were excavated and no remains were found inside the holes, so no estimate of their probable age is possible. Unlike the essentially solar and lunar criteria possibly adopted for the alignments in Trinitapoli, it is possible that some stars of the Centaurus-Crux group were used by the community of Ordona, and we cannot exclude the presence of the effects of precession on the alignments. The spectacular region of the sky of the Centaurus-Crux group was pointed out by several astronomers and scholars of the past, and its diffused light effect, discussed by G.V. Schiaparelli a century ago, could have been of some importance for ancient civilizations.
Keywords: Prehistory, Bronze Age, hypogea, sanctuaries

1. Introduction

In the Italian region of Apulia (Puglia) there are many interesting archaeological sites dating from the Neolithic, Roman and Medieval epochs, and some finds date back to the Palaeolithic period. Beginning in the 5th millennium BC, for more than three thousand years the farmers living on the wide plain of the Daunia region in northern Apulia dug *hypogea* and holes for ritual and funerary purposes. To date, two sanctuaries have been found in Trinitapoli and Ordona (Figure 1).

Figure 1. Southern Italy and the location of the archaeological sites of northern Apulia discussed in the present work.

In the Bronze Age sanctuary of Trinitapoli there are some *hypogea*, but the most impressive phenomenon is the presence of rows with hundreds of holes dug in the stratum of calcareous rock located below some tens of centimetres of topsoil. The area affected by this phenomenon is very large, as shown by the most recent excavations. The characteristics of the holes and the remains found inside indicate a ritual use and exclude agricultural or post hole use. An obvious suspicion is that the holes were dug just for tree plantation, e.g., vines as described by the Latin writer Columella in *De re rustica* (1st century AD). However, several features, such as the geometry of the holes, their short spacing, the age of the prehistoric remains found in some of them and this prehistoric epoch being too early for systematic tree planting, tend to exclude such an interpretation. The astronomical analysis suggests that the inhabitants may have adopted a few specific astronomical orientations for the alignments of the rows, and these results were discussed at the SEAC 2008 meeting (Tunzi et al., 2009; for further details, see also Tunzi et al., 2010).

2. Ordona

Last year a new Neolithic sanctuary was discovered in Ordona, about thirty kilometres west of Trinitapoli. The occasion was the installation of a cable connecting the wind turbines of a wind farm. Only two small areas were excavated, named Ponterotto 1 and Ponterotto 2 (Figure 2). Straight rows of holes dug quite carefully were found in both areas. The distance between the circular holes (with a diameter of about thirty centimetres) is generally less than one metre, and that between the rows more than three metres. The larger of the two excavated areas is on the order of 20 metres; the two areas are located rather far apart, about 900 metres. The remains found in some *hypogea* have been dated to the mid-5th millennium BC. The holes, however, contained just soil, and for the present there is no reliable estimate of their true age. The real extent of the sanctuary is not known but it may cover a very large area, as in the case of Trinitapoli, and it is possible it was similarly used for many centuries.

The orientation of the rows is between 194° and 213° south-west (between 14° and 33° north-east). Here we present a preliminary analysis, in which we focus on the south-west orientation, as the astronomical phenomena in the north-east direction appear less interesting. The azimuth values suggest that, unlike the essentially solar and lunar criteria adopted for the alignments in Trinitapoli, it is possible that some specific stars were used by the community of Ordona. We suspect that these stars belong to the group of Centaurus-Crux. At the time of the *hypogea*, the stars of this group set behind the nearby mountains of Deliceto; during the 3rd millennium BC the direction of the setting could be that indicated by the rows of Ponterotto 2. A millennium later, the setting direction could be indicated by the rows of the other area, Ponterotto 1. This group of stars was visible in Southern Italy during the Neolithic-Bronze Age beginning from October to the end of spring (present day calendar, Figure 3). For example, in mid October there was the heliacal rising of γ Crucis while that of α Centauri occurred some days later. The acronychal setting of the stars (just before the sunrise) occurred at the time of the winter solstice, and the heliacal setting occurred some days before the summer solstice.

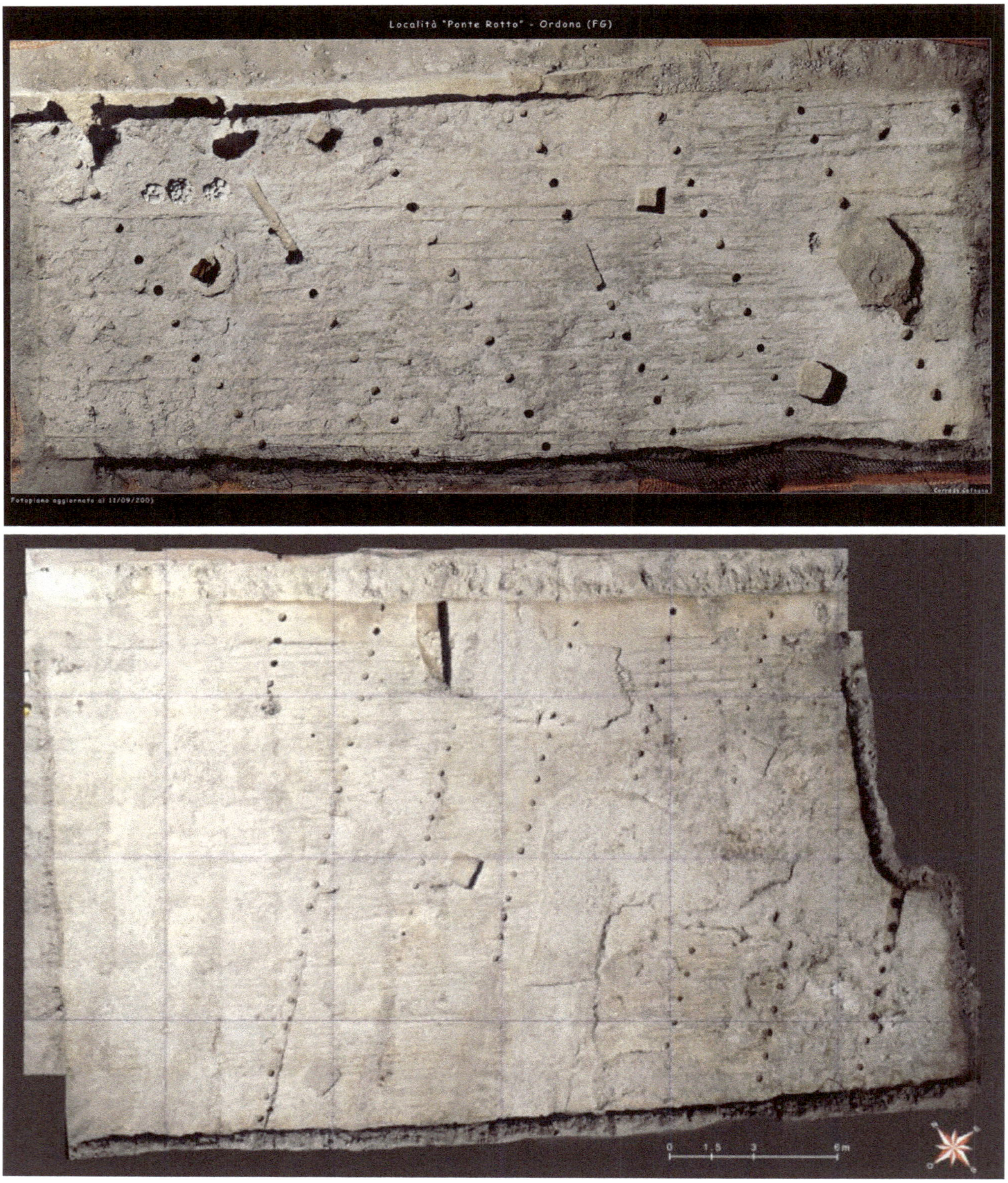

Figure 2. Orthophoto of the two areas of Ponterotto 1 (upper panel) and Ponterotto 2 (lower panel) of Ordona. The holes appear as 'dots'. The separation between two holes is less than one metre, and the distance between the rows is about three metres. The compass in the lower right corner indicates qualitatively the north direction.

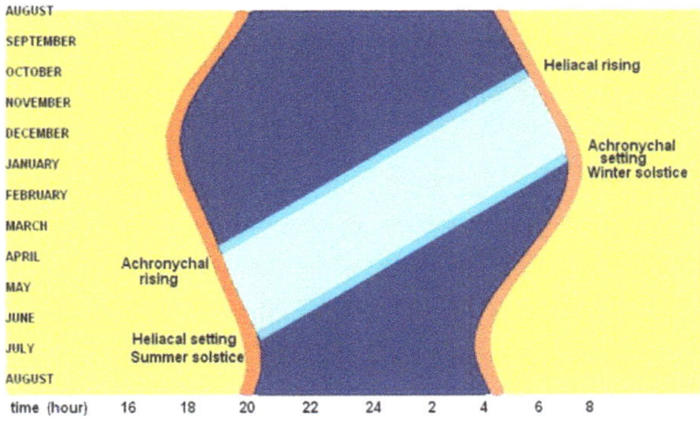

Figure 3. Qualitative plot of the visibility of the Centaurus-Crux stars during the Neolithic-Bronze Age in southern Italy. The visibility is indicated by the band in light colour, while the darkest colour indicates night time (from evening to morning) along the year (the month is shown on the left).

One should note that this group of stars was visible during wintertime. We cannot exclude for the present that the analysis of the alignments shows the effects of precession. In particular, this may be suggested by the progressively slightly different orientations of the rows in the area of Ponterotto 1. Assuming α Centauri as the target star of the people of that time, the holes of Ponterotto 2 should have been dug at the middle of the 3rd millennium BC, and those of Ponterotto 1 from the middle to the end of the 2nd millennium BC (Figure 4). Of course, this is just a preliminary analysis, and further archaeological discoveries could change this scenario.

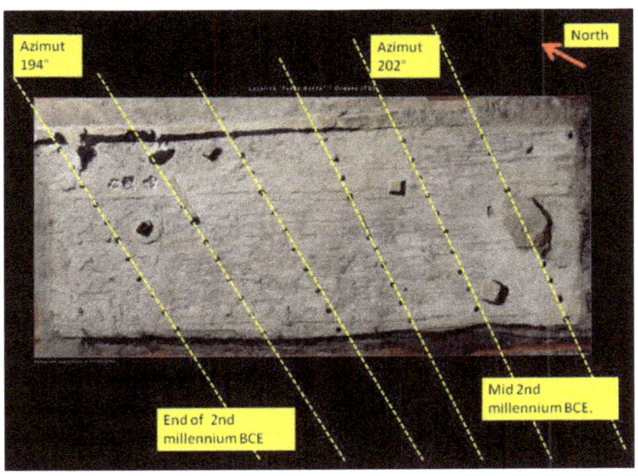

Figure 4. Orientation and hypothetical dates of the hole rows of the site of Ponterotto 1 (Ordona).

3. The Centaurus-Crux region

In support of the Centaurus-Cross interpretation, we recall that Hoskin (2001, 42-51) pointed out the possible importance of these stars for the sanctuaries of the Mediterranean basin. However, it may be that he tacitly assumed, as we did, that the targets of the inhabitants were probably the brightest stars defining the constellations. This seems rather obvious, but this is just an assumption. More than one hundred years ago, Schiaparelli published an interesting study on the astronomy in the Old Testament, which was quickly translated into German and English (Schiaparelli, 1903) and which contains a potentially important discussion regarding this issue. As I will try to show, it could put this issue literally 'in a new light'.

Schiaparelli discussed carefully the identification of some constellations and asterisms in the Old Testament, according to different ancient versions and translations of the Bible. A verse in the Book of Job (Job 9, 9) quotes the Bear (*hasc*), Orion (*kesil*) and the Pleiades (*kimah*). Moreover, the last two words of the verse, *chadre teman*, were translated by Schiaparelli as 'chambers of the south', that is, the verse could be translated 'who made the Bear and Orion, the Pleiades and the chambers of the south', as in the New Revised Standard Version (1989) of the Bible. Schiaparelli pointed out that the word *chadre* should indicate the innermost and private chambers of a building, while *teman* indicated both the right and the southern direction. Schiaparelli probably was exceedingly enthusiastic when he declared that the author of the Book of Job wanted (and not just 'wished' as in the English translation of his book) to indicate by these two words some brilliant southern constellation, since no one is able to know what the author of the Book of Job really wanted or wished to indicate. Schiaparelli discussed at length this spectacular case, though he never travelled beyond Europe and therefore he never saw the southern constellations. What is the reason for his enthusiasm? He described the richness of both bright and faint visible stars in a region of the southern sky from α Argus (Canopus) to α Centauri, taking into account the studies of the astronomers and scholars of his time. The star α Argus is today α Carinae, since Argus has been divided into Carina, Vela and Puppis. This part of the sky is 'the splendour of the southern heavens', as declared by von Humboldt (1858, 146-147). Moreover, Schiaparelli quoted his own study on the distribution of visible stars, and he wrote that this part of the sky produces in the atmosphere a sort of twilight illumination. Indeed he took from von Humboldt's book an impressive description (reported in English even in the original Italian edition of his book) of the effect of the increase of diffuse light when the Southern Cross has risen:

> Such is the general blaze of star-light near the Cross, from that part of the sky, that a person is immediately made aware of its having risen above the horizon, though he should not be at the time looking at the heavens, by the increase of general illumination of the atmosphere, resembling the effect of young Moon.

Von Humboldt's source was a note on α Centaury written by W. S. Jacob, an engineer and amateur astronomer who was working in India. In 1848, Jacob sent that note to his friend Piazzi Smyth, who presented the work to the Royal Society and then published it (Piazzi Smyth, 1849); in the same year Jacob was appointed director of the astronomical observatory of Madras. In the note it is also possible to read that the 'excessive splendour is caused not only by the profusion of first, second, and third magnitude stars in the neighbourhood, but by the extraordinary general breadth and brightness of the Milky Way thereabouts'. Note that 170 years ago the real structure of the Milky Way was not known. Jacob thought he was looking at the southern part of the vast ring of the Milky Way surrounding us, and he writes that 'the superior brightness of so a large proportion of the stars is then naturally accounted for by the greater proximity to us'. That is, we were not in the centre of the ring of the Milky Way, but closer to its southern part.

This is the testimony of astronomers of two centuries ago about a spectacular light effect. Two centuries ago the night sky was probably still as dark as many centuries and millennia before. For example, the street lighting in the towns for most of the eighteenth century still used gas lamps and not electricity; in Bombay in India such gas lamps did not exist until about 1863. Therefore we think that the 'Jacob effect' could have some importance for archaeoastronomy, since such a spectacular sky would have been visible in the Mediterranean basin until the end of the Bronze Age before disappearing owing to precession. That is, during the long nights in wintertime, when the Moon was not visible or not yet risen, the local inhabitants could enjoy for many hours a sort of twilight (Figure 5).

Figure 5. A simulation of the sky visible in Palestine during biblical times (1st millennium BC; program STELLARIUM). The brightest stars near the horizon to the south-east are α and β Centauri and those of the Crux.

4. Simulations

The question now is how to deal with this issue. Does one need to go to lonely southern places with the darkest skies in order to verify it today? This is not as easy today as it was two centuries ago, when the effect was visible very easily even from the centre of a town. For example, today in the best observing sites, that is those with the largest telescopes and the darkest skies, astronomers live in strongly illuminated rooms and it takes a lot of time to accustom the eyes to the dark. To spend a lot of time outside the buildings to accustom the eyes is rather uncomfortable in such places, owing to cool and often windy weather. More generally, however, subjective opinions should be avoided and therefore we think that the only reliable approach is to make accurate simulations.

The simulations require the use of modern star catalogues containing many million stars and the application of some astrophysics, since we have to take into account Rayleigh scattering in the atmosphere and other physical effects. This is a long-term project, and we show here just one simple example (Figure 6). We have projected TYCHO2 (Hog et al., 2000) star brightness using IDL. It is possible to see that the brightest part of the Milky Way is located indeed at southern declinations, near the Southern Cross. When we include about 100 million stars of the UCAC3 catalogue (Zacharias et al., 2010; $V < 16$), the brightness of the Sagittarius region is similar to that of the Crux region, and one would expect an analogous Jacob effect even in this case. Since this second case was not mentioned, a study is needed that take carefully into account the geographical position of the observer, the date and time.

It may be that we are a bit too bold, but our preliminary conclusion is the following: when we archaeoastronomers today talk about what the original inhabitants were actually looking at in the sky, we are probably to some extent talking for the sake of talking. This is another reason for being humble before claiming that these people were interested in a specific celestial object. In our case, it could be possible that, even if the inhabitants used some specific stellar targets for the orientations, they were not interested in the bright stars but in the global light effect.

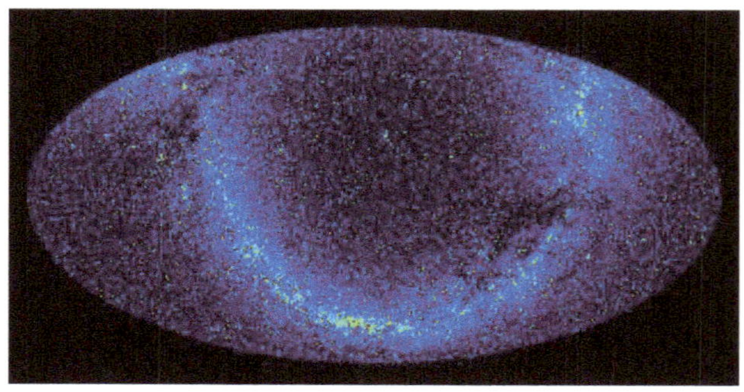

Figure 6. Mollweide (equal area) projection of the brightness of about 2.5 million stars in the TYCHO 2 catalogue ($V < 12$ mag). Abscissae: right ascension; ordinatae: declination. Each point is one square degree. The brightest part of the Milky Way is that of the Centaurus-Crux region (lower part of the figure).

References

Hog, E., Fabricius, C., Makarov, V. V., Urban, S., Corbin, T., Wycoff, G., Bastien, U., Schwekendiek, P and Wicenec, A. 2000. The Tycho-2 Catalogue of the 2.5 Million Brightest Stars, Astronomy and Astrophysics 355, L27.

Hoskin, M. 2001. Tombs, temples and their orientations. A New Perspective on Mediterranean Prehistory. Bognor Regis, Ocarina Books.

Piazzi Smyth, C. 1849. Notice of the orbit of the binary star α Centauri, as recently determined by Captain W.S. Jacob. Transactions of the Royal Society Edinburgh XVI, 445.

Schiaparelli, G. V. 1903. L'astronomia nell'Antico Testamento. Milano, Hoepli.

Tunzi, A. M., Lo Zupone, M., Antonello, E., Polcaro, V. F., Ruggieri, F. 2009. The 'Madonna di Loreto' Bronze Age Sanctuary and its Stone Calendar. In J. A. Rubino-Martin, J. A. Belmonte, F. Prada and A. Alberdi (eds.), Cosmology Across Cultures Astronomical Society of the Pacific Series 409, 375.

Tunzi, A. M., Lo Zupone, M., Antonello, E., Polcaro, V. F. and Ruggieri, F. 2010. Il santuario dell'età del Bronzo di Trinitapoli. Il Calendario di Pietra. In M. Incerti (ed.), Mensura Caeli. Territorio, città, architetture, strumenti. Ferrara, UnifePress, 249.

Von Humboldt, A. 1858. Cosmos. A sketch of a physical description of the Universe, Vol. 3. English translation by E.C. Otté. New York, Harper & Brothers.

Zacharias, N., Finch, C., Girard, T., Hambly, N., Wycoff, G. et al. 2010. The Third U.S. Naval Observatory CCD Astrograph Catalog (UCAC3). The Astronomical Journal 139, 2184.

ARCHAEOASTRONOMICAL WORLD FROM ROMANIA

IHARKA SZÜCS-CSILLIK, ALEXANDRA COMŞA AND ZOIA MAXIM

Abstract: In Romania many researchers have interdisciplinary preoccupations, tackling the studied matter from many points of view. Such an interdisciplinary research group was established in Cluj-Napoca. This archaeoastronomical research investigated Romanian territory by using astronomical orientations. Some of the archaeoastronomical studied places, like sanctuaries, fortresses and necropolises of the Neolithic are found in Banat, Transylvania, Moldova and Muntenia. We enlisted just the well-known and published historical monuments: the Neolithic sanctuary from Parta (Banat Culture), the Neolithic settlements from Baia-Hamangia (Hamangia Culture), Cernica (Boian Culture, Dudeşti Culture), Grădiştea Ulmilor (Gumelniţa Culture), Iclod (Iclod Cultural Aspect), Cucuteni (Cucuteni Culture), and necropoleis from the Dacian world at Sarmizegetusa-Regia (Romanian Stonehenge). We present these locations, culture and the archaeoastronomical results obtained.
Keywords: Neolithic, sanctuary, necropolis, alignment, solstice, equinox, solar arc

Introduction

Romanian archaeoastronomy is the study of how ancient people interpreted what they saw in the night sky and how they integrated their observations into their mythologies, religions and rituals. Once they realized they could determine, e.g., when the opportune planting and harvesting time had come, sky observation became important. The main objective of our investigation is to prove that the celestial phenomenon had a certain impact upon the different cultures and the manner in which this effect is reflected within the archaeological finds. Archaeoastronomy can help provide the measurement (e.g., astronomical orientation etc., including the timing) and the study of symbols, as well as the study of people's relation with the supernatural world. It is difficult to interpret the preserved signs and symbols, often impossible for today's researchers. All these are evidence of what a great help archaeoastronomical research is in the reconstruction of culture.

Research in Romania

On Romanian territory many Neolithic civilizations existed. We briefly present the well-known Neolithic sites, including the Neolithic shrine from Parţa, the Neolithic cemeteries from Iclod, Cernica and Grădiştea Ulmilor, the Neolithic settlements from Cucuteni, Baia Hamangia, and the Romanian ancient observatory, namely Sarmizegetusa-Regia.

SETTLEMENT: CUCUTENI

The Cucuteni-Trypillia culture is a late Neolithic archaeological culture that flourished between 4500 BC and 3000 BC in the Dniester-Dnieper region and in modern-day Romania, Moldova, and Ukraine.

The culture was named after Cucuteni (geographical latitude: 47°16' north), Iaşi county, Romania, and Trypilia, Ukraine, where the first objects associated with this culture were discovered in 1884. The excavations started in 1909.

The Cucuteni culture has been called the first urban culture in Europe. The settlements were usually located on a plateau, fortified with earthworks and ditches. Agriculture is attested, as well as livestock breeding, mainly cattle.
The pottery is connected to the Linear Pottery culture. Extant figurines excavated at the Cucuteni sites are thought to represent the Mother Goddess (Mantu 1994, 225).

One a vase we can find a complex geometric decoration that might transmit an astronomical meaning: the four circles with a cross in the center could represent the four phases of the Moon, each circle being decorated with a horn. Similarly decorated vessels were found in Romania (Frumuşica, Târgu Ocna, Valea Lupului), in the Republic of Moldova and in the Ukraine.

Figure 1. Phases of the Moon (Cucuteni culture), according to Mantu (1994, 226).

SETTLEMENT: BAIA HAMANGIA

The Hamangia culture began around 5200 BC and lasted until around 4500 BC; it was a Middle Neolithic culture that covered Dobroudja, the right bank of the Danube in Munteia and northeast Bulgaria. It is named after the site of Baia-Hamangia (geographical latitude: 44°43' north).

Pottery figurines are extremely stylized and show standing naked faceless women with emphasized breasts and buttocks. There are flexed or supine inhumations in cemeteries. Grave-gifts include flint, worked shells, bone tools and shell-ornaments.

Interesting from an astronomical viewpoint is that the Hamangia culture used the bull symbol, as well as human representations in pairs (man-woman), as seen in the well-known 'The Thinker' and 'The Sitting woman' figurines. The Sun passes through the Taurus (bull) constellation from mid-May to late June (spring, fertility and fecundity).

These two figurines are considered masterpieces of the Neolithic art.

Figure 2. The Thinker and the Sitting woman from Hamangia, according to Berciu (1966, 40).

Painted vessels with complex geometrical patterns based on spiral-motifs are typical with this culture. The shapes include pots and wide bowls. Settlements consist of rectangular houses with one or two rooms. They are normally arranged on a rectangular grid. Settlements are located along the coast, at the coast of lakes, on the lower and middle river-terraces, sometimes in caves.

SANCTUARY: PARŢA

The excavations were started by Joachim Miloja in 1931 and were continued by another archaeologist (Csillik et al. 2000, 115). It belongs to the Banat culture from the period 4600-4200 BC, and it is located near Timişoara (geographical latitude: 45°45' north), Romania.

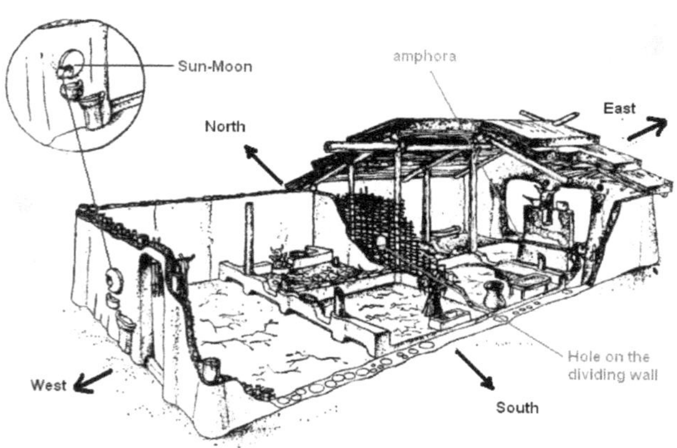

Figure 3. The Neolithic shrine from Parţa.

The size of the sanctuary was 11.6m x 6m x 1.75m and its longest axis was oriented in an east-west direction (Chiş et al. 2000, 12). The sanctuary was divided into two rooms by a wall. On this dividing-wall was a round window-like opening, approximately 35cm in diameter at 1m above the floor and 2.25m from the northern wall. In the centre of the sanctuary was found a big altar table that was 25cm thick and extended 2.5m in length in both rooms. The sanctuary had a circular opening on its western wall, probably representing the Sun, approximately 30cm in diameter, 1.5m away from the northern wall. Surrounding this circular opening through which the light entered the sanctuary, in the exterior part of the wall, an earthen Moon-shaped adornment about 8cm thick was fastened. Fixed against the same wall, under the hole, there was a cup and a grinder. The grain offerings brought to the Sun and Moon divinities probably represent a cult of fecundity and fertility.

On a clay socket stood an idol bust or bull-statue (1.7m high, near the eastern wall and 3m away from the northern wall), probably representing the cult of the bull. The statue was formed of a divine couple: The Bull God and the Great Mother God. In the western part, guarded by two posts, an altar-table was laid. Inside the altar, several flint blades that were used for sacrifices were discovered. In the northwest corner of the socket (0.9m high) was an amphora, probably containing liquid and food devoted to the deities (Lazarovici 2006, 5).

In the northwest corner of the sanctuary was a loom with seven clay weights, illuminated in winter.

Near the northern wall on the altar table there were a number of bull skull ornaments. This was the place where the people brought their meat offerings, illuminated in spring and in autumn.

The place for the grain offerings was on the altar-table, near the southern wall, illuminated in summer.

From the astronomical viewpoint, we proved that the light of sunset entered the sanctuary on the western wall through the Sun-Moon circular-opening and through the hole in the dividing wall and illuminated the northwest corner of the socket, where the amphora was located.

Furthermore, we proved that at the winter solstice, the sunlight illuminated the loom; on some days in spring and autumn, the sunlight entered the hole in the dividing wall and fell upon the socket. During the summer, the Sun illuminated the place on the altar table where the grain offerings were brought (Csillik et al. 2000, 115).

Very interesting is the fact that the vernal point in 4600 BC (Neolithic) was in the constellation of Taurus (Bull) and in spring-autumn the Sun illuminated the socket of the bull-statue (cult of fecundity and fertility).

From these facts we can infer that the Neolithic people from the sanctuary of Parţa had a very good knowledge of the sky (Lazarovici et al. 2002, 10).

Today, one can find the reconstructed sanctuary of Parţa in the Banat Museum in Timişoara.

Figure 4. The illuminated Neolithic shrine from Parţa (model by Radu Zapotinschi).

SANCTUARY: SARMIZEGETUSA-REGIA

Here stood one who studied the waxing and waning of the moon, while still another regarded the labors of the sun and observed how those bodies which were hastening to go toward the east are whirled around and borne back to the west by the rotation of the heavens. (Jordanes, *Getica*, paragraph 69)

Legends mention, about the year 3000 BC, the existence of a great empire, ruled by the priestess of the Uranian Sun, Dacia–Dochia. She was simultaneously the supreme judge and military leader.

On the Grădiştea hill is the ancient Sarmizegetusa, the biggest Dacian known fortress. The fortress has three hectares between its walls, with an irregular plane. But Sarmizegetusa Regia did not have primary strategic importance. The ruin complex that was named The Sacred Precincts is 100m from the Eastern Gate of the Sarmizegetusa fortress. To that place we are guided by a large road paved with limestone slabs, delimited by a short stone wall. The landscape is impressive. In the middle of the secular forest, the eye spots the monumental traces of Dacian sanctuaries covering two terraces.

Several Romanian astronomers did research on this fabulous Dacian place. Many astronomical hypotheses are born from this plateau of the Romanian Stonehenge.

Sarmizegetusa Regia (geographical latitude, 45°37'36'' north; geographical longitude, 23°18'62'' east; altitude, 990m), located in the sacred Orăştie Mountains of Romania, consists of the remains of a Fortified City and a Sacred Area. The city dates from 82 BC to 107 AD, during the reign of Decebal, the last of the Dacian kings.

It was the most important Dacian religious, military and political center of Dacia during this period.

The archaeological ruins clearly illustrate the importance of geometry in designing the city; rectangular and polygonal structures were common, along with circular sacred spaces. The small round sanctuary could be used as a calendar. This sanctuary has 114 items: 13 slabs and 101 poles. The 101 poles are divided by the 13 slabs in 13 groups as it follows: 8 groups of 8 slabs, 1 group of 7 poles, 3 groups of 8 poles, 1 group of 6 poles. One pole equals one day and one slab is a mark for a week or a year. You can start at any pole and one year means three complete rotations plus eight weeks. You mark that slab and continue numbering. After 13 years (4748 days) all slabs are marked and the Dacian calendar should be one day behind the Gregorian calendar (an average 365.23 day year).

We calculated the solar arc for Sarmizegetusa Regia in 100 BC, the result was an azimuthal interval 56°-123°. We found these solstice points in the big round sanctuary. This emphasizes that ancient Dacian people had a very comprehensive geometrical and astronomical view (Stănescu 1999, 32).

The big round sanctuary could be a calendar corrector (365.24 days in a Dacian year).

The exterior circle has 104 slabs forming a perfect circle. The next circle touching the exterior circle and has 210 items: 180 poles divided by 30 slabs in 30 groups of 6 poles. The third interior circle has 68 poles arranged like this: 17 Poles - 4 Slabs - 18 Poles - 3 Slabs - 16 Poles - 4 Slabs - 17 Poles - 4 Slabs. The interior altar has 72 items: 68 poles organized in 2 groups each group divided like this: 13 poles - 2 slabs - 21 poles - 2 slabs. The stalls from the altar and the ones from the third circle form two perpendicular lines (a cross), which is orientated perfectly to the cardinal points and shows the solstice points too. This sanctuary can predict lunar and solar eclipses.

Figure 5. Sarmizegetusa Regia.

Sarmizegetusa Regia illustrates perfectly the level of astronomical knowledge of our ancestors concerning the universe, the time, the seasons and the geographic and astronomic orientation (Szücs-Csillik 2010, 198). The sanctuary has a calendar system still insufficiently studied, but the Andesite Sun is a jewel of universal culture (Comşa 1991, 10), perfectly orientated to the north direction (theodolite, gnomon, Sundial, Sun sign, golden ratio).

NECROPOLIS: CERNICA

An important moment for the Cernica area is the discovery of the Neolithic necropolis (4600-4200 BC) in the neighborhood of the former Iezerul cloister.

To date, on the Muntenia territory, this is one of the greatest necropolis of the Neolithic age, discovered and investigated by Gheorghe Cantacuzino. The monograph (378 graves) from the Cernica Neolithic necropolis was published by Eugen Comşa and Gheorghe Cantacuzino (2001, 18).

This necropolis was accidentally found in 1961, on the occasion of the systematic excavations from Cernica in the Căldăraru village, on the western bank of Cernica Lake. In Cernica there was practiced a form of solar cult: sunrise and sunset probably were observed within the limits of the burial ritual. From aligned skeletons, rates of 92.11% are also comprised in the western area of annual oscillation of the Sun in azimuth. Twenty six skeletons are out of the solar arc, but they are close to the winter and summer solstice points. The reason can be a miscalculation, or an act of expulsion from the community, for reasons we do not know at this moment. The Gaussian distribution of the skeletons from the Cernica necropolis shows a west-east orientation, with an apex at spring-autumn. This fact shows that the mortality rate grows in spring and autumn. This death-rate can be connected with epidemics (influenza, hepatitis, encephalitis) or food shortages (the end of winter).

The anthropological studies made on the Cernica necropolis showed that the Mediterranean anthropological type was mostly frequent within the Neolithic population, then followed by the Protoeuropid, the Alpine and Nordic ones (Comşa 2006, 150).

The skeletons' orientation in the Cernica cemetery backs up the archaeoastronomical hypothesis: in Neolithic times, the orientation of skeletons was towards the sunrise, or sunset on the day of burial, or death, of the individual.

The Neolithic people had possibly made a social differentiation regarding the burial location in the cemetery (in the middle it was the rich, healthy, protected person), and in the orientation of the skeletons (derived from another population, not native, foreign).

The main occupations of all Boian communities were agriculture and animal husbandry, suitable to the geographical environment of the plain, so that they could easily develop a solar calendar.

The solar cult that subsequently appeared and was also used for the burials was created by a systematic observation of the Sun.

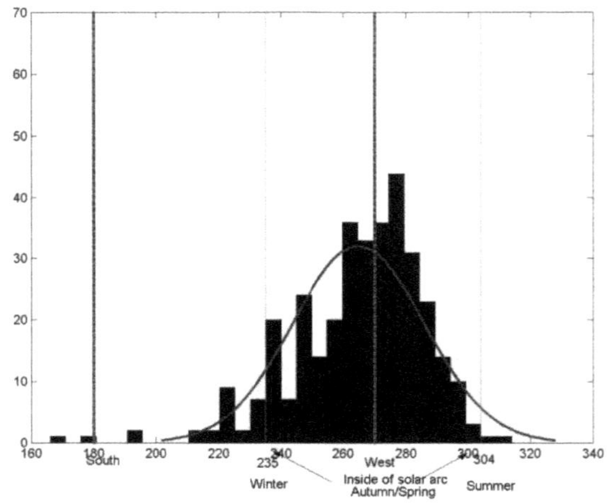

Figure 6. The astronomical orientation of the skeletons from Cernica.

Four pregnant woman skeletons (graves: M158 [242°], M251 [264°)] M256 [280°], M303 [284°]) were found in the Cernica cemetery. Each skeleton is inside the solar arc and is not isolated within the cemetery.

Three on the belly skeletons (graves: M149 [258°], M237A [260°], M318 [260°]) were also discovered in the Cernica necropolis. Each skeleton is inside the solar arc and is not isolated in the necropolis. It is very interesting that none of the skeletons have a funerary inventory. Probably, the Neolithic people buried the dead person on the belly to immobilize its spirit into the pit, to prevent it from disturbing the living people.

Using mathematical and astronomical calculations, we showed that the people of the Dudeşti and Boian culture made their burials at sunrise (sunset) and aligned them towards the Sun. In burials, the dead had their legs towards the sunrise direction.

So, we can admit the existence of a special cult. The purpose of the practice may be a last desperate trial to resurrect the dead person to 'life', the light of the Sun feeding with energy the 'resurrection' moment.

NECROPOLIS: ICLOD

The Neolithic burial site Iclod (4200 BC) was discovered by Márton Roska at the beginning of the 20th century. The dead people were buried in supine position, facing the sunrise direction. Inside the graves, in the later stage of the Iclod culture, there were found tools made of stone, bone and obsidian, near the skeletons.

We determined the Sun azimuth for Iclod, corresponding to summer and winter solstice (Wittmann 1979, 130). We found within an accuracy of 1° that the Sun azimuth measured from the north is 53° for the summer solstice and 126° for the winter solstice.

We proved that 72% of the graves found there are orientated within the rigorous limits of the annual oscillation of sunrise azimuth (Maxim et al. 2002, 20).

The grave whose orientation is out of the annual oscillation of sunrise azimuth belongs to the last stage of

Iclod, when the eastern orientation was replaced with a northern one.

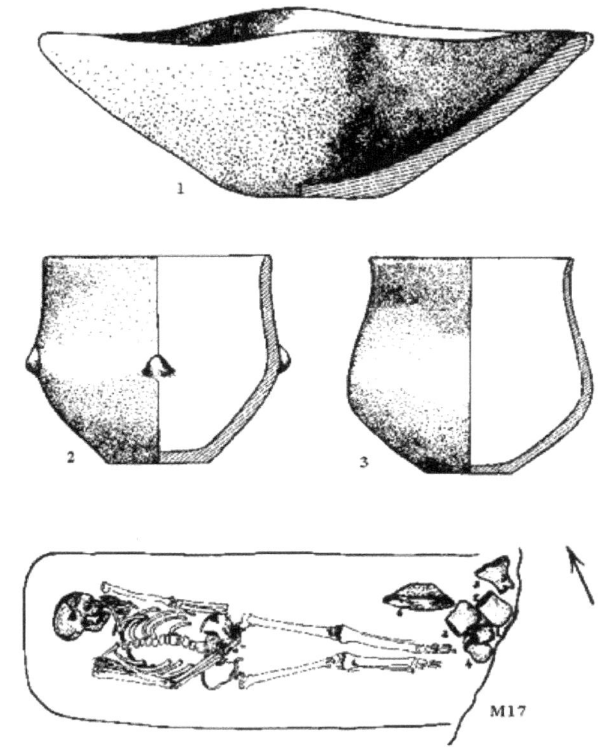

Figure 7. Grave M17 from Iclod.

NECROPOLIS: GRĂDIŞTEA ULMILOR

In Vărăşti village (geographical latitude: 44°14' north), on Grădiştea Ulmilor of the former Boian Lake, one of the biggest necropoleis belonging to the Gumelniţa culture (4000 BC) was discovered.

The necropolis, as space intended for the dead of community, show a stage of population stabilization in this area and, therefore, a stage of more obvious intervention over the environment (Comşa 1995, 56).

In this necropolis the dead persons were grouped in places specially arranged for them (outside the settlements).

Within the excavated areas there were uncovered 118 graves belonging to the Gumelniţa culture settlement. Of these graves, 80 were for adults, 35 for children and 3 for teenagers. There have been found also a few graves overlapping, pointing to two burying stages.

The pits of the graves were usually not identified but in two cases, their shapes were irregularly oval.

Most skeletons were flexed on the left side, orientated towards the sunrise direction between azimuthal interval 70°-117°. In most cases, the hands were bent at the elbows and laid with the palms in front of the individuals' faces.

The funerary inventory was not too rich, consisting of flint tools, copper pins with rhombic, two-lobe or rhombic plate shaped heads, biconical, or flat clay artifacts, bone tips, shell pearls, amber beads, and a few golden artifacts.

The burying ritual was in a flexed position, from a slight, then moderate and up to a very pronounced one, usually on the left side, while the offerings occurrence is usually scarce.

The details of the ritual (position of arms, the various categories of offering and their distribution within the grave, the use, or non-use of ochre, the degree of flexing in the case of this position, the shape of the pit, etc.) do not seem to have been submitted to rigorous standards. We should mention the presence of children's graves under and among dwellings, some of the skeletons bearing various traits that led to the hypothesis of ritual sacrifices.

Conclusion

Archaeoastronomy uses different methods from archaeology, anthropology, astronomy, statistics and probability in order to determine ancient civilizations' preoccupations. Because these methods are miscellaneous and use data from different sources, the collecting and processing process is a long lasting one. Using mathematical and astronomical methods, we can determine, for example, the solstice points for necropoleis, shrines and other historical buildings.

In the future, we plan to use our numerical programs for a study into other Neolithic necropoleis and buildings in the Carpathian Basin.

References

Berciu, D. 1966. Hamangia Culture. Bucharest, Academiei Romane Press.

Chiş, D., Oproiu, T., Csillik, I., and Lazarovici, G. 2000. Astronomical Orientations at Parta. Arheometrie 13, 12-14.

Comşa, A. 2006. The interrelations between burial orientations and astronomy in the Balkan region. Analele Banatului 14, 149-176.

Comşa, E. 1978. Contribution à l'étude de la culture Cris en Moldavie. Dacia 22, 9-36.

Comşa, E. 1995. Gumelnitean necropolis from Vărăşti. Analele Banatului 4, 55-193.

Comşa, E., and Cantacuzino, Gh. 2001. Neolithic necropolis from Cernica. Bucharest, Academiei Romane Press.

Comşa, M. 1991. The "Stone Sun" from Sarmizegetusa Regia. Lucrari interdisciplinare 18, 10-15.

Csillik, I., Oproiu, T., Chiş, D., Maxim, Z., and Lazarovici, G. 2000. Archaeoastronomy in Transylvania. PADEU 11, 113-118.

Lazarovici, G., Chis, D., Oproiu, T., and Csillik, I. 2002. The neolithic shrine at Parta. "Unwritted Messages" from the Carpathian Basin. Konkoly Observatory Monographs 4, 7-18.

Lazarovici, G., and Lazarovici, M. 2006. A home altar at Gura Baciului. Analele Banatului 14, 1-12.

Mantu, C. M., Botezatu, D., and Kromer, B. 1994. Une tombe double à inhumation de l'établissement de type Cucuteni de Scanteia. Prehhistorie Europeenne 6, 225-241.

Maxim, Z., Chis, D., Oproiu, T., and Csillik, I. 2002. The astronomical aspects of the orientation of the graves in the burial site of Iclod. "Unwritted Messages" from the Carpathian Basin. Konkoly Observatory Monographs 4, 19-29.

Stănescu, F. 1999. Dacian Sanctuaries. The archaeometrical and archaeoastronomical analysis. Sibiu, "Lucian Blaga" University Press.

Szücs-Csillik, I., Comşa, A., and Maxim, Z. 2010. Archaeoastronomy in Romania. Romanian Astronomical Journal 20, 197-200.

Wittman, A. 1979. The obliquity of the ecliptic. Astronomy and Astrophysics 73, 129-131.

MEGALITHIC CULTURES

EQUINOCTIAL FULL MOON MODELS AND NON-GAUSSIANITY: PORTUGUESE DOLMENS AS A TEST CASE

FABIO SILVA

Abstract: There is a growing body of evidence supporting Equinoctial Full Moon alignments in megalithic Iberia where previously these were being interpreted as solar. The author's own work surveying Neolithic dolmens in Central Portugal is now shown to exhibit spring and Autumn Full Moon alignments for different hydrographical regions. Even though these Equinoctial Full Moon events have been well defined, modelling them with simulated data to obtain theoretical predictions for alignment distributions can be trickier. Some issues, like their inherent non-gaussianity, are identified and discussed. Both updated and new theoretical models for possible Equinoctial Full Moon alignments are also presented. The introduced models will also permit, using statistics on a high number of measurements, to discern meaning within Equinoctial Full Moon alignments by identifying whether there was a preference for alignments to eclipsed full moons or not.
Keywords: megalithic astronomy, lunar alignments, spring full moon, autumn full moon, megalithic equinox, Neolithic, Portugal

Introduction

Since C. Marciano da Silva introduced the Spring Full Moon (henceforth SFM) concept to account for alignments to the so-called 'megalithic equinox' (Da Silva, 2004) a growing number of Neolithic tombs have been shown to be aligned to this astronomical event, and some also to its autumnal counterpart. Dolmens in Alentejo and more generally in Southwest Iberia were the first to be shown to be orientated towards the SFM (Da Silva, 2004). Using Bayesian analysis, it was found that the megalithic enclosures of Alentejo are also very likely to be orientated towards the Autumn Full Moon (henceforth AFM) (Pimenta et al., forthcoming). Recently, César González and Belmonte (2010) have done a systematic statistical analysis of most of the available data for monuments in the Iberian Peninsula and found that Da Silva's SFM model, for Alentejan and other neighbouring groups, was indeed a better fit than the solar models.

Equinoctial Full Moons (henceforth EFMs), to generalize the term, are temporal events in which the rise (and set) positions of both full Moon and Sun have swapped places. In summer the Sun rises at a northeastern azimuth (in the northern hemisphere), travels high in the sky and sets in a northwestern azimuth. The full moon, on the other hand, rises at a southeastern azimuth, travels low in the night sky and sets at a southwestern azimuth. In winter the sun rises in the southeast and travels low in the sky, whereas the full moon rises northeast and travels high in the sky. The two celestial bodies change their 'position' at the AFM and will change again at the SFM. In essence, the Sun and full Moon change places, relative to the celestial equator, at an EFM. This effect is more pronounced the closer to the equator it is observed, as the east-zenith-west line (the celestial equator) is a clear-cut division of the abode of each of the two luminaries depending on the season.

In this paper, current EFM distribution models available in the literature are critically reviewed. These have been treated as bell-shaped when there is no evidence for it, whether a priori or a posteriori. A proper treatment of simulation results is introduced in the first section. In section two these distributions are compared to data on Neolithic dolmens of central Portugal, and alignments to both the SFM and AFM are identified. The question of meaning is discussed in section three, where the eclipsed EFM hypothesis is put to the test. Because Portuguese dolmen data suggests megalith builders were doing no such selection, ethnographic evidence for yearly EFM rites and myths is briefly introduced. A full treatment of the subject is left for a future work, now in progress. A table with the range and peak values for all the presented distributions is given and the importance and application of the paper's main points are discussed in the last section.

EFM Modelling

EFM DECLINATION DISTRIBUTIONS

By simulating moonrises for a period of 110 years for the current epoch, Da Silva (2004) obtained a theoretical distribution for the SFM with a minimum at 85° of azimuth, an average value of 97.3°, and a maximum at 110°, for the latitude of Évora, Portugal. As noted by that author a similar distribution for the AFM can be constructed from first principles. This is shown in Fig 1 in light grey.

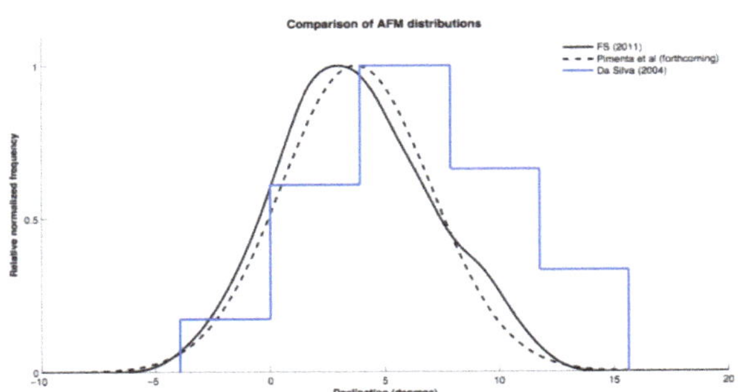

Figure 1. Declination histograms simulated for Autumn Full Moonrise. The approach described in the text is represented by the black solid curve. Pimenta et al.'s (forthcoming) normal distribution is represented by the dashed curve, and the original histogram from Da Silva's (2004) paper on the Spring Full Moon is represented in light grey.

Pimenta et al. (forthcoming), using the Alcyone Ephemeris software package, simulated both the SFM and AFM moonrises for a period of about 428 years in the current epoch, for which there is absolute confidence in the lunar orbital parameters. This distribution includes a much bigger sample of years than the previous one, which means there is greater confidence in its statistical significance. It also includes a bigger sample of the standstill cycle, which is of relevance as will be shown.

These authors have kindly provided us with the results of this simulation, which will henceforth be used.

NON-GAUSSIANITY IN LUNAR MODELS

Pimenta et al. (forthcoming) use the simulated distributions as priors to test megalithic enclosure alignment data from south Portugal using Bayesian analysis. They have, however, obtained the average and standard deviation of the simulated declinations to model a Gaussian curve (the grey dashed line in Figure 1), using it then as a prior to test the data against. Da Silva has also said that the SFM's azimuths 'exhibit a bell shaped distribution' in his recent paper (2010). But there is no a priori reason why this distribution should be Gaussian shaped. This is an event that involves both solar and lunar mechanics that combine in a very non-linear way. This is made evident by the fact that such full Moons don't occur on the same date (on a solar calendar) every year. This is why a distribution is needed to account for the range of possibilities, as well as the differences in likelihood. Such distribution needs not be bell-shaped (Gaussian) as it is being driven by the celestial mechanics and not observations.

Each simulated moonrise should instead be treated as a theoretical prediction for a value behind an observation. This amounts to saying that, if our goal is to get a moonrise distribution that is to be compared to measured histograms, the errors inherent in any observation need to be taken into account. A bell-shaped distribution should be created for each individual value coming out of the simulation. The declination value would be the peak of the distribution, which would have a spread (standard deviation) directly related to the observational errors. All these individual bell-shapes would then be added to get the total distribution, which need not be Gaussian. For astronomical events where there is a unique declination at any given epoch, for example the winter solstice sunrise, this amounts to a single Gaussian. But other more complex events, like EFM's, would exhibit more complicated distributions. The inherent error that is reflected by the standard deviation has to be estimated and should reflect both simulational uncertainty, observational error, choice of lunar limb and refraction issues as well as alignment issues that arise when using rough stones (see the discussion on the aptly-titled 'problem of precision' in Appendix A of Silva, 2010). The shape of the EFM distribution depends heavily on this value. For too high values the distribution will 'fatten' and flatten out, whereas for too small values it will not be 'smooth'. For the rest of the paper a standard deviation of one degree of declination was chosen, which is high enough to accommodate Schaefer and Liller's (1990) uncertainty in the declination due to refraction issues alone.

In the EFM case this has to be done for each declination value obtained from the simulation. This is shown, for the AFM moonrise, as the black curve in Figure 1. There is a considerable difference between this curve and Pimenta et al.'s Gaussian curve (the dashed grey one): the distribution's peak is shifted by a degree. This is because the original distribution is not bell shaped but has a small bump on the side. A similar effect occurs for the SFM moonrise, even though the peak's shift is smaller in that case.

It is feasible, now, to fit the actual curve using Gaussian fitting techniques. It turns out that it can be fitted to good accuracy using two to five Gaussians, which would fully describe each EFM distribution. However, it is still not obvious if and how this would be of use within archaeoastronomy so it won't be discussed presently.

Because the simulations were done for the current epoch (1620-2047 AD) adjustments are needed in order to correct the distribution to any given epoch. According to the latest simulations (Laskar, 1986), already cited by Pimenta et al. (forthcoming), the adjustments should be no greater than 0.4° for 1000 BC, 0.6° for 3000 BC, 0.7° for 5000 BC and 0.8° for 8000 BC, which should be subtracted for the SFM values, and added to the AFM values (see Figure 4). These are only approximations and while proper EFM simulations for such epochs are needed, they are nevertheless good indicators of what to expect.

EFM Alignments in Central Portugal

A recent survey (Silva, 2010) of Middle to Late Neolithic (c. 4000-3000 BC) megalithic tombs, in a region of central Portugal delimited by the Mondego River to the south and the Douro River to the north, has shown a scatter in the declinations of the main axial orientations (see Figure 6 in Silva, 2010). The data of this survey was in good agreement with a previous one conducted by Michael Hoskin (1998; 2001) in which it was concluded that the tombs were orientated towards sunrise at the date of their construction (Senna-Martinez et al., 1997). However, this author suggested that this simplistic interpretation presents a narrow view of the available data, disregarding outliers that suggest lunar, and even stellar, alignments. Another fact that Hoskin and his colleagues' interpretation could not account for was the preference of Mondego river basin tombs towards negative declination orientations, whereas tombs in all other nearby hydrographical areas, that is in the Vouga, Paiva, Torto and Coa river basins, prefer positive values for their declinations (Silva, 2010).

Consolidating the dataset acquired by this author with the one reproduced by Hoskin in his book (2001), declination histograms for the tombs' orientation can be constructed. Data from the more recent survey was given prevalence, which is to say that data from Hoskin's survey was only used for tombs that were not measured recently. Each measurement was attributed a Gaussian distribution using the measured declination as its mean value and a standard deviation that tried to account for both precision (see appendix A in Silva, 2010) and observers' errors. In comparing declination values for the 18 tombs that were encompassed by both surveys, a value of 3 degrees of declination for the standard deviation was arrived at. This value was given to all measurements in the consolidated dataset.

The resulting histograms (Figure 2) have been divided for tombs in the Mondego (Figure 2, top) and other river basins (Figure 2, bottom). The sample sizes (25 and 30 respectively) are enough to give some statistical significance to the analysis of the relative frequencies normalized to the average value. The declination sign preference previously mentioned is now made explicit by the fact that almost no peaks are present in the non-preferred areas of each set of tombs, except for some barely significant peaks with frequency values very close to the average. In other words, the statistically significant peaks for Mondego tombs indeed have negative declinations, and the reverse is true for the combination of Vouga, Paiva, Torto and Coa tombs.

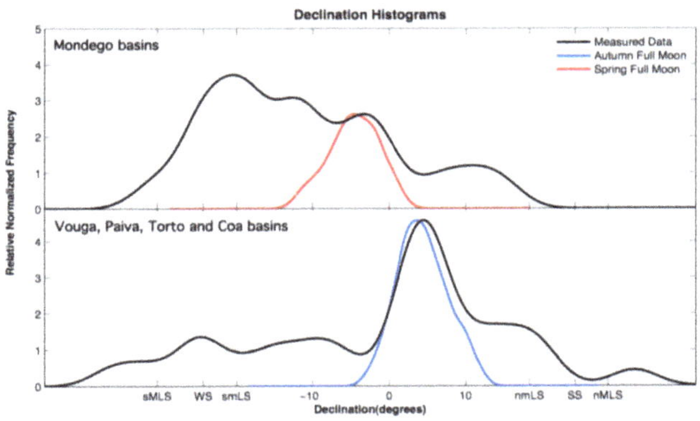

Figure 2. Declination histograms for measured data of the consolidated survey (black lines, see text), for dolmens of the Mondego river basin (above), and other local river basins (below). The expected distribution for Spring Full Moonrise and Autumn Full Moonrise are also represented (greyscale).

The highest peak in the Mondego histogram is very close to the expected declination peak for an alignment with a southern minor lunar standstill, and there is as well a plateau close to the declination for the northern minor lunar standstill in the histogram for the other river basins. Alignments with lunar standstills have indeed been suggested previously (Silva, 2010), and seem to be quite common alignments in megalithic monuments (Ruggles, 1999; Sims, 2006). There is also a barely significant peak at the winter solstice declination for tombs in the other basins, another common alignment in megalithic Europe. There are however quite significant peaks in both histograms for values in between the mentioned solar and lunar extremes and zero degrees of declination. One of which is, in fact, the highest peak of the distribution.

These peaks correspond closely, in both peak value and range, to the expected value for alignments with the Spring Full Moonrise for the Mondego tombs and the Autumn Full Moonrise for tombs in the other nearby basins. This clear-cut difference between the two sets of tombs encourages their independent analysis and might even suggest different purposes or builders for each group. This hypothesis, which gains further credence from architectural and archaeological considerations, will be dealt with in a future work.

EFM's and Meaning

The question of meaning is an imposing one in prehistoric archaeoastronomy. Whereas the importance of both solstices and standstills can and has been quickly ascribed to their visual uniqueness, the importance of an Equinoctial Full Moon event hasn't been considered properly. One is quick to note the importance of the spring season, the renewal of nature, etc., and argue the naturalness of a spring rite. However, one question still remains unanswered: why, then, perform the rites at the EFM, instead of at the actual equinox, or the second full moon following the equinox when the spring is well under way? What is so unique about the EFM that makes it a special, meaningful time?

BEYOND THE FULL MOON PARADIGM

A possible explanation for this is the fact that lunar eclipses occur at these times on very specific years. Whereas usually eclipse dates fall near to the solstices, on standstill years they always occur on the full moon closest to the equinoxes. This is because, at this time, the lunar nodes are close to the equinoxes, which is, in fact, the cause for the extreme lunar declinations that are called standstills (Morrison, 1980). The existence of alignments to both EFM and lunar standstills in close proximity and in monuments of the same style and of the same period, as in the Portuguese case above, is suggestive that such a connection might already have existed in the Neolithic mind.

For a given place, lunar eclipses are much more common than solar eclipses. This is because when one occurs it is visible during the night across the whole hemisphere. They are also more varied in the sense that an eclipsed full moon can exhibit very distinct colours, ranging from near invisibility to brownish red or bright orange. Whether one postulates that pre-literate societies considered specific times on the lunar cycle, like a Dark Moon (Sims 2006; 2007; 2010) or the often mentioned but less ethnographically corroborated Full Moon (Da Silva, 2010), or that the very fact of lunar periodicity was the important thing (Lévi-Strauss 1973; 1978), a lunar eclipse, in which the bright full Moon turns dark or red, would be a momentous time.

Claus Clausen et al. (2008) have explored lunar eclipse alignments for the megalithic sites of Denmark. They have simulated both solar and lunar positions for the Neolithic period and identified the presence of a 'fingerprint feature' in the eclipse histograms. This fingerprint, consisting of two peaks, roughly at 100° and 120° of azimuth, shows up only for western European longitudes and, then again, only for certain centuries. They have identified this fingerprint in the Danish passage grave data they have collected. This technique is, however, not always applicable as, sometimes, one might not have dates accurate to a century, particularly so if dating is reliant only on radiocarbon samples. In fact, one might not even have dates at all, attributing a prehistoric monument to a particular epoch based on architectonical and other archaeological similarities, as is the case for most of the Portuguese dolmens, where radiocarbon dating was only possible for a few of them. There is, nevertheless, another way to test the lunar eclipse hypothesis based on the already mentioned fact that they only occur at these declinations on standstill years.

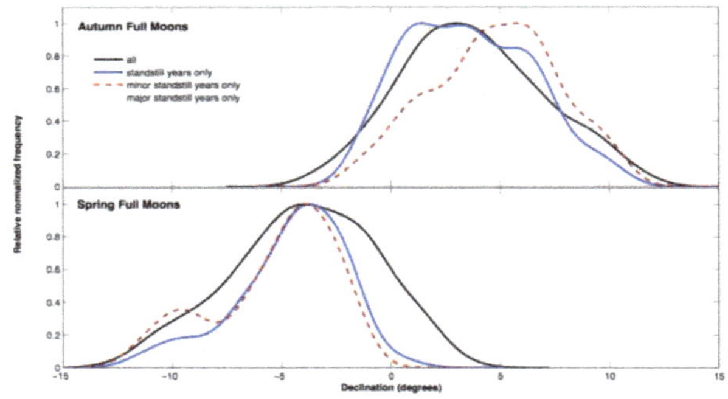

Figure 3. Declination histograms for different models of both Equinoctial Full Moons. The black solid line includes all 460 years of the simulation, the grey solid line includes only the standstill years in the simulated period. The grey dashed and dotted lines include, respectively, only minor and major standstill year data.

From the simulated EFM dataset presented previously, one can select and plot the histogram for the declination of moonrise on standstill years only. Lionel Sims' work on Stonehenge suggests the concept of standstill year to be far more important than the actual, geocentric, lunar declination extreme, which, he notes, is not even observed (Sims, 2007). A standstill year is comprised of the thirteen or so lunations around the declination extreme that would be visible on the upper Grand Trilithon window at Stonehenge as seen from the Heel Stone (Sims, 2006). The standstill extremes of declination always peak at the half-moon closest to the equinoxes, the spring equinox for a major standstill, and the autumn equinox for a minor one (Morrison, 1980; Sims, 2006). This means that the full Moon closest to the standstill extremes is an Equinoctial Full Moon by definition, and it would for sure be included in the standstill year.

There is then interest in distinguishing between alignments to Equinoctial Full Moons on all standstill years (grey line in Figure 3), major standstill years only (dotted line), or minor standstill years only (dashed line), as opposed to a distribution that does not distinguish standstill from interstandstill years (the black curve), i.e., the one that has been considered so far. These have all been renormalized to unity to better represent their differences. The distribution for interstandstill years only is so similar to the 'all years' one that it is not worth representing at this point.

These distributions have very different shapes and all of them exhibit non-gaussian features, like multiple peaks. Such fingerprint features (to borrow the term from Claus Clausen et al., 2008) are not discernible in the consolidated dataset of dolmen orientations for Central Portugal (Figure 1), and therefore one cannot distinguish between these models with this dataset and analysis. It seems feasible, though, that a statistical method appropriate to test broader datasets will permit such distinctions to be made.

The table in Figure 4 shows the declination values for the range and primary peaks of each one of the distributions present in Figure 3, for moonrise on both the SFM and AFM. As is clear from the figure some of the distributions have other relevant features like peaks or plateaus. The values for these are: i) SFM, all standstill years, secondary peak at -9.83°; ii) SFM, minor standstill years only, secondary peak at -9.60°; iii) AFM, all standstill years, secondary peak at 3.23°; iv) AFM, all standstill years, tertiary peak at 5.81°; v) AFM, major standstill years only, secondary peak at 5.42°.

	Spring Full Moon			Autumn Full Moon		
	Min	Peak	Max	Min	Peak	Max
All years	-15.04	-4.01	5.90	-6.39	2.99	14.26
All Standstill	-14.13	-3.76	3.13	-4.39	1.42	13.09
Major only	-10.51	-3.36	3.35	-4.35	1.14	10.35
Minor only	-14.36	-3.89	1.44	-4.35	5.77	13.26

Figure 4. Moonrise Declination values for the different EFM distributions presented in the paper. These include values for the minima and maxima of each distribution as well as its primary peak. Secondary and tertiary peaks values are found in the text.

It should be remembered that the values of the table in Figure 4 and the ones just mentioned are values for simulations in the current epoch and that they should be shifted accordingly when looking for alignments in the past. Another thing to keep in mind when looking at this table is that, close to the equinoxes, the full Moon changes its declination two to three degrees in a single night. Hence if one has western orientations that might potentially be EFM moonset alignments one should take such a shift into account (again subtracting these to the SFM values while adding them to the AFM values). Both of these shifts are, naturally, only approximations and proper simulations with further study are needed, but nevertheless this table provides the basis for current work with EFM alignments.

MYTH, RITUAL AND DEEP STRUCTURE

But let us, for the moment and only briefly, entertain the idea suggested by the data from central Portugal, that EFMs were always meaningful, without any yearly discrimination. It turns out that the ethnography of pre-literate societies seems to support this case. Whereas at equatorial latitudes the solar seasons might not be as important, mainly due to the much more influential climatic dry and wet seasons (Lévi-Strauss, 1969; 1973), at more temperate latitudes the changing of the seasons might have been observed and infused with meaning. In his exhaustive analysis of Native American mythology, the anthropologist and structuralist Claude Lévi-Strauss analyzed certain myths and rites of an 'equinoctial' (his word) character. Such sets of myths are present, in various forms, from Tierra del Fuego to Alaska and the structural analysis of them and associated rites shows an inherent structure that pops out all over the world in rites that happen around the equinoxes (Lévi-Strauss, 1973).

For example, analysing a myth and related rite from the Mandan people of the North American Plains, he remarks on the inherent change of position of the Sun and Moon. He calls this an 'equinoctial concern' because he interprets it, not incorrectly, as a change to happen at the only time when night and day are of equal length: the equinox (Lévi-Strauss, 1978). But it is possible, without detracting from the eminent anthropologist's conclusions, that knowledge of EFMs is here made explicit, as the time the Sun and Moon actually change positions in the sky. In fact, it is possible that EFMs are the ethnographic definition of equinox, besides being the proposed definition for the megalithic equinox.

Conclusions

By going beyond a Gaussian paradigm, where a peak orientation, and sometimes an associated standard deviation, is chosen to represent the theoretical (or predicted) alignment distribution it has been shown how different the actual distributions can be from their Gaussian approximations. This has tremendous implications when these models are compared with measured orientation data as not only the peaks, but also the shape of the distributions, convey knowledge.

To identify an EFM alignment a large sample of orientations might not be needed, as the dolmen data of Figure 2 suggests. But if one wants to go a bit deeper and discern between several models then the shape of the distribution becomes paramount and advanced statistical tools and huge samples will be required. Techniques such as Bayesian Analysis (Pimenta et al., forthcoming), Cluster Analysis and Principal Component Analysis (González Garcia and Belmonte, 2010), which are primed for this sort of job, will certainly be amongst the first ones to be applied.

Evidence for EFMs in the ethnographic records can also be found, but they seem to be in disguise, mostly because EFMs were unknown to the ethnographers who seem to have confused them with the equinoxes. It seems that ethnography, and especially structural analysis, have much to offer archaeoastronomy in general, but particularly in this case. A proper treatment of these aspects is now in preparation.

The case of the Portuguese dolmens serves as an example of the presence of wide distributions of declinations characteristic of EFMs that were previously being interpreted as alignments to sunrise/sun-climb. It is feasible that EFMs and possibly other, yet unknown, 'distribution-type' events might reshape interpretations of some site orientations and unlock many others throughout megalithic Europe and even beyond.

Acknowledgements:

The author would like to thank Kim Malville, Lionel Sims, Cândido Marciano da Silva, Claus Clausen and David Fisher for insightful discussions during the course of this work, as well as the organizers of SEAC2010 for the opportunity to present and publish it. A special thanks goes to Fernando Pimenta for allowing the use and critical discussion of the data he simulated using the Alcyone Ephemeris.

References

Clausen, C., Einicke, O., and Kjaergaard, P. 2008. The Orientation of Danish Passage Graves. Acta Archaeologica 79, 216-229.

Da Silva, C. M. 2004. The Spring Full Moon. Journal for the History of Astronomy 35, 1-5.

Da Silva, C. M. 2010. Neolithic Cosmology: The Equinox and the Spring Full Moon. Journal of Cosmology 9, 2207-2216.

González Garcia, A. C., and Belmonte, J. A. 2010. Statistical Analysis of Megalithic Tomb Orientations in the Iberian Peninsula and Neighbouring Regions. Journal for the History of Astronomy 41, no. 143, 225-38.

Hoskin, M. 1998. Studies in Iberian Archaeoastronomy: (5) Orientations of Megalithic Tombs of Northern and Western Iberia. Journal for the History of Astronomy 29, S39-88.

Hoskin, M. 2001. Tombs, Temples and their Orientations: a new perspective on Mediterranean prehistory. Bognor Regis, Ocarina Books.

Laskar, J. 1986. Secular Terms of Classical Planetary Theories Using the Results of General Theory. Astronomy and Astrophysics 157, 59-70.

Lévi-Strauss, C. 1969. The Raw and The Cooked: Introduction to a Science of Mythology, trans. John and Doreen Weightman. Harmondsworth, Penguin.

Lévi-Strauss, C. 1973. From Honey to Ashes. New York, Harper & Row.

Lévi-Strauss, C. 1978. The Origin of Table Manners. New York, Harper & Row.

Morrison, L. V. 1980. On The Analysis of Megalithic Lunar Sightlines in Scotland. Journal for the History of Astronomy Supplement 11, 65-77.

Pimenta, F., Tirapicos, L., and Smith, A. (forthcoming). A Bayesian Approach to the Orientations of Central Alentejo Megalithic Enclosures. Archaeoastronomy, XXII.

Ruggles, C. 1999. Astronomy in Prehistoric Britain and Ireland. New Haven and London, Yale University Press.

Schaefer, B. E., and Liller, W. 1990. Refraction Near the Horizon. Publications of the Astronomical Society of the Pacific 102, 796-805.

Senna-Martinez, J. C., López-Plaza, M. S., and Hoskin, M. 1997. Territorio, ideología y cultura material en el megalitismo de la plataforma del Mondego (Centro de Portugal). In O Neolítico Atlántico e as Orixes do Megalitismo: Actas del Coloquio Internacional (Santiago de Compostela, 1-6 de Abril de 1996). Santiago de Compostela: Universidade de Santiago de Compostela, 657-676.

Silva, F. 2010. Cosmology and the Neolithic: A New Survey of Neolithic Dolmens in Central Portugal. Journal of Cosmology 9, 2194-2206.

Sims, L. 2006. The 'Solarization' of the Moon: Manipulated Knowledge at Stonehenge. Cambridge Archaeological Journal 16, no. 2, 191-207.

Sims, L. 2007. What is a Lunar Standstill? Problems of Accuracy and Validity in 'The Thom Paradigm'. Mediterranean Archaeology & Archaeometry, Special Issue 6, no. 3, 157-163.

Sims, L. 2010. Coves, Cosmology and Cultural Astronomy. In N. Campion (ed.), Cosmologies: Proceedings of the Seventh Annual Sophia Centre Conference 2009, 4-28. Ceredigion, Sophia Centre Press.

KREISGRABENANLAGEN: EXPRESSIONS OF POWER LINKED TO THE SKY

GEORG ZOTTI AND WOLFGANG NEUBAUER

Abstract: Several archaeologically distinct cultural groups developed a certain class of monumental structures in a rather short period during the Middle Neolithic in Central Europe: *Kreisgrabenanlagen* (KGA; a form of circular enclosures). The astronomical orientation of the entrances into these monuments has been discussed for many years. In a new project we analyze and simulate astronomical aspects for the KGAs in Lower Austria. This work combines previous archaeological surveys by magnetic prospection, excavation results, on-site horizon measurements, virtual reconstruction and astronomical simulation software.
Keywords: Kreisgrabenanlagen, Neolithic circular enclosures, orientation studies, horizon survey, virtual reconstructions

In a short period (c. 4850/4800-4550/4500 BC) of the Middle Neolithic, a certain class of circular monumental ditch systems, *Kreisgrabenanlagen* (KGA), was built in many places in Central Europe. They have been detected by aerial archaeology in Hungary, Slovakia, Czech Republic, Austria, Germany and Poland (Figure 1).

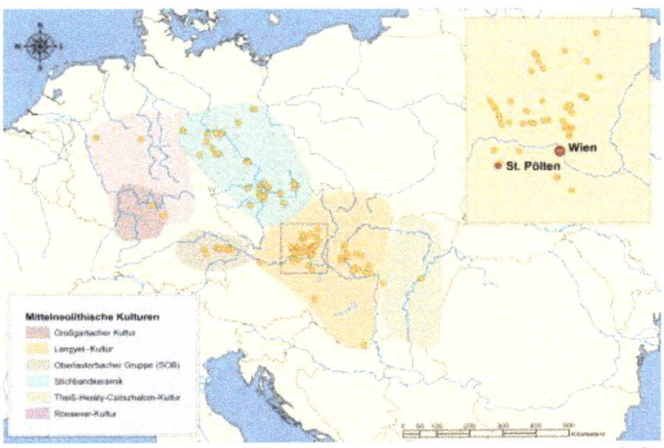

Figure 1. Distribution of Kreisgrabenanlagen and their Neolithic cultures in Europe. (Trnka, 2005a).

The monuments in Austria are characterized by up to four concentric circular V-shaped ditches with diameters typically ranging from 45 to 180m, up to four internal wooden palisades, and at least two opposed entrances. They were an integral but separated part of the associated Middle Neolithic settlements indicating a central-place role with no obvious defensive function. The most likely explanation is their use as some form of ritual place, a place for gatherings, or for certain ceremonies. The area of their distribution spans several archaeologically defined Neolithic cultural groups, so they appear to represent an early transcultural idea (Trnka, 2005a).

Nothing remains visible in today's topography. The structures are typically discovered by aerial photography, where they appear as soil and crop marks. Magnetic prospection has become the method of choice to gain a fast but still reliable survey for the large areas, when excavation is too costly (Figure 2).

Astronomical Aspects of *Kreisgrabenanlagen*

It can be frequently shown that the azimuths of the entrances, as seen from the centre, are identical to certain important rising or setting solar azimuths in the course of the solar year (e.g. solstices) (Becker, 1996; Iwaniszewski, 1996; Bertemes and Schlosser, 2004). On the other hand, several KGAs in Slovakia have been associated with lunar standstill orientations (Pavúk and Karlowský, 2004). A typical problem of previously published orientation studies analyzing the azimuth orientations of entrances or palisade gaps is the lack of horizon data and therefore lack of accurate declinations, which may lead to wrong associations of celestial bodies, especially in cases where horizon altitude is significant. Also, the published maps usually suggest flat terrain to researchers who may not have a chance to visit the sites themselves.

About 40 KGAs are known in Austria, and most of them have been magnetically surveyed since the 1990s (Melichar and Neubauer, 2010). A first investigation of these data supported the idea of solar orientation of several entrances, but also indicated that some azimuths–within and outside the range of solar azimuths–may also be explained by stellar orientation (Zotti, 2008, 2010a).

Figure 2. Magnetogram of KGA Pranhartsberg 2. The double ditches are far from perfect circles, and the entrance axes formed by the passages do not intersect close to the KGA's centre. In the northwest entrance just in-line with the main ditch, two magnetic anomalies may indicate postholes pointing towards summer solstice sunset (see also Figure 7).

These results invite the idea of a calendrical use as one further aspect of KGAs. The observation of sunrises and sunsets on certain dates like the cross-quarter days (dates just between solstices and equinoxes, i.e., approximately early November/February as begin/end of a solstice-centred 'Winter' season, and early May/August as begin/end of a solstice-centred 'Summer' season) in the course of the year, or heliacal risings or settings of certain stars in that period, may have been linked to processes on

Earth, like the beginning and end of the agricultural year, or festivals related to livestock breeding. However, this first study was based on flat archaeological interpretation maps without consideration of local topography, and horizon altitudes were only derived from digital elevation models, not based on field measurements.

Kreisgrabenanlagen as expressions of Power

We can only vaguely guess the state of beliefs of the KGA builders. Clearly, a whole group of people had to work together under the command of a main architect. The creation of astronomically motivated entrances required knowledge about and systematic observation of celestial phenomena as well as the power over a sufficiently large population–a power which may have linked the celestial processes with earthly authority. The astronomical aspects indicate that the monuments had a calendrical function and were probably associated with ritual and festive events. They might have been used for the legitimation of cultural or social patterns through natural phenomena. Because these social habits required such dramatic support, it may be supposed that they were novel. It is reasonable to assume the KGAs functioned in part in the legitimation of territorial ownership by new social groups and non-local traditions. That is, the ritual or functional idea behind the KGAs, or their expression of the power and might of individuals or groups, may have played a vital role for the formation of new cultural and/or political identities (Timothy Taylor, personal communication).

Surveying and Modelling of *Kreisgrabenanlagen*

The existing archaeological data consist mostly of georeferenced geomagnetic surveys and archaeological interpretation maps which show the outline of the filled ditches and palisades, and results from excavations where available. Typically, archaeological excavation reports or magnetograms show 2D maps or lists of orientation azimuths neglecting the visible horizon and the specific topographic situation of the respective site. The recently published overview of the Austrian KGAs (Melichar and Neubauer, 2010) includes topographical information, but no horizon data. In our current project (ASTROSIM), we investigate the potential astronomical orientations in much more detail than was previously possible (Zotti, 2010b).

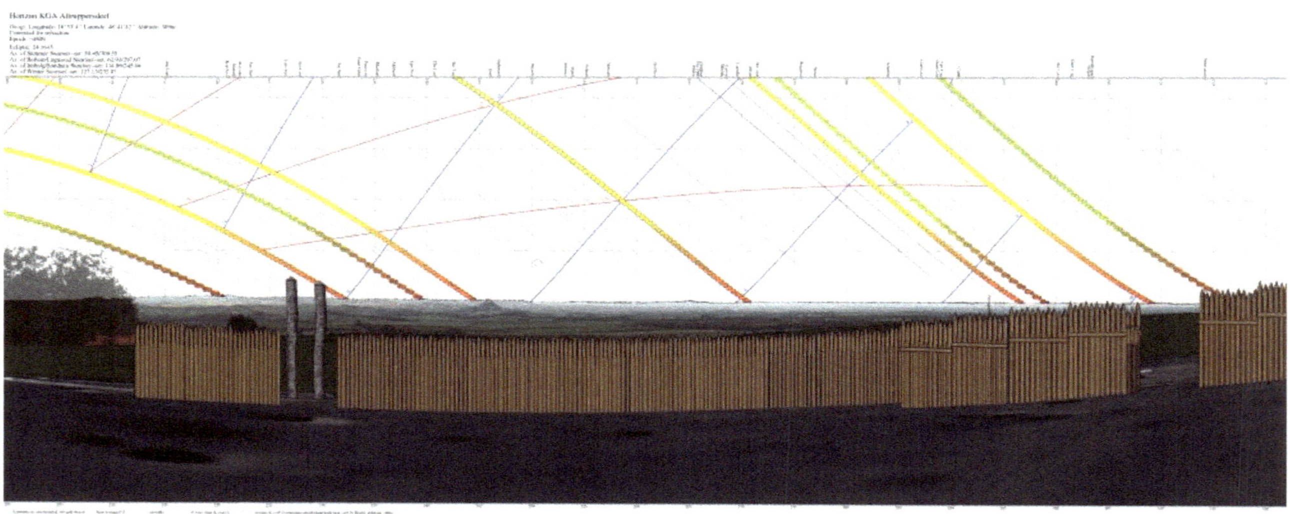

Figure 3. Part of the horizon panorama of KGA Altruppersdorf. This site overlooks a distant south horizon (partially blocked by a nearby tree, in the left edge of this part), the terrain slopes upwards towards north. A conspicuous hill (left of centre), visible even from inside the KGA if the palisade (reconstructed here) was not excessively high, almost coincides with winter cross-quarter sunset (chain of Sun circles). A lack of such landmarks at other sites however suggests a coincidental alignment.

Figure 4. The surveyed horizon line is completed with a carefully adjusted panorama photography and astronomical diagram of diurnal tracks. These panoramas show the frequently elevated horizon, current treelines and buildings (Rosenburg is near centre in this case), and can be directly evaluated, but also used as background in virtual reconstructions and in appropriate astronomical simulation software.

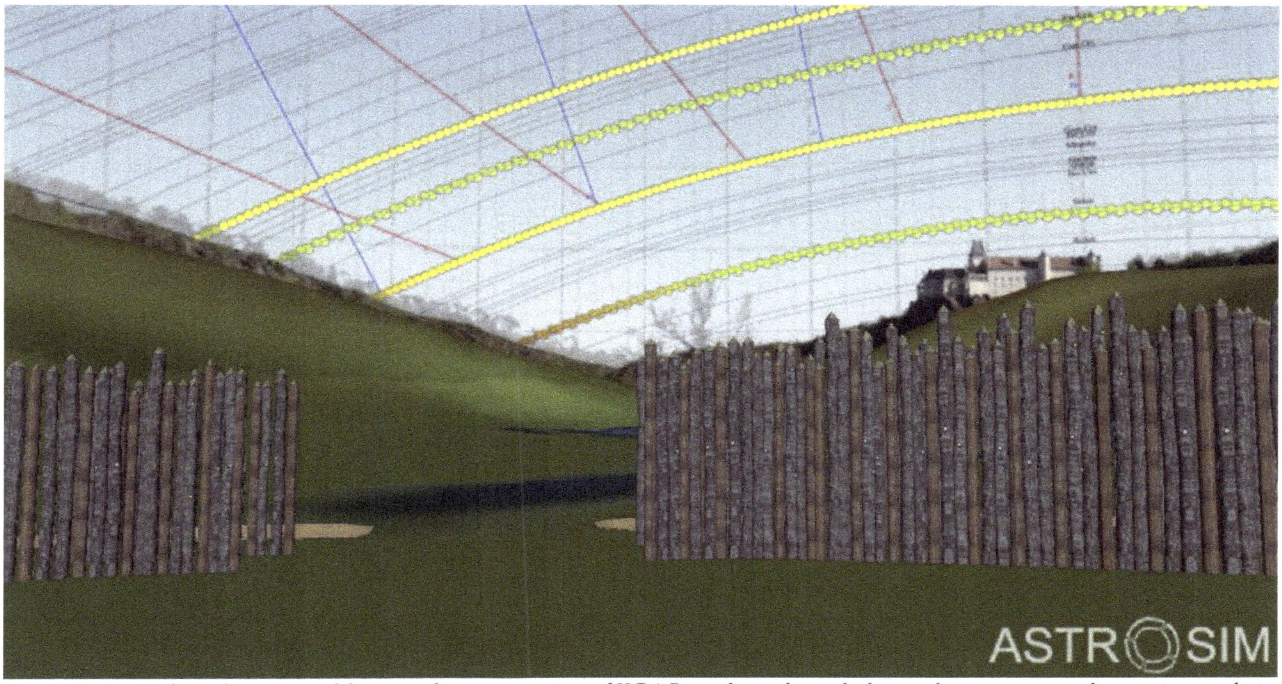

Figure 5. A view from the centre of the virtual reconstruction of KGA Rosenburg through the southeast entrance shows winter solstice sunrise over the mountains (third chain of circles from left) at an altitude of almost 7 degrees, in line with the left edge, but not the centre of the entrance. Omission of such horizon altitudes influence the result of previous studies. However, the rather wide entrance also aligns centrally with the lunar major standstill (lowest chain of circles). This shows the ambiguity of even such an approach, in cases where the entrances are not narrow enough to allow sharp determination of azimuths.

For the astronomical investigation of the KGAs in Lower Austria, we have re-visited all 33 sites where magnetic data is available with a total station to add a survey of horizon altitudes, on which an aligned photographic horizon panorama is carefully fitted. These panoramas, now fitting the surveyed line to within 1-2 arcminutes, are combined with latitude-dependent diagrams showing astronomical data like diurnal tracks of brighter stars or solstice and lunistice curves for the respective epoch. These combined diagrams can now be directly used to read declinations or celestial objects *on the landscape horizon* for the azimuths in question (Figures 3 and 4). The first visit to a site (Altruppersdorf) suggested also looking for conspicuous hills in the astronomically significant directions. At this site, Staatzer Berg, a pyramidal hill, coincides almost with the direction of winter cross-quarter sunset (Figure 3). If such hills or notches on the horizon would have appeared more frequently in our surveys, they could have indicated a motivation for selecting such sites to build a KGA. However, no other place shows such a feature.

Especially for the KGAs in the Kamp valley, e.g. near Rosenburg or Kamegg, the horizon altitude caused by the surrounding mountains certainly cannot be neglected and clearly influences the astronomical result (Figures 4 and 5).

In addition, the panoramas provide an astronomically correct visual background for virtual reconstructions of the archaeological structures.

Our three-dimensional virtual reconstructions are based on the magnetograms projected onto a digital elevation model of 25m resolution, with ditches carved out and palisades re-erected where their traces are detectable in the magnetograms.

The spatial resolution of the magnetograms is 0.125m x 0.5m, thus ± 0.25m can be assumed as maximum lateral displacement error, verified by various excavations carried out on basis of the archaeological interpretation of the magnetic images. The width of the entrance in the palisades, and therefore accuracy by which a solar date could have been observed, is also frequently uncertain due to the bad state of preservation. We therefore cannot hope to verify high-accuracy alignments involving long lines between edges of assumed former wooden structures, but given the large sizes of the ditches, this technique should allow us to study whether astronomically relevant azimuths lie within the entrances or gaps in the palisades as seen from the KGA centres or from one entrance through another.

Our models allow virtual walks through the architecture *in their landscape*, and allow a much better evaluation of the potential significance of astronomical or terrestrial targets for the entrances in question. It is noteworthy to mention that artificial horizon lines derived from single spots for available digital elevation models sometimes are not accurate enough, especially in cases where the horizon is close, so measurements taken on-site remain an essential part of work for reliable results.

Another potentially interesting aspect for simulation is shadow effects caused by entrances through the palisades.

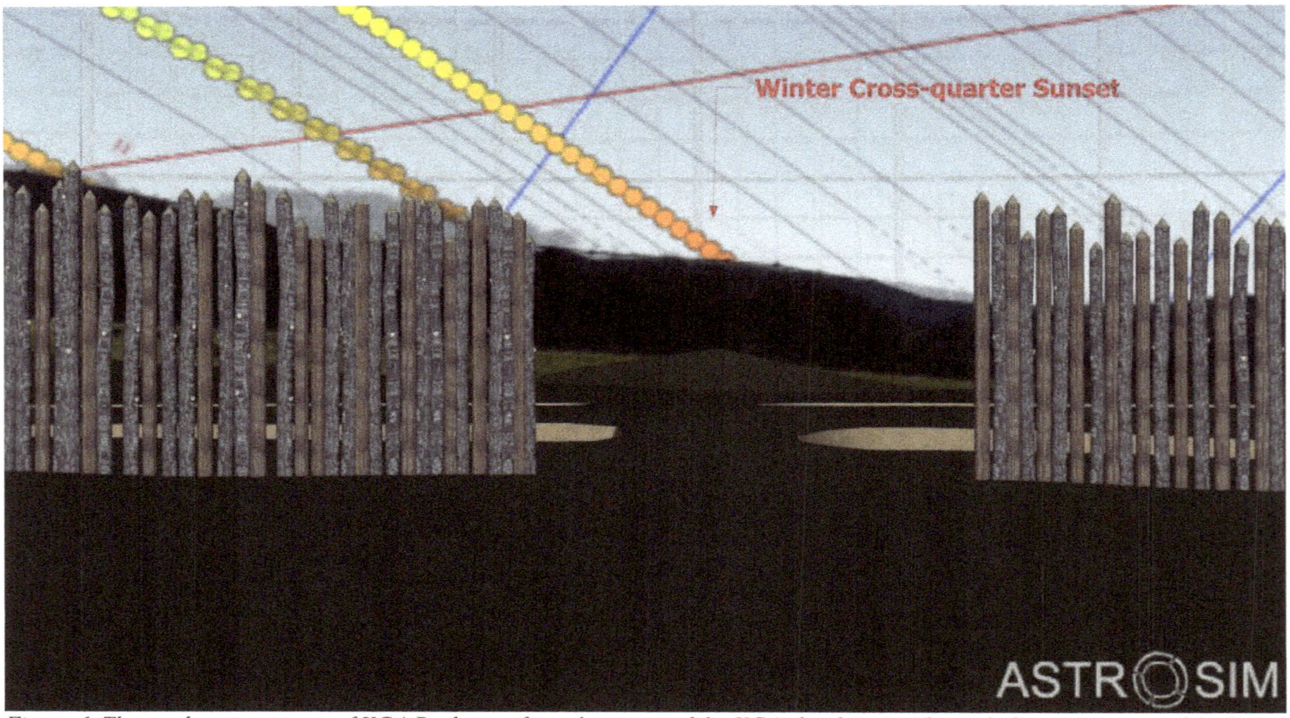

Figure 6. The southwest entrance of KGA Puch seen from the centre of the KGA closely coincides with the setting point of the Sun at the winter cross-quarter days on the elevated horizon. The opposing entrance (not shown) points towards sunrise on the summer cross-quarter days on a low far horizon. There, the palisade traces are better preserved and indicate a narrower entrance.

Simulating and animating the first or last rays of the sun being cast into the inner area at astronomically significant dates, e.g. the solstices, may help to discover such intentions of the builders.

A custom-made panorama export option for the modelling software allows us to create static foreground panoramas for selected observer locations to be used with appropriate astronomical simulation software (Figure 8). A further step will be the creation of a direct walk- through mode for the astronomical simulation program, so that celestial simulation can easily be combined with views from any point around the building structures

First Results

For KGAs with two circular and radial connecting ditches which form 'entrance passages', it has been observed earlier that their axes do not meet in the geometrical centre of the KGA (Pásztor, 2008; Zotti, 2008). In those cases, it seems more likely that the axes of the entrance passages, or the radial ditches, should be followed, and not views from a central location. For single-ditch KGAs with completely eroded traces of palisades, only central places can be assumed, and any results will clearly be less reliable, given that the inner area of KGAs shows no preserved archaeological structures which could have acted as back sights. For such badly preserved sites, we try to define the geometric centre of an idealised circle from the magnetogram and then investigate whether astronomically relevant directions lie within the opening angles of the entrances as seen from there, or from locations within 1-2 steps from this centre if it is not well-defined. (For a circle of 50m diameter, 1m of lateral displacement causes an azimuth error of 2.3°, but typically just in those KGAs entrance orientation cannot be narrowed down to better than several degrees width.)

The orientation of the entrances at KGA Puch towards the rising summer cross-quarter and setting winter cross-quarter days can be confirmed also with the measured horizon. However, at many locations the exact endpoints of the palisades is no longer visible, so that the entrances appear too wide to be able to pinpoint a single day, casting doubts on the previous claims about marking a single day in the solar year, but rather supporting the idea of symbolic orientation.

The creation of virtual reconstructions on top of the magnetograms also leads to potential explanations of some magnetic anomalies. For example, in the northwestern entrance passage of KGA Pranhartsberg 2 (Figure 2), two magnetic anomalies might indicate postholes which, if indicating posts contemporary with the KGA, would have accurately pointed towards summer solstice sunset (Figure 7). Of course, this result can only be an interpretative suggestion and should be verified by excavation.

In the previous study, it was assumed that ultimately all entrances might be associated with celestial objects. All entrances outside the solar and lunar arcs had therefore to be associated with stars. The result had to remain somewhat inconclusive for lack of horizon data, especially in the northern and southern horizons where the diurnal tracks of the stars intersect the horizon in a flat angle. The site visits now show that local topography, especially terrain slope, also must have been an important factor for the entrance orientation. The case of the heavily eroded KGA Gauderndorf at least seems now clear. This KGA has two entrances towards south-southeast and north-northwest, both well outside even the lunar range. However, the connecting line of the entrances closely follows the slope of the rather steep terrain, which invites the simple designation of 'upper' and 'lower' entrances, completely free of celestial associations. It seems therefore that not all KGAs include celestial elements, and certainly astronomical observation was not the most important motivation for their construction.

Figure 7. KGA Pranhartsberg 2. Two magnetic anomalies in the north-west entrance, shown here with potential posts in place, are perfectly in-line with summer solstice sunset.

It is especially noteworthy that the palisades of the KGAs which were erected in sloping terrain could most likely not have formed a closed horizon line, so that parts of the surrounding terrain, like the opposing hill behind the valley, remained visible above the assumed height of the palisade (see also Figure 3). This also means that some entrances which lead down the slope fail to intersect the sky, and again might not be associated with celestial objects at all. Figure 8 shows the case for KGA Steinabrunn, where we previously had suggested a potential association of the south-western entrance with a setting of the star Rigel. However, this entrance lies considerably lower than an observer close to the centre of the KGA, so that the palisade or single posts in the entrance area would have to be more than 4m high to hide the horizon and visually enclose a star (if this would have been the intention). At least this stellar association, although correct in azimuth, therefore now appears less likely. The opposing, upwards-leading entrance can on the other hand be confirmed to coincide with the rising azimuth of Deneb, as previously stated.

Conclusions

We cannot give a final result at this stage of the project. Many more models need to be built and evaluated, and some more simulation software is still under development. This work should be completed later during 2011.

Figure 8. A view of the Greater Dog, Orion and Taurus (with Pleiades at right), as they would have appeared side-by-side on the horizon in 4700 BC. The south-western entrance in KGA Steinabrunn, although indeed sharing the setting azimuth of the star Rigel (lower centre) as suggested earlier, is too low to visually intersect the horizon. In this reconstruction, the palisade is 2.4m high. To intersect the horizon as seen from this central location, it would have to be more than 4m high. (Simulation with Stellarium, an open-source desktop planetarium).

Acknowledgements:

We are grateful to Timothy Taylor, Gerhard Trnka and Alex Gibson for fruitful discussions on the topic.
This work has been supported by the Austrian Science Fund (FWF), project P 21208-G19 (ASTROSIM).

References

Becker, H. 1996. Kultplätze, Sonnentempel und Kalenderbauten aus dem 5. Jahrtausend vor Chr. – Die mittelneolithischen Kreisanlagen in Niederbayern. Arbeitshefte des Bayrischen Landesamtes für Denkmalpflege, No. 59.

Bertemes, F., and Schlosser, W. 2004. Der Kreisgraben von Goseck und seine astronomischen Bezüge. In Harald Meller (ed.), Der Geschmiedete Himmel -- Die weite Welt im Herzen Europas vor 3600 Jahren, 48-51. Stuttgart, Theiss.

Daim, F., and Neubauer, W. (eds). 2005. Geheimnisvolle Kreisgräben - Niederösterreichische Landesausstellung 2005 (Exhibition Catalog). Horn and Wien, Verlag Berger.

Iwaniszewski S. 1996. Neolithic and Eneolithic Structures in Central Europe: Calendric-Astronomical Implications. In W. Schlosser (ed.), Proceedings of the Second SEAC Conference 1994. Bochum.

Melichar, P., and Neubauer, W. (eds). 2010. Mittelneolithische Kreisgrabenanlagen in Niederösterreich. Geophysikalisch-archäologische Prospektion - ein interdisziplinäres Forschungsprojekt. Mitteilungen der prähistorischen Kommission der Österreichischen Akademie der Wissenschaften, Wien.

Pásztor, E. 2008. Megjegyzések a Lengyeli Kultúra Körárkainak Tájolásához (with German abstract: Bemerkungen zur Orientierung der Kreisgräbern der Lengyel-Kultur). Archaeologiai Értesítő 133, 5-20.

Pavúk, J., and Karlovský, V. 2004. Orientácia rondelov lengyelskej kultúry na smery vysokého a nízkeho Mesiaca Orientation of Lengyel rondels to directions of High and Low Moon). In Slovenská Archeológia Vol. LII(2), 211-280.

Trnka G. 2005a. Kreise und Kulturen – Kreisgrabenanlagen in Mitteleuropa. In F. Daim and W. Neubauer (eds). 2005, Geheimnisvolle Kreisgräben - Niederösterreichische Landesausstellung 2005 (Exhibition Catalog), 10-18. Horn and Wien, Verlag Berger.

Trnka G. 2005b. Katalog der mittelneolithischen Kreisgrabenanlagen. In F. Daim and W. Neubauer (eds). 2005, Geheimnisvolle Kreisgräben - Niederösterreichische Landesausstellung 2005 (Exhibition Catalog), 244-247. Horn and Wien, Verlag Berger.

Zotti G. 2008. Zur astronomischen Interpretation ausgezeichneter Richtungen der Kreisgrabenanlagen Niederösterreichs. In Wilfried Menghin (ed.), Acta Praehistorica et Archaeologica 40, 61-67. Berlin, Museum für Vor- und Frühgeschichte.

Zotti G. 2010a. Astronomische Aspekte der Kreisgrabenanlagen in Niederösterreich. In Melichar and Neubauer 2010, 136-167.

Zotti G. 2010b. Astronomical Orientation of Neolithic Circular Ditch Systems (Kreisgrabenanlagen). In L. Pigatto and V. Zanini (eds), Astronomy and its instruments before and after Galileo. Proceedings of the Joint Symposium International Astronomical Union - INAF Osservatorio Astronomico di Padova, Italy, held in Venice, San Servolo Island, Italy, 28 September-2 October 2009. Padova, CLEUP.

RE-STRUCTURING THE WORLD OF SCOTTISH MEGALITHIC SITES AND ANIMATING ASTRONOMICAL PHENOMENA THROUGH 3D COMPUTERISATION

DAVID FISHER

Abstract: The research that is being conducted combines the disciplines of Geographic Information Systems (GIS), computerized 3-dimensional modelling, astronomical formulae (Meeus, 1998; Chapront-Touzé and Chapront, 1991; Bretagnon and Simon, 1986) and mathematically accurate light-ray tracing in order to re-structure the world as seen by the builders of Scottish megalithic sites. This technique permits the tentative straightening of orthostats that are leaning from their perceived upright positions, re-instating stones that have fallen or known to be missing, and allows for the visual investigation into astronomical phenomena without site disturbance or the constraints of weather or limited site visitation. This cross-discipline approach permits experimentation through the course of a year or over the millennia; the output allows the observer to experience, by visualising, what might have driven some aspects of our ancestors' reasoning and possibly their reasons for creating such sites. The results allow for the expansion of the empirical database to aid archaeologists and anthropologists in the identification of cultural traits, or even drive true re-construction of disrupted archaeological sites to substantiate or refute astronomical findings. This paper outlines briefly the steps involved in the creation of computerized models of megaliths and their positioning within the landscape. This is followed by the main focus of the research: animations that illustrate preliminary findings of new perspectives on how the megaliths were viewed; that stellar observations were very much a consideration as well as solar and lunar observations; enabling the setting of a new timeline for Ballochroy; and addressing both the sequence of construction as well as a new date range for construction. Please note that, for the sake of this paper, the animations given during the symposium have been converted to single images.
Keywords: dating, orientation, simulation. tectonics, isostasis, experimental

1. Introduction

Many megalithic sites in Britain have no associated archaeological artefacts to determine the dates of construction; the orientation of the monoliths, the landscape within which they reside and celestial events have been used for that purpose (Thom, 1967). The original site dating was performed before the sciences of plate tectonics (Altamimi, Sillard and Boucher, 2002) and glacial isostatic rebound (Smith, Fretwell, Cullingford and Firth 2006, 949-972; Shennan, Peltier, Drummond and Horton 2002, 397-408) had been formulated. These time-related changes (116m in site movement and 6-8m in isostatic rebound over 5000 years) have been incorporated into the modelling tools to determine impact, if any, on the astronomical orientation of the megaliths.

Due in part to the lack of archaeological evidence that could be associated with the dating of megalithic sites in Scotland, Alexander Thom, while surveying some sites, took the opportunity to consider dating them by measuring the orientation of the stones to the 'setting position' of the solar sphere on the horizon. As an example, Ballochroy is shown in Figure 1.

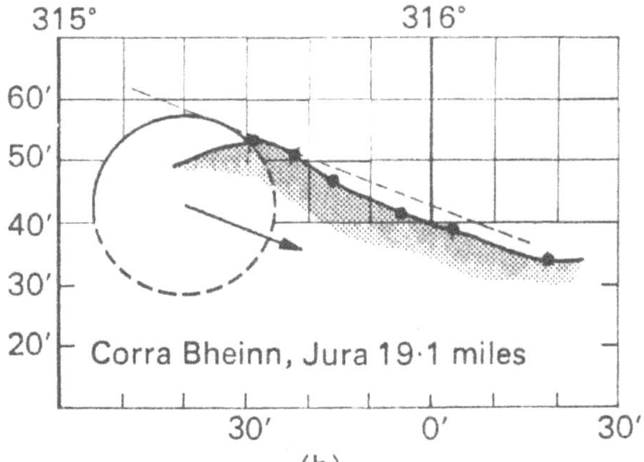

Figure 1. Ballochroy summer solstice sunset (Thom, 1967).

Thom established a date for Ballochroy to ~1770 BC where he determined the summer solstice sun to disappear behind the distant mountain and a sliver to reappear, shown by the dashed line. The dates determined by Thom in this manner have established themselves in the archaeological records; however, at the time of establishing these dates certain land movement sciences, those of:

- Plate tectonics and ....
- Glacio-isostatic rebound, were in their infancy.

Two questions that came to mind were: first, 'would these combined actions of tectonics and glacio-isostatic rebound have any implication on dating sites' and secondly 'could 3-dimensional computer modelling of the megalith worlds be used to determine such impact, if any, from these land movements. In conjunction with the aforementioned effort would it be possible to expand the archaeological and cultural knowledge of eons past by recreating the environment of the time.'

Plate tectonic motion

There are considered to be seven major plates upon which the continents 'ride'; these are illustrated in the following diagram. These plates are in constant motion either moving away or into each other, as indicated by the arrows in the diagram.

The arrows not only show the direction in which the plates move, in addition they give the average distance travelled in a year, this distance being indicated by the length of the arrow. It can also be seen by the direction of the arrows in the diagram that some of the plates appear to rotate. This rotation is more pronounced and is easily seen in the North America plate. The point about which they rotate, referred to as the Euler Pole, can be determined. The Eurasian plate rotation is not so pronounced; in addition its Euler Pole is somewhat more difficult to determine due to the influences of the other plates around it. This rotation is pertinent to the issue at hand—that of megaliths and distant foresights. Due to the amount of separation the foresight has from the megaliths being examined, a different distance of rotation for each location would result. There are several determinations for each plate for the location of their respective Euler Poles. The point selected for this experimentation provides an angular rotation of 0.261° per million years. (Altamimi, Sillard and Boucher, 2002). The Eurasia plate upon which the British Isles is seated (top left of the diagram) for the past 5500 years has, by calculation, rotated approximately 116 metres (380 feet) to the northeast.

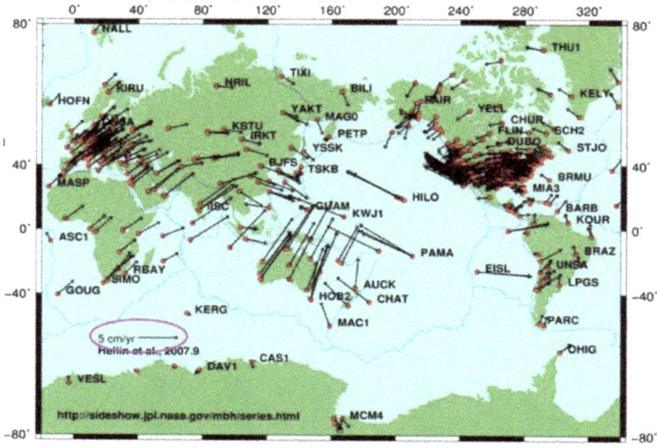

Figure 2. Map of major plates.

Glacio-Isostatic Rebound

In conjunction with the tectonic movement the landmass of Scotland has risen due to glacio-isostatic rebound over the course of the past several millennia since the last ice age. The following figure, illustrates the uplift since the Holocene Storegga Slide tsunami (~6100 BC). The area of focus for this research is highlighted in green, which illustrates the computed uplift in the zone over the past 5500 years of ~6-8m (~19-26 feet) in isostatic rebound.

As you can see in Figure 3, there are two blue zones, which indicate sites that have not risen but potentially sunk, and may be considered for research at a later date.

Mathematical models could be generated resulting in extensive volumes of data in order to test these concepts of plate tectonics and isostasis, which may well provide answers to a degree, but the result of such modelling would fail in several ways. For example:

- These mathematical tables are not tactile and therefore they are not conducive to touch, look and feel (exploration that best allows archaeologists to examine artefacts).
- In addition, mathematical models do not allow for effectively visualizing the impact of the events (for non-mathematicians, that is).

To address these shortcomings, I conceived and constructed a computerised approach to expand archaeological investigation by re-constructing the celestial world of Scottish megalithic sites, within a determined date range of 5500-3500 BC. My initial steps to this end were to:

- Construct 3-dimensional computer models to re-create the megalithic sites;
- utilize Geographic Information Systems (GIS) to generate 3 dimensional topographical landscapes;
- place the sites within their computerized landscape;
- ignoring previously specified dates, I performed a comprehensive review of the environment and determined an acceptable date range within which to conduct the experimentation; and ...
- combined these data with astronomical computations.

With such an undertaking I did not wish to restrict such work to only examine the land movement; by employing these same tools I was able to expand the methodology to enable me to conduct reasonably accurate investigations of the sites that:

- Effectively visualized what our ancestors saw.
- Test 'what-if' statements about megalithic sites and celestial orientations.
- Address the possibility that computerized 3D reconstruction could confirm or redefine dates?
- Address the original concept of landmass movement over the last 6000 years.
- Address the question posited regarding glacio-isostasis and plate tectonics shift: do these movements' impact celestial alignments as stated?
- If so, do these movements re-date the time of construction?

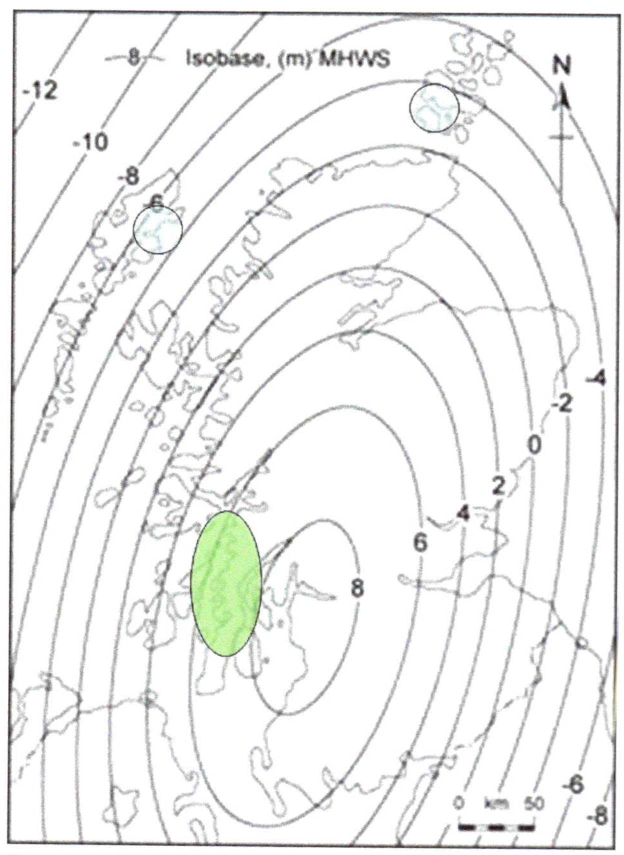

· *Figure 3. Glacial isobase map. Prof. David E. Smith and Peter Fretwell (personal correspondence).*

Any findings from the above investigation would illustrate the need for more thorough physical site investigation and an additional comprehensive field survey.

Site Survey

Astronomical orientations of previous site surveys were selected by 'aligning' stone surfaces with horizon features, which are subjective by their very nature. Instead, my research was guided by allowing the 'natural landscape' and astronomical positioning of the reconstruction model to dictate objective perspectives of the site. The approach was to conduct the following survey activities:

- Global Positioning—a reading was taken for each orthostat.
- Compass bearings—were assessed for each face of a stone, employing a constructors spirit level laid against the face of the stone, and the compass against the spirit level, effectively providing a flat surface for a more objective reading.
- Dimensions— of each face were measured
- Distances and bearings—measurements between stones were taken
- Photograph— of each face
- Photographs—of horizon

Employing the survey data, an accurately scaled 3D wireframe of each stone was constructed and then over-wrapped with the photographs taken onsite. The effect is a realistic looking re-creation of the stone, which can within the model be moved into an upright position, or as in one instance be stood upright from its current prone position.

Determine the Landscape

The landscape was generated by acquiring low scale Digital Terrain Model 'tiles' (1:10000, 10km x 10km DTM's) from the British ordnance survey. These were then converted through a sequence of steps resulting in a 3-dimensional landscape. The 3-dimensional models of the megaliths were then positioned within the landscape using the GPS readings and orientation bearings taken during the field survey. The formulae to generate the tectonic and isostatic land motion was incorporated into each tile (Bretagnon and Simon, 1986; Smith, Fretwell, Cullingford and Firth 2006, 949-972), followed by the corrections for grid north to true north convergence.

The last action to perform prior to running the simulations was the development of the software to accurately place the astronomical elements of the sun, moon, planets and stars into their correct positions to span 6000 years. This was done by employing the formulas of Meeus (1998), Chapront-Touze and Chapront (1991) and Bretagnon and Simon (1986) to within an accuracy of 2-arcseconds over the 6000 year period.

Figure 4. Nether Largie Stone 5 completed 3-D model.

Findings to Date

Some of the preliminary findings will be illustrated using two familiar Scottish sites in Argyll:

1. Ballochroy and
2. Nether Largie

In addition, I will include one site not previously explored, where no preconceived perspectives exist, as a control; this site is recorded as:

3. Dunamuck Farm, (across the main road from the village of Bridgend).

The first simulations of the sites were conducted for Ballochroy. Preliminary findings of the 'what-if' questions posited indicated several new and intriguing perspectives such as:

- An initial model of the summer solstice sunset at Ballochroy was run for 1770 BC (see Figure 5) and the slight edge of the sun can just been seen, appearing along the edge of the hill over Jura.
- Modelling has demonstrated that by 1750 BC the solar event at Ballochroy ceased to occur, causing one to change perspective on what might be the actual date of construction. It also means that if the site was

Figure 5. Model generated image of the summer solstice sunset at Ballachroy 2800 BC.
NB: the presentation made at the symposium was comprised of computer-generated animations. For the purpose of this paper those animations have been reconfigured into not so attractive single image.

- constructed as Thom estimates in 1770 BC ± 50 it only had a short duration of usefulness.
- The simulations have also demonstrated that the event happened from the beginning of the date range selected, that is, 3500 BC. Therefore, dating the site by the solar sunset at the time of the summer solstice is not a viable dating mechanism for this site.

Other views of the sun over the Paps of Jura are more viable, but are not presented in this paper as time would not permit.

Slanting Tops

While conducting the site surveys it was observed that each site had at least one stone with a top that slanted. The general response to this observation in the past has been that the tops had broken off over time. If that were the case, the slanting tops should be irregular in shape when compared from one site to the next, but I observed that there appeared to be uniformity amongst them.

Conducting simulations revealed an aspect to these sites that heretofore had not been known: shaping the tops of some of the stones appears to be a deliberate act, as the following computer-simulation-generated image illustrates.

Figure 6. Southern minor lunar standstill.
NB: the short Stone C has indeed broken.

The first simulation for a slanting top (shown in Figure 6) was a result of a random run with no specific viewing position being selected. As can been seen in Figure 6, the phenomenon of the lunar orb running down the sloping top seems to infer that the slanted tops do indeed have a specific purpose, positing the question—could this be pure coincidence? Simulation runs at others sites would determine if it was a coincidence or not.

Winter Solstice Sunset at Ballochroy

The winter solstice sunset illustrated by Thom (1966, 1-59) and reiterated by others (e.g., Hadingham, 1976) demonstrated the viewing perspective as being along the western edge of the stones. Whereas the simulations indicate that in order for the viewer to gain a more advantageous viewing point of the setting of the winter solstice sun, it is indeed the eastern edge where the Sun's final setting is seen to touch the central stone, in addition to the island of Cara in the distance (See Figure 7). This viewing position also causes the Sun to appear as if it is setting over the cist.

Figure 7. Winter solstice sunset 3500 BC.

Ballochroy's Third Stone Anomaly

This cist to the southeast of the stone row is the remains of a cairn of which the top has been removed sometime over the past 300 years. In academic circles, there has been much debate, as to 'what came first, the cairn or the stones?' The cairn's cist is just visible to the left of the stones (see Figure 7). If it were the cairn that was constructed first, then the setting sun on the winter solstice could not have been witnessed. If, as we have seen, the winter solstice was important, and the cairn came after the erection of the stones, then a replacement observation would be required—and there is one.

No discernible orientations for the Sun rising on the days surrounding the winter solstice at Ballochroy could be found in the simulations conducted for the site. Therefore, at this site, the only means open to me to address this question, was to simulate stellar events, in particular the heliacal rising or setting at the time of the winter solstice.

Earlier in the morning, prior to morning twilight, β Aquarius had risen at a point on the horizon which, if the observer positions themselves so that the point of rising is made to appear at the junction of the southern edge of the centre orthostat and the horizon, they are positioned to determine if the day is that of the winter solstice. As the morning progresses and twilight begins, α Aquarius would be witnessed to rise at the same junction point of the horizon and the centre stone, although this is somewhat difficult to perceive in Figure 8. Some minutes later, as they both rise in the morning sky, β Aquarius (3.1 mag) sits at the pinnacle of the needle of Stone C (to the right). The image in Figure 9 would occur just prior to daylight. If daylight occurs prior to β Aquarius reaching the pinnacle of Stone C or the star travels beyond the apex it is not the day of the solstice.

Figure 8. Heliacal rising of α and β Aquarius on the winter solstice morning.

Figure 9. Heliacal marker for winter solstice.

If the viewing location were fixed by a marker, then this event would only be valid for 100 years or thereabouts, but as the viewing location is self determining and moves over time, the window of opportunity for observation then expands to a period of approximately 800 years—determined by the transition of dark-to-twilight-to-daylight.

The presence of the third stone has been considered somewhat of an anomaly, with no apparent understandable intent or purpose, a stone that has no apparent role in the arrangement. I believe that this research may proffer a viable hypothesis explaining the presence of this third stone.

To whit—in completing the main lunar events at this site, an additional phenomenon became evident—that of the Moon's southern major standstill.

Standing with one's back to the central stone and looking along the eastern edge of the third stone, the Moon can be seen to rise then for two hours, progress along the horizon (almost like a ball rolling down a hill), eventually setting at the point on the horizon that is in alignment with the southeast face of the third stone (see Figure 10). Not only do these phenomena provide an explanation for the third stone but they also lock this stone into the arrangement.

Figure 10. Moon at major southern standstill.

Verification of Phenomena at Another Site

One obvious question is, are these stellar and sloping top phenomena peculiar to Ballochroy and therefore potentially coincidental? Or can they be verified as events that can be observed at other sites? To test the question, a site called Dunamuck, comprising of two groupings of stones, which is situated approximately 5km south of Nether Largie, was selected, because it is considered to have little or no alignments to taint the test.

The first test conducted at the Dunamuck site was on the southeast group of stones, consisting of two stones, one with an almost flat top and the other sloping. The stone with the flat top appears to have had its top shaped to replicate the 'plateau' in the distance. When the observer positions himself so that the 'flat top' aligns with the contour of the horizon at the time of the summer solstice, the Sun can then be observed to rise, and thence ride along the slanting slope of the second stone (see Figure 11).

Figure 11. Summer solstice sunrise at Dunamuck.

There is another group of stones at Dunamuck, two of which were still standing at the time of the site survey— a third was laying flat on the ground. The stone on the ground and one of the standing stones (fallen since 2007 observation) each have a slanting top—begging the question: what is the rationale or purpose for having two stones with the same slanting top within a close proximity of each other?

The stone on the ground was well enough exposed to obtain accurate measurements, and within the modelling software it could be 'stood upright' to a very good approximation of its original position. The only explanation I could discern regarding two orthostats, each with a slanting top stood next to each other, is illustrated in Figure 12.

Figure 12. Autumnal equinox at Dunamuck B.

The black stripes in the figure indicate dark night at the time of simulation, changing to the dark blue of morning twilight. The plain grey stone in the centre is the fallen stone stood upright. What is being shown is Aldebaran riding down the top of the left most stone, whilst the Pleiades ride down the top of the centre stone. This phenomenon only occurs at the time of the autumnal equinox.

Simulation runs were conducted for the whole year over the course of two millennia, whilst viewing the stones from both sides. This is the only stellar event that works for this pair of orthostats. Plus, once the stars have 'dropped' off the end of the stones, if the onlooker moves to the opposite side of the orthostats the autumnal equinoctial sunrise can be observed to ride up the slanting tops.

Time does not permit other interesting solar phenomena at the site but it is safe to say that the stones of Dunamuck are the solar sister site to Nether Largie's lunar orientations.

Verification has been tested that both sloping tops and stellar phenomena are of astronomical consideration for megalithic sites.

Nether Largie

For this paper there is one final site in which the preliminary findings also prove to be most enlightening, and that is Nether Largie. To maintain consistency I will continue to use Thom's work as a reference; as such his plan for Nether Largie is given in Figure 13. During the site survey a fifth stone was located at the site at the Group Q stones, so a minor addition has been made to the diagram indicated by the blue dot (surrounded in the diagram with a red circle).

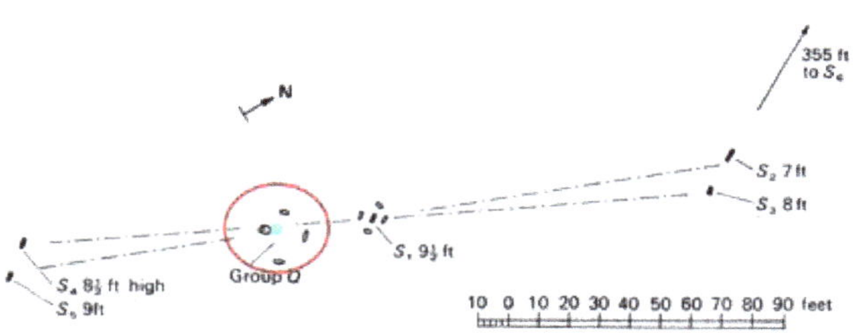

Figure 13. Nether Largie plan.

A systematic series of simulations were run against each stone for the primary solar and lunar events; however, no solar events of any kind could be discerned (quarter days as yet have not been tested). However, in the simulation process some exceedingly interesting data were uncovered. I refer to the Q stones as they relate to their topographical locations (east, west, etc.). When positioned at Q east and looking southwest toward stones 4 and 5 on the day before the southern major standstill, the moon can be seen to set between them in a Col in the distant horizon (see Figure 14). On the actual day of the standstill the moon slides along the ridge, seemingly pointed to by stone 5, and sets into the side of the distant mountain (the path is indicated by the red arrow in Figure 15).

Figure 14. Day BEFORE southern major standstill.

Figure 15. Day OF southern major standstill.

Yet again, this site presents us with another stone with a slanting top, this time the Central Stone, referred to as S1.

By standing to the north of the stone at a point that places the lower right 'corner' of the slanting top at a niche in the horizon (see Figure 16—offset in the diagram for the sake of illustration) the southern minor standstill moon, as it sets, rides down the slanting top and sets in the niche. By standing on the opposite side of this stone the reciprocal event of the northern major moonrise can be witnessed to ride up the sloping top.

Figure 16. Nether Largie minor southern standstill over S1.

Figure 17. Northern standstill Moonrise above stone 6.

Modelling affords the opportunity to stand upright stones that have fallen, as at Dunamuck, or to straighten stones that are leaning severely. Stone 6 at Nether Largie is such a leaning stone. When straightened by 26° the top of the stone has a shape that follows the contour of the distant hill, and at the time of the northern major standstill the Moon may be observed to rise along both (as shown in Figure 17).

Nine more lunar oriented viewpoints for Nether Largie have been uncovered via these simulations. These additional orientations demonstrate that at the time of site construction, foliage was not present to obscure the view. This absence of foliage could be a matter of low tree coverage at the time, or tree clearance was conducted to allow the events to be witnessed.

An interesting aspect that is shown in the Figure 18 for Nether Largie is that just about every observed event happens in two's. In those instances that do not (the northern and southern minor moonsets), it leads one to ponder what is missing from the site that these minor events are not recorded.

To Summarise the Findings

- The paths through the sky of the solar and lunar orbs were 'observed' more so than 'horizon events', by the use of true or 'created horizons'.
- The slanting tops are a deliberate innovation of our ancestors and not breakage of stones over time.
- The actual shape of the stones and their orientation indicate that stellar events may also have been observed at more than one site.
- We can adjust the dates of construction for Dunamuck and Ballochroy to a bounded range of 3100 BC –2200 BC (older than 1770 BC).

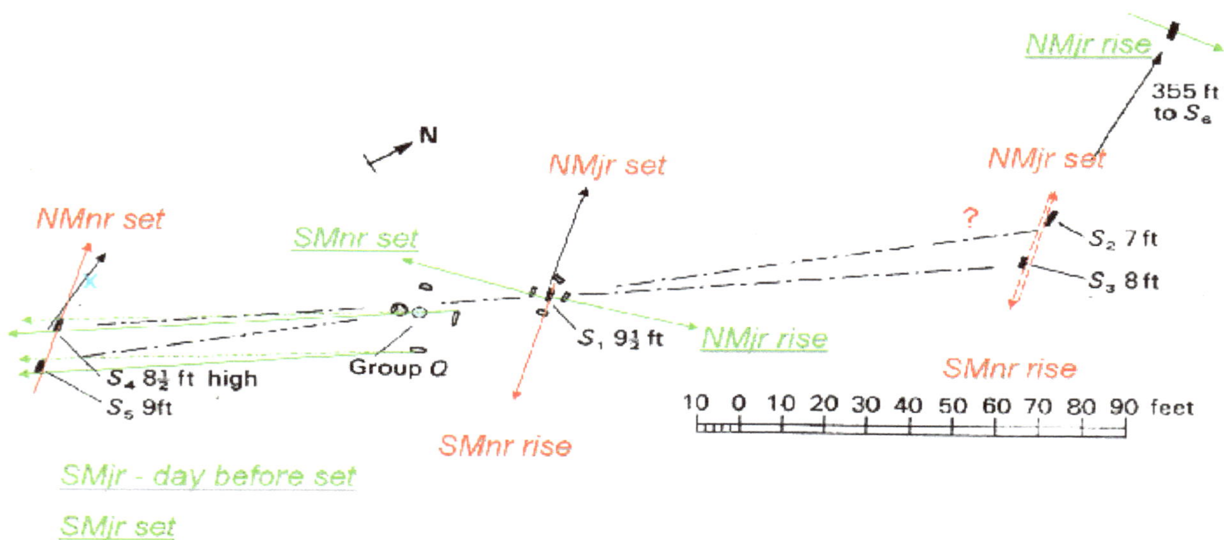

Figure 18. Additional Lunar orientations at Nether Largie.
NB: The dashed red lines in Figure 17 associated with stones 2 and 3 indicate that these orientations cannot be confirmed with a secondary reference due to the twisted lean of stone.

- Until now we have omitted half the viewing perspectives of the stones.
- We should start to consider stars and maybe even planetary connections with the megaliths.
- The model can reflect the stones as they stood.
- NB: One of the Stones at Dunamuck has fallen in the last 2 years! Seems like a great time to arrange a dig to investigate the hole that's left!

Cultural Implications

These ancient people were not content to be mere external witnesses of timed events, as if consulting a watch; it was the phenomena that intrigued them as they deliberately structured their monuments in the following manner:

- By marrying sky, earth and man together as one, they 'PARTICIPATED WITHIN' the event.
- CONTROLLING the event, by shifting the observation away from the natural horizon to a horizon they created themselves.
- The DUALITY of Nether Largie is staggering.
- What cultural implications can be derived from such findings?

Date ranges for Nether Largie concur with currently held beliefs, and are association with the dated cairns in the area. Ballochroy on the other hand may well be contemporaneous with the cairns, already evaluated across the Kintyre Peninsula, that are dated some 1500 years earlier than previously hypothesised.

3-Dimensional Modelling Conclusions

3D modelling has demonstrated that it offers several evaluative techniques of on-site investigation, to add to the archaeologist's investigative tool-kit. In addition:

- It provides us with the ability to record the site as is.
- To straighten stones that lean.
- Replace those that have fallen over or are missing. As stated earlier, since surveying Dunamuck another stone has fallen, but I have photographs and modelling for future reference.
- Effectively conduct investigation into astronomical phenomena and provide further supportive material into the decision whether or not an archaeological dig would be justified to reinforce our knowledge base.
- 3D modelling enables us to continue our investigations without the complications and discomfort of adverse weather conditions, in field expenses (food and lodgings and travel expenses), or restricted views due to tree growth. In a few months with 3D modelling, we can view complete cycles for a site—lunar, solar and even stellar events.

3D modelling has opened a new interpretation of 'phenomena observation' at megalithic sites, at least for stone rows. This technique enables us to experience a new perspective, not only on how to view the stones themselves, but just as importantly, on how they were viewed by our Scottish ancestors, and how we must now re-view the cultural implications regarding the structures and how they were employed.

In closing I would like to point out that my observations of the impact of plate tectonic and glacio-isostasis movement for the Scottish sites is only a matter of seconds, in both time and azimuth of a horizon observation, which is negligible in the greater scheme of things. These findings are a great relief—for archaeoastronomers studying British prehistoric sites— not having to take such factors into consideration!

References

Altamimi, Z., Sillard, P. and Boucher, C. 2002. ITRF2000: A new release of the International Terrestrial Reference Frame for earth science applications. Journal of Geophysical Research: Solid Earth (1978-2012) 107, no. B10.

Bretagnon, Pierre and Simon, Jean-Louis. 1986. Planetary Programs and Tables from -4000 to +2800. Richmond VA, William Bell Inc.

Chapront-Touzé, Michelle and Chapront, Jean. 1991, Lunar Tables and Programs from 4000 BC to AD 8000. Richmond VA, William Bell Inc.

Hadingham, E. 1976. Circles and Standing Stones. London, Book Club Association.

Meeus, Jean. 1998. Astronomical Algorithms 2nd ed. Richmond VA, William Bell Inc.

Shennan, I., Peltier, W.R., Drummond, R. and Horton, B. 2002. Global to local scale parameters determining relative sea-level changes and the post-glacial isostatic adjustment of Great Britain. Quaternary Science Reviews 21, no.1, 397-408.

Smith, D.E., Fretwell, P., Cullingford, R.A. and Firth, C.R. 2006. Towards improved empirical isobase models of Holocene land uplift for mainland Scotland, UK. Philosophical Transactions of the Royal Society A: Mathematical, Physical and Engineering Sciences 364, no. 1841, 949-972.

Thom, A. 1966. Megalithic Astronomy: Indications in Standing Stones. Vistas in Astronomy 7, 1-59.

Thom, A. 1967. Megalithic sites in Britain. Oxford, Clarendon Press.

RECUMBENT STONE CIRCLES: THEORY OVERVIEW BASED ON FIELDWORK CONDUCTED AT THREE SITES

LIZ HENTY

Abstract: Surveys were conducted at three Recumbent Stone Circles in North East Scotland. A variety of surveying techniques were used and assessed for reliability. A review of Thom, Ruggles and Burl investigated the archaeoastronomical theory that the recumbent stone is aligned to the setting of the major or minor standstill moon, an arc which includes the winter solstice setting. The major findings of this project were based on horizon astronomy. Given the altitude of the horizon behind the recumbent, the astronomical events would not have been visible in their true positions. Importantly, the three sites had alignments to horizon features opposite the recumbent, not previously discovered. At Tomnagorn the recumbent is opposite Bennachie, a locally revered landmark. At Sunhoney and Midmar Kirk the recumbent is opposite the Barmekin of Echt, a cattle-fold. This ties in with the theory that the megalith builders were cattle herders. Additionally there is some evidence of symbolism associated with a bull cult. This research, based on an interdisciplinary approach, suggests cosmological meaning was derived from a marriage between archaeoastronomy and sacred geography.
Keywords: Recumbent Stone Circles, North East Scotland, Surveying methods, Horizon astronomy, Symbolism, Archaeoastronomy, Sacred Geography

Figure 1. Panoramic view of Sunhoney Recumbent Stone Circle showing the recumbent stone and flankers and the angle of the horizon behind it.

Introduction

This paper reviews archaeoastronomical theory pertaining to the megalithic Recumbent Stone Circles found in North East Scotland. It includes a detailed study of three such circles, which, by locating the researcher in the place of the ancient sky-watchers, asks whether phenomenology can add useful content to the academic research carried out by theorists such as Thom (1967), Ruggles (1999) and Burl (2000). Fieldwork methodology which drives the astronomical conclusions is also examined. Additionally, it enquires whether the builders may have combined astronomy with other factors such as landscape, symbolism and economic culture to create a religious worldview that encompassed both earth and sky.

General Background

Britain is home to a rich collection of stone circles built in the megalithic period of the Neolithic and early Bronze Ages. The scarcity of archaeological artefacts and the absence of written material means that most theories, including dating, remain speculative, yet scholars agree that astronomical events had an influence on their construction. Archaeoastronomy seeks to discover the impact of astronomy on culture and whether this is incorporated in the buildings by alignment, orientation, or in symbolism derived from the sky.

In North East Scotland, in an area of 80 by 50km spreading outwards from Aberdeen, there is a distinct type of stone circle built between 2500 and 1750 BC (Ruggles and Burl 1985, S25). These circles are called Recumbent Stone Circles (RSC), as their outstanding characteristic is a large recumbent stone 'with an average weight of 24 tons' (Shepherd and Shepherd 1996, 151) set between two tall flanking pillars. Although each site probably supported only about twenty people, the size and weight of the stones indicates that communities worked together, presupposing a stable culture (Burl 1999, 7). According to Burl (1999, 220), of the original estimate of 200-300 circles, 175 were known in 1911. Of these, Shepherd and Shepherd (1996, 151) say only 99 remain that are probable, of which 74 can be accepted as definite. These Recumbent Stone Circles have been researched extensively, principally by Thom, Ruggles and Burl.

Thom's detailed theories on megalithic precision astronomy have been widely debated, particularly by Ruggles (1999) in a further scientific survey. Despite Ruggles' view that progress is better made by systematic surveys of a large sample of related sites, this paper suggests that a detailed study of a few sites can bring a qualitative depth to the quantitative statistics derived from larger samples. The three sites examined were Sunhoney,

Midmar Kirk and Tomnagorn, all located along a 7km line, 22km west of Aberdeen. The choice of sites was purely subjective, governed by their proximity to Aberdeen and their apparent linear siting, despite Barnatt's opinion (1989, 172) that RSCs are randomly distributed.

Fieldwork Methodology

Initial fieldwork was commenced prior to academic research to avoid being influenced. The same methods were used to survey all three sites and as the study was primarily to look for archaeoastronomical alignments, the varying internal features were ignored. Latitude and longitude were established by converting eastings and northings read from OS Landranger Map 38 (http://gps.ordnancesurvey.co.uk/etrs89geo_natgrid.asp). These results were checked using a Magellan hand-held GPS unit. Using a magnetic compass, east-west and north-south lines were pegged across the circle to establish an approximate centre. This centre was adjusted by measuring the resulting radii with a cloth tape. Dimensions from the centre to the left hand edge of each stone were marked on a rough plan. Measurements around the perimeter included the widths of the stones and the distance between them. This measurement gives a difference in the final perimeter value from that computed from the diameter using *pi*=3.1416. Both methods can only give approximate values because the straight measure does not compensate for the curve of the line and the *pi* calculation is only accurate for a true circle.

Plans were drawn to scale from the field measurements. Thom's plans do not appear to have been challenged; indeed Ruggles and Burl (1985, S58) acknowledge 'Thom's accurate site plans'. Consequently the research site plans were superimposed on Thom's (1980) plans for comparison. MacKie (1977, 30) reveals that Thom used a best fit circle drawn from averaging the radii to find the mean radius for the plan circle. Additionally, Thom and his son (1988, 143) did not measure the perimeters by tape but calculated them from measured diameters so their plans appear to be circular. Using dotted lines on the research site plans to link the stones shows a different picture. All three circles appear to be flattened. The line of the recumbent and flankers does enhance this effect but at Tomnagorn and Sunhoney the north-east quadrant has a mean radius respectively of 3-4 feet longer than the radii of the other quadrants. The Midmar Kirk site has been reconstructed and stones replaced, so its circularity does not negate the above findings. Burl (1970, 75) has speculated that the sites were orientated towards sunrise. Summer solstice rising occurs in the north-east quadrant.

The main difficulty was establishing true north. Firstly north was found using a magnetic, prismatic compass and the results adjusted from magnetic north to true north using the current (2010) correction of 3° 10′ west. The readings did not correspond to Thom's north point. The GPS compass readings were erratic and unreliable in the circles, presumably because of the granite stones. The only method left was to establish a north/south line using the Sun. At noon in the northern hemisphere the Sun culminates due south at 0° longitude. Local noon depends on the angular distance from 0° and is calculated by using a scale of 15° to one hour in time, subtracting the time for longitudes west of 0° and adjusting for daylight saving. Having timed local noon, a line through the circle centre was pegged using the shadow cast from bamboo canes. This method compared favourably to the compass reading. A further check was made of the longitude GPS measurements along the line. The final stage of the fieldwork was measuring the azimuths or angles of the stones from the circle centre using both a large circular protractor and the prismatic compass. As measured against the scaled plan, these angles were unreliable so the help of a trained surveyor with a theodolite was enlisted. The distances and angles obtained from the theodolite readings were consistent with the final site plans. Significant rising and setting azimuths calculated for 2000 BC were added to the plans.

At every stage of the fieldwork there was room for error, from the stretch of the cloth tape to the difficulty of holding the heavy theodolite ranging rod completely vertical. Though every attempt was made to ensure accuracy, mainly by employing a variety of methods, there is probably a margin of error of at least 5° throughout. This finding raises questions about the absolute accuracy of past surveys.

Astronomy and Archaeoastronomy

Astronomical measurements are based upon the cardinal directions, all azimuths being measured from 0° at north. Archaeoastronomy places emphasis on the solar, equinoctial east and west points and the lunar risings and settings in the arcs of approximately 30° (north-east, south-east, south-west and north-west) described by the lunar standstills. The solstice risings and settings are at the centre of these arcs. Thom remarked (1967, 95) that it would have been difficult to measure north with no pole star. As the above events do not occur at north we cannot assume that north was significant for the builders. Even using a gnomon to show a north/south shadow at local apparent noon implies anachronistic knowledge. The absence of exact alignments to the cardinal directions at the sites suggests that other points were more important.

Archaeoastronomical theory has shifted in the decades following Thom's initial surveys. As the orientations of the recumbent and flankers fall between the south-west and the south-south-east, it was believed that they were tied to the summer lunar standstills. Yet Barnatt says (1989, 29) that no common pattern to major or minor lunar standstills was found. Additionally, in midsummer, because of the tilt of the ecliptic, the Moon's maximum altitude does not reach more than 4°, only visible on a level horizon. Burl (1999, 13) revised his theory so that the recumbent was laid in line with the full Moon, 'not as it rose or set' but so it could be seen in 'its full glory'. Research using Stellarium planetarium software showed that, at the sites' latitude, in midsummer the time between the Moon's rising and setting is only around four hours. It skims along the horizon and because the horizon altitude, as viewed from the circles' centre, is around 8° at both Midmar Kirk and Sunhoney (Figure 1), it cannot be visible. At Tomnagorn the horizon altitude is 3°, so only the rim would be visible. Therefore, the astronomical events most capable of observation in the south-west are the setting of the winter solstitial sun and the passage of the winter moon which would pass high over the recumbent. Planetarium research showed that the *visible* winter solstice sunset occurs some 26° south of the recumbent centre. Further research (beyond the scope of this study), based on horizon astronomy, could establish where in the circle the astronomical events are actually *visible* both at midsummer and midwinter. This might discover new targets in the circle for astronomical events.

Sims (2006, 5-8) has suggested that a double alignment to the south-west, which combines winter solstice sunset with southern lunar standstill, reflects 'a *ritual* pairing'. However this pairing could not be observed because the Moon at standstill would be a dark moon. Only in years between standstills could a full Moon occur at the winter solstice. Some evidence was found to back Sims' theory of ritual pairing. At Sunhoney, stone 5 (Figure 2) is tall and pointed with a *lunar* alignment (major standstill rising) and diagonally opposite is its smaller mirror image (Figure 3) with a *solar* alignment (summer solstice setting).

Figure 2. Stone 5 at Sunhoney RSC is aligned to the southern major lunar standstill rising.

Barnatt's findings (1989, 29) are that prehistoric astronomy was of relatively low precision yet 'orientations highlight the impressiveness of astronomical events'. Ruggles (1999, 213) details the recumbent azimuths for 58 sites and their variation implies orientation is more likely than precise alignment. If the theory that these circles are aligned to the Moon's southern lunistice settings in the south-south-west is correct, then the angular distances measured from the circle centre to the recumbent centre should all be similar. Ruggles gives the centre line azimuth of the recumbents at Sunhoney and Midmar Kirk as both being 231° degrees and Tomnagorn as 202.5°. The fieldwork results were 232.5°, 235.18° and 221.4° respectively. Differences in location or date of construction cannot explain this variation. Indeed, fieldwork showed that the three circles in this study are within 1°5′ of the same line of latitude. If the siting of the stones was astronomically precise, then their positions would be within this variation.

Figure 3. At Sunhoney RSC, Stone 11, diagonally opposite stone 5, is aligned to the summer solstice setting.

The question of precision is more a *current* concern. We need to plan the sites accurately, measure horizon altitude and plot the precise azimuths of the stones to calculate where the major stations of the Sun and Moon are in relation to them. Such hindsight can be used to theorise on the original placements but it seems highly unlikely that such accuracy was observed by the builders.

Archaeoastronomy, Sacred Geography and Symbolism

It was not until the third site visit to Tomnagorn that a discovery was made (only possible on a totally clear day) that has not been mentioned in other studies. In the far distance, and opposite the recumbent, is the Mither Tap, the summit of Bennachie. Slightly obscured now by an intervening tree, a position a few steps to the left gives a spectacular view of a landscape feature which is locally revered (Figure 4). Eliade wrote (1987, 26) that 'sacred space implies a hierophany, an irruption of the sacred that results in detaching a territory from the surrounding cosmic milieu'. Elsewhere he said (1958, 99) that 'the symbolic and religious significance of mountains is endless'. Having placed myself in the landscape I had witnessed a hierophany. MacGregor (2002, 154-5) notes an uncanny resemblance between the cleft in the Mither Tap and the image of a recumbent and its flanking pillars and goes on to say 'the nature of the experience at the top of Bennachie provided the source of inspiration for the form of the RSC tradition'.

Figure 4. The view opposite the recumbent at Tomnagorn RSC showing the Mither Tap of Bennachie in the distance.

Culture which draws its symbolism from the sky is astrological. Looking at the fixed stars which were prominent in the period of megalithic building we find that the Pleiades rose due east and set due west. Given that these stars represent the shoulder of the Bull in the constellation of Taurus, the sites were examined for evidence of symbolism relating to the Bull. Certainly the recumbent and its curved, upward reaching flankers give the impression of a bull's head with its horns pointing to the sky. There is some academic backing for this symbolic approach. As there were few cattle, and no sheep or horses, Burl (1999, 13) suggests cattle would have been prized. Milne (1912, xix) suggests that people lived in 'large and substantially walled folds to contain themselves and their cattle'. The hill fort at the Barmekin of Echt, opposite the recumbent at Sunhoney was one such cattle-fold. Sims (2006, 1) says the current view is that the monuments were built by 'cattle herders who continued to hunt and sometimes planted'. Burl (2000, 71) records that Celtic customs 'like the horned god date back at least to the Bronze Age when the stag, and perhaps the bull, were representative of the sun'. Lacaille (1930, 226), researching a bull cult in Scotland found evidence of bull or ox bones in a Bronze Age cairn within this area. Trevarthen (2002) has associated the annual cycle of cattle-breeding with the ritual year at megalithic monuments in Inverness-shire. This suggests a clan-based totemic culture, which lends itself to the social, structural functionalist explanation of religion as favoured by Durkheim (2008).

Archaeoastronomical opinion is that ritual was closely tied to astronomic events. By concentrating on the sky alone, similar landscape correspondences have been largely overlooked. Yet there is the symbolism of the sacred space itself, discrete in its location and set apart from the surrounding mundane world. Eliade in particular has explored this sacred/profane dichotomy as well as discussing the sacred properties of the stones themselves. Though the megalithic builders may have had what Thom and Burl (1980, 1) describe as 'an empirical knowledge of astronomical phenomena', ritual was part of a complete cosmology which combined knowledge of the sky with that of the earthly landscape, seasons, agriculture and economic needs. The discovery of landscape alignments *opposite* the recumbents at Tomnagorn and Sunhoney concurs with Bradley's suggestion (1993, 45) that many alignments may be found within a site, not just astronomical but also topographical. MacGregor agrees (2002, 142) that most sites have not been considered in terms of their relationship with the wider landscape. Our cultural limitations and categorisations may blind us to the prehistoric cosmos, described by Bradley (1998, 109) as that 'perception of space that extends outwards from the individual and upwards into the sky'. Ruggles and Burl concluded (1985, S47) that lunar orientation and the presence of a conspicuous hill top seem to have been separate goals.

Indeed, Ruggles and Burl (1985, S58) agree that 'evidently there is no simple all-embracing astronomical explanation for the Recumbent Stone Circles'; a view that this research supports. McCluskey adds (1981, 119), 'perhaps it is time to temper the present emphasis on using quantitative astronomy and statistical analyses to infer the scientific motives of prehistoric man'. A comparative interdisciplinary approach combined with qualitative phenomenology may be a better way forward, though Ruggles warns (1999, 11) 'by dismissing prehistoric astronomy altogether we may be just as guilty of projecting our own prejudices into the past as by recreating it in our own image'.

Conclusion

The astronomical conclusions for any site depend on an accurate survey. This research found that every aspect of surveying is prone to a margin of error. Similarly, conclusions relating to precise azimuths depend not only on establishing the circle centre and true north but also on the horizon properties for each astronomic event. Discoveries in the field led to a tentative theory that the horizon opposite the recumbent may be as significant as the horizon behind it. This could be an avenue for further research. Phenomenological fieldwork and ethnography suggested alternative theories. Firstly, astrological symbolism may have influenced ritual and secondly, a stable land-based economy based on cattle-herding may have driven symbolic religious practices. Overall, the conclusion is that archaeoastronomy theory alone is not sufficient to explain the predominance and orientation of the RSCs; other factors may have played an important part in their layout. This research, based on an interdisciplinary approach suggests cosmological meaning was derived from a marriage between archaeoastronomy, symbolism and sacred geography.

References

Barnatt, John. 1989. Stone Circles of Britain: Taxonomic and distributional analyses and a catalogue of sites in England, Scotland and Wales, Part I. Oxford, British Archaeological Reports, British Series 215, (i).

Bradley, Richard. 1993. Altering the Earth: The Origins of Monuments in Britain and Continental Europe. Monograph Series 8, The Rhind Lectures 1991-92. Edinburgh, Society of Antiquaries of Scotland.

Bradley, Richard. 1998. The Significance of Monuments: On the shaping of human experience in Neolithic and Bronze Age Europe. London, Routledge.

Burl, H. A. W. 1970. The Recumbent Stone Circles of North-East Scotland. Proceedings of the Society of Antiquaries of Scotland 102 (1969-1970), 56-81.

Burl, Aubrey. 1999. Writing in Milligan, Max. Circles of Stone: The Prehistoric Rings of Britain and Ireland with text by Aubrey Burl. London, The Harvill Press.

Burl, Aubrey. 2000. The Stone Circles of Britain, Ireland and Brittany. New Haven and London, Yale University Press.

Durkheim, Emile. [1912] 2008. The Elementary Forms of Religious Life. Translated by Carol Cosman, abridged with an Introduction and Notes by Mark S. Cladis. Oxford, Oxford University Press.

Eliade, Mircea. [1957] 1987. The Sacred and the Profane. USA, Harcourt Books.

Eliade, Mircea. 1958. Patterns in Comparative Religion. Translated by Rosemary Sheed. London, Sheed and Ward.

Lacaille, A. D. 1930. The Bull in Scottish Folklore Place-Names, and Archaeology. Folklore 41, no. 3, 221-248.

MacGregor, Gavin. 2002. Making Monuments Out of Mountains: The Role of Colour and Texture in the Constitution of Meaning and Identity at Recumbent Stone Circles. In Andrew Jones and Gavin MacGregor (eds), Colouring the Past, Ch. 7, 141-158. Oxford, Berg.

MacKie, Euan W. 1977. Science and Society in Prehistoric Britain. New York, St. Martin's Press.

McCluskey, Stephen C. 1981. Comment on the paper Stone Age Science in Britain? by Alvar Ellegard. Current Anthropology 22, no. 2.

Milne, John. 1912. Celtic Place-Names in Aberdeenshire. Aberdeen, Carnegie Trust.

Ruggles, Clive. 1999. Astronomy in Prehistoric Britain and Ireland. New Haven and London, Yale University Press.

Ruggles, C. L. N. and Burl, H. A. W. 1985. A New Study of the Aberdeenshire Recumbent Stone Circles, 2: Interpretation. Archaeoastronomy No 8, Journal for the History of Astronomy, xvi, supplement to Volume 16, S25-S60.

Shepherd, Ian and Shepherd, A. G. 1996. Aberdeen and North-East Scotland. HM Stationary Office.

Sims, Lionel. 2006. Lighting up Dark Moon: ethnographic templates for testing paired alignments on the Sun and the Moon. Available at:
http://homepages.uel.ac.uk/L.D.Sims/Lighting%20Up%20Dark%20Moon.pdf (1 February 2010).

Thom, Alexander. 1967. Megalithic Sites in Britain. Oxford, Oxford University Press.

Thom, A. and Thom, A. S. 1980. Megalithic Rings. Collated with archaeological notes by A. Burl. Oxford, British Archaeological Reports, British Series 81.

Thom, A. and Thom, A. S. 1988. The metrology and geometry of Megalithic Man. In Ruggles, C. L. N. (ed.), Records in Stone: Papers in memory of Alexander Thom, Ch. 5. Cambridge, Cambridge University Press.

Trevarthen, David. 2002. Holy Cows: Natural Precursors to a Ritual Year? Cosmos 18, 35-41.

POWER OF CALENDAR AND CLOCKS

CALENDARS AS SYMBOLS OF POWER

SONJA DRAXLER AND MAX E. LIPPITSCH

Abstract: Calendars may be regarded as systems to organize time. Our innate biological clock, synchronized by periodic environmental changes, makes it natural to align this system along astronomical phenomena like the alternation of night and day, changes in the appearance of the Moon, or variations of solar altitude over the seasons. Thus already in the earliest civilizations time management became connected with astronomy. Synchronized action of groups of humans enhanced productivity, but also served as a means for improving social coherence, especially the collective celebration of festivities that are essential for human societies. Calendars therefore at every time had the particular purpose to fix the festive days. Solemnities also served to improve the loyalty of tribesmen to their chiefs, and hence astronomy obtained political significance. Years were counted after the reign of princes, rulers decreed the date when taxes had to be paid, and priests the dates of sacrifices. Calendars became symbols and tools of power. Power needs publicity. Hence the public display of calendars in various forms dates back to early times.
Keywords: calendar reform, astronomy, Greek calendar, Roman calendar, Julian calendar, Gregorian calendar, calendar table, Stift Rein

1. Introduction

From the early beginnings to the present day astronomical sciences and everyday culture have influenced each other strongly. One of the topics of particular importance was the reckoning of time. Calendars have been developed throughout history for religious, social and political purposes, using the astronomical and physical techniques of the respective time (Hameter, 2005). The development and use of calendars have always been connected with power. First it was the power of small groups acting synchronously, later the power of priests and rulers.

At the beginning it was sufficient just to keep track of the seasons. Farmers needed to know when to sow and when to harvest. Synchronized actions of humans enhanced productivity, but also served as a means for social coherence, especially the collective celebration of festivities, which is essential for human societies. Calendars had the particular purpose to fix the festive days. Years were counted after the reign of kings, and priests ordered the dates of sacrifices. This shows very well that calendars became a symbol of power.

2. Ancient observatories and calendars

While the tracking of lunar phases was obviously performed as early as 25,000 BC (e.g., the Blanchard Bone), calendars based on the course of the Sun seem to be connected with the development of larger social communities, as proven by the construction of sun observatories. An example for one of the earliest of these is the Goseck circle, a Neolithic structure in Germany. It consists of concentric ditches built more than 6000 years ago. At the winter solstice observers at the centre would have seen the Sun rise and set through the southeast and southwest gates (Reichenberger, 2003).

Another example of an early calendar is the so-called Palermo Stone, a large fragment of a stele known as the Royal Annals of the Old Kingdom, written in the 25th century BC. The original stone was more than 2m long, about 0.61m wide and 6.5cm thick; so far only parts of it have been excavated at the archaeological site in Memphis. The stone gives a list of kings of ancient Egypt before and after Menes, with regal years and notations of events like religious festivals, information on taxation and warfare, and on the level of the Nile flood.

To develop useful calendars it is necessary to have precise star observations. One of the first star catalogues written in cuneiform script is the MUL.APIN list, from about 1000 BC. It is based on earlier Babylonian star catalogues, the so-called *Three Stars Each* lists, and gives the names of all major stars and constellations as well as rising, setting and culmination dates.

One of the first written documents of European literature referring to temporal markers is Hesiod's poem *Works and days*. This didactic poem includes practical instructions on agriculture, sailing, social and religious activities as well as an almanac describing favourable and unfavourable days for special activities. Several parts contain astronomical information like the rising and setting of stars and constellations, and the significance of celestial objects for humans. Hesiod's work contains the earliest recorded European mention of the star Sirius:

When the piercing power and sultry heat of the sun abate, and almighty Zeus sends the autumn rains, and men's flesh comes to feel far easier, for then the star Sirius passes over the heads of men, ... (Hesiod 1914, 414-417).

In addition Hesiod described the time of the rising of the Pleiades as the time for harvesting, and the time in which Arcturus rose in the morning as the time for the grape harvest.

3. From the Greek and Roman to the Gregorian calendar

The roots of our modern calendar reach back to the Roman and the Greek calendars. However, in classical Greece there was no uniform calendar. One reason for that was the fact that there was no common state reigned by a king like in Egypt but a lot of small city-states, so-called *polis*. Each city-state had its own calendar with some common features: they all were lunisolar with twelve months and a periodic intercalation of a thirteenth month. The month in which the year began, as well as the names of months, differed among the states, which caused major problems in dating events. So for example Athens started the year in summer, Sparta in fall and Delos in winter. Sometimes years were also recorded after the person in power or after Olympiads which were held every four years.

In Athens in the 5th and 4th centuries BC, different calendars were used simultaneously, mainly to fix days for special purposes. So for example a festive calendar of twelve months was based on the cycle of the moon, and then there was a democratic state calendar with ten arbitrary months, and an agricultural calendar using the rising of stars to mark seasons. The Attic calendar was in the control of governmental officers, who were not astronomers. Manipulation of the calendar was customary; if a festive day fell on a day needed for an assembly meeting, then an extra day could be inserted. Aristophanes describes the problems in his comedy called *Clouds* from 408 BC:

> … you do not observe the days at all correctly, but confuse them up and down; … the gods … are defrauded of their dinner, and depart home, not having met with the regular feast according to the number of the days. … (Aristophanes 1907, 614-619).

Nevertheless, the problems connected with the different calendar systems were not based on a lack of astronomical knowledge. One of the oldest known astronomical devices of the 2nd century BC, the Antikythera mechanism (Freeth, 2006), demonstrates clearly the high level of astronomy in Greece. The mechanism consists of gears in a highly complex arrangement, and shows the Egyptian calendar or Sothic year, the zodiac, the date, position of the Sun and Moon, phases of the Moon, a parapegma to mark the rising or setting of stars, the Metonic, Callippic and Saros cycles, and much more. Today the device is displayed in the Bronze Collection of the National Archaeological Museum of Athens, together with reconstructions of the mechanism.

The ancient Roman calendar changed its form several times. In antiquity it was believed that the first calendar was invented by Romulus, the founder of Rome, about 753 BC. This calendar had ten months with 304 days and 61 days of winter not assigned to any month. The calendar was reformed by Numa Pompilius, the second king of Rome, about 713 BC to get a year with 355 days. To keep the calendar year aligned with the solar year a leap month was added from time to time. In 304 BC Gnaeus Flavius, a pontifical secretary, introduced in Rome the custom of publishing in the Forum tables engraved in stone with information on the *dies fasti*, the days when legal business was allowed, together with references to victories and triumphs. The list can be regarded as an origin of the written *calendarium*. *Fasti* in ancient Rome were detailed records or plans of all days showing all official and religious events. A fragmentary fresco of a Roman calendar, the *Fasti Antiates Maiores* (Figure 1), was found at the ruins of Nero's villa in Antium. On the top the abbreviated names of the months are shown, and the letter K, which means Kalends, the first day of the month corresponding to the new moon.

Special days are marked, like for example Non (*Nones*), indicating the first quarter of the moon, or Ides (*Eidus)* for the full moon. After the calendar reform ascribed to Numa Pompilius, in March, May, July and October *Nones* were fixed to the seventh and Ides to the 15th day, and in all other months to the fifth and 13th day, respectively. At the bottom the total number of days in each month is given. The last column shows the intercalary month. The days are listed vertically and are represented by a letter from A to H, representing a week with eight days. Every ninth day is identified as a market day.

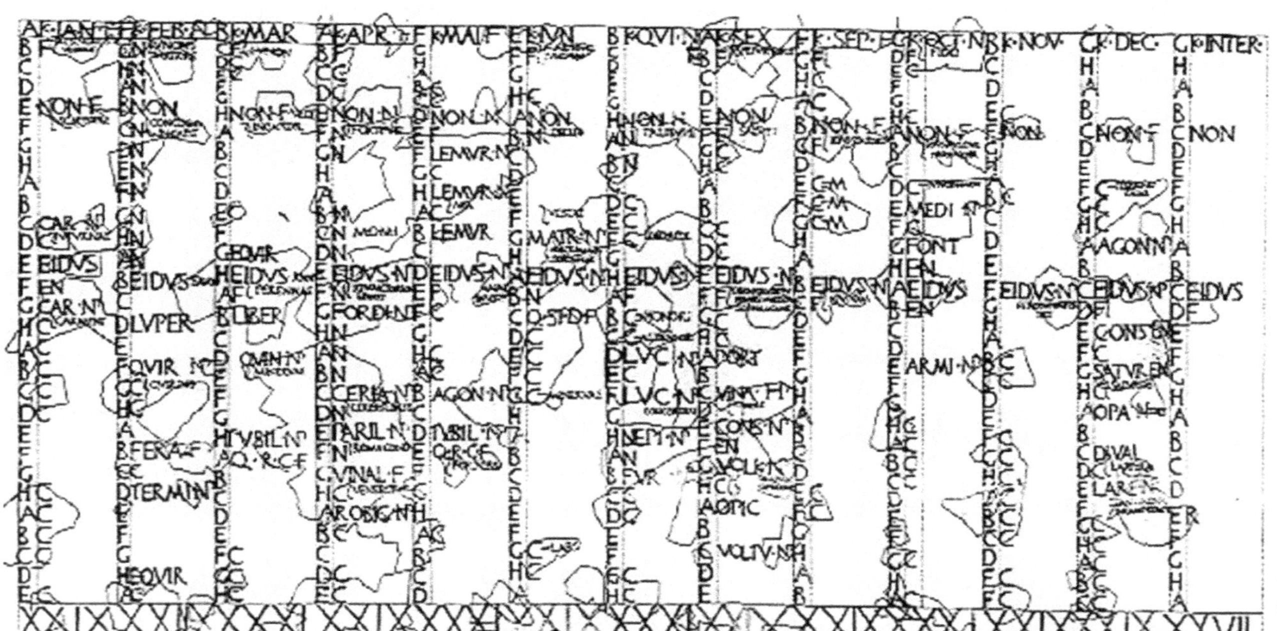

Figure 1: Fasti Antiates Maiores,. Miniature of a fragmentary fresco of a pre-Julian Roman calendar. Source: en.wikipedia.org; public domain.

An additional letter for each day shows its special status: F means *dies fasti,* days on which civil and legal business was allowed; N, *dies nefasti*, days in which these activities were prohibited; C, *dies comitiales*, days when public assemblies could convene and vote. There were also days in which activity either was restricted in some way or avoided as unlucky; days following Kalends, Nones and Ides were regarded as being inauspicious for any new undertaking. The Roman writer Marcus Terentius Varro describes that fact in his work *On the Latin language* (1938), which is mainly dedicated to Cicero. In book VI he treats words denoting time-ideas:

Dies postridie Kalendas, Nonas, Idus appellati atri, quod per eos dies nihil novi inciperent.
The days next after the Kalends, the Nones, and the Ides, were called atri 'black', because on these days they might not start anything new (Varro 1938, VI 29).

The disadvantage of the republican Roman calendar was that the insertion of an intercalary month to keep the solar and lunar year in pace was not subject to an objective rule but left to a consortium of priests to decide. If it was managed correctly, the Roman year could stay roughly aligned with the tropical year. On average, intercalation should have happened every second or third year but sometimes the Pontifex maximus used the right to lengthen a year when his allies were in power or to shorten it when an opponent was ruling. As a consequence the calendar drifted rapidly out of alignment with the sun. This was one of the reasons for Julius Caesar's plan to reform the calendar, which was intended to correct the existing problems permanently by creating a calendar that remained aligned to the Sun without any human intervention. Caesar invited the astronomer Sosigenes of Alexandria to elaborate the details of a new calendar. As a result a year with 365 days was created, based on twelve months and a leap day in February every four years. This new calendar was introduced by Julius Caesar in 46 BC. A Latin poem in six books, called *Fasti*, was written by Publius Ovidius Naso (2004), to illustrate the official calendar published by Julius Caesar. The poem describes the Roman calendar, explaining the origins and the customs of important Roman festivals together with astronomical notes. Each book discusses one month of the Roman calendar. It is not clear if Ovid never finished the poem due to his banishment by Augustus in 8 AD, or if the second half is simply lost. In the introduction to Book III he describes the calendar reform of Julius Caesar.

Sed tamen errabant etiam tunc tempora, donec Caesaris in multis haec quoque cura fuit. ...
Ille moras solis, quibus in sua signa rediret, traditur exactis disposuisse notis. Is decies senos tercentum et quinque diebus Junxit, et e pleno tempora quarta die.
But the calendar was still erratic down to the time when Caesar took it, and many other things, in hand.
...
He is said to have drawn up an exact table of the periods in which the sun returns to its previous signs. He added sixty-five days to three hundred, and then added a fourth part of a whole day (Ovid 2004, Book III, Introduction, 155, 161)

Although the new calendar was much simpler than the pre-Julian calendar, it created errors in adding the leap day. Instead of every four years the pontifices added a leap day every three years. This is described by Macrobius in his work *Saturnalia* (1852). It is written as conversations held by a group of friends during holidays at the house of a Roman aristocrat. Amongst others in the first book, Macrobius discusses the calendar and the problem of the wrong insertion of leap days, which was solved by Augustus.

Hic error sex et triginta annis permansit: quibus annis intercalati sunt dies duodecim, cum debuerint intercalari novem. Sed hunc quoque errorem sero deprehensum correxit Augustus, ...
This error continued for 36 years, by which time 12 intercalary days had been inserted instead of the number actually due, namely, nine. But, when this error was at length recognized, it too was corrected, by an order of Augustus, ... (Macrobius 1852, I, 14.14)

Augustus corrected Julius Caesar's reform by skipping the wrong leap days in order to realign the year. In addition he renamed the month Sextilis to Augustus, and he brought the first obelisk to Rome, the Horologium or Solar Clock of Augustus, which was erected in the Campus Martius in Rome. The obelisk was brought from Heliopolis, Egypt, and used as a meridian instrument. Today this obelisk can be visited in front of the Italian parliament in Piazza Montecitorio in Rome. The obelisk is described by Pliny the Elder in his Natural History:

Ei, qui est in campo, divus Augustus addidit mirabilem usum ad deprendendas solis umbras dierumque ac noctium ita magnitudines, strato lapide ad longitudinem obelisci, cui par fieret umbra brumae confectae die sexta hora ...
The one that has been erected in the Campus Martius has been applied to a singular purpose by the late Emperor Augustus; that of marking the shadows projected by the sun, and so measuring the length of the days and nights. With this object, a stone pavement was laid, the extreme length of which corresponded exactly with the length of the shadow thrown by the obelisk at the sixth hour ... (Pliny the Elder date?, XXXVI Chap. 15)

However, the influence of the respective rulers on the calendar did not only effect the duration of the year. The calendar was also used to make public the days when people had to pay their taxes, and to show the festive days. Over time the number of new festive days rapidly increased, as we can read in Tacitus' Annals:

Ob haec consal[ut]atus imperator Nero, ..., utque inter festos referretur dies, quo patrata victoria, quo nuntiata, quo relatum de ea esset, aliaque in eandem formam decernuntur, ...
For all this Nero was unanimously saluted emperor, ..., and among the holy days were to be included the day on

which the victory was won, that on which it was announced, and that on which the motion was brought forward. ... (Tacitus date?, 12, 41, 5)

In some cases ceremonies were held lasting up to 20 days, as for example after Caesar's victory in the Gallic Wars. This is the reason Cassius recommended cancelling some of these festive days.

... ut C. Cassius ... si pro benignitate fortunae dis grates agerentur, ne totum quidem annum supplicationibus sufficere disseruerit, eoque oportere dividi sacros et negotiosos dies, quis divina colerent et humana non impedirent
Other proposals too of a like kind were carried, on a scale so extravagant, that Caius Cassius .... argued that if the gods were to be thanked for the bountiful favors of fortune, even a whole year would not suffice for thanksgivings, and therefore there ought to be a classification of sacred and business-days, that so they might observe divine ordinances and yet not interfere with human affairs. . (Tacitus date?, 13, 41, 5)

Figure 2: Calendar table in Stift Rein. Photo courtesy of Dr. Roth.

Caesar's calendar reform was not absolutely perfect. The year was about 11 minutes too long, therefore the vernal equinox shifted each year by this small amount; so for example after 128 years it shifted by 1 day. Such differences were too obvious to be ignored. Between the seventh and the 16th century many scientists requested calendar reform, including the Venerable Bede, Roger Bacon and Nicolaus Cusanus. However, scientists at that time did not have enough power to cause a change. Therefore it lasted until 24 February 1582, when Pope Gregory XIII introduced a new calendar with his papal bull called *Inter Gravissimas*. In comparison to the Julian calendar the Gregorian one improved the approximation to the mean solar year to 365.2425 days compared to 365.25. To fix the Easter date according to the Council of Nicaea ten days were dropped in October 1582.

The adoption of the calendar proceeded very slowly. Only four countries adopted the calendar immediately in October 1582: Spain, Portugal, the Polish-Lithuanian Commonwealth and Italy. It was again a question of power, in this case mainly religious power. The Latin phrase *cuius regio, eius religio* was changed into *cuius regio, eius calendarium*. Many Protestant countries initially objected to adopting a Catholic invention. In Styria, for example, the conversion was ordered for 15 October 1583 by the state parliament, but the local agencies as well as the city government of Graz refused to change (Rüpke, 2006). However, in December 1583 Archduke Karl II claimed the immediate adoption of the new calendar. The Protestant countries of Germany adopted the calendar in 1700, Russia in 1918, Greece in 1923, and Turkey in 1926. During the period between 1582, when the first countries adopted the Gregorian calendar, and 1923, when the last European country adopted it, it was often necessary to indicate the date of some event in both the Julian and the Gregorian calendar.

4. Calendar table in Stift Rein

In Stift Rein, a monastery in the north of Graz, a very interesting calendar showing both Julian and Gregorian dates can be visited (Figure 2): a big stone calendar table, with a diameter of 152 cm, gives information for the years 1600 to 1800 and data for 73,000 days. In addition a lot of astronomical and religious information can be found on the table. This sophisticated work was created in 1607 by Andreas Pleninger, an organist and stone etcher from Regensburg. The stone table belonged to the Roman Emperor Ferdinand II, who maybe wanted to show that he wielded power over Catholics as well as Protestants by showing all dates in the Gregorian as well as the Julian calendar.

In the centre of the table (Figure 3) there is an armillary sphere. Next there are the gods of the day with the names of the day, and then numbers showing sunrise and length of the day, and beautiful pictures of the zodiac as well as of the work connected with a certain month, followed by the lengths of the respective months.

Going from the centre outwards there is an eternal calendar for the 365 days of the year (Figure 4). Here astronomical facts like the rise or decline of special stars, the letter of the day depending on the respective weekday, or the saints of the day are given. Finally the eternal calendar is bordered by a circle of holes. The holes were used to span a thread to the centre to find all the information for a special day of the year on one line.

Figure 3: Centre of the Calendar in Stift Rein. Photo courtesy of Dr. Roth

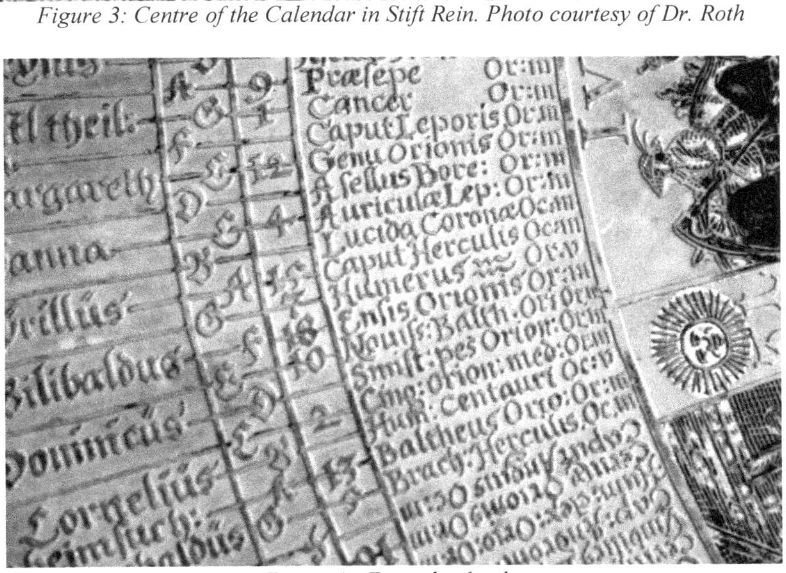

Figure 4: Eternal calendar

On the other side of the holes there are two calendars, one for Julian dates, the *calendarium stylo veteri*, and one for Gregorian dates, the *calendarium stylo novo*, both spanning the years 1600 to 1800. In both calendars all major Christian holidays are shown (Figures 5 and 6). The Julian calendar starts with the year 1600, *Jarzal nach Christi Geburt*. Then there are the Dominical letters, *Sonntagsbuchstabe*, and the additional letter for the leap year, the golden number, which gives the year's position in the 19 years Metonic cycle, and much more. The Gregorian calendar, starting with the year on the top, shows the same festive days; only the golden number is displaced by the epact number, which gives the number of days after the last new moon for the first of January.

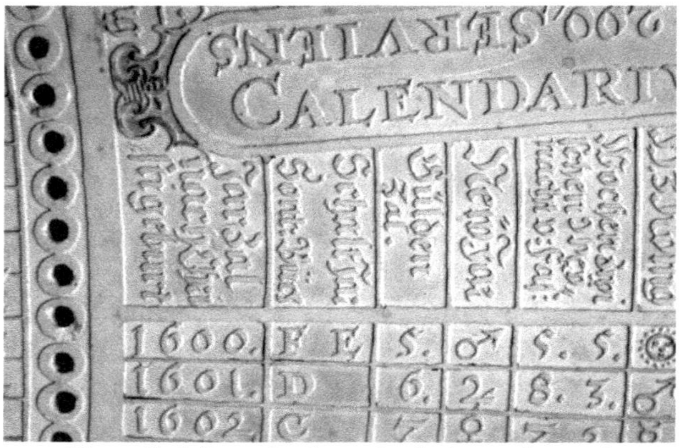

Figure 5: Julian calendar, calendarium stylo veteri.

Figure 6: Gregorian calendar, calendarium stylo novo.

Today the Gregorian calendar is the internationally accepted civil calendar. Nevertheless, since the last papal reform, several proposals have been offered to make the calendar more regular. One was the French Republican Calendar proposed during the French Revolution and used by the French government for about 12 years from 1793 to 1805. The calendar had 12 months with three weeks of 10 days. The remaining five or six days were placed at the end of each year. In the 1950s the UN favoured a perpetual World Calendar with four equal quarters of 91 days, 13 weeks or three months. The remaining days stand outside of the week, a main reason that religious groups oppose adoption. Other reforms involved changing the numbering of the years, like the Holocene calendar, starting with year 1 at 10,000 BC. Other groups tried to make an astronomical proposal and favoured to start the year on the 21st of December or March. All these proposals were dominated by questions of power: power carried out by political or religious groups, trying to enhance their influence.

Acknowledgements:

We appreciate very much the excellent cooperation with Stift Rein.

References

Aristophanes. Aristophanes Comoediae, eds. F.W. Hall and W.M. Geldart, vol. 2, 1907. Oxford, Clarendon Press. Translated into English by William James Hickie, ed., http://www.perseus.tufts.edu (accessed ?).

Freeth, Tony, Bitsakis, Y., Moussas, X., Seiradakis, J. H., Tselikas, H. Mangou, H., Zafeiropoulou, M. et al. 2006. Decoding the ancient Greek astronomical calculator known as the Antikythera Mechanism. Nature 444, 587-591.

Hameter, W. (ed.). 2005 Ideologisierte Zeit. Kalender und Zeitvorstellungen im Abendland von der Antike bis zur Neuzeit. CITY?, Studienverlag.

Hesiod, Works and Days. The Homeric Hymns and Homerica, trans. Hugh G. Evelyn-White (1914), 414-417. London and Cambridge, MA., Harvard University Press.

Macrobius. Saturnalia. Latin text by Ludwig von Jan (1852). Quendlingburg and Leipzig, Gottfried Bass.

Ovid. Fastorum, Libri VI. Latin text at The Project Gutenberg. http://gutenberg.org/cache/epub/8738/pg8738.html (accessed February 27 2015).

Pliny the Elder. : Naturalis History XXXVI Chap. 15

Reichenberger, Alfred. Presseinformation 07.08.2003 Goseck. http://www.praehist.uni-halle.de/goseck/goseck.pdf (accessed February 27 2015).

Rüpke, Jörg. 2006. Zeit und Fest. Eine Kulturgeschichte des Kalenders. City?, Verlag C.H. Beck.

Tacitus. Annalen. Reclam, Philipp, jun. GmbH, Verlag (1986)

Varro. On the Latin language, Books V- VII, trans. Roland G. Kent (1938). Cambridge MA., Harvard University Press. Reproduced 1993, Loeb Classical Library 333.

Wastler, Joseph. 1887. Die Technik der Steinätzung und deren Künstler in der Steiermark im 17. und 18. Jahrhundert. Mitt. Centralkomm. z. Erf. u. Erh. d. Kunstdenkmäler, NF 13.

ASTRONOMICAL CLOCKS – REPRESENTATIONS OF POWER

GUDRUN WOLFSCHMIDT

Abstract: The astronomical clocks started with the Astrarium, made by Giovanni di Dondi (1318–1389) for Padua in 1364. Soon these clocks spread to different European countries. Three types of astronomical clocks can be distinguished: the clocks in the Free Imperial Cities in middle Europe (e.g., Esslingen), in the Hanseatic cities (e.g., Rostock, Stralsund), and the moon clocks (e.g., Nuremberg); the highlights are Prague and Strasbourg. Astronomical clocks were built in order to inform and impress the people.
Keywords: Astronomical Clocks, Astrolabes, Moon Clocks, clockwork universe

1. From Mechanical to Astronomical Clocks

Mechanical clocks with verge and foliot escapement are one of the great inventions of medieval times. Mechanical clocks are also mentioned in the *Divine Comedy* of Dante. Famous astronomical clocks in cathedrals in England date back to the 14th century (Beeson, 1971): there was the Clock of Abbot Richard of Wallingford in St. Alban's Abbey, around 1330. 'Richard's intention must surely have been to demonstrate to the townspeople that the Abbey was not merely a powerful institution but also one with intimate connections to the celestial realms.' (Whyte, 2011). The oldest surviving clock in England is at Salisbury Cathedral, dating from 1386; the clock of Wells Cathedral (1392) in England is preserved in the Science Museum in London (North, 1976). More clocks were produced in the 15th century in order to show the power of the church (Exeter in 1423, Ottery St. Mary in Devon, and Wimborne Minster in Dorset). Early reports of mechanical clocks are included in the work *Libros del Saber de Astronomia* (Books of Wisdom of Astronomy) (1276/77), which was submitted at the court of the astronomically interested King Alfonso X of Castile (1221-1284). A weight driven mercury clockwork is described, which rotates an astrolabe. Soon the mechanical clocks spread to different European countries, in France, e.g., in Lyon, Bourges, and Rouen. These early devices struck only the hours and did not have hands or a dial. What were the reasons for the construction? 'Most of the first clocks were not so much chronometers as exhibitions of the pattern of the cosmos. [...] Clearly the origins of the mechanical clock lie in a complex realm of monumental planetaria, equatoria, and geared astrolabes.' (White 1966, 122-123). They show much more than just the time, the praying hours or the date on a calendar disk. In addition there is a lot of decoration and a variety of astronomical indications.

The mechanical weight driven clock with strike was supplemented in the late Middle Ages by astronomical functions. Its universality represents the emerging mechanical worldview, the clockwork universe (Maurice and Mayr, 1980). Astronomical clocks are impressive artefacts: besides the time, they also show the date, the phases of the moon and calendar information, as well as the motion of the Sun through the zodiac, times of sunrise and sunset, and sometimes also the planet's orbits. Many astronomical clocks have not survived, like the Astrarium of Giovanni di Dondi (1318-1389) in Padova (1364), or were changed considerably during the centuries or brought to museums. By the mid-14th century, elaborate astronomical clocks were being built. They were described by their designers in such detail that we have been able to reconstruct working models in modern times, as in the case of Dondi's Astrarium (models exist in Florence and Paris, see Figure 1). In Italy astronomical clocks can be found on clock towers in Padova (1599-1605), in Mantua (1472) and in Cremona (1583/88)–the largest medieval clock in Europe.

Figure 1. Astrarium of Giovanni di Dondi (1318-1389) in Padova (1364); model in Paris. Photo: Gudrun Wolfschmidt.

World Heritage Sites connected to astronomical clocks are Lyon, Prague, Roskilde, Split, Zytglogge tower (1530) in Bern, Switzerland, Strasbourg and Venice. In Italy you find astronomical clocks normally not in cathedrals but connected to town halls. The powerful Doge and the Signoria of Venice did not allow the chapter of San Marco to built an astronomical clock and erected a clock tower just next to San Marco. The famous *Torre dell'Orologi* on Piazza San Marco in Venice (Figure 2) was designed by Mauro Codussi in 1496-1499; crowned by two Mori striking the hours, it shows the old Italian system of reckoning 24 hours from sunset to sunset; a similar example is in Brescia, Lombardy (1546/74). Also in Paris a *Tour d'Horloge* was erected in 1370; the king ordered that all clocks had to show his time. In Nuremberg towers for striking the clocks were erected: in 1388 the towers of the two main churches, Sebald and Lorenz in the north and south, in 1440 the White Tower and Laufer strike tower of the west and east side of the inner (old) city fortification. It is interesting that in the beginning a clock was not watched but listened to.

In particular, the construction of the 2nd Strasbourg Cathedral Clock, designed by Conrad Dasypodius (1531-1601) and built in 1572-1574 by Isaac and Josia Habrecht, should be given as an instructive example (Oestmann, 1993). It was initiated by the city government directly after the introduction of the Reformation in the Free Imperial City under the direction of Protestant scholars, whose attitude was influenced by humanism. Funds were provided in order to decorate the Cathedral with an astronomical clock. The newly founded academy (1556) wanted to show the ability of the mathematical professor in constructing a mechanized model of the cosmos. It represents the synthesis of the most advanced scientific knowledge of the era in the domains of astronomy, mathematics and technology. In 1842 the mechanism was replaced by a new clock made by Jean Baptiste Schwilgué (1776-1856). As a demonstration model for the astronomical processes the monumental clock of Strasbourg Cathedral shows not only the wisely arranged and well-regulated cosmos, but also, on a metaphorical level, the social constitution of the Imperial City. It was a sign for the right to control exclusively the Cathedral as Protestant city government.

2. Types of Astronomical Clocks

In middle Europe one can distinguish three types:

- The astronomical clocks in churches in Hanseatic cities around the Baltic Sea have an astrolabe as a dial (usually without the zodiac or the dragon pointer); in addition, they have a detailed calendar with names of the saints of the day, festive days, epacts etc.
- The astronomical clocks at town halls in southern 'German' Imperial cities and northern Italy are characterized by a large astrolabe with a zodiac and possess, besides the hour hand pointer, a Sun, Moon and sometimes also dragon pointers.
- The lunar clocks are used to display mainly the time and the indication of the moon phases, sometimes accompanied by other figures.

2.1 Astronomical Clocks in Churches and Cathedrals in Hanseatic Cities

The astronomical clocks in churches in Hanseatic cities (Schukowski, 2006; here you will find all the details and dates) have an astrolabe as a dial; in addition, a large calendar disk provides a lot of information. The clock in St. Mary's church in Rostock (Schukowski, 1992/2009; Schukowski and Helms, 2012) was built in 1379, renewed by Hans Düringer of Thorn in 1472 with a calendar disk, a Renaissance case (1642), Apostles parade and carillon; it is one of the view clocks of the Hanseatic type which is still functioning with the late medieval clockwork.

Figure 2. Torre dell'Orologi on Piazza San Marco in Venice, designed by Mauro Codussi (1499). Photo: Gudrun Wolfschmidt.

There are more examples of Hanseatic clocks in Lübeck: (St. Mary, 1405, renewed in 1561/66 and destroyed in 1942; the calendar disk of 1405 in St.-Annen-Museum, a new clock was constructed between 1955 and 1976); Lübeck (cathedral, 1628, survived); Bad Doberan (Münster Doberan, made by Nicolaus Lillienveld in 1390, only the dial survived); Stralsund (Nikolai church, made by Nicolaus Lillienveld in 1394, cf. Figure 8); Lund, Sweden (around 1424, restored in 1923); Danzig, today Gdansk, Poland (St. Mary, Hans Düringer, 1463/70, restored in 1997); Stendal (St. Mary, 1411); and finally Münster (St. Paulus Cathedral, the oldest in Western Germany, made in 1408-1534, renewed in 1540/42 by Dietrich Tzwyvel, restored in 1932 and 1951), here at 12 o'clock you find the three Magi revolving around Mary and Jesus instead of the parade of the Apostles.

2.2 Astronomical Clocks at Town Halls in Southern 'German' Imperial Cities

The astronomical clocks at town halls show the self-confidence and pride of the Imperial Cities. They are characterized by a large astrolabe with a zodiac and possess besides the hour hand pointer a Sun, Moon and dragon pointer. Most examples are in southwest Germany: Ulm (1520, 1580/81 Isaak Habrecht, 1944 destroyed), Heilbonn (Hans Paulus 1525, Isaac Habrecht 1580, clockmaker Hörz of Ulm 1896-1944 destroyed), Esslingen (1581/89), Tübingen (Johannes Stöffler, 1511, cf. Figure 9), and Ochsenfurt.

Figure 3. Astronomical clock Prague (Praha) (1410/1490): upper clock. Photo: Gudrun Wolfschmidt.

2.3 Astronomical clock Prague (Praha)

The astronomical clock in Prague is a mixture of both types. In 1410 Jan Šindel from the Charles University in Prague constructed in cooperation with the clockmaker Nicolaus of Kadaň the outstanding astronomical clock (*orloj*) at the Town Hall in Prague (Figure 3), measuring about six meters in height. The bottom clock (around 1490) contains a calendar. In the 17th century allegorical statues with moving parts were added, and in the 19th century the figures of the Twelve Apostles playing their roles at the chime of every hour of daylight. The upper clock displays the Sun, Moon and star time, sunrise and sunset (*aurora, ortvs, crepvscvlvm, occasvs*), and the time in the modern way (24 equal hours) but also the old Italian hours (twelve day-time and twelve night-time hours with different lengths in summer and winter). In addition there are several allegorical figures which are moving. In the 17th century more moving figures were added and in the 19th century the chime of Apostles (1865-1866). The lower clock, made in 1486, contains a calendar disk with lockets representing the months (Figure 4). A new calendar disc was installed by Josef Mánes in 1866. The clock was heavily damaged in World War II. The Prague clock is unique in being one of the oldest with original clockwork, having been in operation from its beginning to the present time for six centuries–and even the astronomical dial shaped like an astrolabe survived in the original form (Hadravová and Hadrava, 2005). So you find many features typical for clocks in churches but it is a clock at a town hall (see also in Switzerland Winterthur, Sion and Solothurn).

Figure 4. Astronomical clock Prague (Praha) (1410/90): lower clock, calendar and pointer. Photo: Gudrun Wolfschmidt.

2.4 Moon Clocks–Nuremberg, Church of Our Lady

Lunar clocks are used to display mainly the time and the indication of the moon phases, sometimes accompanied by other figures (cf., the Zytglogge tower in Bern and Fronwagturm in Schaffhausen, Switzerland, or the clock towers of the Castle of the Pomeranian Dukes in Stettin / Szczecin, Poland, 1693, and e.g., Plauen, Görlitz, Marburg, Jena, Kaufbeuren and Mindelheim Jesuit College).

In Riga, Latvia, there is an astronomical clock at the House of the Blackheads (Latvian, *Melngalvju nams*; German, *Schwarzhäupterhaus*, Figure 5). This is a building situated on the town hall square. The original building was erected during the first third of the 14th century for the Brotherhood of the Blackheads Guild, a guild for unmarried German merchants in Riga. Major works were done in the years 1580 and 1886, adding most of the ornamentations, and the clock was reconstructed from 1995 to 1999. In this case, not only did the church and the city government want to show power by constructing a clock, but also the rich merchants of the Hansa.

Figure 5. Astronomical clock at the House of the Blackheads in Riga, Latvia. Photo: Friedhelm Beichler.

Nuremberg is the final and best example of an astronomical clock (Wolfschmidt, 2010) where astronomy is linked very clearly to power in the late Middle Ages. You recognize first, above the dial, the blue and golden ball, indicating the moon phases. At the top in the small iron tower you see, like in Venice, the two figures, dressed in Turkish costume, striking the bell with a hammer. The mechanism of the astronomical clock (Figure 6) was designed and completed by the locksmith Jörg Heuss (1506/09); the copper figures were made by Sebastian Lindenast the Elder. The figure programme included a figure with a sundial and a figure with an hourglass–besides the trombone, flute, piper, drummer and bell bat. During World War II, the clock was stored in the Nuremberg art bunker and thus preserved until today. At noon–twelve o'clock–a herald appears and rings a bell, followed by a figure, the conductor, and two trombone players, a flute and a drummer piper. As a highlight, the seven Prince Electors with the imperial regalia appear and chime three times the Emperor, Charles IV (1316-1378), who is holding a sceptre and raising the hand in salute. In 1424, the German Imperial Regalia was transferred to Nuremberg for permanent safekeeping, at the behest of King Sigismund (1368-1437), Holy Roman Emperor from 1433 until 1437. The Church of Our Lady is the historical place where the Imperial Regalia, insignia and the Reich relics were shown–once a year–to the people from the west balcony below the astronomical clock. The so-called 'Männleinlaufen' parade of the seven Prince Electors, the chime, was donated by the City of Nuremberg in memory of Emperor Charles IV to commemorate the Golden Bulla from 1356. In this Golden Bulla the modalities of the Imperial Election are established; in addition it was stated that every emperor had to hold his first *Reichstag* (Imperial Diet or Parliament) in Nuremberg. Through economic privileges, Charles IV supported Nuremberg. This is the reason that the government of Nuremberg built, above the main portal of the Church of Our Lady, the electors chime around the emperor in this artistic astronomical clock as a real statement of power.

Figure 6. Nuremberg, Church of our Lady (Frauenkirche), Moon clock, Emperor Charles IV and chime of the seven Prince Electors. Photo: Gudrun Wolfschmidt.

2.5 Clocks in Universities and Science Museums

We have seen that in the case of Hanseatic clocks you find a chime of the Apostles or the three Magi. There are two very interesting examples–university clocks–where you find a chime of professors in the clock. One is in Szeged University, Hungary (Figure 7). The other example is the Clock in *Collegium Maius*, Cracow (Kraków), Poland. The present clock has been restored four times in the history of *Collegium Maius* since the 15th century. The university shows its ability and power in presenting this clock. Also the 'Deutsches Museum–Masterpieces of Science and Technology' in Munich presents an astronomical clock (1932) with the time, weekday and corresponding planet, the month, the position of the Sun in the zodiac and the Moon phases.

3. Conclusion: Astronomical Clocks – Representation of Power

Astronomical clocks are very impressive artefacts not only in the medieval times but also nowadays. They are a remarkable cultural heritage in a scientific, technological and artistic way. What were the reasons for their construction? They show much more than just the time, the praying hours or the date on a calendar disk. In addition there is a lot of decoration and a variety of astronomical and astrological indications. In 1407 the chapter of the Cathedral in Chartres erected a mechanical clock, in order to increase the authority of the church and to attract visitors through spectacular works.

Figure 7. University Clock in Szeged, Hungary, with the chime of professors (upper part). Photo: Gudrun Wolfschmidt.

This astonishing craftsmanship outshone everything that has been accomplished before and presented a spectacular work for the glory of the church. The monumental clock of Strasbourg Cathedral–as a demonstration model for the astronomical processes–shows not only the wisely arranged and well-regulated cosmos, but also, on a metaphorical level, the social constitution of the Imperial City. It was a sign for the Protestant city government to have the right to exclusively control the Cathedral. It is a similar case to Venice. Not only churches and city councils erected astronomical clocks in order to demonstrate craftsmanship, cosmopolitanism and power but also the wealthy merchants and universities wanted to have a clock to show their influence and scientific knowledge. Finally in Nuremberg at the Church of Our Lady, where the Imperial Regalia have been displayed to the public since 1423, an astronomical clock was donated in memory of Emperor Charles IV. It presents the chime: the Holy Roman Emperor is shown seated with the seven Prince-Electors surrounding him–a real demonstration of astronomy and power.

Science is power–this motto of Francis Bacon (1561-1626) from the Early Modern Time became a programme for the whole new science; mathematics was used for measuring and structuring the world. By means of the mechanical clock the old temporal hours (*horae inequales*) disappeared and the new hours of equal lengths were introduced; this transition changed our awareness of time and had an important impact on social and economic life. The cosmos was seen as a large celestial clockwork. The clock was a symbol for innovation, knowledge, richness and prestige. This was the reason that everybody–the church, the imperial cities, or influential people and institutions–wanted to have an astronomical clock in order to have power by being a ruler over time.

Figure 8. Astronomical Clock in the Cathedral in the Hansa City Stralsund. Photo: Gudrun Wolfschmidt.

Figure 9. Astronomical Clock, Town Hall in Tübingen. Photo: Gudrun Wolfschmidt.

References

Beeson, C. F. C. 1971. English Church Clocks 1280–1850: Their History and Classification. London, Antiquarian Horological Society.

Hadravová, Alena, and Hadrava, Petr. 2005. Astronomy in Medieval Prague. In Gudrun Wolfschmidt and Martin Šolc, Astronomy in and around Prague. Acta Universitatis Carolinae – Mathematica et Physica, Vol. 46, Supplementum.

Maurice, Klaus, and Mayr, Otto (eds). 1980. The Clockwork Universe: German Clocks and Automata 1550–1650. New York, Smithsonian Institution, Neale Watson Academic Publications.

Oestmann, Günther. 1993. Die astronomische Uhr des Straßburger Münsters: Funktion und Bedeutung eines Kosmos-Modells des 16. Jahrhunderts. Stuttgart, Verlag für Geschichte der Naturwissenschaften und Technik.

Schukowski, Manfred. 1992/2009 (2nd edition). Die Astronomische Uhr in St. Marien zu Rostock. Königstein, Verlag Langewiesche (Die Blauen Bücher).

Schukowski, Manfred. 2006. Wunderuhren - Astronomische Uhren in Kirchen der Hansezeit. Schwerin, Thomas Helms Verlag.

Schukowski, Manfred, und Thomas Helms. Sonne, Mond und zwölf Apostel. Die Astronomische Uhr in der Marienkirche zu Rostock. Schwerin, Thomas Helms Verlag 2012.

White, Lynn Jr. 1996. Medieval Technology and Social Change. Oxford, Oxford University Press.

Whyte, N. 2011. The astronomical clock of Richard of Wallingford. Available at: http://www.nicholaswhyte.info/row.htm (Accessed 27 January 2011).

Wolfschmidt, Gudrun (ed.). 2010. Astronomie in Nürnberg. Nuncius Hamburgensis – Beiträge zur Geschichte der Naturwissenschaften, Band 3. Hamburg, tredition science.

Chalcolithic / Bronze Age / Iron Age Cultures

LUNI-SOLAR SYMBOLISM IN AN ARTEFACT FROM BULGARIA

VESSELINA KOLEVA

Abstract: A small bone artefact found in the prehistoric settlement mound near Karnobat, Southeast Bulgaria, is presented. It has the shape of a stylized anthropomorphic figurine and is decorated with holes, hollows and notches incised on the face side and the edges. The analysis of the object suggests the presence of lunar and solar symbolism with possible time-measuring functions. The article draws parallels between different archaeological and ethnographical finds from Europe.
Keywords: Hollow bone figurine, Karnobat settlement mound, lunar synodic cycle, sun-dial.

Introduction

The heavenly bodies have always attracted the attention of people. Their rhythmic motions became basic measurement units of the first calendars. The sunlight and the shadows it casts are the basis of the first time measurement tools–the sundials. The alternation of day and night, of warm and cold seasons and the mysterious lunar transformations give rise to the veneration and the worship of the Sun and the Moon. Their most easily recognized and obvious symbols are the circle and the crescent, or the horns of a cow or a bull.

There are a wide variety of notions and rites related to the Moon. The very old notion that the women's menstrual cycle is associated with the lunar month has been preserved to this day; the pregnancy period is also measured in lunar months. In many folk beliefs the Moon is the celestial body that represents the life cycle with its three phases of birth, life and death. The Moon can take on feminine attributes and so the waxing Moon symbolizes the Maiden, the full Moon–the Mother and the waning Moon–the Crone. According to an etiological myth preserved in the Gotse Delchev region in the Western Rhodopes the Mesechina (a folk name of the Moon) is a woman (maiden): 'Look how clear she is in the beginning. She has washed and wiped herself, everything is gone. From the 24th day on she is somehow muddy. This is when her monthly period begins. And when it is over, she comes out so clear'. (Boneva 1994, 11). There is a belief that weddings should be held when the Moon 'gets older', i.e, at least seven or 15 days after the new moon. Some see in its dark spots animal and human figures, but most often they are thought to depict a human face. Observant people from the village of Vasil Levski, Central Bulgaria, see the moon spots as 'notches' and while the Moon is 'waxing' they determine how many 'nights' old the Moon is (Stamenova 1986, 295).

The written records and oral ethnographic data are reliable sources for research of the astronomical knowledge in the past. Beyond them, however, we can only rely on the archaeological artefacts–scarce and scattered in time and space. But, as Mircea Eliade says, although cultures differ in style they are comparable in terms of images and symbols, and the universality of the archetypes allows us to reach deep into man's spiritual world and his surroundings (Eliade 2002 168). Along with the comparison of images, morphological and quantitative analysis of the presented artefact is necessary in order to verify an archaeoastronomical hypothesis, and this will be our research approach here.

The bone artefact from Karnobat, Bulgaria

The object (Figure 1) was found in a prehistoric settlement mound near the town of Karnobat, Southeast Bulgaria (latitude 42° 39.7′ N, longitude 26° 58.5′ E, 200m above mean sea level). The height of the mound is 12m and its diameter at the base is 120m. In the 1920s Atanas Ignatiev (1880-1948)–a teacher in history in the local school–conducted together with his pupils the first archaeological investigation of the mound (Momchilov 2002, 7). Over a span of 20 years, every time after ploughing near the mound, as well as after planned shallow excavations were carried out, a large number of 'Neolithic' finds were brought to the school museum (Ignatiev 2002, 49). After a long pause, new archaeological research was carried out in 2006 (Boyadzhiev et al., 2007).

Figure 1. The face side (A) and the back side (B) of the hollow bone figurine from the Karnobat settlement mound. Published with permission of the Historical Museum in Karnobat. Photos by V. Koleva.

The cultural layers show that the mound was inhabited starting in the Neolithic Age, all through the Eneolithic and until the beginning of the Bronze Age. The artefacts collected by Ignatiev made up the foundation of the archaeological collection of the local museum. The majority of them are dated to the end of the 5th millennium BC and belong to the Kodzhadermen-Gumelnitsa-Karanovo VI cultural complex (Boyadzhiev et al. 2007, 22; Ignatiev 2002, 57).

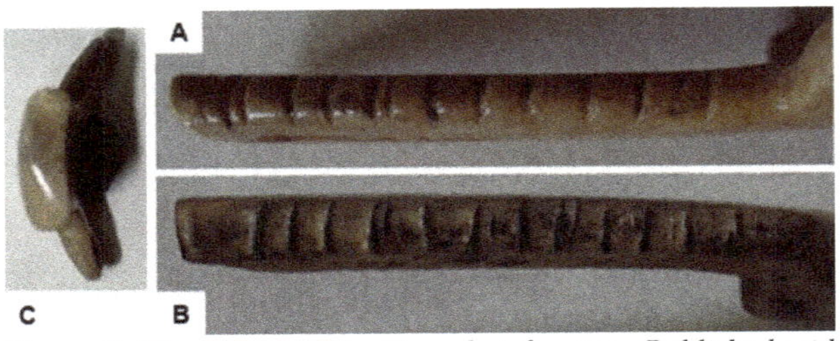

Figure 2. The edges of the Karnobat figurine. Published with permission of the Historical Museum in Karnobat. Photos by V. Koleva.

The object that attracted our attention with its shape and incisions has an inventory number 108. It is exhibited in the prehistoric section of the museum and has not been published to date. Ignatiev described it as 'a cultic scoop-shaped object made of bone'. It is of flaxen colour and is carved out of the diaphysis of a long bone of a large ruminant (such as bull or aurochs). The latter is the conclusion of Professor Nikolay Spassov, a palaeontologist at the National Museum of Natural History in Sofia.

The artefact is well preserved. Its surface is well polished and the notches and holes are incised precisely and almost symmetrically with respect to the long axis. The object consists of two parts–a wide semi-circle and a narrow rectangular 'handle'. Its total length is 8.3cm. The length of the 'handle' is around 5.3cm and its maximal width is 1.8cm. The maximal length and width of the oval part are 3.1cm and 4.1cm, respectively. The thickness of the bone varies from 0.3 to 0.6cm. The diameters of the holes are 0.2 and 0.3cm.

The number of the various incisions is of great importance to our study. On the face side (Figure 1) there are four small holes and two hollows incised in the semi-circle part. On the edges of the rectangular part there are legibly incised notches–12 on the left and 13 on the right side. The notches are 3 to 5mm away from one another and only the distance between the first and the second as well as between the fifth and the sixth notches on the right is 1.5mm. When looking at the edges sideways (Figure 2) there are one to two shorter and shallower incisions between some of the main notches on (A) and (B). There are more short notches on edge (B).

Interpretation

The 'scoop-shaped cultic' object–as its discoverer calls it–resembles also a mushroom or a phallus, as well as a highly stylized anthropomorphic figurine. The latter could be viewed in two ways. One way is to see it as a head and a very schematic body without limbs and distinguishing sex features (Figure 1A).

We suppose that on the 'face' of the figurine (the hollow side of the 'head') the two pits at the top are the 'eyes'. The small hole between them could mark an important place on the forehead such as 'a third eye'; the big hole is the 'mouth' and the two small side-holes are the 'ears'. The other way to see the object is if we turn it upside down and with its backside towards us-it looks like a female body with a plump lower part (Figure 1B). The three small holes outline an enormous pubic triangle and the big hole could mark the uterus. The upper part of the body is represented very schematically.

The presence of groups of signs on the anthropomorphic figure reminds us that counting with the help of the body parts is an archetypal and universal practice (Georges 2005, 58-62). The fingers and toes, hand and foot joints, the ears, eyes, nose and mouth are used to count not only objects but also time units. Due to the limitations in human perception and memory, counting aids, such as notches on bone or stone, appear as far back as in the Upper Palaeolithic (Georges 2005, 45, 59 Figure 1.35; 61 Figure 1.37), or maybe even much earlier (Rincon, 2004).

Each type of incision is strictly distributed on a different part of the presented artefact, so it could embody a different meaning and a different function. A good reason for us to check for a time measuring function of the signs is their total number (31) which is close to the duration of the lunar synodic month (29.53 days). Thus the artefact could be a body-shaped device used to count out the days in a Moon cycle. Let's say that there is a one-to-one correspondence between the lunar phases and the parts of the human body. Then the body turns into a tool to count out time periods of 29 or 30 days.

The body of a standing man was also the first gnomon. And using the shadow of a gnomon, man was able to define the cardinal directions and to divide the day into parts and the year into seasons as far back as antiquity.

A good illustration of an ancient body part counting practice could be the famous 'Adorant' found in the Geißenklösterle cave near Blaubeuren, Germany, and dated between 35,000-32,000 BC. The anthropomorphic figure with raised arms is depicted as a bas-relief on the face side of a rectangular plate (3.8cm x 1.4cm x 0.45cm) made of mammoth ivory. There is a set of six or seven incisions on its left hand. The total number of notches on the four edges framing the figure and in the four columns on the back side is 88, which is close to the duration of three lunar months. The back side 'contains tiny spots of manganese and ochre, which must have been applied intentionally' (Rappenglück 2010, 20 Figure 1.2.1) and therefore can be an additional proof of the counting function of the signs.

Other images that can be interpreted as patterns for space orientation and time measurement using the Moon and the human body are the drawings found in the caves Canchal de Mahoma and Abri de las Viñas in Spain, and dated c. 9000 BC (Marshack 1964, 743). They depict a schematic anthropomorphic figure with dots and crescents around it, whose total number corresponds to a full lunar cycle.

Possible functions

The bone artefact possesses morphological and quantitative (arithmetical and geometrical) characteristics that can turn it into something more than an amulet or cultic object, even if it has been used as a ritual spoon. It could have had some specific practical usages.

Luni-solar calendar

The total number of signs is 31. On the periphery, i.e., without the big hole (the 'mouth') (Figure 1A) there are 30 or 29 signs that can be counted out. This coincides with the duration of the lunar synodic month (29.53 days). Counting can start and end with the endmost notches on the lower part of the 'body', corresponding to the first and the last Moon crescent. In this case the full Moon will coincide with the small hole on the 'forehead', and the days before and after the full Moon–with the signs on the semi-circle. An additional function of the notches on the edges could be to count out months. In a luni-solar calendar their number is 12 in a common year and 13 in an embolismic year. The short shallow notches on the edges could possibly be auxiliary signs, namely, for remarkable months or days.

The signs on our artefact are handy for counting just like the day notches on the edges of the pocket bone calendar shown on Figure 3B. It was used by hunters from Yakutia, Russia, until the 19th century. The day notches are grouped in months and special signs mark the important feasts in the Orthodox liturgical calendar.

Sundial

The object could be used also as a time-measuring device. Let's place the figurine horizontally and parallel to the meridian, now let's stick vertically a wooden peg into the big hole ('the mouth'). When the object is oriented with the 'head' to the south then the shadow falls over the 'body' of the figurine (Figure 3A), and the noon shadow lies along the axis of symmetry. The length of the noon shadow varies during the year. We did some calculations for the latitude of the mound $\varphi = 42.66°$ and the inclination of the ecliptic $\varepsilon = 24.15°$ for the epoch 4500 BC (Wood 1981, 86 Table 4.1). With a 2.0cm long peg, the length of the shadow at noon in the days of the summer solstice will be 0.7cm from the base of the gnomon and will reach the line that connects the notches on the edges closest to the gnomon. On the days of the equinoxes the shadow will be 1.8cm long, reaching the midpoint between the fourth and fifth notches. On the winter solstice it will be 4.7cm long and will reach the line connecting the farmost notches.

There is a type of sundial, called by ancient writers 'analemma', that only shows the height of the Sun at noon, every day of the year, by the length of the gnomon shadow (Sawyer 1994, 7.4). So the Karnobat artefact can also be interpreted as an ancient analemma.

If we turn the object 180°, i.e., place it with the 'body' to the south, its semi-circular part will become the face of a sundial, on which the gnomon shadow will be moving clockwise–from the 'right ear' in the morning, through 'the third eye' at noon up to the 'left ear' in the afternoon when the Sun is exactly in the east, south and west respectively. On the equinox days the shadow at noon will reach the 'third eye' (the distance between the base of the gnomon and the edge of the hole is c. 2.0cm). The hollows and the holes can also serve to separate the light part of the day. They are located at about ± 30° and ± 90° respectively in relation to the gnomon shadow at noon (the left and right part of the 'face' are slightly asymmetrical, probably due to imprecise handwork).

The sundial itself can be used to find the exact north-south bearing. Furthermore, some angular observations can be used to estimate the phase and the age of the Moon. For instance, if you place the oval part in front of your eyes, holding the rectangular part like a handle perpendicular to the ground, the Moon disk with its angular diameter of 0.5° will fill up the small (2mm) and the large (3mm) holes when the distance to the eye is 23 and 34cm, respectively, i.e., at a distance which corresponds to the distance of clear vision. If fitted with a thread in the 'forehead' opening, the object could be worn as a pendant.

The so-called *Pastirska sončna ura* ('Shepherd's sundial') (Figure 3C) could also be worn as a pendant. This peculiar portable sundial was used until the first half of the 20th century in the Alpine regions of Slovenia (Cavc, 2007; Ceferin, 2000). Much earlier analogues of the Slovenian sundials are some analemmatic sundials fitted on the back of miniature nocturnal dials accurately made in France in the 16th century (see the Website of the Museo Galileo in Florencia, Italy).

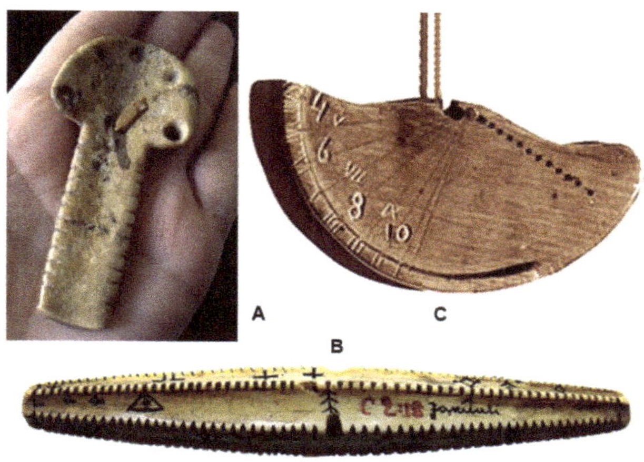

Figure 3. (A) The Eneolithic bone figurine from Karnobat as a possible time measuring tool. Length 8.3cm. Photomontage by D. Kolev. (B) Yakutian bone calendar used by the Siberian hunters in 19th century. The important Orthodox feasts in September and October can be seen. Length c. 12cm. With the permission of the Estonian Literary Museum, Tartu, Estonia. Photo by N. Sivkov. (C) Shepherd's sundial used till 20th century in Slovenia. Semi-circular shape. Length c. 12cm. Vlasto Kopač collection (Cavc, 2007).

The gnomon of the Slovenian Shepherd's sundial is also stuck perpendicular to the face and a number of holes form the so-called date line. The date line on the sundial shown on Figure 3C consists of 13 holes and the peg is moved from one hole to the next every 14 days in the course of the year. This regular time scale adjusts the time depending on the Sun's declination during the year–from the summer to the winter solstice and then backwards. The date line of another Slovenian sundial of the kind (Ceferin, 2000) has two rows of 17 holes each and one additional hole. The holes were possibly meant for the peg to be stuck into each of them approximately every 10 days.

The time during the day is measured using the hour lines on the left part of the sundial, and short notches, or hour points, on the periphery of the right-hand side. The time before noon on the inner hour circle and the time after noon on the outer one is monitored by placing the sundial vertically or horizontally, respectively. The date line itself is symmetrical to the leftmost hour line for VIII (PM) on the left.

Parallels

So far the hollow bone figurine from Karnobat does not have an exact analogue in Bulgaria. Therefore we have tried to find parallels with similar characteristics. Following the advice of archaeologists, we searched for analogues synchronous to the Karnobat artefact among the so-called hollow idols (Figure 4A) (Website of the Archaeological Museum in Varna, Bulgaria).

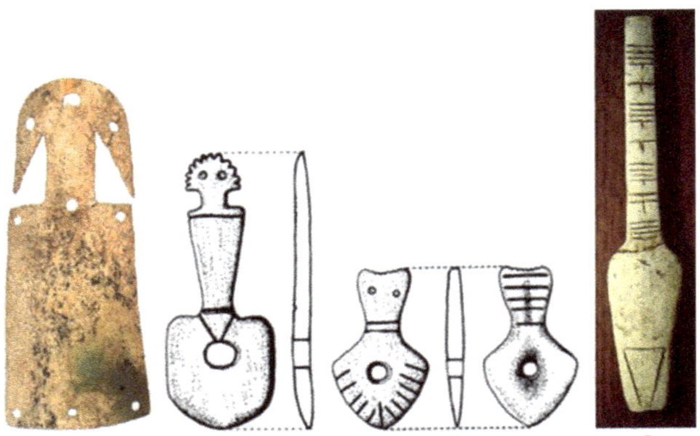

Figure 4. (A) Hollow bone figurine (H = 18.7cm), 4500-4000 BC, woman's grave № 4, Varna, Bulgaria (Website of the Archaeological Museum in Varna). (B) and (C) Bone pendants, 4500-4000 BC, Igeşti, Romania (Sztancs et al. 2010, 145,146). (D) Flat bone figurine (H = 13.6cm), 15,000 BC, Mezhirichi, Ukraine (Pidoplichko 1976, 205 Figure 80). (The images are not of the same scale)

One of the initial assumptions was that the Karnobat object could be a fragment of a bigger hollow idol. Fortunately, there are no visible traces of it being broken or modified (Figure 2C). So far some 40 hollow bone idols and only one made of marble have been found in Bulgaria and Romania. All the idols belong to the Gumelnitsa–Karanovo VI culture. Their dimensions vary between 5 and 22cm. There are holes in groups of three to seven drilled into the trapezoid-shaped body and into the head. According to some interpretations these idols symbolize solar or lunar deities (Voinea, 2008). Quite in line with our analysis of the Karnobat artefact, we can also suggest some practical usage of these idols as time-measuring tools. Lunar symbolism can be found in the unusual crescent-like shape of the head, or rather headdress. (Compare with the semi-circular or half-moon shaped 'head' of the Karnobat figurine.) So if we consider the idols as anthropomorphic Moon symbols, then the holes on the different zones of their bodies can mark Moon phases of special importance. Most of the idols were discovered in dwellings, and eight were found in symbolic graves (cenotaphs) of the Varna necropolis (Ivanov, 1991). The two types of locations can also point at two different possible functions: a utilitarian and a cultic one.

Among the anthropomorphic idols (Precucuteni and Cucuteni-Ariuşd cultures, 4500-4000 BC) there are two bone pendants (Figure 4BC) found in Igeşti, Romania, that we think bear considerable similarity to the Karnobat artefact in terms of morphology and amount of signs (Sztancs et al. 2010, 145 Figure 3-14; 146 Figure 4-3). We consider the geometric proportions and the round opening are also suitable for gnomonic time measuring.

Another example is the female figurine (Figure 4D) from a mammoth tusk found in a mammoth bone dwelling of the Upper Palaeolithic site near Mezhirichi, Ukraine, and dated c. 15,000 BC. The length of the object is 13.6cm, and the maximum width and thickness are 2.5cm and 1.2cm respectively. On the plump lower part three long notches outline the pubic triangle filled with a hardly discernible netlike ornament (Pidoplichko 1976, 203 Table 22, 205 Figure 80). There are 27 notches (20 long horizontal and 7 short vertical) incised on the face side of the prolonged upper part of the body. We suggest that the notches mark a period of 27 days–the period when the Moon is visible–and this is related to the worship of the Moon as a female deity. The notches are arranged in a way as if a sort of an arithmetical notation is used. It could be based on finger counting because all groups contain one short and one to four long notches.

Conclusions

The hollow bone figurine from Karnobat might seem to have unclear functions and uses at first sight, but when astronomical and mathematical interpretation is applied, it reveals the presence of lunar and solar symbolism with possible time-measuring functions.

Manly P. Hall says that 'the oldest, the most profound, the most universal of all symbols is the human body' (Hall 1928, 73). Typically for the people of all epochs, the useful and wise knowledge, which helps them to survive and develop, is often hidden behind anthropomorphic images, figures and signs. The anthropomorphic shape and female features of the Karnobat artefact combined with the amounts of signs commensurate with the lunar synodic cycle suggest the worship of a lunar deity related to the notions of fertility and immortality in the nature and in the human society. The possible function of a portable time-measuring device also corresponds to the cultural level, the agricultural cult and the mobility of the inhabitants of the Balkan Peninsula during the Eneolithic.

Acknowledgements:

The author is grateful to Dr Dimcho Momchilov (director) and Rositsa Hristova (archaeologist) from the Karnobat Historical Museum for the opportunity to study the artefact as well as for their assistance; to Prof. Ana Raduncheva and Dr Yavor Boyadzhiev (National Institute of Archaeology and Museum, BAS) for the useful discussions; to Prof. Tamila M. Potemkina (Institute of Archaeology, RAS) for the valuable literature provided, as well as to Dr Mare Kõiva and Andres Kuperianov (Estonian Literary Museum in Tartu, Estonia), and Svetlana Koleva from the University of Sofia for the suggested folklore and ethnographic examples.

The present research is part of a scientific project funded by the Institute of Astronomy and National Astronomical Observatory of the Bulgarian Academy of Sciences.

References

Boneva, T. 1994. Naroden svetogled. In R. Popov and S. Grebenarova (eds), Rodopi. Tradicionna narodna duhovna i socialnonormativna kultura, 7-50. Sofia, Prof. Marin Drinov Academic Publishing House.

Boyadzhiev, Y., Boyadzhiev, K., Gyurova, M. 2007. Arheologicheski prouchvanija na selistnata mogila Karnobat. In H. Popov et al. (eds), Arheologicheski otkritiya i razkopki prez 2006, 21-23. Sofia, Faber.

Cavc, A. 2007. Pastirjevo znanje. O času. In Velika planina [Elektronski vir]. Fikfak, J. (ed.). Ljubljana, ZRC SAZU. Available at: http://odmev.zrc-sazu.si/planina/PastirjevoZnanje.htm (accessed 18 April 2015).

Ceferin, A. 2000. Shepherd's sundial. Solar pocket watch. Available at: http://www.thezaurus.com/?/webzine/shepherds_sundial/ (accessed 18 April 2015).

Eliade, M. 2002. Obrazi i simvoli. Sofia, Prozorets.

Georges, I. 2005. Enciklopedichna istoria na tsifrite, Choveshkiyat razum, opisan chrez chislata i smyataneto, Tom 1. Sofia, Elementi.

Ignatiev, A. 2002. Predistoricheskata selistna mogila pri gara Karnobat. Istoriya i kultura na Karnobatskiya kray 4, 49-70.

Ivanov, I. S. 1991. Der Bestattungsritus in der chalkolithischen Nekropole von Varna (mit einem Katalog der wichtigsten Gräber). In J. Lichardus (Hrg.), Die Kupferzeit als historische Epoche. Symposium Saarbrücken und Ortzenhausen 6.-13.11.1988. Teil 1, Bonn 1991, 125-149.

Hall, M. P. 1928. The Human Body in Symbolism. In The secret teachings of all ages. An encyclopedic outline of Masonic, Hermetic, Qabbalistic and Rosicrucian symbolical philosophy. San Francisco, H. S. Crocker Company.

Marshack, A. 1964. Lunar notation on Upper Paleolithic remains. Science 146, 743-745.

Momchilov, D. 2002. Atanas Ignatiev Karaivanov (1880-1948). Istoriya i kultura na Karnobatskiya kray 4, 7-14.

Pidoplichko, I. G. 1976. Mezhirichskie zhilista iz kostey mamonta. Kiev, Naukova dumka.

Rappenglück, M. 2010. Earlier prehistory. Case Study 1.2: The decorated plate of the Geißenklösterle, Germany. In C. Ruggles and M. Cotte (eds), Heritage Sites of Astronomy and Archaeoastronomy in the context of the UNESCO World Heritage Convention, 19-21. Paris, ICOMOS and IAU.

Rincon, P. 2004. Early human marks are 'symbols'. Available at: http://news.bbc.co.uk/2/hi/science/nature/3512470.stm (accessed 18 April 2015).

Sawyer, F. W. 1994. Of Analemmas, Mean Time and the Analemmatic Sundial. Available at: http://citeseerx.ist.psu.edu/viewdoc/download?doi=10.1.1.142.3875&rep=rep1&type=pdf (accessed 18 April 2015).

Stamenova, Zh. 1986. Naroden mirogled. In G. Mihaylova et al. (eds), Plovdivski kray. Etnografski i ezikovi prouchvaniya, 293-309. Sofia, Prof. Marin Drinov Academic Publishing House.

Sztancs, D.-M. et al. 2010. Fiches typologiques de l'industrie osseuse de Roumanie. I. Préhistoire. 1. Idole / pendeloque / amulette anthropomorphe énéolithique en bois de cerf de Păuleni-Ciuc, dep. de harghita. Acta Terrae Septemcastrensis 9, 121-150.

Voinea, V. 2008. About figurines 'en violon' within the civilization Gumelniţa–Karanovo VI. Sprawozdania Archeologiczne 60, 73–101.

Website of the Archaeological Museum in Varna, Bulgaria. Available at: http://www.amvarna.com/gindex.php?lang=1&lid=3&slid=17&show=8 (accessed 18 April 2015).

Website of the Museo Galileo in Florencia, Italy. Available at: http://catalogue.museogalileo.it/gallery/NocturnalInv2501_n01.html (accessed 18 April 2015).

Wood, J. E. 1981. Sun, Moon, and Standing Stones. Moskva, Mir.

ASTRONOMY, RELIGION AND THE STRUCTURE OF SOCIETY IN PREHISTORIC FINLAND

MARIANNA RIDDERSTAD

Abstract: Archaeoastronomical studies of Finnish prehistoric monuments have shown that the occurrence of astronomically-orientated structures is connected to the existence of different social strata in societies from the Subneolithic to the late Iron Age. From the Bronze Age on, celestial symbolism is seen in the metalwork, often in objects that can be connected to high social status. The reasons for these relations are probably related to the profane leadership gaining religious authority provided by astronomical and calendrical knowledge. Only in the Iron Age is it possible to gain insight into the astronomically-motivated religious beliefs and rites of all social classes. From the literary sources on Finnish pre-Christian mythology and the archaeological evidence from the Iron Age to the beginning of the historical period it can be deduced that the cult of the sun goddess was an important part of Finnish Iron Age religion.

Keywords: astronomical orientations, prehistoric society, giants' churches, cairns, sun goddess

Introduction

Archaeoastronomical research carried out in Finland since 2008 has already established that astronomical practices were an important part of the religious life from the Subneolithic period to the late Iron Age, from c. 3000 BC to c. 1000 AD (Okkonen and Ridderstad, 2009; Ridderstad and Okkonen, 2010; Ridderstad, 2010 and 2011). It seems that in all these periods, constructing astronomically-oriented structures was connected to the social structure of the society in question.

Only from the late Iron Age on is there information available on the actual religious beliefs that motivated astronomical practices. This information is in the form of folk epics, runic poetry, fairy tales, recorded beliefs, sayings and incantations, and is, therefore, also subject to interpretation. While our knowledge on Stone Age and Bronze Age astronomy mostly reflects the significance of astronomical practices to the upper classes of the societies in question, in the Iron Age, one is able to gain insight into the astronomically-related beliefs and practices of all social strata.

In the following, I will review the astronomical practices of the different periods and the significance of astronomy in the prehistory of Finland. Finally, I will summarize the ancient practices which still live on in the folk tradition of the Finns.

Subneolithic

Astronomical orientations first appear in Finland in the Subneolithic period (about 3500-1800 BC), when agriculture first arrived in the region. The evidence, however, is not from southern Finland, where agriculture was first practiced by the Corded Ware culture, but from Ostrobothnia, a region inhabited by cultures using the Pöljä and Kierikki asbestos ceramic ware. The geographical line of contact between the Corded Ware culture and the Ostrobothnian cultures was located approximately in the present-day Vaasa region.

The Ostrobothnians were hunter-gatherers, mostly seal-hunters, living by and off the sea. The annual temperatures were higher than today, and the seal-hunting was very productive. The hunters produced seal train oil, probably trading their product with nearby communities which had already taken on farming (see Okkonen 2003, 219-226). The first evidence of agriculture in the north is from Puolanka in 2300 BC (Vuorela 2002, 84-87).

In this period, for the first time in the region, signs of social stratification start to appear: some dwellings were considerably larger than others (Okkonen 2003, 219-226). Also the wealthy Kierikkisaari trading site is dated to this period. The accumulation of wealth seems to have been uneven, although the communities were still small and rather equal.

The appearance of the so-called giants' churches (hereinafter GCs) in this period may be related to new religious concepts, partly obtained as a result of cultural contacts related to trading relations (Okkonen 2003, 223-225). The GCs are monumental (some over 60m long) stone enclosures with rectangular or round walls, usually built on high locations. The walls have openings, 'gates', and there are often large cairns in the immediate vicinity of the gates. The GCs were built on islands or near the shoreline, and they originally had wide views over the sea and the surrounding terrain.

While the actual use of the GCs is not known, most of them, especially the largest ones, have orientations to the sunsets and sunrises of the main solar dates of the year (Okkonen and Ridderstad, 2009; Ridderstad and Okkonen, 2010). The three largest GCs all have orientations to the winter and summer solstice solar events.

The GCs are often surrounded by cairns of different sizes, suggested to be graves (Okkonen, 1998). It is clear that not all members of the society could have been buried in the cairns, which leads to the conclusion that there was social inequality among the builders of the GCs.

The varying sizes of the GCs and the concentration of the largest ones in certain areas, sometimes only a few kilometers away from the next site, may reflect a competitive situation between the nearby communities (Okkonen 2003, 226).

The appearance of astronomically-oriented GCs in the period, when the former nomadic style of life changed into co-existence in permanent villages all year round, may be related to the new religious concepts and socially organized rites needed to strengthen social ties between the inhabitants. Directing some of the abundant resources into communal building projects, which resulted in a more beneficial religious life for all of the participants, may have made it easier to accept the leadership of the leaders of the project in other areas of the life of the community.

The building of the GCs stopped and the flourishing culture of their seal-hunting builders came to an end around 1800 BC, probably due to the climate change related to the end of the warm Atlantic period.

Bronze Age

When the Bronze Age monumental cairns started to appear on the coast of Finland, from the Ostrobothnia to the eastern side of the Gulf of Finland, the former dwelling sites of the GC builders had already moved away from the coast due to the post-glacial rebound.

The Finnish Bronze Age culture has so many similarities with the Scandinavian Bronze Age culture that it has been seen as a representative of the same cultural sphere. In the early Bronze Age, the cairns were monumental, the largest ones being over 35m in diameter. Towards the end of the period and in the early Iron Age, the cairn size got smaller. Most cairns were oval or round, but there were also rectangular, triangular and other shapes. An important subtype was the so-called long cairn. In the Bronze Age, a special type of long cairn appeared: cairns built into the form of a boat. There are several types, some more elaborate than others, but all have a wide 'stern' and a thinner 'bow', often with one larger stone marking both the front and the back end of the structure (Ridderstad, 2010).

In contrast to the Subneolithic cairns, almost all Bronze Age cairns are built of red stones (Ridderstad, 2010). This may provide one way to separate Subneolithic cairns from the similar-looking Bronze Ages ones.

All Bronze Age cairns are situated on high locations, looking towards the sea or other water routes of the time. Most of the cairns are situated on southern slopes of the terrain, or on the edges of ridges, so that the views are towards the sunrises and sunsets of late autumn and winter (Ridderstad, 2010).

All the stone-boat cairns measured in 2010 were 'sailing' towards the sunset (Ridderstad 2010). The shape, the red color and the direction towards the sunset over the water may be related to the belief in the land of the dead beyond the sea, and also to the belief of the solar boat. Perhaps it was believed that the solar boat itself would sail the souls of the dead to the Netherworld.

Not all cairns were equally placed: if the site contained more than one cairn, the largest ones were usually on higher ground than the smaller ones (Ridderstad, 2010; see also Kuusela et al., 2010). This, and the existence of several small cairns immediately around the monumental ones, were probably related to the unequal status of the persons buried in the cairns. In all of Europe, the Bronze Age saw the rise of the warrior class (Kristiansen and Larsson, 2005); it is probable that also in Finland the local leaders were honored with burials invoking the status symbolism of great warrior heroes and seafarers.

Figure 1. Image on a Bronze Age brooch.

Also in Finland, the Bronze Age jewelry and weaponry show decoration with celestial symbols: different types of spirals, star shapes and solar symbols can be seen (Figure 1; for the spiral as a solar symbol, see Ridderstad, 2009). Of special interest are the decorative spiraling shapes resembling labyrinths, since the earliest stone labyrinths of Sweden and Finland have been suggested to date to the Bronze Age.

Iron Age

Several different burial types appear in Finland during the Iron Age. Most of these graveyards have not yet been analyzed from the archaeoastronomical point of view. However, many of the cemeteries were situated on high locations and on the southern slopes of terrain.

The cairn building continued to the Viking Age, but the cairns got smaller and often became mixed with earth. A novelty was the placing of a large boulder in the middle of the cairn. The Iron Age cairns so far investigated with archaeoastronomical methods show orientations similar to the Bronze Age ones (Ridderstad, 2010).

The metal objects of the period show solar and stellar symbolism similar to the Bronze Age (Figure 2). From the Younger Roman Period onward, intense contacts with the Balts are reflected in the jewelry: images of the Sun, Moon and stars are frequent. Those elaborate decorative metal objects were probably important status symbols for their bearers.

It is known that the Iron Age solar deity of the Finns and the Sami was female, Päivätär. However, not much is known of her cult, which has led to speculations about the effect of Christianization on the Finnish Iron Age pantheon. Siikala (2002, 24) has suggested that the cult of Päivätär was replaced by the cult of Virgin Mary, *Maaria* in Finnish (cf., Mary as the Queen of Heaven, and the symbolism of 'Maria in Sole'). From runic poetry and incantations it can be deduced that Päivätär-Maaria had many similarities with the sun goddess Saule of the Balts (Ridderstad, 2011).

Figure 2. Celestial symbolism in the Finnish Iron Age jewelry.

Figure 3. Solar symbols in Finnish runic calendars.

It can be seen from the locations of the most important Christian feast days in the Medieval liturgical calendar, and from the folk beliefs and rituals associated with those days, that much of their significance had been drawn from the pre-Christian festivals celebrated on the eight main solar dates of the year (see Figure 3; Ridderstad, 2011). This is especially true for Christmas, and for the days of the Virgin Mary and the Holy Cross. The most important of the pre-Christian festivals had probably been associated with the winter and summer solstices, the beginning of May (May Day), and the beginning of November (*kekri*, the ancient Finnish New Year).

In the earliest Medieval church murals of Finland, calendrical themes dominated (Stigell, 1974). Many originally pagan themes are found in the imagery, with their meaning Christianized. Solar symbolism is a dominating theme, including the symbolism of the division of the solar year into eight parts (Figure 4; Ridderstad, 2011). Many of the solar symbols and the symbols for the wheel of the year presented in the paintings have close parallels both in the calendars of the time and in the symbolism of the Iron Age jewelry (compare Figures 2, 3 and 4). Therefore, it can be suggested that much of the Iron Age solar and calendrical symbolism continued into the historical age in its new, Christianized form.

Figure 4. Symbolism of the Sun and the eight-divided wheel of the year in early Medieval Finnish church murals.

In Figure 5, a church mural located in the Sipoo church in Southern Finland, which depicts a woman in a labyrinth, is presented. Most of the stone labyrinths of Finland have been dated to the Iron Age, and were used in historical times for the dancing games of young people, which they held in spring time and early summer (Kraft, 1985). In the games, a young girl was 'imprisoned' in the center of the stone labyrinth and had to be freed by the young man, who reached her first running or dancing through the stone pathways of the structure. The games have been interpreted as remnants of Iron Age, or perhaps even earlier, spring fertility rites (Kraft, 1985). The 'Lady in the Labyrinth' has an obvious parallel in the Lady of the Labyrinth of Knossos, the Bronze Age Minoan solar goddess. Thus, she was probably a personification of the sun goddess, who was liberated in the spring (Ridderstad, 2011). Both her calendrical significance and her identification with the Christian 'Maria in Sole' explains why her image was painted on a church wall in Medieval Finland.

The Helkajuhla festival of Ritvala of Sääksmäki, southern Finland, which has been celebrated since times immemorial, has many similarities with the ancient Baltic Midsummer celebration (see Vaiškunas, 2003). Nowadays Helkajuhla is celebrated on the day of Pentecost, but in the 19th century it was celebrated every Sunday from the day of the Ascension of Jesus to the Midsummer (Gottlund, 1832). In Finland, the Midsummer, *mittumaari* in Finnish, was largely a festival of Maaria. The central participants

Figure 5. 'The Lady in the Labyrinth'. An early Medieval painting in the church of Sipoo.

of Helkajuhla are young girls, who walk a cross-shaped route in the village singing Helkavirsi songs, which are Kalevala-metred Medieval ballads with some clearly pre-Christian elements embedded. The girls sing, for example, of an elk, which drools, and out of the saliva grows a giant tree, a symbol of fertility. Finally, the procession walks to the Helkavuori hill, and the girls dance around a bonfire; previously, they used to leap over it, too. It was believed in Ritvala that if the annual celebrations ever ceased, the fields of Ritvala would never bear fruit again. The present form of the festival is probably Medieval, but the original Helkajuhla predates Christianity. The origins of the Helkajuhla are in the same kinds of spring and early summer fertility rites as the labyrinth-dances. Originally, the rituals of the festival have probably included rites for both the solar and the weather god, possibly a *hieros gamos* between the sun goddess Päivätär and the weather god Ukko (cf., the Baltic myth of Saule's and Perkunas' wedding; Ridderstad, 2011 and references therein). The original ending of the festival in Midsummer is reflected in the songs: they include the symbolism of cutting down the world tree, a giant oak, which was believed to grow from midsummer to midwinter and to gradually cover the light of the sun (see Lintrop, 1999). The preservation of the festival in Sääksmäki can be explained by the fact that, as late as in the end of the 14th century, the pre-Christian faith was still strong in the region.

The largest Iron Age hill fort of Finland, the Rapola hill fort, is located only 5km from the Helkavuori of Ritvala. It has been suggested that some of the hill forts, including Rapola, may have been places of ritual activity (Luoto, 1999). The archaeoastronomical investigation of the hill forts is still ongoing. In the summer of 2010, measurements were carried out in the Kauttua hill fort in Eura, and an orientation of a large standing stone towards the winter solstice sunset was detected (Ridderstad, 2011).

Conclusion, with a view to the present day

It is concluded that the appearance of astronomically-oriented monuments seems to be related to the appearance of social stratification in the Subneolithic period. The trend of connecting astronomical orientations and celestial symbols with a high social status is seen also in the Bronze and Iron Ages, although cultural continuity between the different periods cannot be established for certain. It is probable that the religious authority associated with the phenomena of the celestial realm was connected with an elite status in the societies of all periods and is, subsequently, seen in the burial practices and the most conspicuous remains of material culture.

In the Iron Age it is possible to gain insight into the actual content of the astronomically-motivated religious beliefs and rites of all social classes. From the literary sources on Finnish pre-Christian mythology and the archaeological evidence from the Iron Age to the beginning of the historical period it can be deduced that the cult of a female solar deity was an important part of the Iron Age religion. It is likely that Iron Age mythology and religion had parts that originated in the Bronze Age and even earlier, but it is almost impossible to establish continuity for any specific belief or cultural meme; the most likely candidates are the division of the solar year into eight and the rituals related to labyrinths.

The Finnish Christmas *joulu* and the Midsummer celebrations never quite lost their Iron Age character. Also May Day, the ancient spring festival celebrated in the solar mid-quarter day of early May, has always retained its pagan form. The rites and beliefs surrounding the ancient Finnish New Year, *kekri*, became part of All Saints' Day. The Christian Easter absorbed much of the significance of the ancient festival of the coming of spring, the spring equinox. These festivals are still widely celebrated; others have lost their significance. Bonfires are lit on Midsummer, Easter, May Day, in the end of May and near All Saints' Day (*kekri*). In these festivities, the traditions of the ancient solar worship still live on.

References

Gottlund, C. A. 1832. Suomen Kansan Vanhoja Runoja IX, 1, 84-87. Available at: http://skvr.fi (Accessed 31 May 2015).

Kraft, J. 1985. The Goddess in the Labyrinth. Religionsvetenskapliga skrifter 11. Åbo, Åbo Akademi.

Kristiansen, K., and Larsson, T. B. 2005. The rise of Bronze Age society: travels, transmissions and transformations. Cambridge, Cambridge University Press.

Kuusela, J.-M., Vaneeckhout, S., and Okkonen, J. 2010. Places of importance and social communication: studying the Pre-Roman cairn field of Viirikallio in Laihia, Finland. Estonian Journal of Archaeology 14, 2, 22-39.

Lintrop, A. 1999. The Great Oak, the Weaving Maidens and the Red Boat, not to Mention a Lost Brush. Folklore 11.

Luoto, J. 1999. Sääksmäen Rapola – Hämeen muinaislinnojen ongelma, in U. Lähdesmäki, M. Meriluoto-Jaakkola and T.-L. Soininen (eds.), Masunni, Kirjoituksia Tampereelta ja Pirkanmaalta 3, 121-130. Tampereen museoiden julkaisuja 48. Tampere.

Okkonen, J. 1998. Muinaiset kivirakennelmat Keski- ja Pohjois-Pohjanmaalla. Licenciate thesis, University of Oulu.

Okkonen, J. 2003. Jättiläisen hautoja ja hirveitä kiviröykkiöitä – Pohjanmaan muinaisten kivirakennelmien arkeologiaa. Acta Universitatis Ouluensis B52. University of Oulu.

Okkonen, J., and Ridderstad, M. 2009. Jätinkirkkojen aurinkosuuntauksia, in J. Ikäheimo and S. Lipponen (eds.), Ei kiveäkään kääntämättä – Juhlakirja Pentti Koivuselle, 129-136. Pentti Koivusen juhlakirjatoimikunta, Tornion kirjapaino.

Ridderstad, M. 2009. Evidence of Minoan astronomy and calendrical practices. Available at: http://arxiv.org/ftp/arxiv/papers/0910/0910.4801.pdf.

Ridderstad, M. 2015. Orientations of the Bronze and Iron Age cairns in Finland. Fennoscandia Archaeologica.

Ridderstad, M. 2016. Sääksmäen Ritvalan Helkajuhlan alkuperä ja aurinkokultti rautakauden Suomessa. In preparation.

Ridderstad, M., and Okkonen, J. 2015. Orientations of the Giant's Churches in Ostrobothnia, Finland, in Proceedings of SEAC 2009, From Alexandria to Al-Iskandariya, astronomy and culture in the ancient Mediterranean and beyond, 25-31 October 2009, Alexandria, Egypt. Accepted.

Siikala, A.-L. 2002. What Myths Tell about Past Finno-Ugric Modes of Thinking, in A.-L. Siikala (ed.), Myth and Mentality, Studies in Folklore and Popular Thought, 15–32. Studia Fennica Folkloristica 8. Helsinki, Finnish Literature Society.

Stigell, A.-L. 1974. Kyrkans tecken och årets gång : tideräkningen och Finlands primitiva medeltidsmålningar. Suomen muinaismuistoyhdistyksen aikakauskirja 77. Helsingfors, Finska fornminnesföreningen.

Vaiškunas, J. 2003. Some aspects of Lithuanian folk observation of the sun during the summer solstice period, in Calendars, Symbols, and Orientations, in M. Blomberg, P. E. Blomberg and G. Henriksson (eds.), Legacies of Astronomy in Culture, Proceedings of SEAC 2001, Stockholm, 27-30 August, 2001, 33-38. Uppsala Astronomical Observatory Report 59. Uppsala.

Vuorela, I. 2002 Luonnon kerrostumat säilövät menneisyyttä, in R. Grünthal (ed.), Ennen, muinoin, Miten menneisyyttämme tutkitaan, 76–92. Helsinki, SKS.

ASTRONOMY AND THE POWER: THE SINGULAR BUILDING OF *TURÓ DEL CALVARI* (VILALBA DELS ARCS, TARRAGONA)

MANUEL PÉREZ GUTIÉRREZ, DAVID BEA CASTAÑO, JORDI DILOLI FONS AND SAMUEL SARDÀ SEUMA

Abstract: The archaeological interventions executed between 1999 and 2004 on the site of Turó del Calvari in Vilalba dels Arcs (Terra Alta, Tarragona) have brought to light a protohistoric building that has been interpreted as one of the earliest enclosures of power in operation during the Early Iron Age in the northeast of the Iberian Peninsula. The exceptional construction and geometric nature of the structure and the materials recovered show the existence in that time of a complete series of mechanisms of representation and control of the territory based on the use of the liturgy and feast as an element of political cohesion of an emergent elite, within a process that reproduces on scale a Mediterranean cultural system in an indigenous space. In 2009, the accomplishment of observations with astronomical and topographic aims showed four unpublished aspects in the excavation. These aspects reveal, in a special topographic situation, an exquisite geometry, an intentional divine proportionality and two precise astronomic orientations: to the setting of Arcturus and to the sunset at the summer solstice.
Keywords: Astronomy, archaeoastronomy, calendars, power, singular building, cult place, wine, commensality, feasts

Introduction

In 1999, as often happens in archeology, chance brought to light the remains of a building near the Ebro River, on land that, according to Hecataeus of Miletus, belonged to the Ileraugales of the preiberian period and that several centuries later became the Ilercavones, whose capital was in the city of Íbera, the current town of Tortosa. The building has been established to be from 590 to 550 BC (Early Iron Age). The excavation that was carried out between 1999 and 2004 yielded interesting and peculiar archaeological material, both of local and Phoenician manufacture, that has been described as exceptional in the pre-Phoenician world. No other item has been able to match it to date (Castaño and Fons, 2005; Blázquez, 2008).

Figure 1. The singular building of the Turó del Calvari and its immediate surroundings.

Separated from any other settlement of its time (the closest settlement, by the name of Coll del Moro de Gandesa, Terra Alta, Tarragona, is located more than 15km away), no domestic or common item was found in it. Only elements of a cultic, ritual and liturgical type were found, which somehow prove the existence of a series of power and control mechanisms based on the use of liturgy and banquets as a political and social cohesive element for an emerging elite and within a process that reproduces on scale an orientalizing cultural system (Sardá, 2007 and 2008).

This absence of what we could consider 'normal' elements inside and around the building, along with the existence of others not common at all, was what caused the archeological site to be named the 'Singular Building' of the Turó del Calvari (since it was next to Calvary crossings on a hill, or turó). The execution of a series of astronomical and topographic observations in 2009 did confirm its singularity with new contributions that confirm how special this building is, something already suspected by the excavators.

The topography of the 740km^2 of the Terra Alta (High land), on the south shore of the low Ebro where our building is located, is formed by a succession of big slopes and gullies that lead, mostly in a northerly direction, to the river that has always shaped this area. We have to consider, therefore, that this is an undulating territory, with an average altitude of about 350m above sea level, in which there are no predominant elevations. Our building rises between two of these gullies, flanked on three of its four sides by steep slopes that protect it.

First Singularity: Topographic Alignment

The first singularity that the construction shows is that its main axis notably matches the direction of the dividing slope on which is sitting, north-northwest. At first, investigators thought this was done in order to camouflage the building with its surroundings so it would go unnoticed. This would allow for the thesis that the topographic orientation is not casual, but deliberate.

However, there are two circumstances that seem to contradict this thesis. The first circumstance is the size of the building: the height of the two floors that it supposedly had (established through archaeological procedures) was around 10m. The 1.5m of width of the base matches this proposal, since less than half of that would be necessary to build the equivalent to one floor.

Figure 2. On the ground, view of terrain and the building from the NNW and detail of Turó del Calvari.

approach

The second circumstance is that the building was enhanced by building it on a circular podium of 2m in height, which raised it as well as protected it: this feature would only allow an the building from the main entrance and not from the sides. The overall height that it must have had would not allow for it to go unnoticed from considerable distance, since the topography of the terrain does allow for it to be sighted from many kilometers in every direction.

Second Singularity: Geometry

The topographical survey of the building showed a second singularity: an exquisite geometry which undoubtedly reveals a certain planning before the construction. Every dimension in the building is proportional to the measurement standard used to build it, which is the foot, 0.295m at the time (Calvo, 2006; Moret, 2008; Benlloch, 2009; Panchón et al., 2002). The deviation of the different modules of the different elements in the building is almost unnoticeable, as is the deviation from the measurement standard in each one of them: not more than a few millimeters.

The actual space of the building is determined by two circumferences tangent to each other, with a radius of 10 feet each and with the center on the main axis and the two adjacent tangents. This provides a bi-apsidial useful surface of 40 by 20 feet (about 12m by 6m). The thickness of the wall (a homogeneous 5 feet or 1.5m) allows for the outer limit to be obtained following the layout of two circumferences of 15 feet, each of whose centers match the ones of the aforementioned circumferences, closing the building with the mentioned outer circumferences and the two outer tangents. The dimensions of the construction are therefore 50 by 30 feet (about 15m by 9m).

The interior of the building is divided in two rooms by a wall of adobe of about ¾ of a foot of thickness, which is in accordance to the normalized size for adobe block. It stands on a low stone base, with a width of about one and a half feet. It seems the wall was not higher than one meter, since its only purpose was to elevate the floor in the southern room. This wall made the northern room 30 feet long, and the southern room 20 feet long. The actual width for both rooms was 20 feet.

Third Singularity: Proportionality

The third singular characteristic appears when comparing the above-mentioned measurements. The actual measurements of the building (40x20 feet) are in the proportion 2:1, whereas the external and main dimensions are in the proportion 5:3, just like the small altar, located in the corner of the northern wall, and the wall, whose dimensions are exactly 10 times less than the building. The lengths of the two rooms separated by the wall, of 25 and 15 feet, are also in the proportion 5:3. The radii of the circumferences used in the design of the building are in the proportion 3:2. The estimated proportion between the base and the dividing wall is 3:1 (strictly speaking it is a 10:3 proportion).

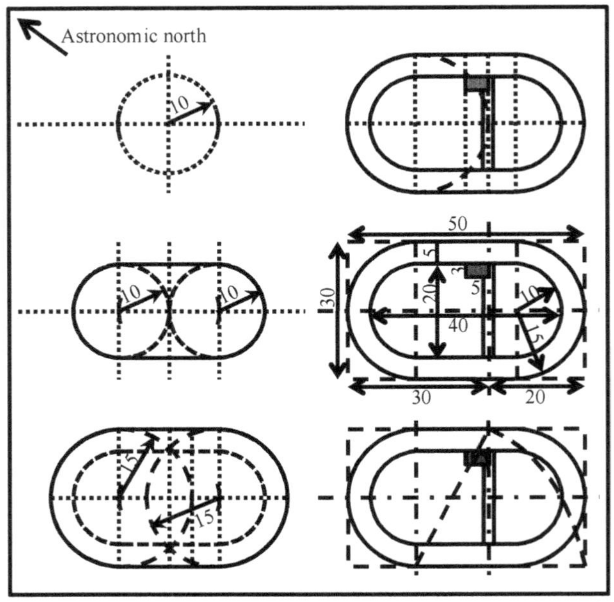

Figure 3. The plain of construction, the dimensions and the proportionality of Turó del Calvari. Dimensions are proportional to a foot of 0.29m.

It is plain to see that both the dimensions and the proportionality amongst them are, at the very least, quite singular. If we take the foot as the standard unit, all of the dimensions are a multiple of 5 (the fingers in one hand), except those of the interior wall which, as mentioned, are measured in a different kind of foot. Also, all relations of proportionality are basic and they can all be achieved with the prime numbers of the first 5 digits that are also the first terms in the Fibonacci series, that is: 1, 2, 3 and 5. These provide the proportions 2:1, 3:1, 3:2 and 5:3. It should also be considered that the main proportion (which is the proportion of the building itself, the altar and the distribution of the rooms) is 5:3, which is also the first approximation that can be obtained with a rational number to the golden ratio, used in many constructions along centuries, since it makes easier the planning of any project with the standard ruler and compass used by builders. It was Euclid who proved the irrationality of the golden ratio, golden proportion or divine proportion, called φ in honor to Phidias, architect of the Parthenon and whose sculptures were and are still considered some of the most beautiful of ancient times.

If the height of the building was known, we could also discuss its proportionality, but obviously we can only speculate in the absence of this information. For the height to be in golden proportion with the dimensions of the base (or the width or the length), it should be either the minor segment for the width (5.4m, 18 feet), the width of the building (9m, 30 feet), its length (15m, 50 feet) or it would have to become the biggest segment of the proportion with the length, and it would have measured an improbable 25m (83.3 feet). The excavation site provided us with information of the existence of two floors in the building, based on the remains of some framework in the top floor and the basis for a column that most likely would hold the framework. Considering this, the construction would rise to either 18 or 30 feet high.

Fourth Singularity: Astral and Solar Directions

Finally, there is a fourth singularity shown by astronomical observation, which confirms the special and peculiar use for this building. The determination of the astronomic meridian by Sun observation together with the topographic mapping allowed us to obtain the astronomic azimuth of the elements of the construction. These azimuths (both the main axis and the interior and exterior sides of the walls) are almost identical and provide an orientation of the axis of the building of 316.5° with respect to the north. A topographic and photographic mapping of the local horizon tells us there is no topographic feature of significance in any direction in general, and in the direction of the axis of the building in particular. In fact, the west horizon and north horizon are practically horizontal. The zenithal distances of the horizon in the Arcturus setting is 90° 21'. It was also confirmed that the position in which the direction of the building intersects with the local horizon is more northern that the sunrise of the summer solstice (azimuth 302°, declination, 23° 43'). It is even more northern that the major standstill of the Moon at the winter solstice (azimuth 311°). The zenithal distance is 90° 30' in both cases. However, a simulation of the celestial sphere and its movements during the time and place of construction of the building shows that the main axis of the building points to the place in the horizon where the setting of the star Arcturus (declination for the building time is 34° 10'), which is the third brightest star on the sky and the one that verifies its rising and setting closest to the North Pole. As stated in 1718 by Edmund Halley, the movement of Arcturus is quite pronounced (a bit more than 2 inches yearly) which, together with the precession of the equinoxes, means that the variation of the azimuth of the setting of the star is half a sexagesimal minute yearly, a considerable value that has reached nowadays 22°.

In addition, we can observe that the minor axis of the Turó del Calvari seems directed towards the setting of the southern major lunar standstill, that occurs in the summer solstice. Of course, if there was a window in the southern wall, the moonrays would fall onto the altar on this summer day.

The orientation towards a known star, used since ancient times by many nations and cultures, does not seem accidental. In our case, there are up to three references prior to the construction of the building. Homer mentions them in *The Odyssey*, along with Orion and both Ursas, providing the basic rules of Phoenician navigation when Calypso tells Odysseus how to return home. It is also mentioned by Hesiod in *Works and Days*, where he uses the heliacal rising and setting of Arcturus as indicators for the dates of pruning, harvesting and storing of wine. As we all know, it was Phoenician merchants who first brought wine into all Mediterranean countries. Finally, it is clearly linked to the grapevine and the wine in the ancient Attican tale that tells how Dionysius, son of Zeus and Semele (daughter of Cadmus, legendary Phoenician hero, founder of Tebas and introducer of the alphabet in Greece), granted this gift to the human kind. Received by Icarius, he brought it to his neighbors who, after drinking it and getting drunk, killed him. His dog Mera looked for his daughter Erigone who, upon discovering the corpse, committed suicide. Dionysius granted them a space on heaven: Icarius as Boötes, Erigone as Virgo and Mera as Sirius or Procyon.

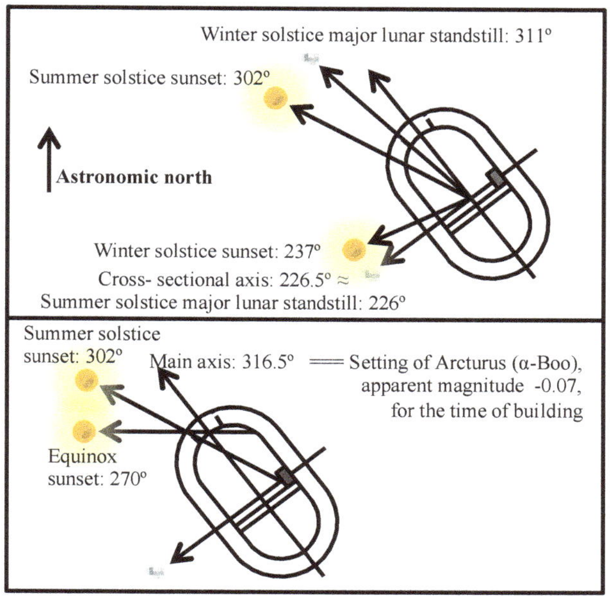

Figure 4. The astronomical orientations in the Turó del Calvari.

There is another astronomic singularity that confirms the interest in the Sun by the settlements in the Iberian Peninsula, as well as in many other places (Pérez 2007; 2009; 2010). The location of the main entrance (at the northernmost point of the building) makes it impossible for sunbeams to enter the house except at a very specific time of the year: from the vernal equinox to the autumnal equinox, and always at times close to the sunset. At the moment of maximum declination, the sunset happens at the northernmost position of the year. There is only one structure (believed to be a small altar or a base used for one of the mobile altars found during the excavation) that is lit by the sunshine, only for a brief moment during the summer solstice. The Sun never reaches the back of the building, where the smallest room is; this room was most likely elevated, in the shape of a small stage. We find therefore an altar that only gets hit by sunbeams during the summer solstice, unlike most of the Punic-Phoenician structures in the Iberian Peninsula, which are facing the sunrise of the summer solstice.

At this point we should return to the discussion about the golden ratio (5:3) of the building. Aratus, in the 3rd century BC (probably while collecting information going back beyond the 1st millennium BC, and therefore prior to the time of our building), described the motion of the Sun on the celestial sphere. In particular, he explained the seasons and the difference in the length of day and night that occurs through the year. Specifically, when the Sun was placed in Cancer, it reached its maximum declination and summer began. 'The circle [the Tropic of Cancer] was divided into eight parts, five of which occurred during daytime and the remaining three beneath the horizon' (Aratus, *Phaenomena*, 498). In other words, as the summer begins and the Sun illuminates the inner altar, the ratio between the time that the Sun spends above the horizon

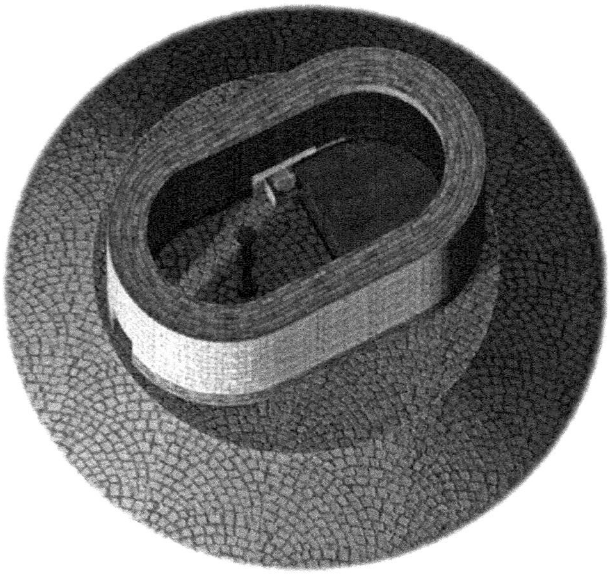

Figure 5. The sunrays going into the building and lighting the altar during the summer solstice sunset.

and below the horizon is also 5:3. Although it we can not say for certain that the builders of the Turó del Calvari possessed this knowledge, everything that we find in the building, both tangible and intangible, suggests this was very probably the case.

Summary

In short, we find ourselves with a building from the First Iron Age (pre-Iberian). It was clearly build by natives, but with a strong oriental influence on the house itself and on all the furniture and accessories found in the excavation. All the materials found at the site, both indigenous and Phoenician, surprised the researchers to the extent that they called it simply the 'Singular Building'.

The dimensions of this construction (considering also the idea of elevating it on a circular base) make this building simply astonishing. Moreover, the great deal of care put into the exquisite geometric design leads us to believe there was a planning unmatched at that time and place, considering also the amazing use of the measurement standards (the feet of 0.295 meters as well as the Attic foot) which allows every dimension to be a multiple of that standard and also a multiple of 5. And last but not least, there is the proportionality among all its different measurements, which makes the building something pleasing to the eye, something simply beautiful, although the reasons why are still foreign to us.

On the other hand, the idea that the orientation of the building may be casual is dismissed once it is considered as a whole. The relationship between Arcturus/Boötes and the wine, the grapevine and the gastronomic element is proven, although it raises a question, not trivial at all, that is left unanswered. There does not seem to be any difficulty in directing the building towards the summer solstice sunset, as it happens with many Punic-Phoenician altars, and the topography should not have been a problem. However, it was built in a disposition that is similar to the position of the Sun that is found in many places in the Iberian Peninsula: the summer solstice sunset. The way the structure is designed and built, including the astral and solar orientation, is simply unique.

Conclusions

If Greek culture and all its knowledge had been present on the coasts surrounding the mouth of the River Ebro, the Iberia of Hecataeus, the standard of measurement (the Attic foot), the geometry and proportionality and the adoption of stories, legends and astronomical knowledge would not have been surprising. Our problem, and the delicate dilemma we are faced with, is that there is no archeological evidence or written sources that confirm the presence of Greek settlers at the end of the 7th century BC, either at the mouth of the river Ebro or in the excavation of our building.

In order to try to solve this dilemma, we raise the following questions that might shed some light on this problem:

First, there are strong indications that Homer (probably related to the nobility) was born in the Ionian region of Asia. There he might have obtained all the information regarding the Mediterranean Sea and the tricks to navigate from the Phoenicians and the Canaanites, who had been doing it for centuries.

Secondly, as mentioned before, wine was brought from the Western Mediterranean by the Phoenicians. In Spain, there is a record of wine production in the areas of Castellón and Valencia during the time our building was built.

Lastly, the well-known Attican legend Icarius and Erigone, associated with Dionysius (a Mycenaean god probably inherited from Semitic settlers, recognized by the Greek as grandson of the Phoenician hero Cadmus) could very well have been adopted precisely with the introduction of the wine and vineyard in the Attican peninsula.

Then, in finishing, there is still a question unresolved: how it is possible that so many Greek aspects appear in a place that had not yet been colonized by them?

References

Aveni, A. 2002. Empires of time: calendars, clocks and cultures. Boulder CO, University Press of Colorado.

Benlloch, P. Olmos. 2009. Aproximació a la metrología ibérica a Catalunya (segles V-II aC). Revista d'Arqueologia de Ponent 19, 51-74.

Blázquez J. M. 2008. Últimas aportaciones a la presencia fenicia y cartaginesa en Occidente. Arquitectura y urbanismo. Gerión, 26-2: 9-73.

Calvo, J.C. 2006. Sistemas metrológicos prerromanos en la península ibérica. STUDIUM. Revista de Humanidades, 12.: 35-55.

Carrasco, J. L. Escacena. 2009. La Égersis de Melqart. Hipótesis sobre una teología solar cananea. Complutum 20, no. 2, 95-120.

Castaño, David Bea, and Fons, Jordi Diloli. 2005. Elements de representació durant la Primera Edat del Ferro al curs inferior de l'Ebre: el recinte del Turó del Calvari (Vilalba dels Arcs, Terra Alta). Revista d'Arqueologia de Ponent 15, 179-198.

Evans, J. 1998. The history and practice of ancient astronomy. Oxford, Oxford University Press.

Moret, P. 2008. Recherches historiques et archéologiques sur l'ibérie Antique. Mémoire d'habilitation. PhD dissertation, Toulouse, Université de Toulouse-Le Mirail.

Panchón, R. F., and Manzano, F. 2002. Metrología en las civilizaciones de Mesopotamia, Egipto, Fenicia, Israel, Grecia, Cartago, Roma y otras culturas de la antigüedad. XIV Congreso Internacional de Ingeniería Gráfica. Santander, España, 5-7 de junio de 2002.

Pérez Gutiérrez, M. 2007. Astronomía en la Edad del Hierro peninsular: orientaciones astronómicas en los castros celtas de la provincia de Ávila. Tésis doctoral inédita. Universidad de Salamanca.

Pérez Gutiérrez, M. 2009. Astronomía y Geometría en la Vettonia. Complutum 20-2: 141-164.

Pérez Gutiérrez, M. 2010. Astronomía en los castros celtas de la provincia de Ávila. Institución Gran Duque de Alba. Diputación provincial de Ávila.

Prados Torreira, L. 1994. Los santuarios ibéricos. Apuntes para el desarrollo de una arqueología del culto. Trabajos de prehistoria, 51, no. 1, 127-140.

Sánchez Moreno, A. 2005. Santuarios ibéricos en la Bastetania. Arqueología y Territorio 2, 65-80.

Sardà, S. 2007. Els materials ceràmics del Turó del Calvari (Vilalba dels Arcs, Terra Alta). Pràctiques de consum ritual a la primera meitat del segle VI anE. Trabajo de Investigación, DEA, Universitat Rovira i Virgili.

Sardà, S. 2008. Servir el vino. Algunas observaciones sobre la adopción del oinochoe en el curso inferior del Ebro (s. VII-VI a. C.). Trabajos de prehistoria 65, no. 2, 95-115.

Valdés, M. 2001. El proceso de sinecismo del Ática: cultos, mitos y rituales en la <<primera polis>> de Atenas. Gerión 19, 127-197.

Ventura, F. 2008. Orientations of the Phoenician and Punic shaft tombs of Malta, in C. Esteban and J. A., Belmonte (eds), Astronomy and cultural diversity, the Proceedings of the International Conference OXFORD VI and SEAC 99. OACIMC.

PRECISE ASTRONOMICAL MEASUREMENTS OF ANCIENT DACIAN SITES WITHIN THE PYTHAGOREAN MEGA-TRIANGLE SARMIZEGETUSA-REGIA-RETEZAT-PARÂNG

FRANZ KEREK AND FLORIN STANESCU

Abstract: Sarmizegetusa-Regia in the Southern Carpathians mountain range was the capital city of the Dacian kingdom (84 BC-106 AD) but its outlandish location and enigmatic sanctuaries make it more likely it was a sacred destination. The Dacian faith was a Pythagorean religion and we propose that Sarmizegetusa was sited according to precise astronomical alignments and the Pythagorean doctrines of well-ordered harmonic structures in heaven and on the earth. Sarmizegetusa is at the right-angle vertex of a mega triangle formed with two prominent peaks of the surrounding Retezat and Parâng Mountains, an assumption confirmed here in detail by precise satellite and GPS data. The fortress hill in Sarmizegetusa is on the winter sunset alignment towards Retezat and this celestial event can be observed from this place. A recent find of massive gold bracelets in Sarmizegetusa also argues for its status as a sacred-ceremonial destination. We identified two exactly dimensioned straightedges on the vertices of the mega-triangle, presumably constructed by Dacian astronomers to enhance by triangulation the precision of their azimuth angle measurements. The 160m edge of the truncated Retezat pyramid was probably arranged by man-made flattening of the peak. Tracks of similar constructions can be identified around the twin peaks Piatra Taiata and Coasta lui Rus in the Parâng Mountains.
Keywords: Dacians, Sarmizegetusa-Regia, Retezat, Parâng, Pythagorean, mega-triangle, gold

Introduction

The ancient kingdom of the Dacians (called Getae by Hellenes) comprised closely the territory of modern-day Romania (Figure 1). Its capital Sarmizegetusa-Regia was erected in the Southern Carpathians under the rule of King Burebista (82-44 BC) and his holy priest Deceneus.

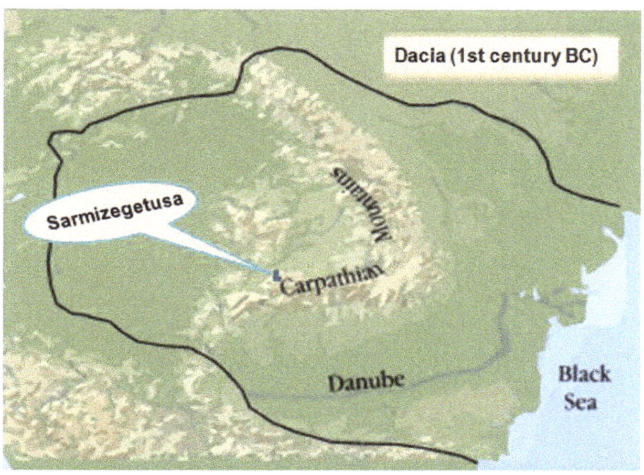

Figure 1.Territory of the Dacian kingdom.

During two centuries (82 BC-106 AD) of rivalry, several Roman Emperors planned to conquer the wealthy Dacian kingdom and its legendary gold mines. In his first Dacian campaign (101-102 AD) the Emperor Trajan could not decisively defeat King Decebalus. In preparation of the next campaign, Trajan´s engineer Apollodorus of Damascus built in 104 AD a bridge across the Danube (Serban, 2009). The victorious second campaign (105-106 AD) ended with the conquest of the capital Sarmizegetusa-Regia and Dacia was annexed as an imperial province.

Impressive pictures of Trajan´s epic Dacian wars are artistically presented on the 125 reliefs of Trajan´s Column, erected in 113 AD in Rome (Davies, 1997). The 190m long marble frieze winds up around the shaft 23 times and portrays Trajan´s Dacian campaigns. The lower half provides scenes of the first campaign while the upper half presents the second campaign and siege in 106 AD. The column mainly depicts the Roman army in various military activities and portrays the Emperor addressing the soldiers, commanding military operations, or receiving legates of the Dacians (Florescu, 1969).

Many scenes on Trajan´s Column illustrate the redoubtable heroism of the Dacian defenders. Pictures with their strong fortresses attest to their high level architectural and military achievements. Scenes with Dacians saving their gold riches relate to the immense capture by Trajan´s army of an estimated 165 tons of gold and 330 tons of silver.

Sarmizegetusa the sacred capital

It remains enigmatic why Sarmizegetusa, the capital of the Dacian kingdom, was built on the slope of the Dealul Grădiştei Mountain. The location was not optimal for an administrative capital and was militarily vulnerable. Our first paper (Kerek, 1994) suggested that the location of Sarmizegetusa-Regia was selected primarily by religious-astronomical criteria. Accordingly, this was a holy, god-designated place further strengthened by the surrounding fortresses and it was heroically defended against the Romans.

Figure 2. Stone sanctuaries in Sarmizegetusa.

The presumably sacred destination of Sarmizegetusa-Regia is convincingly supported by the presence on the site of three circular sanctuaries and six rectangular constructions with arrays of column bases (Figure 2). The large circular sanctuary is 29.4m in diameter and comprises an external circle of 104 precisely shaped massive stone blocks and an adjacent circle with 30 series of 6+1 stone pillars. The main axis of the internal horseshoe construction of 64 timber posts is aligned to sunrise at the midwinter solstice. Also notable is the massive andesite-sun measuring 2.76m in diameter and divided into 10 segments with an internal disc with a 1.46m radius. Its destination as sacrifice altar is suggested but not confirmed.

Pythagorean location of Sarmizegetusa

To identify the criteria which probably decided the siting of Sarmizegetusa, the doctrines of the Getaeo-Dacian faith are to be considered. This was a Pythagorean religion and the ancient historians Herodotus (Histories IV, 94-6) and Strabon mentioned that Zalmoxis the God of Getae was a disciple of the great Pythagoras (580-500 BC).

Contrary to his modern image of mathematician, Pythagoras was renowned in ancient times as a philosopher and religion founder. The theorem that bears his name was already known by the Babylonians (Ferguson, 2008a). Pythagoras propagated it to the Greeks since it provided brilliant proof for his ‘all things known have numbers’ theory. Pythagorean teachings on immortality, which involved rigorous self-discipline and restrictions on eating beans and drinking wine, were adopted by the Dacians (Eliade 1972, 55-67).

Pythagoras discovered that harmonious sounds depend on ratios of small integers (Ferguson, 2008b). The harmonic order of the planets, called the cosmos, was considered similar to regular solids (sphere, cube). Pythagoreans searched passionately for well-ordered harmonic systems on the Earth and in heaven. This belief could have motivated Dacian priests to find a god-designated holy location for their capital. We identified (Kerek, 1994) that Sarmizegetusa-Regia is at the right angle vertex of a mega-triangle built with two prominent peaks in the Retezat and Parâng Mountains (Figure 3). One side of the triangle is an astronomical alignment: viewed from Sarmizegetusa at the winter solstice, the Sun sets behind the Retezat peak. This event marks the start of a new year and is celebrated in all cultures.

The detailed analysis of the mega-triangle (Kerek, 2007) revealed that Sarmizegetusa is at the 90° intersection of the winter sunset alignment towards Retezat with the straight line towards the twin peaks Coasta lui Rus and Piatra Taiata at midpoint of Parâng Mountains (Figure 3).

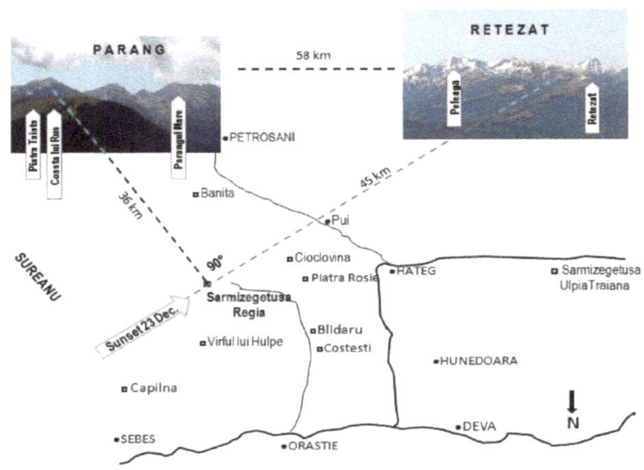

Figure 3. The Sarmizegetusa-Retezat-Parâng mega-triangle (▫) Dacian fortresses, (•) present day localities.

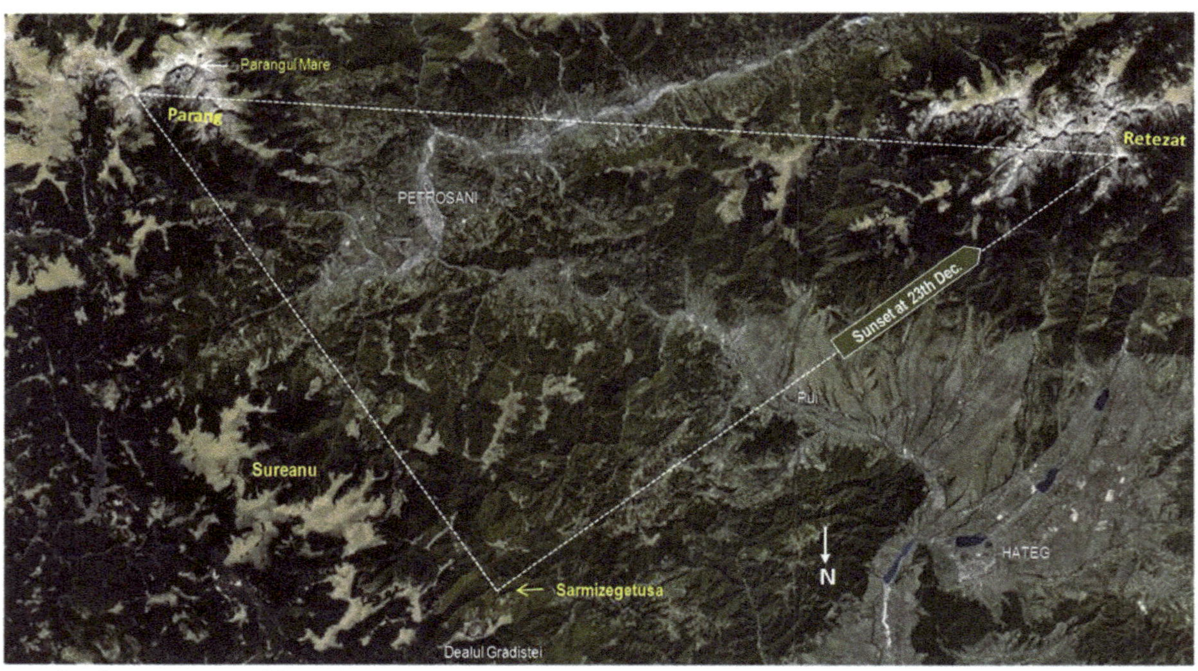

Figure 4. Satellite map image of the Pythagorean mega-triangle Sarmizegetusa-Retezat-Parâng in the Southern Carpathian Mountains with the assumed involvement of the sacred Dacian capital Sarmizegetusa Regia.

In this sense the sacred capital Sarmizegetusa was erected at a location in Pythagorean harmony with two prominent mountain peaks of the region, i.e., in Retezat and Parâng.

The satellite picture (Figure 4) confirms that the midwinter sunset alignment towards Retezat peak intersects in Sarmizegetusa at 90°, the straight line towards the midpoint of the Parâng Mountains. Interestingly this line closely follows the eastern ridge of the W-shaped Parâng Mountains and is nearly perpendicular on the main Parâng ridge from midpoint to the Parângul-Mare peak (2519m).

The fortress hill observatory

Sarmizegetusa Regia (Figure 5) consists of the sacred zone with sanctuaries, the area of the fortress and the civil settlement. Three circular sanctuaries (1-3) and ruins of six rectangular constructions (temples) (4-9) were identified in the sacred zone but their exact destination is less elucidated. The area of the sanctuaries is connected to the fortress with a broad stone-paved road (10) with well-preserved fragments. The dimensions and very fine execution of this road suggest a ceremonial destination.

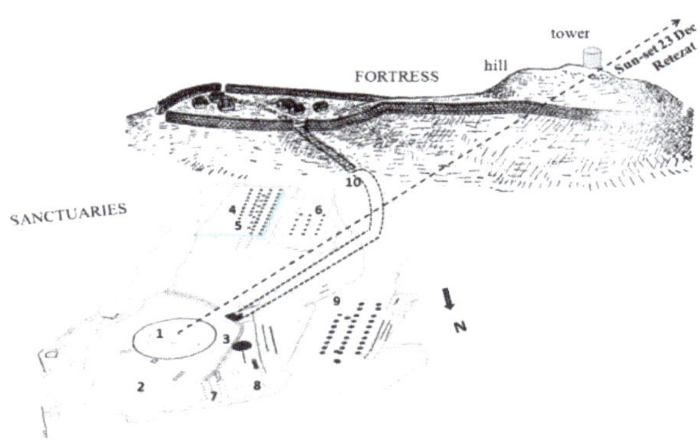

Figure 5. Fortress and sanctuaries in Sarmizegetusa-Regia.

For archaeoastronomical investigations the fortress area was hitherto ignored in comparison with the zone of the sanctuaries. Historians (Glodariu, 1983) assumed that this wall-enclosed citadel was a military destination used for the protection of the civil settlement and the sanctuaries.

The highest point of this, a hill inside the fortress walls that is probably man-made, lays 35-40m above the sanctuaries. Presently the knoll is visible only winter time when it is less obscured by the foliage of the trees. As shown by the drawing made in 1847 (Figure 6), the uncovered hill dominated Sarmizegetusa and was designated by Ackner (Wollmann, 1982) as the 'Akropolis at Gradistea Muncelui'. Ruins of a stone construction were found by Ackner on the hill. More recent diggings (Glodariu, 1996) confirmed the ruins on the hill and assumed they belonged to a military watch tower.

In a recent paper Stanescu (2001) reconsidered the possible astronomical use of the hill with special concern for the observation of the Pleiades. Actually, the hill—with 1038m altitude—is the highest point of the capital Sarmizegetusa. A 10-12m-high tower on the hill enabled sky watching over a broader horizon in comparison with the restricted outlook from the sanctuaries (Ruggles, 2005).

In this sense the fortress hill, considered hitherto to be only a military destination, could have an astronomical role.

We disclose now the finding that Retezat Mountain is visible from the fortress hill and thus the impressive midwinter sunsets behind this peak could be observed directly from the sacred capital.

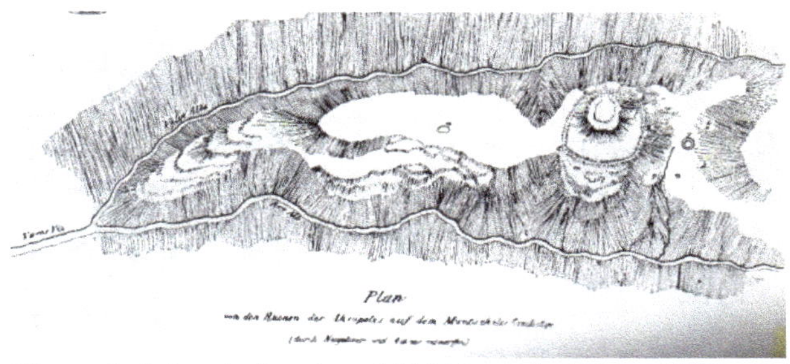

Figure 6. Ackner's drawing of 1847 from the fortress hill.

High resolution Google-Earth Pro and Garmin GPSmap 60CSx data places the top of the fortress hill at 45°37.32′N and 23°18.50′E. This point lies on the midwinter sunset alignment towards Retezat peak (Figure 5) similar to the midpoint of the large circular sanctuary with the coordinates 45°37.37′N and 23°18.64′E. The tower on the hill served for the observation of the sunset and the midpoint of the large circular sanctuary was positioned on an extension of this line. The broad paved road, which starts from the circular sanctuary, follows in its first part the same alignment (see 10 in Figure 5).

The turn of the year and further major celestial events, e.g., Moon- or solar eclipses were solemnly celebrated by Dacian priests. Festive ceremonies with processions along the sacred road from the hill to the sanctuaries are realistic scenarios. The discovery in 2000 of 16 gold bracelets weighing 0.6-1.5kg near the sacred zone (Constantinescu, 2010) hints at the pomp of Dacian religious processions (Figure 7).

Figure 7. Gold bracelets recently found in Sarmizegetusa.

Analogue to other cultures the ceremonies in Sarmizegetusa were aimed to demonstrate the divine power of the priesthood. The exact prediction of vivid celestial events such as eclipses of the Sun or Moon provided optimal opportunities for the ceremonies. Identification of the fortress hill tower as the observation point of the celestial events gives a profound meaning to the broad stone paved procession road to the area of sanctuaries.

Dacian astronomer-priests were involved in the observation, measurement and calculation of the celestial events. According to Jordanes (1972, 41) Gaeto-Dacian priests were skilled astronomers with advanced understanding of the movements of the Sun, Moon, planets and stars. Jordanes credited the holy priest Deceneus with establishing the relative size of the Sun and the Moon.

Astronomical observations and measurements were most likely performed on the Dealul Grâdiştei plateau at 1550m altitude with a roundabout clear horizon. This plateau is 500m above the sanctuaries (985m) at 2.7km distance with possible optical signalling. Geographical coordinates of the plateau midpoint are: 45°38.60′N and 23°18.85′E. The area of the Dealul Muncelului plateaus is recognised easily on the satellite image on Figure 4 by the characteristic salamander-like contours of the zone. It is assumed that the data from the observation plateau were processed at the temples of the sacred and used by the priests for the records and prediction of celestial events.

Precise azimuth values

Dacian priests (astronomers and land-surveyors) used for angle measurements the instruments groma and dioptra (Drachmann 1969, 241-247) of Roman engineers. However, for the distances of 36-45km within the Dacian mega-triangle the precision of these instruments was not satisfactory.

We assume that Dacian priests had improved the precision of azimuth angle measurement by triangulation using straightedges of the known length between marker points around prominent peaks on mountainous horizon.

An appropriate straightedge was provided by the top edge of the pyramid frustum of the Retezat peak at 2482m altitude (Figure 8). For precise assessment of azimuth values, the southern margin RTS and the northern margin RTN of the Retezat edge are considered separately.

Figure 8. The truncated pyramid formed Retezat peak.

Geographic coordinates of the RTS margin of the edge are 45°22.75′N and 22°50.85′E. The coordinates of the northern margin RTN are 45°22.85′N and 22°50.98′E. The azimuth angle values calculated from the fortress hill in Sarmizegetusa (N->E) are of 233.04° to RTS and 233.38° to RTN margins of the Retezat edge. The calculated difference between the azimuth values is 233.38°-233.04°=0.34° which correspond to 1/3 degree.

The high-resolution satellite map of the geographical midpoint of the Parâng Mountains reveals the existence at this place of two nearby mountain peaks. The geographic coordinates of the 2301m high Coasta lui Rus (PCR) are 45°21.38′N; 23°35.42′E and that of the 2298m high Piatra Taiata (PPT) peak are 45°21.30′N and 23°35.42′E.

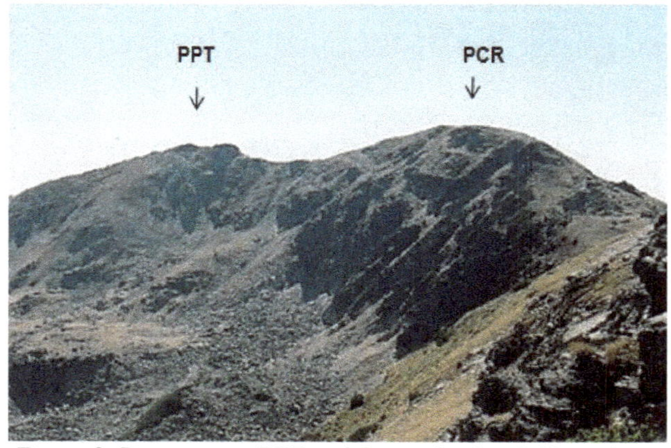

Figure 9. Piatra taiata PPT and Coasta lui Rus PCR peaks.

From Sarmizegetusa the alignment towards the PPT peak has an azimuth angle value of A=143.38° while the alignment to the PCR peak is A=143.65°, both calculated clockwise N->E from the fortress hill tower. From the resulting 0.27° difference between azimuths the calculated length of the PPT-PCR edge perpendicular to the Sarmizegetusa alignment is 175m. The actual distance between the Parâng twin peaks is 340m but this line closes 45° with the Sarmizegetusa alignment.

We suggest that the straightedges described here were the results of impressive constructive efforts. The assumed flattening of the Retezat peak aimed to provide the truncated pyramid frustum with the upper edge of the desired dimension. Viewing the truncated Retezat edge from Sarmizegetusa (45 km) it corresponds to 1/3 degree compared with 1/2 degree angular diameter of the Sun. Dacian constructors were definitely able to perform this impressive effort. The name Retezat means cut off in Romanian and according to the legend the flat top resulted when the original mountain peak was cut off by the sword of a giant fighting against the dragon.

Examination of the zone of the twin peaks reveals some very strange, probably man-made stone features. Even the name Piatra taiata means literally sliced stone in Romanian and suggests human intervention.

The scheme in Figure 10 summarizes the results of precise azimuth angle measurements within the Pythagorean mega-triangle Sarmizegetusa-Retezat-Parâng.

The important conclusion of the azimuth angle data shown on Fig. 10 is that the resulting difference between

233.39°–143.38°=90.01° which confirms with high precision the right angle of the Sarmizegetusa-Retezat-Parâng triangle.

The mean angular diameter of the Sun and Moon are accidentally the same, i.e., 0.53°, corresponding roughly to a half degree. The visual diameter of the Sun at D= 45km distance is of d=420m calculated for 0.53° angular diameter. For the azimuth difference of 0.34° between RTS and RTN it results in a 260m distance between these marker points.

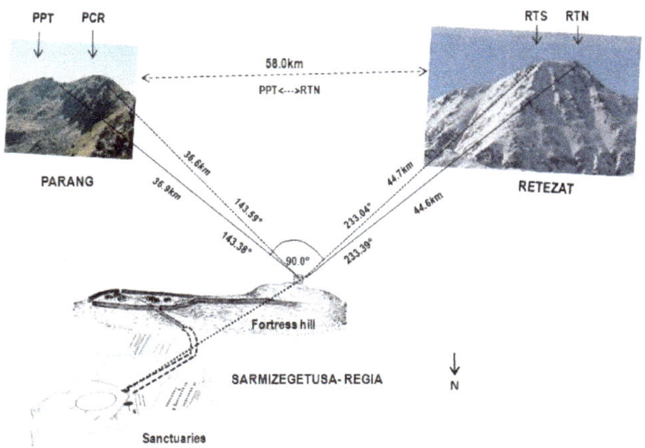

Figure 10. Azimuth and distance values within the triangle.

Sunset at December solstice

By watching from Sarmizegetusa the sunset at winter solstice the 0.53° angular diameter of the Sun is to be compared with the edge of the Retezat corresponding to 0.34° angular diameter from this distance. It is less probable that the Dacians could measure from the capital the very small daily variation of the sunset azimuth values. More precise observation of the sunset is possible from the Federi-hill above the locality of Pui (Figure 4) at half the distance of the Sarmizegetusa-Retezat alignment.

Animation of the sunset at the winter solstice is depicted in Figure 11, which suggests the impressive event that can be followed by the turn of the year.

The exact sunset azimuth data were calculated from Sarmizegetusa for December of the years 10 AD and 2010 AD, respectively. The results provided by the US navy algorithm (www.ssrb.noaa.gov) are presented on diagram of Figure 12. Azimuth corrections due to the 2482m altitude of Retezat are included and the altitude difference is diminished by 156m due to the curvature of the Earth at 45km distance.

Figure 12. Animation of December sunset at Retezat peak.

The diagram on Figure 12 confirms the winter sunset azimuth value of 233.6° during the Dacian era and 234.1° in the year 2010. The difference is caused by the change of the obliquity of the ecliptic in 2000 years from 23°40′ to 23°26′.

The excellent correlation of the measured and calculated azimuth values is noted. The very small range shift (233.8°-233.6°) of the sunset azimuth values at the winter solstice (15-25 December) suggests that the observation period with probable ceremonies lasted for several days.

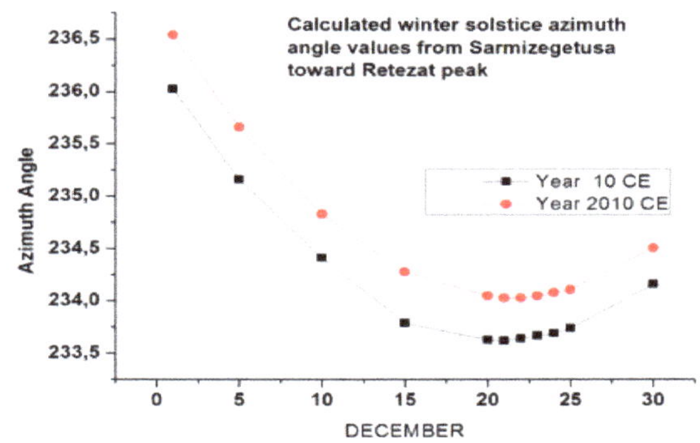

Figure 11. December sunset in years 10 AD and 2010 AD.

Conclusion

The location of the Dacian capital Sarmizegetusa-Regia in Pythagorean harmony to the prominent mountain peaks is confirmed here by precise measurements and suggests further insights into the high level astronomic culture of ancient Dacians.

Acknowledgements:

The authors are indebted to George D. Suciu for the very helpful discussions and proofreading of the manuscript.

References

Constantinescu, B. Vasilescu, A. Radtke, M. and Reinholz, U. 2010. Micro-SR-XRF studies for archaeological gold identification. Applied Physics A 99, no. 2, 383-389.

Davies, P. J. E. 1997. Trajans´s Column and the Art of Commemoration. American Journal of Archaeology 101, 41-65.

Drachmann, A. G. 1969. A Detail of Heron´s Dioptra. Centaurus 13, no. 3, 241-247.

Eliade, M. 1972. Zalmoxis the Vanishing God. Chicago, University Press, 55-67.

Ferguson, K. 2008a. The Music of Pythagoras. New York, Walker Company, 76-93.

Ferguson, K. 2008b. The Music of Pythagoras. New York, Walker Company, 62-75.

Florescu, F. B. 1969. Trajan´s Column. Bucharest and Habelt Verlag Bonn, Editura Academiei.

Glodariu, I. 1983. The Dacian architecture. Cluj-Napoca, Editura Dacia. 97-100.

Glodariu, I. Iaroslavischi, E. Rusu-Pescaru, A. and Stanescu F. 1996. Sarmizegetusa Regia: Capitala Daciei Preromane. Acta Musei Devensis, 133-135.

Herodotus. 1991. Histories, Book IV, 94-96. Munich, DT-Verlag, 352-353.

Jordanes, 1972. The origin and deeds of the Goths. Köln, Insel Verlag, Chap. XI, 41-43.

Kerek, F.1994. Pythagorean Topography of the Dacian Sacred Capital. In W. Schlosser (ed.), Proceedings of the 2nd SEAC (European Society for Astronomy in Culture) Conference Bochum, 57-68.

Kerek, F. 2007. The Dacian capital Sarmizegetusa Regia was sited according to precise astronomic alignments and Pythagorean doctrines. British Archaeological Reports International Series 1647, 35-41.

Ruggles, C. L. N. 2005. Ancient Astronomy: an encyclopedia of cosmologies and myth. Santa Barbara CA, ABC Clio, 370.

Serban, M. 2009. Trajan's Bridge over the Danube. The International Journal of Nautical Archaeology 38, no. 2, 331-342.

Stanescu, F. 2001. Establishing astronomical directions in the Dacian sanctuaries. Studii de Istorie Antica, 325-333.

Wollmann, V. 1982. Johann Michael Ackner, Leben und Werk. Cluj-Klausenburg, Dacia Verlag. 258-261.

ORIENTATION IN THE LANDSCAPE OF OPEN AIR ROCK ART IN THE MOUNTAINS BETWEEN THE ALVA AND CEIRA RIVERS: THE PODOMORPH CARVINGS

FERNANDO PIMENTA, NUNO RIBEIRO, ANDREW SMITH AND LUÍS TIRAPICOS

Abstract: In this paper we analyze the orientation in the landscape of 50 open-air rock art sites with 459 carvings representing segmented oblong circles and podomorphs, or carved feet, located in the mountains between the Alva and Ceira rivers, in central northern Portugal. In our interpretation we propose a possible association with a fertility ritual taking place around the full Moon nearest to the Summer Solstice.
Keywords: lunar alignments, Summer Full Moon, fertility rituals, rock art, Portugal

Introduction

The region under analysis is situated between the rivers Ceira and Alva, which run parallel to each other, and whose sources are situated in 'Serra da Estrela' ('Star Mountain'), the second highest mountain in Portugal. In this hydrographical basin, which is rich in gold, silver, tin, lead and zinc deposits, the Portuguese Association of Archaeological Research (APIA) found more than 700 engraved slabs and two painted stone slabs, clustered in 11 areas within a 20 by 20km region. They were situated along two sets of mountain peaks, one with a north-south orientation, and the other, crossing the former, with an east-west orientation in a coastal to inland direction (Lousã Mountains / Açor Mountains). These mountain ridges were used as natural pathways for thousands of years: for communication, trade, mining and transhumance (the seasonal movement of shepherds and their flocks, sometimes journeying from more than 200km away). Some of these routes survived until the last century under the names as 'Via do Sal' (the Salt Route) or 'Estrada Real' (the Royal Road) (Ribeiro et al., 2010b).

Figure 1. Location of 719 open air rock art sites (ruler in meters).

Most of the carved stone slabs appear in large clusters, concentrated on the mountain ridges along these routes, located in passages that, in some cases, could only have been used during the spring or summer months. Those routes can be connected to transhumance pathways going towards the Tagus and the Upper Alentejo region (Ribeiro et al., 2010b).

Figure 2. Pereiro 9, one engraved slab from a cluster of 54 sites located on one mountain ridge.

Another class of carved stones, in the same region, appear in isolated sites in V-shaped valleys, dominating a river, and contain large numbers of petroglyphs.

Figure 3. Entreáguas, a richly carved site located in a V-shaped valley, facing a river.

Some of the sites discovered appear associated with lithic industry artefacts or remains of mining activities. This territory was probably used in the Palaeolithic as observation and hunting grounds. The very poor soils, composed mainly of shale and gravelly land, did not provide the necessary conditions, required in early agriculture, to support large sedentary communities so pastoralism, hunting and transhumance were probably the only options. Seasonal occupancy of the mountain tops for mining activities probably started in the Final Bronze Age, together with increasing trading activities (Ribeiro et al., 2009).

The region's landscape, with mountains that hinder passage, helps separate regions and isolate the few communities that lie in the valleys. The rock art probably played an important role in the delimitation of cultural boundaries between the resident population and the itinerant shepherds, miners and traders who seasonally shared the area, either as marks of passage or territory possession, some of them probably associated with rituals of propitiation or power (Ribeiro, 2006).

The archaeological survey

Some petroglyphs can be dated from the Palaeolithic, through the Bronze and Iron Ages, until modern times. Several thousands of individual motifs were identified: groups of incised points and cup marks, crosses, podomorphs, segmented oblong circles, circles and spirals, anthropomorphs, zoomorphs and tree forms.

Figure 4. Different carved motifs (one cross, four podomorphs, one segmented oblong circle with a cup mark, one circle, one spiral, three anthropomorphs, two zoomorphs).

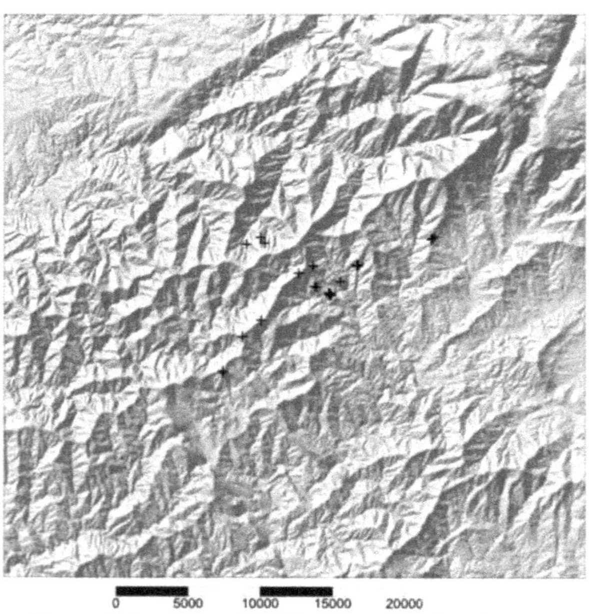

Figure 5. Location of the 50 sites analyzed in this paper

After terrain preparation, the carvings were copied onto plastic sheets, using different colours to represent different layers. The tridimensional slab form was measured with a level theodolite and the readings, together with the magnetic north indication, were printed on the same plastic. Survey date and GPS location were recorded for each site. Franson CoordTrans software was used later to convert the grid data. After washing, the sheets were cut and digitized in a large scanner and finally treated in Adobe Photoshop for reconstruction and colouring.

Methodology used in the archaeoastronomy project

From the various petroglyphs identified, we decided to select the segmented oblong circles and podomorphs since they provided a symmetry axis and a 'walking' direction as well as being common to most of the surveyed areas. The selection based on the predominance of the podomorphs in the different areas resulted in 50 sites from seven areas, with a total of 459 podomorphs, with a broad range of possible datings.

Figure 6. Location of the seven areas in which the 50 sites are clustered (ruler in meters).

The methodology used for the archaeoastronomical work was divided in three main classes:

1) ORIENTATION OF THE SITE IN THE LANDSCAPE

A 25m grid Digital Terrain Model and dedicated GIS software developed by one of the authors was used to provide information, independent from vegetation or cloud obstruction, for each site: a) maximum slope; b) azimuth of maximum slope (aspect); c) distance to horizon profile; d) horizon elevation profile; e) azimuth direction of the highest distant peak; and f) a feature detection algorithm was devised in an attempt to search, in a consistent way, for possible common horizon marks in the above-calculated profiles.

Using the elevation and distance profiles, this algorithm detects local maximums and minimums as well as plane differences in a 2° azimuth sliding window using numerical differentiation, filtered over an 8° azimuth sliding window, in 0.01° azimuth steps. The declination of each possible horizon mark is also recorded.

2) STONE SLAB ORIENTATION

Each plan was rotated to geographical north, using IrfanView and the magnetic declination data correction taken from GeoMag software. Calibration was performed by Bergsoft Image Measurement software. A tridimensional irregular grid was then produced by Didger software from the calibrated surveyed slab data. Surfer software was used to perform the gridding process by kriging, to produce a 3D slab model and to calculate the slope and aspect in each node of the grid slab. We used the median of slope and aspect as result, in order to have robust statistics against outliers.

3) PODOMORPH ORIENTATION

We used Bergsoft Image Measurement software again to perform podomorph angle measurement over the rotated plans. Starting from the process of carving a podomorph until the final measured angles, it is expected there will be a large uncertainty which we estimate to be about 5-10° in azimuth.

Results and analysis

The results for the orientation of the site in the landscape are presented in the diagrams from Figures 7 to 10.

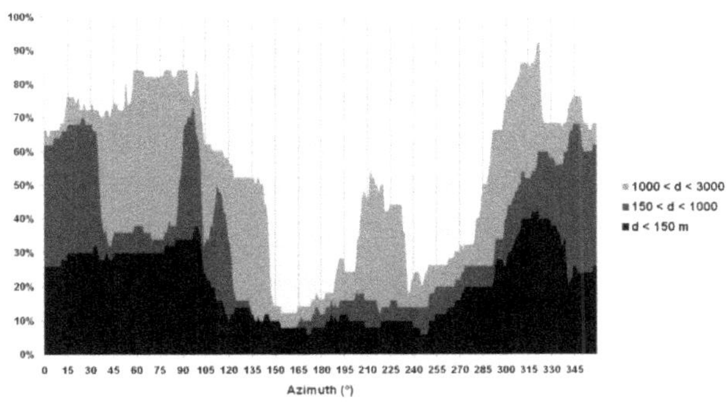

Figure 7. Distance to horizon distribution.

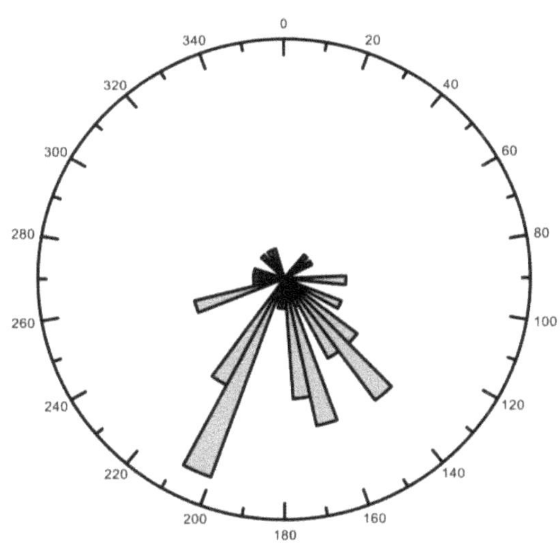

Figure 8. Site azimuth of maximum slope.

From Figure 7 we can conclude that there is a preference for an open and distant horizon mainly to the southeast; from Figure 8 we see that the sites are facing southwest-southeast; Figure 9 reveals that for 50% of the sites there is a possible horizon mark at 24° of declination that can be associated with the setting Sun around the summer solstice; and finally, there are two main directions for the distant highest peak visible from the sites at the southwest and northwest.

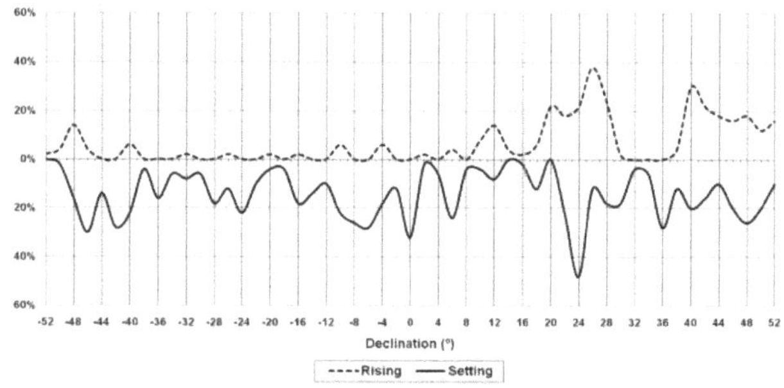

Figure 9. Distribution in declination of possible horizon marks (in 2° bins).

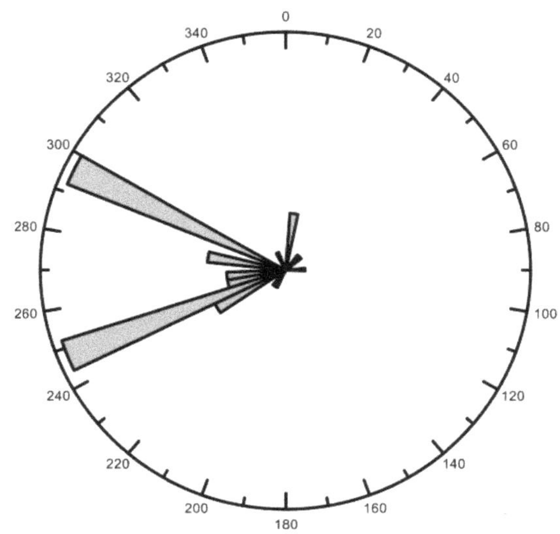

Figure 10. Distribution for the azimuth of distant highest peak.

The results for the stone slab orientations, presented in Figure 11, reveal that most of the slabs are facing south, in a similar way to the sites.

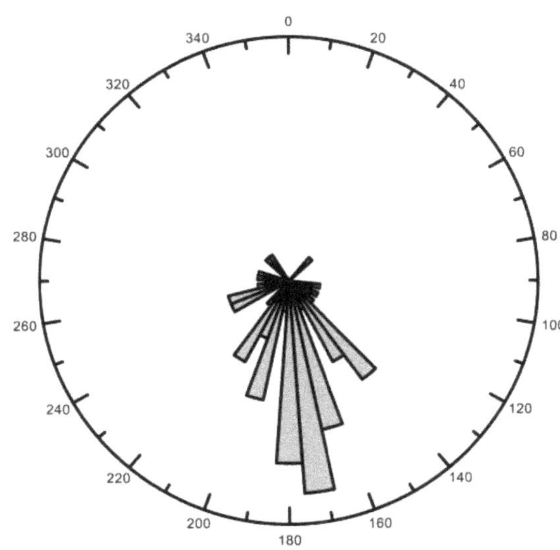

Figure 11. Stone slab azimuth of maximum slope.

Finally, the results for the podomorph orientations are presented in Figure 12 for the azimuths and in Figure 13 for the declinations.

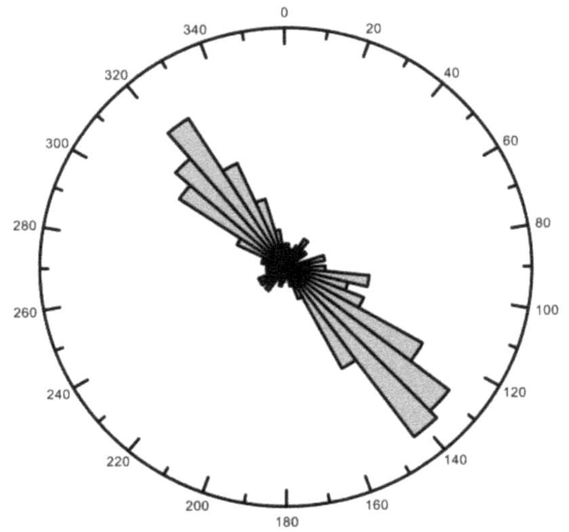

Figure 12. Podomorph azimuth distribution.

Discussion

The distribution of podomorph azimuths in Figure 12 shows a very clear pattern. It is not possible to explain this northwest-southeast pattern by topographic motivations such as the direction to a highest distant peak, direction of the mountain ridge, direction of pathways or inter-visibility between sites.

The consistent horizon mark shown in Figure 9 reveals a possible association with the setting Sun around the summer solstice, but those marks correspond to azimuths between 294° and 300°, which are further south than the northwest podomorph azimuth peak shown in Figure 12. This peak, and the respective positive declination, corresponds to azimuths that are blocked by a near horizon.

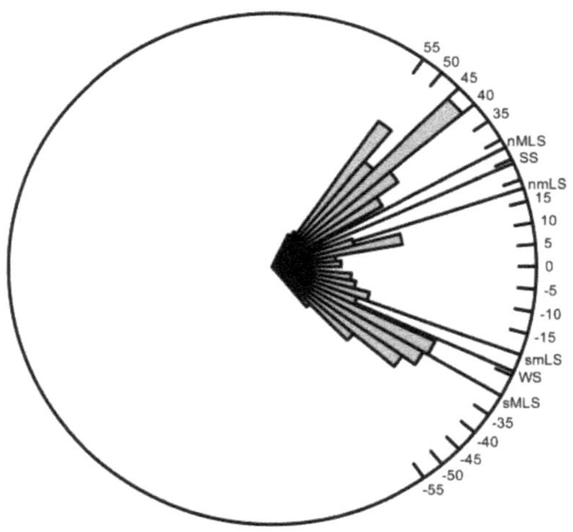

Figure 13. Podomorph declination distribution with indication of the solstices and major and minor lunar standstills.

The south-facing preference shown by the sites and the stone slabs as well as a distant horizon to southeast seem to give strength to the southeast podomorph azimuth peak. The corresponding distribution of negative declinations, shown in Figure 13, given the estimated uncertainties, is very close to the southernmost possible lunar declinations.

There was a seasonal occupancy of the mountaintop areas, where the sites are located, in spring and summer. Transhumance, for example, started around the summer solstice and ended in August or September. Our hypothesis is that those sites were associated with an annual interest in the full Moon rising around the summer solstice, when the full Moon reaches its southernmost declinations. In this case the Sun would set near the common horizon mark shown in Figure 9, roughly in the northwest direction shown in Figure 10 and nearly opposite to the rising Full Moon in the southeast. In several of the studied sites the Sun would set over a river. Is there a special meaning for an observer to be at the centre of a line connecting the summer setting Sun over a river, with the full Moon rising over a hilltop?

If our hypothesis is correct, then it would be expected that the declinations would be distributed between about -20° and -30°, but there is apparently a 5° declination bias to the south in our data. Does this reveal systematic errors in our data? Does this reveal an interest in another event? Can it be the major lunar standstill, occurring each 18.6 years, when, at several sites, the full Moon could be seen 'rolling' over a hilltop, having only the upper limb visible or even showing just a bright glow from behind the mountain? A reassessment of both the sites and our methodology is necessary to reduce uncertainties. A selection based on carvings with similar datings and styles will probably provide additional information.

A possible interest in the summer full Moon was also proposed by the authors for the rich engraved stone slabs that appear isolated in the region, facing a river, in V-shaped valleys (Pimenta et al., 2007; 2009).

Can we find any ethnographic records that could support this hypothesis?

Figure 14. Survey of 'Eira do Piódão' by APIA.

APIA recently surveyed a threshing floor in this region, near Piódão village (Figure 14), which contains over 500 carvings. A threshing floor is a flat surface with a hard floor, where cereals were sieved to separate the grains. Its origin is linked to the advent of agriculture and the development of techniques, tools and specific spaces. A threshing floor also fulfilled a social function in the life of the local population since it provided a place in which certain public ceremonies, such as balls or masses, could take place. In this site we identified 75 podomorphs (included in the present study). Together with incised points and cup marks, the podomorphs are not only one of the three largest groups of motifs present but also some of the oldest carvings made (Ribeiro et al., 2010a). Can we relate the important presence of podomorphs on a threshing floor with a fertility ritual? Are there other ethnographic sources supporting this assumption?

The Moon appears associated to the Mother Goddess in Portuguese popular religions, later integrated into the devotion to St. Mary, and in the organization of Portuguese agricultural calendars. Popular prayers to the Moon were recited facing it, as reported by Espírito Santo (Espírito-Santo, 1984) and also mentioned by Strabo (Strabo, 1923). Several chapels, built on the top of hills, celebrate the 'Senhora do Monte' ('Lady of the Hill'), sometimes represented standing over a lunar crescent. Generally associated with a fecundity ritual and the idea of birth and regeneration, they have in common the myth of a bright appearance over a hill: statues that were worshiped on the altars, after having been buried in caves to protect them from enemies (the Visigoths, Romans, Moors, etc.), return glittering to the surface, normally on the top of a hill. These myths associate carnal dream images of the Mother Goddess with the Earth as a provider and the expression of conflicts between shepherds and farmers. (Espírito-Santo, 1984).

Some examples from these chapels in the region: the hilltop 'Senhora do Mont'Alto' chapel in Arganil (Assumption of Mary feast, 15 August; in the past the return from transhumance feast in Arganil was celebrated together), the sanctuary of 'Senhora das Preces' between Aldeia das Dez and Piódão (feast in the first Sunday of July), and the nearby chapel of 'Senhora das Necessidades' on Colcurinho hilltop (feast on Pentecost Sunday). An ethnographic record of the region (Nobre, 2006) mentions a rite still practiced at the beginning of the last century, probably associated with a fertility ritual, which took place on the night of the summer solstice: young single people would spend the night in the hill until sunrise, after climbing by a trail 'marked by the feet of many pilgrims':

> (...) in that already distant past, young single people drunk water from the fountain of love, on the night of St. John, before climbing to Colcurinho Chapel, so from up high they would see the rising sun. They climbed on foot by a trail between the woods, singing and sometimes dancing (...). This trail is still marked by the feet of many pilgrims, blessed by the fountain (...). I, too, took the same path to Colcurinho, where in the wall of the chapel I wrote love poems, asking for blessings to Our Lady of Needs.

Conclusions

From the archaeoastronomical results and the ethnographic sources, there is some evidence relating the rock art sites with podomorphs to a fertility ritual, possibly facing the rising full Moon, while the Sun sets near the summer solstice. An additional interest in the major lunar standstill may have been integrated.

Nevertheless a reassessment of the sites must be done to reduce systematic errors and give a better estimation for the uncertainties in the podomorph orientation azimuths. This will also provide an opportunity for further investigation directions, such as the relation between the orientation of the carved stone slabs' natural fractures and those of the other nearby stones that were not carved.

We hope that the development of this work will help answer the following questions: How were the sites selected? Why were some rocks chosen to be decorated but not other nearby slabs? What motivations could have triggered the carvings? When might whatever rituals that took place in those sites have been performed?

Acknowledgments:

The authors are grateful to the Instituto Geográfico Português (www.igeo.pt) for the 25m grid DTM data provided. The authors would like also to thank Fábio Silva for his comments.

References

Espírio-Santo, M. 1984. A Religião Popular Portuguesa. A Regra do Jogo. Lisboa, Edições, Lda.

Nobre, C. G. A. 2006. Vide Memorial – Camélias Brancas. Volume I – Edição do autor, 148-149. Coimbra, Impressão Ediliber, Lda.

Pimenta, F., Ribeiro, N., and Tirapicos, L. 2007. Lunar and solar connections at a rock art site in Central Portugal. In Lights and Shadows in Cultural Astronomy, Proceedings of SEAC 2005, Isili, Sardinia 28 June to 3 July, 264-271.

Pimenta, F., Ribeiro, N., Smith, A., and Tirapicos, L. 2009. The Sky and the Landscape of Rock Art in the Ceira and Alva Basins. In Cosmology Across Cultures, ASP Conference Series, Vol. 409, 359-363.

Ribeiro, N. 2006. Open air Rock in the Ceira and Alva River Valleys: Some Symbols. In Proceedings of the XV World Congress, Session WS34 (Lisbon, 4-9 September 2006), 43-49. BAR Internacional Series 1793.

Ribeiro, N., Joaquinito, A., and Pereira, S. 2009. Zoomorphic art in the open air rock art complex of the Ceira and Alva rivers basins and adjacent Unhais river basin– Portugal. IFRAO Congress 2009, GLOBAL Rock ART.

Ribeiro, N., Pereira, A. S., Pimenta F., Joaquinito, A., and Ventura R. 2010a. O Sítio de Arte Rupestre da Eira do Piodão: Um caso de estudo. Arganil – Portugal. Conference Rotas naturais do Centro Interior de Portugal da Pré-História ao Século XX 1° CIAEE, Congresso Iberoamericano de Arqueologia, Etnologia e Etno-história Dourados, Mato Grosso do Sul, Brasil.

Ribeiro, N., Joaquinito, A., and Pereira, S. 2010b. The Symbolism of Open-Air Rock Art at the end of the Upper Palaeolithic in Central Interior Portugal and its Possible Relation With Natural Paths. Pleistocene Art of the World. IFRAO Congress September 2010 – Symposium: Signs, symbols, myth, ideology.

Strabo. 1923. The Geography of Strabo. In Vol. II of the Loeb Classical Library edition, book III, Chapter IV:16, in the public domain.

TOTAL SOLAR ECLIPSES CLOSE TO THE PLEIADES ON THE NEBRA DISK AND SWEDISH ROCK-CARVINGS

GÖRAN HENRIKSSON

Abstract: The Pleiades star cluster is a remarkable object in the night sky. It consists of 6 to 11 relatively faint stars depending on your eyes and the darkness of the sky. A total solar eclipse close to the Pleiades is a very rare event. The Nebra bronze disk is a unique object dated by archaeologists to about 1600 BC from the style of the two swords found together with the disk. The main motifs are a circular disk and a crescent of the same size, identified by most investigators as the Sun and the Moon respectively, and a group of seven round dots identified as the Pleiades. However, the Nebra disk was found by amateurs under non-scientific circumstances and its authenticity has been questioned. It was therefore necessary to use advanced technical methods to prove that it is an original object from the Bronze Age.

In the present paper the crescent is identified as the partial phase of a total solar eclipse 15 minutes after totality. There exists only one total solar eclipse in Nebra—around 1600 BC—and that was the total solar eclipse on 16 April 1505 BC. As a bonus this eclipse took place to the left of the Pleiades, in the same relative position as on the disk, and the Pleiades may have been visible during the total phase. The unique motif on the Nebra disk should in my opinion correspond to an extraordinary experience for the people who made it.

Earlier, the author has identified two total solar eclipses close to the Pleiades that are depicted on the Swedish rock-carvings at Klinta on the island of Öland, on 26 March in 1169 BC, and at Oppeby in Nyköping, at Flyhov in the province of Västergötland and at Hjulatorp in the province of Småland, on 28 March in 1411 BC. The Pleiades are depicted in a similar way as on the Nebra disk. This is not surprising as there were important cultural connections between Germany and southern Sweden during the Bronze Age.
Key words: Nebra disk, Rock-carvings, Pleiades, total solar eclipse, Bronze Age

Introduction

In the Sumerian collection of omen texts, the *Enuma Anu Enlil*, different kinds of ominous risings and settings of the Sun are mentioned, for instance with clouds to the right or left of the Sun, but also that the Sun was surrounded by 'the Stars' (Pleiades). W. van Soldt (1995) has translated all the Tablets that mention *solar omina* and on Tablet 26(27) page 70 we can read: '[If the sun rises and ...] is surrounded by *the Stars*: there will be misfortune in that country'. The only possibility to see stars surrounding the Sun is during a total solar eclipse. But total solar eclipses at a specific site on the earth are very rare events and a total solar eclipse close to a specific group of stars in the sky is an extremely rare event. In fact there was only one total solar eclipse close to the Pleiades, visible in the ancient Babylon, and it took place on 10 March of the Gregorian calendar in 2471 BC, about two hours after sunrise.

During my investigation of the motifs on the Swedish rock-carvings in the 1990s, I had recognized the Pleiades depicted as a narrow collection of cup-marks in a correct relation to the Sun during the total solar eclipse in 1169 BC, visible only on the southern half of the island of Öland and in a narrow zone along the south-eastern coast of the province of Skåne, and in 1411 BC within a narrow zone crossing the middle of the southern part of Sweden, Götaland.

The Pleiades was also one of the main motifs identified in the first article I read about the Nebra disk, written by Christian Weber and published in the German magazine *Focus* (2002). The article was recommended to me by a professor in archaeology in Uppsala because, in his opinion, it proved that people were interested in the sky during the Bronze Age. At that time it was called the bronze disk from Sangerhausen. When I saw the picture of the bronze disk for the first time I independently recognized the Pleiades without reading the caption below.

The discovery of the Nebra disk

Unfortunately, the bronze disk and some other items found together with it, such as two swords, two axes, a bronze chisel and two silver spiral armbands, were not discovered during an excavation by archaeologists, but by treasure hunters using metal detectors. The discovery was made on 4 July in 1999 by two men who sold all the treasure to a collector who quite soon sold it to another collector; the real location was only known by the two men who found it. Later in 1999 all the treasure was offered for sale to the Museum of Prehistory and Early History in Berlin for a half million dollars. When it became clear that it had been found in Sachsen-Anhalt and, according to the German law, belonged to that state, the museum in Berlin could not buy it.

After that archaeologists did not see the bronze disk until it finally resurfaced from the black market in January 2002, when its latest owners persuaded the German news magazine *Focus* to publish an article about it, in order to increase its value. In this article, written by Christian Weber, the original finding place was still unknown and it was called the Bronze disk from Sangerhausen. The Director of the Landesmuseum in Halle, Dr Harald Meller, cooperated successfully with the police and both the bronze disk and the dealers were taken by the police and the remaining items of the treasure were found at the home of one of the dealers. It soon became clear that the style of the swords dated them to roughly 1600 BC. This means that the disk, looted from the same site, was probably of similar age (Meller, 2004b).

The finding place of the bronze disk

The police were later able to trace the previous owners of the disk and one of them agreed to cooperate and showed Meller the looted site on Mittelberg, a large hill near the town of Nebra, in Sachsen-Anhalt. The original looters had discovered the artefacts in an underground small stone chest covered by a stone mound near the top of the hill. A circular wall, c. 1m high and nearly 70m in diameter, surrounded the stone mound. Such structures are found throughout Europe and may have marked prehistoric ceremonial or holy places known as ringed-ditch enclosures. Two similar external walls warned ancient wanderers that they were treading on holy ground.

Today the Mittelberg hill is forested, but during the Bronze Age it would most likely have been logged bare and the ringed-ditch may have enclosed an early observatory according to Meller. The bronze disk of Nebra was stolen from a site known for its archaeological significance. Most stolen artefacts vanish in the black market, but in this case the police and forensic scientists managed to recover the bronze disk and later pinpoint the spot where it originally had been buried. Samples of 113mg of soil found on the disk and 217mg of soil found on one of the swords matched perfectly with samples of soil taken from the looted site on Mittelberg. An archaeological excavation at the bottom of the stone chest, led by Dr Meller, showed that the disk had been placed in a vertical position. No traces of human bones were found. This was a typical sacrifice of valuable artefacts to please the gods in the centre of a holy area surrounded by a ringed-ditch enclosure (Meller, 2004a).

For the trial it was absolutely necessary to identify the site of the crime. The two looters received a four-months and a ten-months sentence, respectively, by a Naumberg court in September 2003. An appeal court raised these to six and twelve months, respectively.

Origin of the metals

An initial analysis of trace elements in the metals by x-ray fluorescence performed by Ernst Pernicka (2004) showed that the copper originated at Bischofshofen in Austria, while the gold was thought to come from the Carpathian Mountains. However, a recent comparison with trace elements in gold from 300 sites all over Europe showed that the gold was collected from the river Carnon in Cornwall, according to a research team with Gregor Borg and Anja Ehser from the University of Halle and Ernst Pernicka from the University of Tübingen (2010). The earliest finds of tin-bronze date to about 2200 BC in central Europe. There is evidence for prehistoric copper mining in, for example, the Alps or mainland Greece, but the provenance of most of the contemporary tin is still an unsolved problem. The tin in the bronze from the Nebra treasure was analysed regarding its tin isotope composition and it was found that the tin-isotope ratio of the Nebra disk fits well with the bulk of investigated tin ores from Cornwall (Haustein, Gillis and Pernicka, 2010).

The date of the disk

It was not possible to get a direct dating of the disk but from its content of lead and the radioactive decay of the isotope ^{210}Pb the scientists got a lower limit of 100 years for the age of the bronze in the disk and it proves that the Nebra disk is not a modern forgery. However, it is most likely that the disk and the other items of the treasure were buried in the stone mound at the same time. The copper in the disk came from the same mine as the other objects and the two magnificent swords can be dated for stylistic reasons to about 1600 BC from a comparison with the well known Hungarian swords from 1650-1500 BC, and the metalwork is very similar to the technique used in Denmark about 1600 BC. It was also possible to make a technical dating of one of the swords by the ^{14}C-method because a small piece of bark from a birch was found in the handle. A sample of 0.6 milligrams of carbon was used and the bark could be dated to somewhat more than 1600 BC (Meller, 2004a).

Summary of facts about the Nebra disk

The Nebra disk was discovered by treasure hunters during the summer of 1999 on a hilltop called Mittelberg near the small village of Nebra in Sachsen-Anhalt. It was found together with other more ordinary items from the Bronze Age in a small stone chest, below the ground surface and covered by a stone mound, at the centre of an earlier unknown cult-place surrounded by a ringed-ditch enclosure originally with a palisade.

It is made of bronze and the diameter of the disk is about 32 cm and its weight is about 2.3kg. The thickness is 4.5mm at the centre and about 1.8mm at the edge, Figure 1. It was found together with two magnificent swords made in Hungary about 1600 BC, two axes, two silver bracelets, two finger rings and one bronze chisel.

The Nebra disk was first considered to be a modern forgery, but a careful technical investigation has now proved that it is an original object made about 1600 BC.

The Sun and the Moon on the Nebra disk?

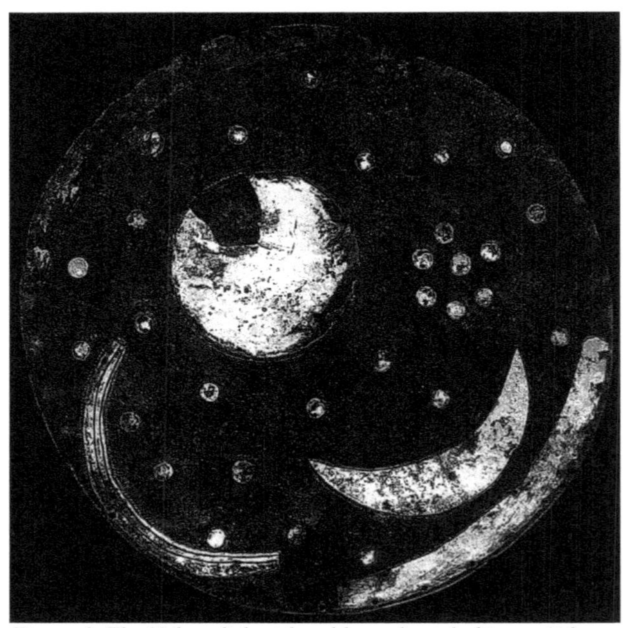

Figure 1. The Nebra disk with golden celestial objects. (After a photo in Focus nr 9, 25 February 2002.)

In the paper by Christian Weber (2002) the round disk was identified as the Sun and the crescent as the Moon. I have also identified the round disk as the Sun, but the crescent cannot be the Moon because it should then be oriented in the opposite direction if it got its light from the Sun. My conclusion is instead that the crescent represents a late phase of a total or almost total solar eclipse.

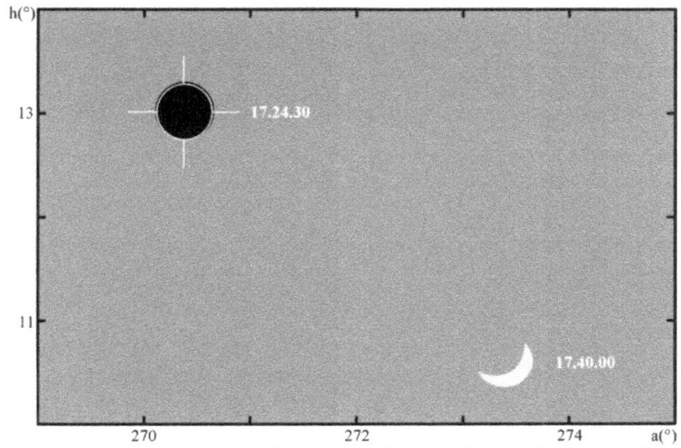

Figure 2. The position of the Sun during the total phase of the solar eclipse on 16 April 1505 BC, at 17.24.30 mean solar time in Nebra, and at the partial phase 15 minutes and 30 seconds later.

The same evening as I read the paper about the bronze disk from Sangerhausen/Nebra, all powerful solar eclipses during the Bronze Age in Sangerhausen were calculated and the only possible solution was the total solar eclipse on 16 April 1505 BC. This date fits well with the archaeological dating of the two extraordinary large swords, dated to about 1600 BC, and other objects that were found together with the bronze disk. There were no alternative total or almost total solar eclipses in the neighbourhood of Nebra during the nearest hundreds of years and the date fitted very well with the Sun's position close to the Pleiades on that date, see Figures 2, 3, 4 and 5.

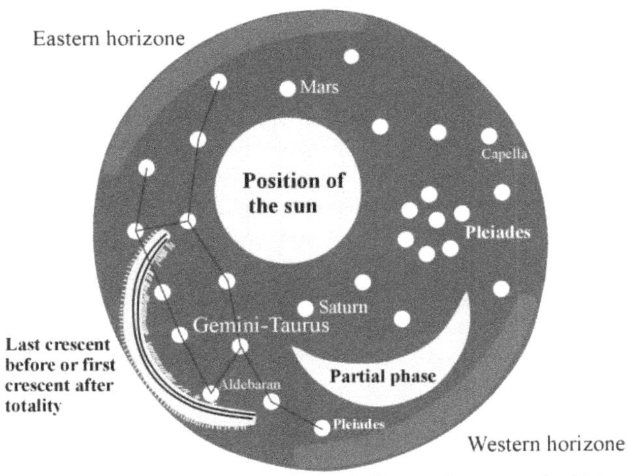

Figure 3. An attempt to identify the different objects on the Nebra disk. The central motif is a total solar eclipse on 16 April in 1505 BC, close to the Pleiades visible at low altitude in the west. The 'ship' is in fact the narrow faint crescent just before and after the total phase.

My first interpretation of the Nebra disk as a depiction of the total solar eclipse in 1505 BC was presented on 6 March 2002 at a lecture in Uppsala. I started to write a paper, but when I read on the Internet that the disk probably was a forgery I decided to wait until the true status of the disk was clarified.

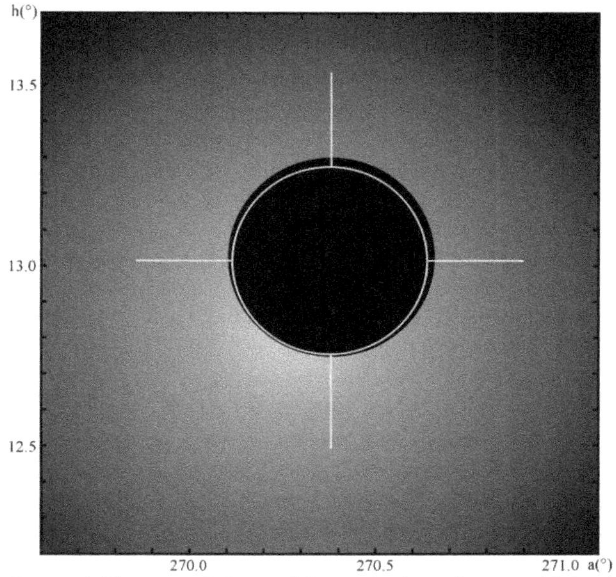

Figure 4. The central phase of the total solar eclipse on 16 April in 1505 BC, at 17.24.30, local mean solar time in Nebra. The magnitude of the eclipse was 1.005.

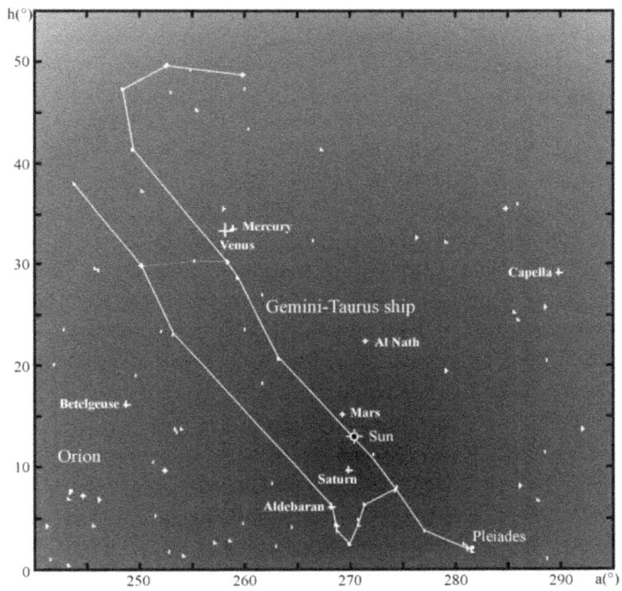

Figure 5. The total solar eclipse on 16/4, 1505 BC, at Nebra. The ship is the Gemini-Taurus ship known from the Swedish rock-carvings. Compare with Figure 3.

'The ship'

Many authors believe that the crescent below the Sun disk is a ship. The shape of the 'ship' has been compared with the ancient Egyptian ships and it has therefore been identified as the Egyptian Sun-ship that carried the Sun across the sky during its daily motion. Some keen-eyed scientists have discovered small rays that radiate from the rail and the keel of the 'ship' and their explanation is that the ship was made of animal skins with some of the fur still left!

At first I found it reasonable to interpret the crescent-like figure below the Sun as one of the six calendar-ships along the ecliptic, in this case the Gemini-Taurus ship that I have identified on the Swedish rock-carvings

Henriksson (1999), Figure 5. This ship may instead have been depicted by the 'stars' to the left in Figure 4. However, the most likely interpretation of the thin crescent with rays is as the faint crescent visible just before or after totality. It is depicted correctly at the lower left side of the Sun, marginally eclipsed by the Moon, Figure 4.

The 'Stars' on the Bronze disk

There exist at least 29 small golden round disks placed in the space between the larger figures. These disks have been interpreted as 'stars' and especially the narrow group of six 'stars' surrounding a seventh 'star' in the middle has been identified as the Pleiades. There have been many attempts to combine the other 'field stars' to make up known constellations, but it has been difficult to convince other researchers that your particular solution is the correct one. It has also been pointed out that the 'stars' may have been distributed in some kind of geometrical pattern.

The first investigator of the disk, Harald Meller (2004b), writes: 'The circular group of seven gold stars on the sky disk is almost certainly a representation of the Pleiades constellation. In early March its disappearance in central Germany signals the beginning of the planting season.' He also discusses the significance of two other main objects on the disk:

> One is shaped like a crescent moon; the other is round, what I first thought to be the sun but now think is the darkened moon of a lunar eclipse. We know that once every decade the Pleiades appears next to the crescent moon, as seems to be depicted on the disk. A lunar eclipse follows seven days later. If Bronze Age astronomers used the disk to predict eclipses, it shows that these so-called barbarians had been studying the heavens for generations.

The 'golden horizons'

Most investigators, including myself, have interpreted the two opposite golden parts along the edge of the disk as the eastern and western horizons.

The German astronomer Wolfhard Schlosser (2002) has measured the angle that the 'golden horizons' make along the periphery and he concluded that it corresponds to the solar arc at Nebra, the angle within which the Sun rises in the east and sets in the west during the year.

This conclusion seems reasonable, but I do not think that the disk has been used as a calendar instrument because it is too small to give accurate directions if it had been used for real measurements. It seems more likely that it had a symbolic function.

Total solar eclipse at Klinta on Öland in 1169 BC

A drawing of this eclipse would naturally indicate the position of the Sun relative to the Pleiades, Figure 6. This eclipse was total in Sweden only on the islands of Öland and Gotland, in the Baltic Sea, and within a small area of south-eastern Scania.

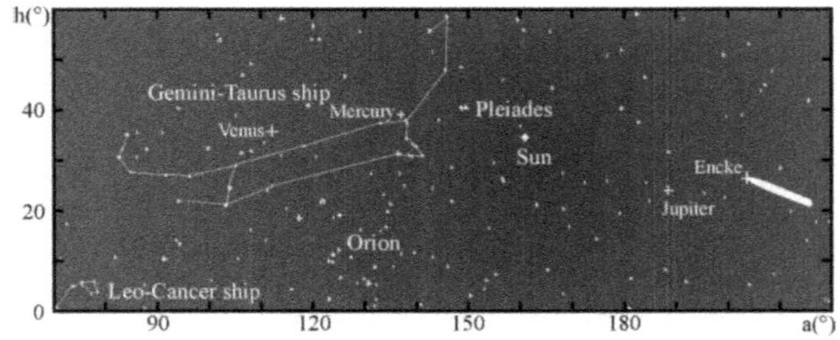

Figure 6. The sky during the total solar eclipse on 26 March 1169 BC from Klinta on the island of Öland.

When I saw, in the dissertation by Göran Burenhult (1973), that the only rock-carving on Öland, the one at Klinta, showed a cluster of cup marks beside a group of concentric circles above a typical ship of the Gemini-Taurus type, it was obvious to identify this rock-carving as a representation of the total eclipse of 1169 BC. However, the published picture is the mirror image of the situation in the sky. This caused me to suspect that Burenhult's picture had been reversed in the process of publication. Upon checking on the real stone at the Historical Museum in Stockholm, my prediction was confirmed, see Figure 7!

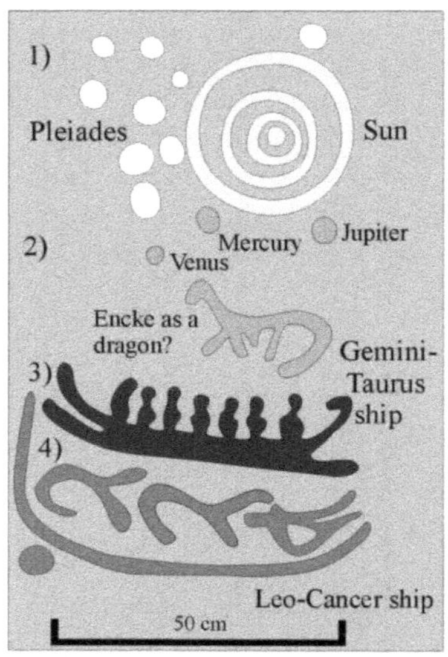

The rock-carving on a standing stone in the cemetery at Klinta, Smedby, Öland. Since the stone is narrow, the scenes had to be placed in four horizontal registers.
From the uppermost down:
1) Sun to the right of the Pleiades
2) Venus, Mercury, Jupiter and the comet Encke (?)
3) Gemini-Taurus ship
4) Leo-Cancer ship
Based on Burenhult 1973, but with the correct orientation.

Figure 7. Rock-carving on a standing stone at Klinta.

Almost total solar eclipse below the Pleiades in 1411 BC, at Flyhov, Oppeby and Hjulatorp

During the powerful solar eclipse at Flyhov on 28 March 1411 BC, the position of the Sun was below the Pleiades. The maximum eclipse took place at 16.28, local mean solar time, when 0.982 of the Sun's diameter was eclipsed. The western sky darkened, but no stars or planets became visible. The people who made the depiction must have known the Sun's position in relation to the Pleiades this day from other people's observations, within the zone of totality, or from some rule for the Sun's motion during the year. If they had seen the Pleiades during the totality they had probably also depicted the planets Mercury, Venus and Mars, see Figures 8, 9, 10 and 11.

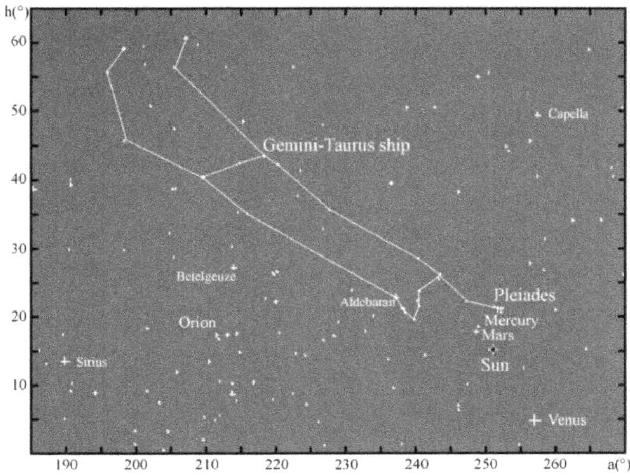

Figure 8. Powerful solar eclipse below the Pleiades on 28 March in 1411 BC at Flyhov in Västergötland.

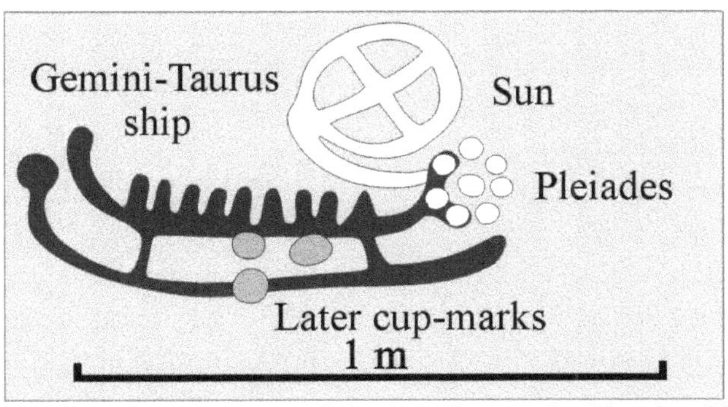

Figure 9. Partial eclipse of the Sun close to the Pleiades, 28 March 1411 BC, on a rock carving at Flyhov, in the parish of Husaby in Västergötland. (After Burenhult, 1973)

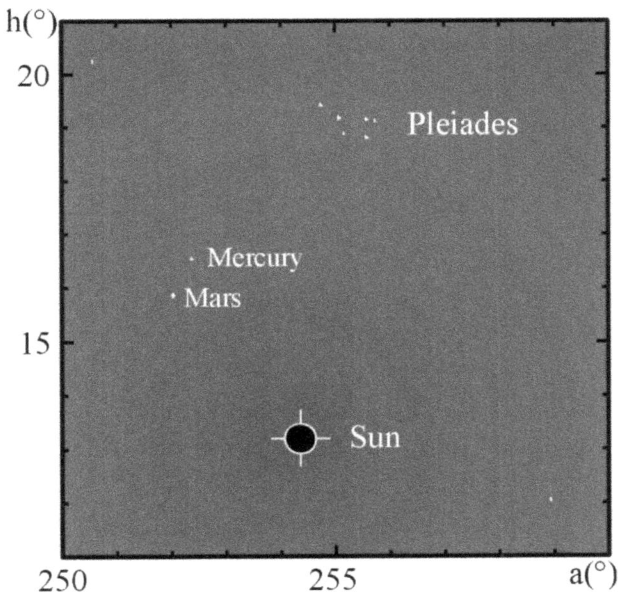

Figure 10. Partial solar eclipse below the Pleiades, m=0.980, 28 March 1411 BC, at 16.43 local mean solar time at Oppeby, Nyköping (m<5.0).

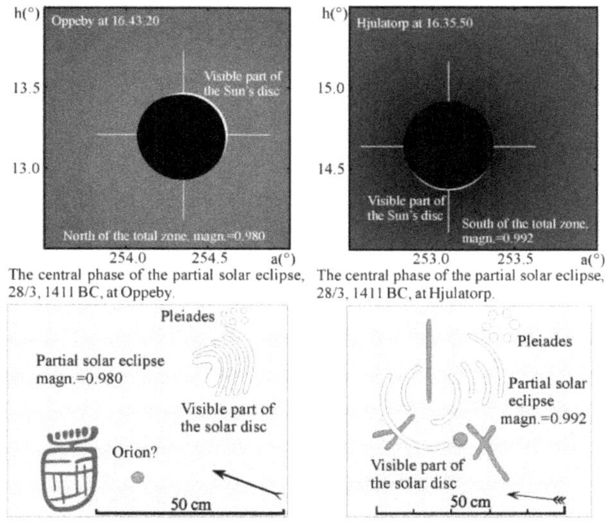

The central phase of the partial solar eclipse, 28/3, 1411 BC, at Oppeby.

The central phase of the partial solar eclipse, 28/3, 1411 BC, at Hjulatorp.

The partial solar eclipse, 28/3, 1411 BC, on a rock-carving at Oppeby, in Nyköping. (After a photo by G. Henriksson 1986.)

The partial solar eclipse, 28/3, 1411 BC, on a rock-carving at Hjulatorp, in the parish of Berg in Småland. (After Kjellmark and Lidsten 1909.)

Figure 11. The almost total solar eclipse below the Pleiades on 28 March in 1411 BC, depicted at Oppeby in Nyköping, left, and at Hjulatorp, in the parish of Berg in Småland, right.

Discussion

Emilia Pásztor and Curt Roslund (2007) consider the motifs on the Nebra disk as individual symbols that can be compared with art and objects from the Bronze Age, but also much later objects used by shamans from Asia and North America. The round disk and the crescent were compared with a picture made by an Inuit artist from Cape Dorset in Canada in 1993, as an example of pictures of the Sun and the Moon by native people. However, this is not the only possible interpretation because there was a partial solar eclipse at this place on 21 May in 1993! They write at the end of the abstract: 'Here the authors offer a subtle interpretation that sees it as the shamanistic device of a local warrior society.' If the Nebra disk had just been an ordinary shamanistic device we should expect to find many similar disks. But the Nebra disk is so far unique and in my opinion it has been constructed and sacrificed together with other valuable objects for a unique religious reason.

A much more relevant comparison can be made with a motif on a Minoan gold ring, found in Mycenae, with a circular disk with thin rays in the sky and to the right a similar crescent as on the Nebra disk. This motif might be a depiction of the total solar eclipse in Knossos on 17 July in 1584 BC.

The first investigator of the Nebra disk, Harald Meller (2004b) thinks that the round disk in the middle is the totally eclipsed moon close to the Pleiades with a later crescent phase. However, total lunar eclipses occur every decade and are not so dramatic and unique as total solar eclipses, when stars and planets become visible at daytime. Unique objects such as the Nebra disk and the Sun disk from Trundholm, in Denmark, have been made for a unique reason such as a total solar eclipse!

Acknowledgements

I want to thank Associate Professor Mary Blomberg, Department for Archaeology and Ancient History, Uppsala University, for her corrections of my English.

References

Borg, G., Ehser, A. and Pernicka, E.. 2010. Himmelsscheibe von Nebra: Das Gold stammt au England. Focus Magazine 12.

Burenhult, G. 1973. The Rock Carvings of Götaland. Acta Archaeologica Lundensia, Issue 8, no. 4.

Henriksson, G. 1999. Prehistoric constellations on Swedish Rock-carvings. In A. Le Beuf and M. Ziółkowski (eds), Actes de la Vème conférence de la SEAC, Gdańsk, 5-8 Septembre 1997 (Światowit supplement series H: Anthropology 2), 155-173.

Haustein, M., Gillis, C. and Pernicka, E. 2010. Tin isotopy—a new method for solving old questions, Archaeometry 52, no. 5, 816-832.

Meller, H. 2004a: Der geschmiedete Himmel: Die weite Welt im Herzen Europas vor 3600 Jahren. Landesmuseum für Vorgeschichte, Sachsen-Anhalt.

Meller, H. 2004b. STAR SEARCH. National Geographic 205, no.1, 76-87.

Pásztor, E. and Roslund, C. 2007. An interpretation of the Nebra disc. Antiquity 81, no. 312, 267-278.

Pernicka, E 2004. Die Naturwissenschaftlicken Unter-suchungen der Himmelsscheibe. In H. Meller (ed.). Der geschmiedete Himmel, 34-7. Stuttgart, Theiss.

Schlosser, W. 2002. Zur astronomischen Deutung der Himmelsscheibe von Nebra. Archäologie in Sachsen-Anhalt 1, 21-3.

Van Soldt, W. H. 1995. Solar Omens of Enuma Anu Enlil: Tablets 23 (24)-29 (30). Istanbul.

Weber, C. 2002. Die Jagd nach den Sternen. FOCUS 9. Available at http://www.focus.de/wissen/natur/archaeologie-die-jagd-nach-den-sternen_aid_207772.html (accessed 15 March 2015).

ARCHAEOASTRONOMICAL ANALYSIS OF THE KARATAEVO FORTRESS SANCTUARY ON THE NORTHERN BLACK SEA COAST

LARISA N. VODOLAZHSKAYA

Abstract: Karataevo Fortress is a unique archaeological site of the Middle Bronze Age. It is located in the western outskirts of Rostov-on-Don. It is a single complex with Liventsovka Fortress. Both fortresses belong to the North Caucasian culture of the Middle Bronze Age. The trench from the base of the rectangular construction was discovered in the eastern part of Karataevo Fortress, close to the central part of the Karataevo-Liventsovka complex. The trench is a trace from the base of the construction. The construction was associated with rituals. The construction-sanctuary was orientated along the rays of the Sun setting on the summer solstice. Light fell into the sanctuary through a doorway in the northwestern wall and reached the central bonfire site at sunset. Noontime rays penetrated the southern doorway and spread toward the north corner and a small bonfire site. The direction between the northeastern and southwestern doorways coincides with the projection of the Milky Way (the galactic equator) on the night of the summer solstice. The sanctuary was used for astronomical observations and for calendar rituals. The rituals were related to seasonal work. They were supposed to ensure fertility and prosperous life for the inhabitants of the fortress and its surrounding settlements.
Keywords: Sanctuary, summer solstice, Milky Way, Karataevo Fortress, North Caucasian culture

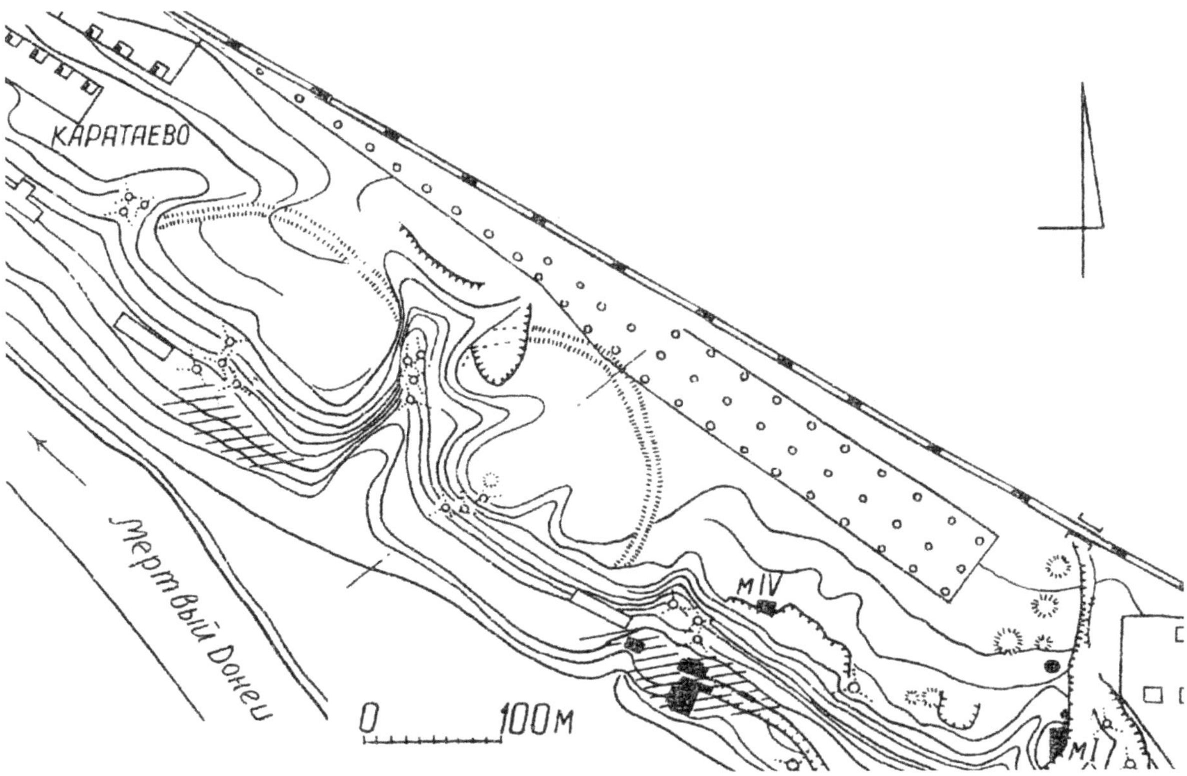

Figure 1. Site plan with fortresses around hamlets Leventsovka and Karataevo (Bratchenko 2006, 35).

Karataevo Fortress is a unique archaeological site of the Middle Bronze Age. It is located in the western outskirts of Rostov-on-Don. It is a single complex with Liventsovka (or sometimes Leventsovka) Fortress (Figure 1). Both fortresses belong to the North Caucasian culture of the Middle Bronze Age (founded in 1800-1700 BC). The trench from the base of the rectangular construction was discovered in the eastern part of Karataevo Fortress, close to the central part of Karataevo-Liventsovka complex (Figure 2). Excavations were conducted during an archaeological expedition led by Dr L. S. Ilyukov. Archaeologists have dated the construction to the Middle Bronze Age. Tombs in its territory were dated to Scythian times (Figure 3). 'N' indicates the direction of True North in Figures 3-8.

The construction size is approximately 13.2 × 8.0m. Gaps made in the middle of long trenches are facing each other. A gap exists in the center of the northwestern trench, too. The width of these gaps is 1.3m. It is likely that the gaps correspond to the doorways into the construction. The big bonfire site (pit of fire) is 2m in diameter and was built in the center. A small bonfire site and slag after the smelting of bronze were discovered near the northern corner of the construction. The building had ritual significance and it was a sanctuary.

Azimuths of sunrises and sunsets and other astronomical objects were calculated for 1700 BC, according to archaeoastronomical analysis. We have calculated the sunrise and sunset azimuths in 1700 BC for the geographical coordinates: 47°13' NL, 39°33' EL, using the program Red Shift (Figure 4). Sun declination was taken to be 23°54'. Magnetic declination was taken to be 6° E on the basis of field measurements.

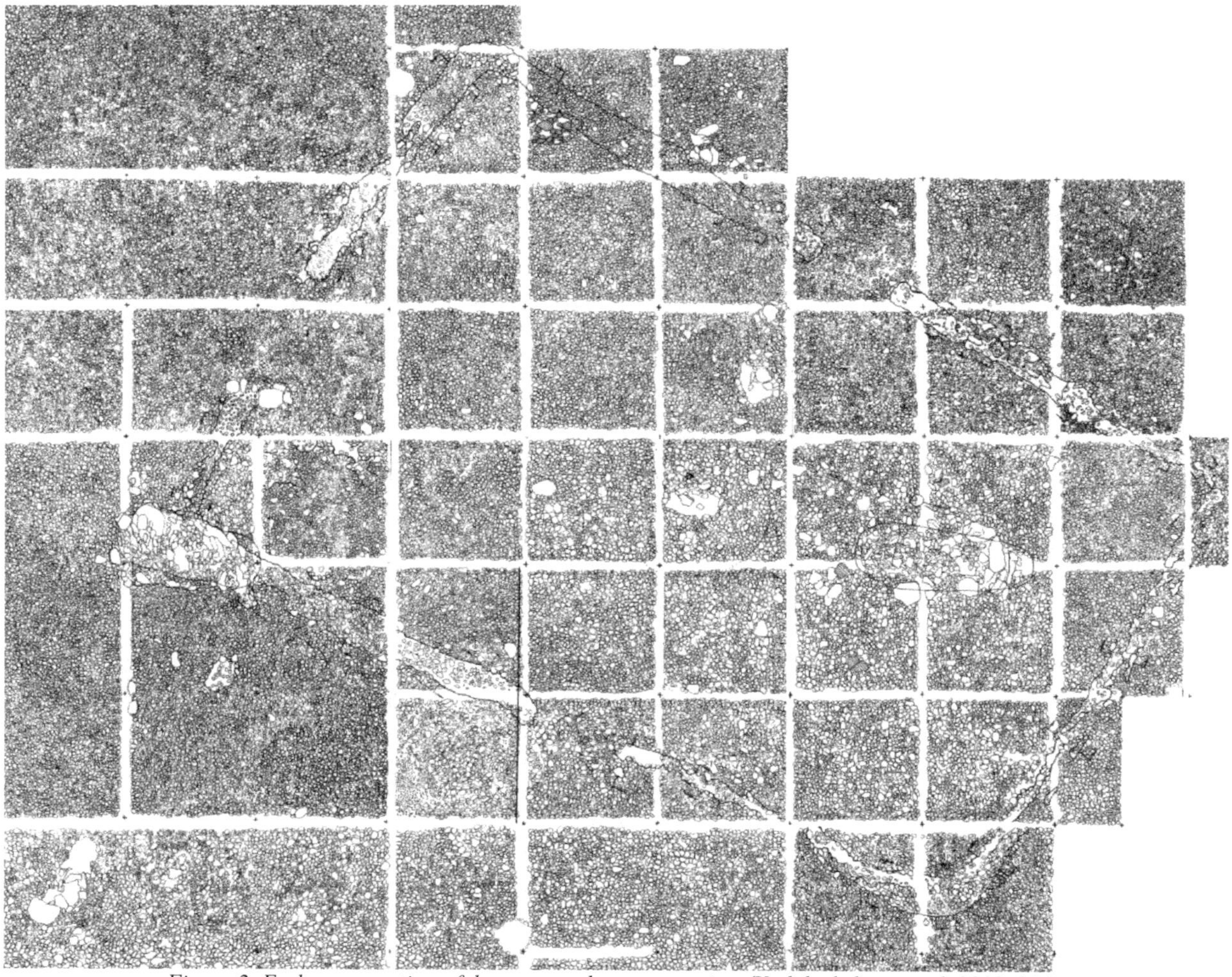

Figure 2. E-plan excavation of the rectangular construction (Vodolazhskaya et al. 2003, 256).

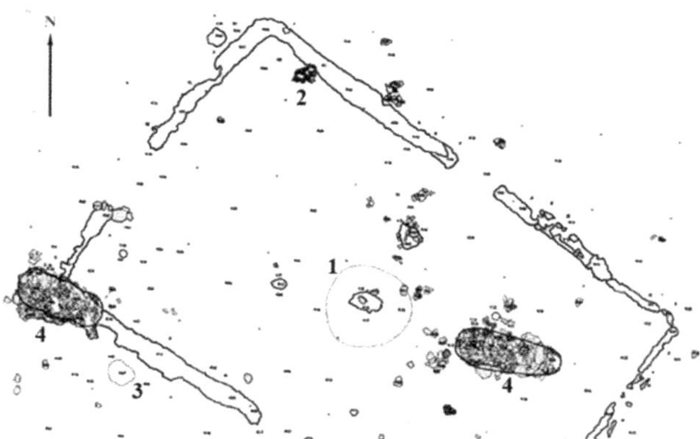

Figure 3. Plan of the trench from the base of the rectangular construction (taking into account the magnetic declination): 1) central bonfire site; 2) small bonfire site in the northern corner of the construction; 3) accumulation of sheep bones; 4) Scythian tombs (Vodolazhskaya et al. 2003, 258).

The sanctuary was orientated along the rays of the sunset on the summer solstice (Figure 5). Light fell into the sanctuary through a doorway in the northwestern wall and reached the central bonfire site at sunset. The sanctuary was associated with the ritual of the summer solstice. Rays of the Sun rising and setting fell in the northeastern doorway and reached the northwestern doorway on the days of the vernal and autumnal equinox (Figure 6). Noontime rays penetrated the southern doorway and spread toward the northern corner and a small bonfire site (Figure 7).

Season	Local true solar time	Azimuth	Date
Sun rise in the summer solstice			
Summer	04h02m	52°38’	8-07-1700 BC
Sun set in the summer solstice			
Summer	19h59m	307°22’	8-07-1700 BC
Sun rise in the winter solstice			
Winter	07h50m	125°51’	3-01-1700 BC
Sun set in the winter solstice			
Winter	16h10m	234°09’	3-01-1700 BC

Figure 4. Sunrise and sunset azimuths in 1700 BC for geographical coordinates: 47°13’NL, 39°33’EL.

The direction between the northeastern and southwestern doorways coincides with the projection of the Milky Way (the galactic equator) on the night of the summer solstice

(Figure 8). The direction of projection is calculated at the end of navigation twilight. The star α Scorpii-Antares becomes visible in the southern doorway to an observer at the central bonfire site, during nautical twilight. The ancient people noticed the touching points of the Milky Way at the horizon related to the cardinal directions, to the points of sunrises and sunsets at the equinoxes and solstices or the passage through the Zenith (Rappenglück 2002, 273).

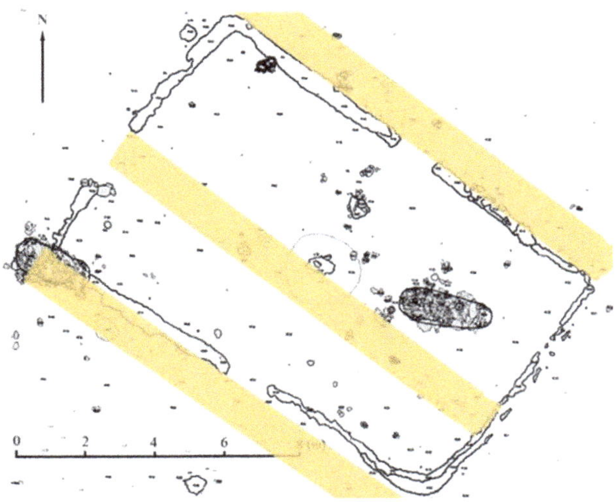

Figure 5. Direction of beams at a sunset on summer solstice.

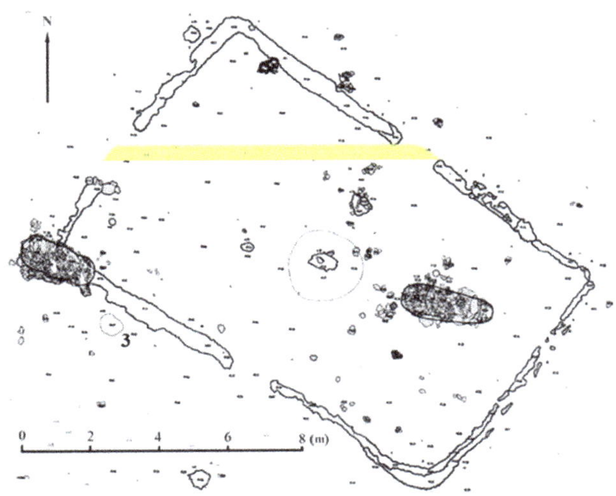

Figure 6. Direction of beams at rising/sunset on days of equinoxes.

The constellation of Scorpius was the symbol of Ishara, the goddess of fertility in the mythology of some peoples of Western Asia (Oelsner et al., 1998, 178). Seasonal holidays related to Ishara were conducted on the summer solstice and the vernal and autumnal equinoxes. It is known that the cult of the great female deity of the barbarians of the Black Sea was subjected to the strong influence of western Asia in ancient times already (Shaub 2007, 80-158).

It is possible that the influence of western Asia is reflected in the metric relations of the Karataevo Fortress sanctuary. Its width is 8m roundly. It is equal to the Sumerian length unit–one '*garush*'. It was used in the markup of arable fields (Kramer 2002, 126-127). The length of the construction (from the corner to the doorway) is equal to about 5.9m, a Sumerian length unit of one '*gar-*(*dy*)'. The inside space of the Karataevo sanctuary (13m x 8m) is equal to 104m^2. This is about three '*sar'*. '*Sar'* is a Sumerian unit of area equal to one '*gar-*(*dy*)' squared (Katz 2007, 71-73).

Direct influence of western Asia on the archaeological cultures of the Northern Black Sea region in the Middle Bronze Age is not proven. But it can be assumed that the influence of west Asia could be mediated through the Caucasus, such as the archaeological Maikop culture and Catacomb culture.

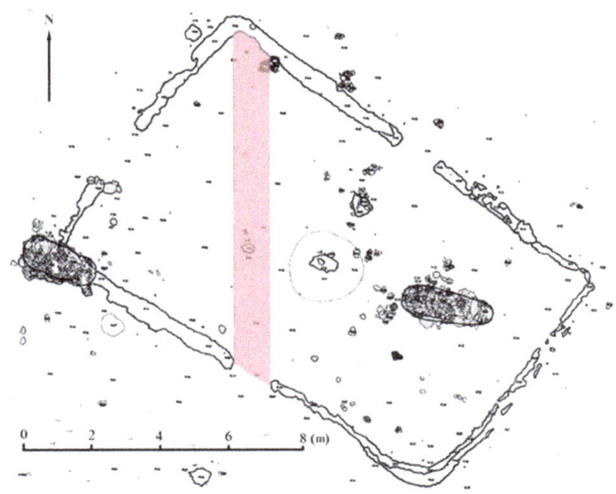

Figure 7. Direction of sun beams at midday.

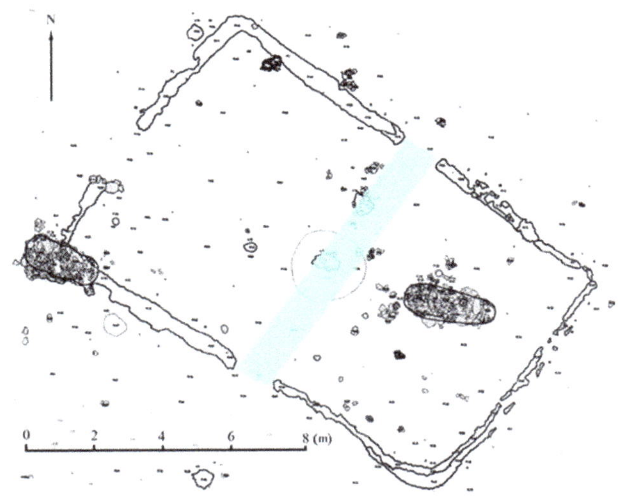

Figure 8. Milky Way projection on summer solstice after sunset.

Thus, for the archaeoastronomical analysis of the Karataevo Fortress sanctuary, patterns were discovered that suggest an affiliation of the sanctuary with the cult of the great female deity, who was of western Asian origin or was influenced by western Asia. The construction-sanctuary was used for astronomical observations and the rituals associated with seasonal work to ensure fertility and the prosperous existence of the inhabitants of the fortress and its surrounding settlements.

References

Bratchenko, S. N. 2006. Materialy i doslidzheniya s arhyeologii Shidnoi Ukraini: Sbornik nauchnyh trudov. Vypusk 6, Lugansk: Izd-vo VNU im. V. Dalya.

Katz, V. J. 2007. The mathematics of Egypt, Mesopotamia, China, India and Islam a sourcebook. Princeton, Princeton University Press.

Kramer, S. 2002. Shumery. Pervaja tsivilizatsija na Zemle. M.: ZAO Tsentrpoligraf.

Oelsner, J., and Horowitz, W. 1997-1998. The 30-Star-Catalogue HS 1897 and The Late Parallel BM 55502, AfO. Bd. 44-45. Archiv für Orientforschung.

Rappenglück, M. A. 2002. The Milky Way: Its concept, function and meaning in ancient cultures // Astronomija drevnih obshchestv: Materialy konf. 'Astronomija drevnih tsivilizatsij' Evropejskogo o-va astronomii v kulture (SEAC) v ramkah Obedinennogo Evropejskogo i Natsionalnogo astronomicheskogo sezda (JENAM). Moskva, 23-27 maja 2000g. Otv. red. T.M.Potemkina, V.N.Obridko. M.: Nauka.

Shaub, I. U. 2007. Mif, kult, ritual v Severnom Prichernomorje (VII-IV vv. do n.e.). - Spb.: Izdatelstvo S.-Peterb. Un-ta. Filologicheskij fakultet SPbGU.

Vodolazhskaya, L. N., Vodolazhsky, D. I., and Ilyukov, L. S. 2003. Metodika komp'yuternoj fiksacii graficheskogo materiala arxeologicheskix raskopok na primere Karataevskoj kreposti. (Technique of computer fixing graphic material of archaeological excavations on the example Karataevo fortress). Informacionnyj byulleten' Associacii Istoriya i komp'yuter. (Association Newsletter History and Computer), No 31, 248-258.

EGYPT, MINOAN CULTURE

STELLAR AND SOLAR COMPONENTS IN ANCIENT EGYPTIAN MYTHOLOGY AND ROYAL IDEOLOGY

ROLF KRAUSS

Abstract: Ancient Egyptian texts of Dynasties I-III refer to the falcon god Horus as a star. The Pyramid Texts of the Old Kingdom and the Coffin Texts of the Middle Kingdom document 'Horus- son-of-Isis' as the morning star and the 'Elder Horus' as the morning star. The Egyptian king embodied the god Horus; thus this stellar model expressed royal ideology of earliest historic Egypt. The origin of the cult of the sun god Ra, which was introduced in Dynasty II, is not known. Very early on, the stellar god Horus was subordinated to the new solar deity. Horus as the morning star was assigned the role of Ra's butler and/or son; by Dynasty IV at the latest, kings had assumed the title 'son of Ra'. The cult of the sun god became the dominant feature of the ideology and religion of ancient Egypt, but the earlier stellar components were preserved, if inconspicuously. Both solar and stellar ideologies endured down to the end of pharaonic times.
Keywords: Egypt, morning star, sun god, royal ideology

Horus and Seth as stellar gods

Since late predynastic times Egyptian kingship was related to the god Horus (Wilkinson 1999, 184f). Kings' official names called them Horus so-and-so; these Horus names were inscribed in a panel atop a depiction of the palace façade which was surmounted by a figure of the falcon Horus (Figure 1). Until the 1950s Egyptologists characterized Horus as a sky god (Erman 1934, 18). But since then it has become clear that the sources for this interpretation are not older than the 1st millennium BC; in earlier periods Horus was a stellar god (Anthes 1959, 185-190; Kahl 2007, 2). The earliest evidence for Horus as a star is provided by the names of royal domains of Dynasties I-III (Kahl 1994, 595f). Some of them could be understood figuratively, such as 'Horus, the star has arisen' or 'Horus, star of the khet', a group of gods. But clearly the name 'Horus, foremost star of the sky' is not metaphorical. It follows that the other domain names which designate Horus a star are to be taken verbatim; thus already at the time of King Anedjib in the mid-First Dynasty Horus is attested as a star. From the time of his predecessor Djet there is a comb bearing that king's Horus name and a depiction of a falcon traversing the sky in a boat (Figure 1; Anthes 1959, 188ff.). On analogy with the later documentation, this bird should be Horus and a star.

But which star did the Egyptians identify as Horus? Later sources, notably the Pyramid Texts (PT) and the Coffin Texts (CT) suggest a tentative answer. The earliest PT occur in the pyramid of Unas, the last king of Dynasty V; the CT are attested centuries later. By contrast to these text corpora, the early written sources consist of short phrases similar to headings. The use of later sources to elucidate the earlier limits any conclusions.

The PT, as well as the CT and other later texts, call Horus a son of Osiris-Orion and Sothis-Isis-Hathor. Osiris and Isis are not attested before Dynasty V. On the other hand, PT 371 identifies Khenti-imentiu (Foremost-of-the-Westerners), the divine patron of the royal necropolis at Abydos from the predynastic onwards, with Osiris who in turn was the god of the dead par excellence beginning in Dynasty V. Possibly Osiris was known as Khenti-imentiu as early as the beginning of Dynasty I (Wilkinson 1999, 288). Djed- and tyet-amulets, emblems associated with Osiris and Isis (respectively) in later times, have been found in Early Dynastic contexts (Wilkinson 1999, 287, 292). According to CT 335, Osiris-Orion had two fledglings named Horus: Mekhenti-irty (Horus Eyeless) and Harendotes (Horus, Protector-of-his-father). The older Horus was born while his father was alive whereas the younger Horus was sired posthumously, after Seth had murdered Osiris. Since it was Horus's mother, under the name Isis, who was credited with raising him, he was also called Horus, son-of-Isis.

Figure 1. Comb with Horus name of king Djet, after Petrie 1925, Pl. XII.

The younger Horus can be identified with the morning star, the planet Venus. In a Demotic list of the planets, Venus is called 'Horus, son-of-Isis, the god of the morning' (Neugebauer 1943, 121). The inscription around the zodiac of Dendera cites Horus, son of Isis, as the star god of the morning (Krauss 2008, 208f). About 850 BC, Horus (though not necessarily the younger Horus) is described as the morning star: 'one sees the king on his chariot like a sparkling star, like Horus of the morning in the starry sky' (Krauss 2008, 209). An ambiguous reference to the star of Horus is 'Horus-Psusennes' in the well known variant of Psusennes II's nomen (c. 950 BC), with the meaning 'Horus, the star has arisen in the city (Thebes)'.

CT 722 describes the god-of-the-morning (= morning star) as a 'sehed star' and a 'beautiful wa'a-star of gold, which has risen from the akhet'. CT 255 expresses the same idea citing 'Horus who ascends as gold (*m nbw*) from upon the lips of the akhet '. Together these two citations identify the morning star as Horus, viz. the younger Horus, since the morning star is the younger Horus in other CT spells. Katja Goebs analyses CT 469 and 1017 as variants of the so-called Cannibal Hymn in the PT (Goebs 2008, 205-393). CT 469 describes the ascent of the god-of-morning-dawn (*netjer hedj-ta*) into the sky, saying: 'My mother Sothis (Sirius), she makes a path for me ... to the place from which Orion comes forth. I find Orion standing on my way, his djam-scepter in his hand, ennobled. I receive it from his hand, that I may be divine with it.' Since Sothis and Isis are identified in PT 33 and 593 (Krauss 1997, 173ff), Horus, the god of the morning dawn, i.e., the morning star, is here the younger Horus.

If the younger Horus is indeed personified in the morning star, the older Horus could be personified in the planet Venus as the evening star. This may be deduced by starting from PT 303. There Osiris is said to have taken the older Horus, as his and Hathor's son, to the pet-sky to let him appear as a great god in the *qebehu* (apparently a part of the pet-sky). CT 167 informs us that the throne of the older Horus as Mekhenti-irty is located in the qebehu. Since the older Horus (Haroeris) is 'in the middle of the upper stars (and) opposite the lower stars' (CT 1143), he is associated with the night sky. Finally, CT 106 asserts that Khenti-irty (Eyeless) is a great sehed-star. The sehed-stars are described in PT as capable of free movement; they are apparently planets like the morning star cited above (Krauss 1997, 254ff).

More pertinent information is implicit in the myth about the eyes of Horus. The older Horus lost an eye in a struggle with Seth. The myth about the injured (and subsequently healed) eye of Horus is already entrenched in the PT and thus should be older than late Dynasty V. Since space is limited, I shall not deal here with the sound eye of the younger Horus (*udjat*, a term which came also to be used for the injured eye of the older Horus). Note, however, that analysis of texts from the 1st millennium BC has led Egyptologists to interpret the fate of the Horus eye as a reflection of the phases of the moon (Erman 1934, 22f).

According to Book of the Dead 137B the Horus eye appears in the evening (Hornung 1979, 270f, 500): 'Spell for lighting the torch for NN: The bright eye of Horus is coming, the shining eye of Horus is coming. ... The one is coming who visits the sky after Ra and before your companion, (o) Ra.' The spell compares the torch, lit for the dead with the eye of Horus which shines in the sky after sunset (after Ra) and before moonrise (before the companion of Ra = moon). If the evening star is Horus's eye, then the period specified corresponds to the second half of the lunar month when the moon rises after sunset.

Several spells in the CT link the injured eye to the night sky, where it is to be found in the retinue of the moon (CT 1151). The eye is 'beneficial in the night' and 'makes fire with its beauty' (CT 1053); ' becoming the fiery eye of Horus' is the title of CT 316. Its fiery nature enables the Horus eye to 'cook' enemies (CT 845). Obviously, the fire of the Horus eye alludes to the particular brilliancy of a certain star. Papyrus Cairo 86637, an almanac spanning a 365 day year which I have identified as 1298/97 BC (Krauss 2002, 193ff), furnishes specific information. The eye of Haroeris is incomplete on day 218, when the sum of its parts is less than half. On day 349 the eye of Haroeris 'has come, is filled and complete, without something missing.' Since the eye was first sighted as akhet eye on day 111, it took about 240 days until its filling was completed. The filling of the eye as reported in the almanac reflects the course of evening visibility of Venus: at first the planet shines with a relatively dim light, reaching maximum brilliancy only after about eight months. The loss of the eye corresponds astronomically to the two-month invisibility of Venus before the planet reappears as the evening star. The role of the moon god who finds the lost eye, brings it back, and heals it (PT 82; CT 249), can be seen as reflecting the frequent encounter of the waxing moon and the evening star.

To sum up: textual sources beginning with the PT identify Horus with the planet Venus (for the Benu or phoenix as another personification of Venus, see below). Besides embodying the planet Venus, Horus in the later sources is a falcon embodying the king. Not all of these ideas are explicit in the early sources. The identification of the early dynastic Horus as the planet Venus in particular and not just as any star is confirmed by his epithet 'foremost star of the sky' which applies first and foremost to Venus among all the stars. The sources of Dynasties I-II are inadequate for differentiating between the younger and the older Horus.

Provided that the Horus king embodied the planet Venus, it can be stated that the royal ideology of Early Dynastic Egypt rested on a stellar model or at the very least had a stellar component. The same ideology also provided the god Seth with a role. The last king of Dynasty II Khasekhemuy claimed to embody Horus and Seth. Seal impressions bearing his name show Horus and Seth, both wearing the double crown of Upper and Lower Egypt, sharing the *serekh* (Figure 2). This iconography implies that the king bears the Title 'Horus and Seth', not just 'Horus'. The king's name itself, meaning 'the two sekhem-powers have arisen', evidently refers to Horus and Seth.

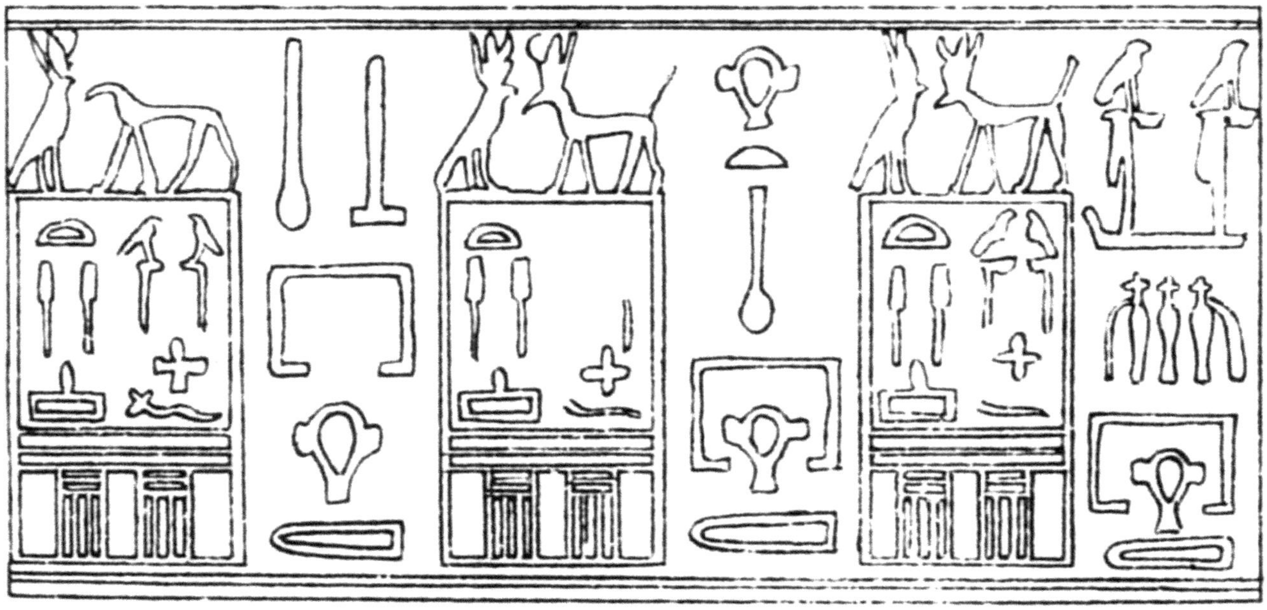

Figure 2. Seal impression with Horus name of King Khasekhemuy, after Petrie 1901, Pl. XXIII.

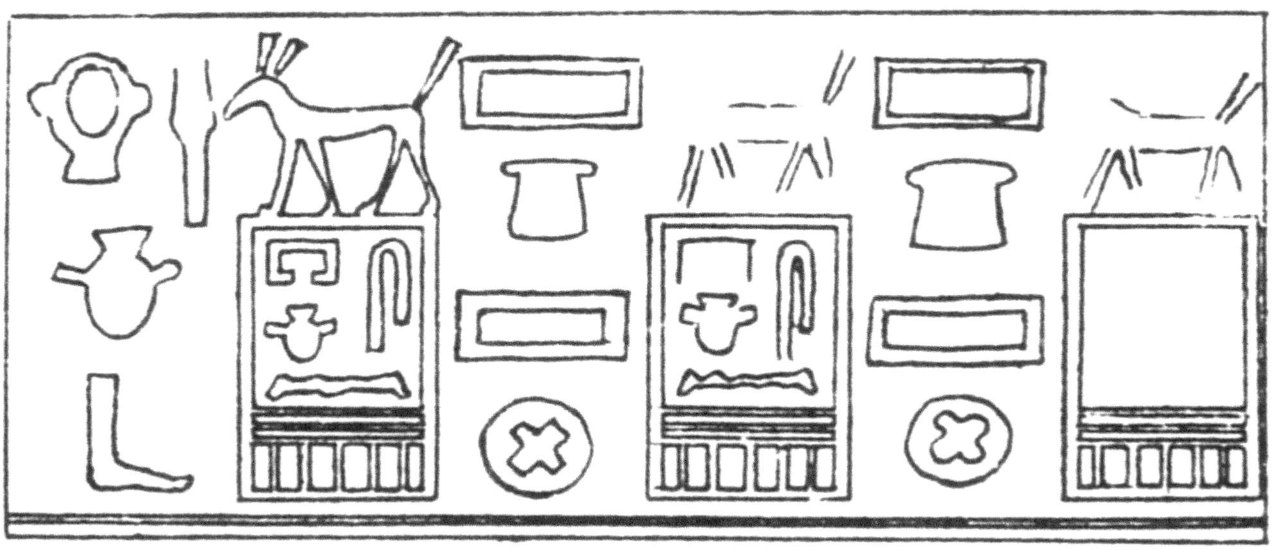

Figure 3. Seal impression with Seth name of king Peribsen, after Petrie 1901, Pl. XXI.

King Peribsen of mid-Dynasty II did not call himself 'Horus' or 'Horus-and-Seth', but rather 'Seth Peribsen'. The writing of his name shows the figure of Seth atop the serekh (Figure 3). The king also honoured Horus alongside Seth, provided his name be understood to mean 'their-heart-has-come-out', with 'heart' the equivalent of sound mind and 'their' referring to Seth and Horus.

There are no assertions about Seth from the Early Dynastic Period comparable to those about Horus. Sources from later times identify Seth with the planet Mercury. The figure of Seth, labelled Segeb, appears in the so-called sky diagrams dating from c. 1500 BC and later (Neugebauer and Parker 1969, 175ff). Segeb as Mercury is last found in Demotic horoscopes. The earliest mention of Segeb as a star is found in CT 1030. According to PT 719 Seth ferries the deceased king over the Winding Waterway which I have explained as the ecliptical belt (Krauss 1997, 236). In astronomical reality only planets and the moon cross over the ecliptical belt; therefore, the authors of the PT must have envisioned Seth as a planet. He ought to be the inner planet Mercury, since the outer planets are later attested as forms of Horus (Neugebauer and Parker 1969, 175ff).

To sum up: Seth of later sources is the planet Mercury, a quadruped of characteristic appearance who embodies in the king like Horus. He is the rival but also sometimes the friend of Horus. There are few correspondences with the Early Dynastic criteria; nevertheless, it is possible that Seth, like his partner Horus, also personified a stellar deity. If so, the likely candidate for Seth as a star ought to be Mercury, both as suitable partner and rival for Venus, the other inner planet. Those ancient Egyptians who elaborated on the roles of Horus and Seth may well have interpreted the appearances and disappearances of Mercury and Venus in the eastern and western sky, the long invisibility, and the dramatic changes in the brilliancy of Venus as evening star as indicative of fighting, injuring, and rivalry, but also of reconciliation and joint action—depending on the political, social, and ritual realities visualized in the heavens. Stellar ideology will have had its roots in the predynastic culture of Upper Egypt and in its presumed Afro-Asiatic (Hamito-Semitic) origin. It is well known for example that Venus was male in southern Arabia, while female in Mesopotamia and Syria, although by Roman times, the Arabs in Syria knew male morning and evening stars (Drijvers 1981, 246). The identification of the ruler with the planet Venus is attested only in Egypt.

Introduction of the solar cult

The cult of the solar god Ra was introduced during the dynasty of Khasekhemuy which experimented with the concept of the rivals Horus and Seth being at peace in the person of the king. Jochem Kahl discovered that one of the names of King Raneb designated him as the embodiment of Weneg (Kahl 2007, 29ff). This being is mentioned in the PT as a son and follower of Ra. If Weneg were already subordinate to Ra as early as Dynasty II, the Horus name Raneb could mean 'Ra is (the) lord', or, more accurately, 'Ra is (my) lord', since at this early period the first person singular possessive pronoun was not written, but implicit. Nebra's predecessor Hetepsekhemuy may have introduced the solar cult, because in his reign the sun god seems to have been present as Akhty (belonging to the akhet-horizon) (Kahl 2007, 62).

According to Kahl, the introduction of Ra to whom Horus was subordinated '...suggests that there was a break in religious tradition and theological frames of reference' (Kahl 2007, 61). Kahl speaks of a change of paradigm with Dynasty II. Indicators of this change which Kahl cited include the abandonment of Abydos as the royal necropolis in favour of Saqqara and a new attitude towards human life, reflected in the discontinuation of practices involving human sacrifice.

Horus's subordination to Ra was achieved by altering his filiation; no longer the son of Orion-Osiris, he became the child of Ra, as expressed, for example in CT 150 and 364. An example of the resulting discrepancies is provided by the cannibalistic texts of the CT and PT. In the CT versions Sothis/Isis is the mother of Horus, which implies that Orion/Osiris fathered him. In the PT version, Atum, a form of the solar god, is the father of King Unas, but his mother is not mentioned by name. Orion is cited in the PT version as 'father of the gods' who gives Unas a writ with which he becomes a great sekhem. By contrast, in CT it is Orion who gives his djam-scepter to his son to cause him to be divine. Thus in the PT version, the ancestry of Horus is solarised and the authentic father role of Osiris-Orion is obfuscated. From Dynasty IV onwards, kings come to be entitled 'son of Ra', paralleling the mythological subordination of Horus to the Sun.

Seth-Mercury was not adopted by Ra, but given a place in the solar barque. In CT 160 Seth opposes a snake when the solar barque sets in the evening. By contrast, the almanac papyrus Cairo 86637 has Seth in the bow of the barque, attacking the snake-like monster Apophis at dawn in the morning (Krauss 2002, 197f). This position in the solar barque is astronomically appropriate, regardless of whether Seth-Mercury appears in the evening or in the morning, since the planet never moves far away from the Sun.

The Egyptians not only subordinated the old stellar gods to the sun god, but also adapted the solar god to the stellar gods. Several hundred years after the introduction of the solar cult, the Egyptians created a syncretistic form of the sun god which associated the falcon, along with the name Horus, with the horizon (akhet) as Horus-Akhty (Horakhte; Ra-Horakhte). It is noteworthy that in the sky diagrams Harakhty is a name of the planet Mars, also known as Horus desher (red Horus).

The origin of the solar cult

The sudden appearance of the solar cult in Dynasty II raises the question of its origin. In other Mediterranean and Near Eastern cultures, the role of solar gods was in no way comparable to their importance in ancient Egypt. (The exceptional role of the sun god in the late Roman Empire is of no concern here; see Drijvers 1981, 247f.) In all probability, the early cultures of Naqada and in the Delta knew some solar deities. But were the rather sophisticated concepts associated with the solar cult rooted in one of the predynastic Egyptian cultures? One such concept was Ma'at—order and truth; personified as the daughter of Ra, 'Ma'at became (in Dynasty II) a norm and political objective' (Kahl 2007, 61).

Another sophisticated concept is the 365-day solar calendar which will have been introduced in the time of Hetepsekhemuy and Raneb; either it belonged originally to the solar cult or was developed subsequent to its adoption. The solar year and the beginning of the civil day in the morning at dawn are unique in the ancient world. There also seems to be a solar concept of time, stressing the continuous temporal existence (eternity) of the sun god, at least according to Egyptologists' interpretation (Assmann 1975, 50f.). The sun god was deemed creator of the world, a concept which is alien to neighbouring cultures. He brought himself into being from the primeval ocean, and then proceeded to create the cosmos with its gods and all else. According to CT 1130 and chapter 175 of the Book of the Dead he shall also eventually dissolve all that he has created. This explanation to account for the world is monistic, to use an anachronistic expression, by contrast to the analysis of some Egyptologists who are of the opinion that the Egyptians of the Early Dynastic Period developed a system of ordering the world in dualities (Beckerath 1984, 13f.).

The solar cult also seemed to have had its own concept of the Beyond. 'Two ancient doctrines of this celestial hereafter have been comingled in the Pyramid Texts: one represents the dead as a star, and the other depicts him as associated with the Sun-god' (Breasted 1912, v). The stellar ideas of the afterlife were threefold: as stars, one group of the dead follow Orion-Osiris in the southern sky, experiencing seasonal death and rebirth; another group of the dead are the Imperishable Stars who live forever in the northern sky, with the heavens divided by the ecliptical belt; finally there is the king who, as the embodiment of Horus, becomes the morning star (Krauss 1997, 277-283). The idea of the exclusive stellar afterlife of the king and the ruling élite (pat) could reflect the extreme stratification of late predynastic society; the doors of the sky were not open to the *rekhyt*, the mass of the population (Krauss 1997, 283f). The solar idea of the afterlife, by contrast, apparently focuses on the solar barque which is manned by the dead as stars. In ancient Mesopotamia and other neighbouring cultures there was nothing comparable to either the Egyptian stellar or solar afterlife. It can be postulated that the new ideas associated with the solar cult originated in Early Dynastic Egypt, independent of external influences. There is no evidence for foreign influence on the material culture at that moment. If the solar cult and the changes associated with it were imported, then they arrived as ideas only, leaving no discernible markers in material culture.

The benben stone is one feature of the solar cult with links to Mediterranean cultures and their holy stones, traditionally called *baetyls*. The benben stone belonged to the indispensable paraphernalia of the solar cult as implied by the existence of a benben also in Amarna, the city which King Akhenaten dedicated to the sun god (ca. 1350 BC). Its earliest attestation is found in PT 600 where Atum-Khepher, a form of the solar god, is described as appearing as the benben stone (*m bnbn*) in the House of the Benu in Heliopolis. The divinity Benu (phoenix), a personification of the planet Venus, was depicted as a heron in the sky diagrams and other sources after c. 1500 BC. It seems possible that the solar god was affiliated from early on with the Benu as Venus. In any case, the Benu as well as Horus was embodied in the planet Venus.

Possibly the source of the solar cult was further afield than Egypt's immediate neighbours—perhaps as far away as the Atlantic megalithic culture which included elements of a solar cult, at least according to some authors (Green 1991, 24-32).

References

Anthes, R. 1959. Egyptian Theology in the Third Millennium BC. Journal of Near Eastern Studies 18, 185-190.

Assmann, J. 1975. Zeit und Ewigkeit im alten Ägypten: ein Beitrag zur Geschichte der Ewigkeit. Heidelberg, Winter.

Beckerath, J. v. 1984. Handbuch der ägyptischen Königsnamen. München, Deutscher Kunstverlag.

Breasted, J. H. 1912. Development of Religion and Thought in Ancient Egypt. New York, Scribner.

Coffin Texts: De Buck, A. and Gardiner, A. H. 1935-1961. The Egyptian Coffin Texts I-VII. Chicago, IL., University of Chicago Press.

Drijvers, H. J. W. 1981. Die Dea Syria und andere syrische Gottheiten, in M. J. Vermaseren (ed.), Die orientalischen Religionen im Römerreich, 241-263. Leiden, Brill.

Erman, A. 1934. Die ägyptische Religion. Berlin, de Gruyter.

Goebs, K. 2008. Crowns in Egyptian Funerary Literature: royalty, rebirth, and destruction. Oxford, Griffith Institute.

Green, M. 1991. The sun-gods of ancient Europe. London, Batsford.

Hornung, E. 1979. Das Totenbuch der Ägypter. Zürich and München, Artemis.

Kahl, J. 1994. Das System der ägyptischen Hieroglyphenschrift in der 0.-3. Dynastie. Wiesbaden, Harrassowitz Verlag.

Kahl, J. 2007. 'Ra is my lord': searching for the rise of the Sun God at the dawn of Egyptian history. Wiesbaden, Harrassowitz Verlag.

Krauss, R. 1997. Astronomische Konzepte und Jenseitsvorstellungen in den Pyramidentexten. Wiesbaden, Harrassowitz.

Krauss, R.. 2002. The Eye of Horus and the Planet Venus: Astronomical and Mythological References, in J. M. Steele and A. Imhausen (eds), Acts of the Conference Under One Sky: Astronomy and Mathematics in the Ancient World, British Museum London, June 25-27 2001, 193-208. Münster, Ugarit-Verlag.

Krauss, R. 2008. Über Horus und die Horus-Augen sowie andere Elemente der altägyptischen Astronomie. Acta Praehistorica et Archaeologica 40, 205-219.

Neugebauer, O. J. 1943. Demotic Horoscopes. Journal of the American Oriental Society 63, 115-126.

Neugebauer, O. J. and Parker, R. A. 1969. Egyptian Astronomical Texts, III: Decans, planets, constellations and zodiacs. Providence, RI, Brown University Press.

Petrie, W. M. F. 1901. The Royal Tombs of the Earliest Dynasties. Part II. London, Egypt Exploration Fund.

Petrie, W. M. F. 1925. Tombs of the Courtiers and Oxyrhynkhos. London, Quaritch.

Pyramid Texts: Sethe, K. 1960. Die altägyptischen Pyramidentexte 1-3, 2nd ed. Hildesheim, Olms.

Wilkinson, T. A. H. 1999. Early Dynastic Egypt. London, Routledge.

THE ELITE AT KNOSSOS AS CUSTODIANS OF THE CALENDAR

GÖRAN HENRIKSSON AND MARY BLOMBERG

Abstract: Among the buildings chosen for our pilot archaeoastronomical project investigating Minoan astronomy were the palace at Knossos, most likely home of the local political leader who also may have been the actual or symbolic leader of all Minoan society, the nearby peak sanctuary on Mount Juktas, and one of the large and stately villas built around the palace and furnished with fine religious equipment, both of the latter probably administered by priests. We found at these places striking similarities in alignments to the Sun, and these suggest to us that the three buildings were part of a system whereby the political and religious elite used orientations to significant yearly risings of the Sun to regulate what became a sophisticated calendar. A corollary is systematic astronomical observations for a long period of time.
Keywords: archaeoastronomy, Bronze Age Crete, Knossos, lunisolar calendar, Minoan calendar, peak sanctuaries, solar calendar

Introduction

The power structure of Minoan society is not completely clear to historians as the script, Linear A, has not been deciphered and its subject matter seems to deal primarily with agricultural accounts. To the archaeologists, despite their long study of the ruins of buildings and communities, and the contents of graves, the members of the Minoan elite are hardly more than shadows. The considerable pictorial remains show no ruler iconography (Davis, 1995). Still, we can assume that there was a political and religious elite.

In our study we employ classical archaeoastronomical methods: using a digital theodolite with observations of the Sun to determine the orientation of the coordinate system and the orientations of foundations to celestial bodies, determining the positions of celestial bodies at the appropriate times in the past using our own computer programs, and calculating the visibility of bright stars. Our study of the 19 buildings selected for our project has introduced us to a group of people who were educated in astronomical knowledge. By piecing together the evidence from 13 of the 15 buildings so far investigated, we have recovered in some detail the advanced calendar developed by this group and understood some of the complex mechanisms involved in its regulation.

Our recent study of the south-east house at Knossos gave us the clue as to the calendric functions that bound the villa with the palace and the peak sanctuary. The palace was the place of political authority, Mount Juktas was, in part, an observatory, no doubt bound to religion, and the south-east house was probably the residence of priests educated in astronomical knowledge. We have published our findings at the palace and on Mount Juktas (Blomberg, Henriksson and Papathanassiou, 2002), but our investigation of the south-east house has thrown new light on the cooperation between these places in evolving and regulating the calendar.

The archaeology of the south-east house

The house was built into the south-eastern slope on the site of the original settlement at Knossos, with many details typical for the Middle Minoan IIIA period (Evans 1902-1903, 3-10; 1921, 425-429). This would have been c. 1750-1680 BC, using the early chronology (Manning, 1995). It is part of the large complex of buildings around the palace and within sight of the peak sanctuary on Mount Juktas (Figure 1).

Figure 1. Knossos. The Kephala Hill and surrounding area. Based on Myers, Myers and Cadogan (1992), with permission.

Access was by a staircase descending from the top of the hill to a lower level. Part of the original plan was the pillar crypt just below the stairs (Figure 2). According to the excavator Sir Arthur Evans, 'there is clear evidence that such pillared crypts fulfilled a religious function'. It was built of fine masonry and contained an exquisite lamp of purple gypsum. Evans said of the lamp that it 'exceeds in richness of design any other object of the kind found within the palace'. There was also a libation table and a pyramidal stand, which probably held a double axe, the most common Minoan symbol. The pillar in the centre of the room has an incised double axe, and Evans drew the parallel between this room and the pillar crypts in the so-called central palace sanctuary where 29 doubles axes were incised into the pillars. Thus we can assume the considerable importance of the house and the small chamber. It was noted that the building is set at a different angle from the eastern wing of the palace, and this was explained as being due to the steep bluff. We propose an archaeoastronomical reason below.

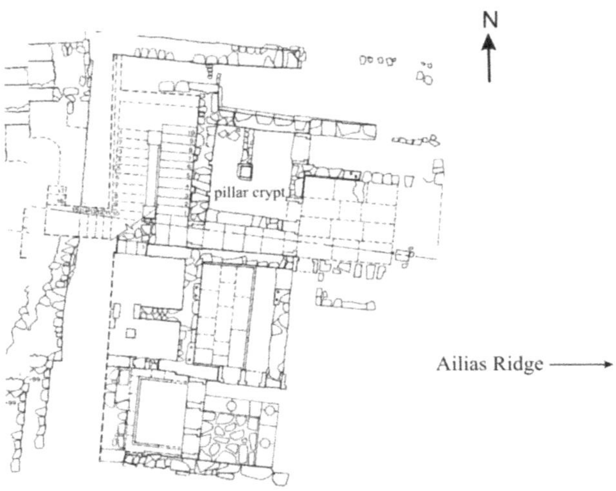

Figure 2. Plan of the south-east house. From Hood and Taylor (1981), with permission.

There is a well-defined niche in the western wall, symmetrical with the northern door (Figure 3). A hole was found below a stone in the south-west corner 'widening below into a cave artificially cut out of the soft rock. It was largely filled with earth that had apparently filtered into it, but, except for a few fragments of the Minoan and Neolithic pottery, nothing was found in it' (Evans 1901-1902, 110; 1902-1903, 3-10). We explain below the significance of the 'hole' in this corner of the pillar crypt.

Figure 3. The pillar crypt from the east. Photo, Mary Blomberg 1996.

The archaeoastronomy of the central palace sanctuary and the peak sanctuary

THE PALACE

We present those details of the orientation of the central palace sanctuary and the peak sanctuary that are relevant for the discussion here. The full account has been published elsewhere (Blomberg, Henriksson and Papathanassiou, 2002).

The central palace sanctuary was part of a much earlier building (Catling, 1973-1974). The present arrangement of rooms was probably the work of the Mycenaeans who converted the western part into a corridor for access to the well-known storerooms beyond (Hallager, 1987), thus accounting for the worn condition of the concave stone (Figure 4), the importance of which is explained below.

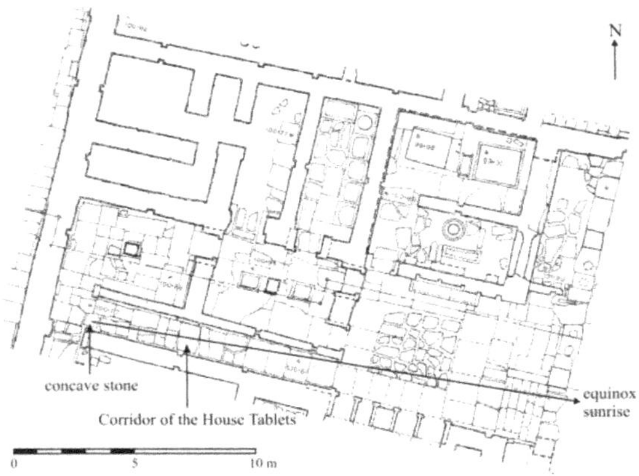

Figure 4. The plan of the central palace sanctuary as excavated. From Hood and Taylor (1981), with permission.

The sanctuary was originally one room (Driessen, 1990) and was orientated so that the upper limb of the Sun at sunrise on the equinoxes appears at the intersection of the Ailias ridge and the northern jamb of the southernmost door (Figure 5).

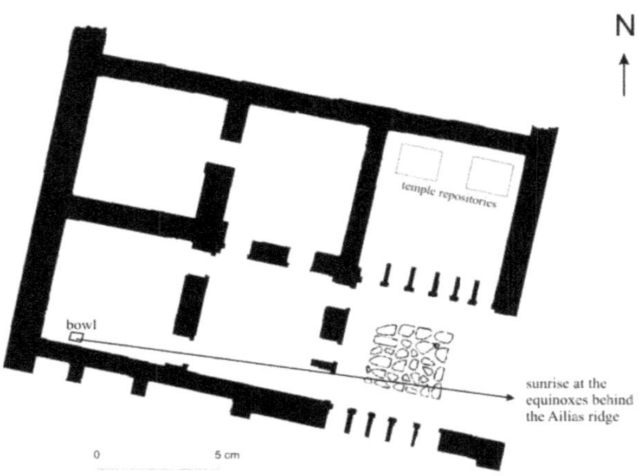

Figure 5. The reconstruction of the original sanctuary (Driessen, 1990). With permission from the author.

Eleven days after the autumn equinox the upper limb of the Sun at sunrise appears at the intersection of the ridge and the southern jamb of the door (Figure 6). The Sun does not appear in the doorway again until 11 days before the spring equinox. In the 11-day interval the rays of the rising Sun strike a much worn concave stone near the western wall (Figure 7). The stone is the remnant of a shallow bowl, the function of which was to cast a reflection on the west wall at the equinoxes (Figure 8). The 11-day interval is crucial for regulating a lunisolar calendar since the lunar year is 11 days shorter than the solar year. If the relevant phase of the Moon appeared in the 11-day interval, this would indicate that it was time to intercalate a lunar month into the calendar so that the lunar and the solar years would have concurrent cycles beginning at regular intervals. This indicates that the Minoans had such a calendar. Months were added three times in eight years and, more exactly, seven times in 19 years.

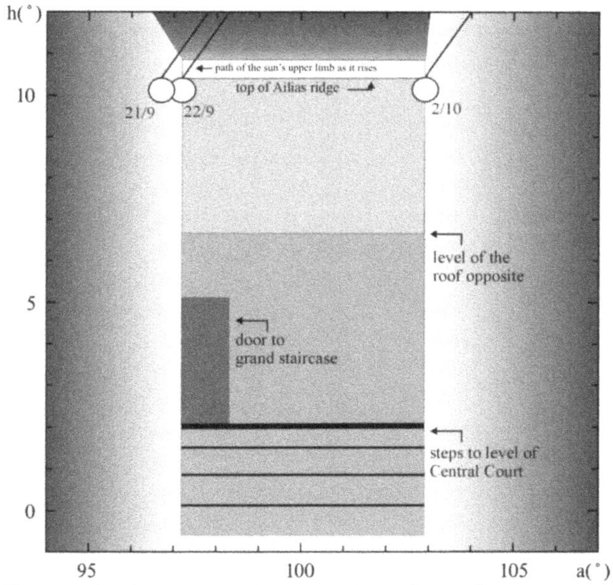

Figure 6. Autumn equinox sunrise from the sanctuary. Reconstruction by Göran Henriksson.

Figure 7. The bowl which casts the reflection. Photo, Göran Henriksson 1996.

Figure 8. The reflection from the Sun's rays. Photo, Mary Blomberg 1996.

The beginning of the year was set with respect to the autumn equinox by the heliacal rising of Arcturus at Petsophas, which was marked by the orientation of one of the walls of the peak sanctuary (Henriksson and Blomberg. 1996). The heliacal rising occurred one lunar month before the autumn equinox so that the rising of the relevant phase of the Moon in the 11-day interval after the heliacal rising would indicate that it was time to intercalate a month. It is this relationship at Petsophas that signifies most clearly that the Minoan new year began at the autumn equinox rather than at the spring equinox. We have evidence of this calendar also at Mount Juktas and at the south-east house (below). The fact that the Hellenistic calendar at Knossos began at the first crescent Moon after the autumn equinox may be a survival of the Minoan calendar (Guarducci, 1945).

The size of the reflection varies each year due to the relationship of the Sun's coordinates at the true equinox. It is larger the closer the Sun is to the true equinox, and the largest reflection, in 1998, was made when sunrise took place two hours from the true equinox (Figure 9). The orientation is dictated by the altitude of the Ailias ridge to the east, which also is responsible for the orientation of the pillar crypt in the south-east house. The nearly horizontal ridge faces the Kephala hill, on which the palace and villas were built.

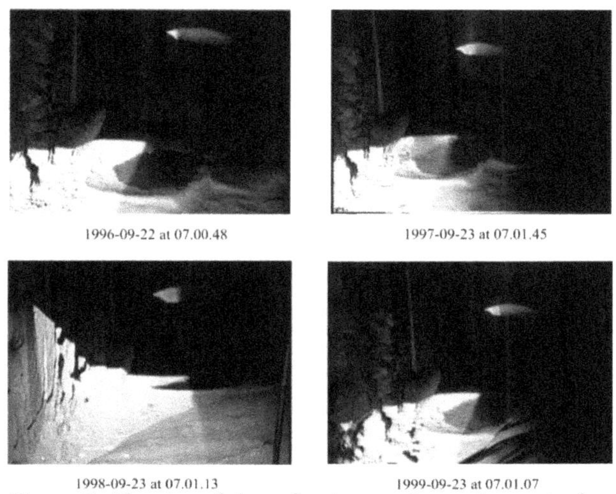

Figure 9. The size of the reflection on the equinoxes in four consecutive years. Photos, Göran Henriksson.

Mount Juktas

Mount Juktas is about 15km southwest of the palace and is a prominent landmark for the whole area. On it is the largest Minoan peak sanctuary, which was built at some time in the late Early Bronze Age, c. 2300-2000 BC. There is a magnificent view towards the sea to the north and the mountains to the east. Such sanctuaries are considered by Minoan archaeologists to have been religious cult places and this is no doubt the case. Relationships between them and celestial phenomena were very important. On the basis of our investigations at all of them that have surviving foundations, we found orientations to the equinoxes, a solstice, the rising or setting of a bright star, or the beginning of a solar month. Some of the sanctuaries had orientations to more than one significant celestial event. Petsophas on the east coast had four. There is no specific orientation from the palace to the peak of Mount Juktas, as is sometimes maintained, although it is prominent on the southern horizon.

The placement of the peak sanctuary on Mt Juktas was such that from the altar the upper limb of the Sun at the equinoxes could be seen to rise in a saddle of the mountain Selena, and 11 days after the autumn equinox it rose where two mountains intersect. Sunrise 11 days after the spring equinox did not occur at a well-marked place (Figure 10). There is also a so-called ramp, the orientation of which is such that the morning of the equinox and the 11th day before and afterwards are indicated in a similar manner as in the central palace sanctuary. The walls of the ramp are related to sunrise in such a way that the ramp is completely illuminated only in the 11-day interval. As at Knossos this interval and its relationship to the autumn equinox would have provided an easy method for regulating a lunisolar calendar, just as it did also at the peak sanctuary on Petsophas.

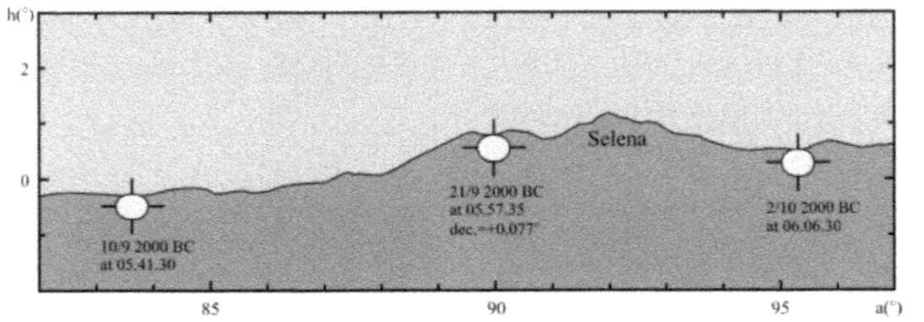

Figure 10. Sunrise opposite Mount Juktas on the autumn equinox and 11 days before and after the equinox. Reconstruction by Göran Henriksson.

The small terracotta figurines of animals and humans, separately made parts of bodies, and other objects found on the peaks can be connected to stars and constellations as we see them used in Aratos' poem *Phainomena*. This would confirm us in our view that the sanctuaries were also used as observatories and astronomical schools where the centuries-old activity of building up knowledge of the motions of the celestial bodies had laid in the hands of a group of learned persons on the most suitable observation places of the island. Future generations were educated there to continue after their teachers. A detailed presentation has been published explaining how these small objects could have been used in teaching the motions of the stars and constellations (Blomberg, 2006). To us it is clear that these sanctuaries were intimately connected to the study of celestial phenomena.

The archaeoastronomy of the pillar crypt in the south-east house

When we measured the orientations of the pillar crypt, we found relationships to the Sun at the equinoxes that are the same as those in the central palace sanctuary. The upper limb of the Sun at sunrise appears at the intersection of the top of the Ailias ridge and the northern door frames of the two doors of the room and illuminates the hole in the southwest corner and the southern corner of the niche (Figure 11). We think that the hole was originally the site of a similar kind of marker as the bowl in the central palace sanctuary for recording the Sun's rays on equinox morning. The Mycenaean successors to the Minoans may have destroyed it, perhaps thinking that it concealed the sanctuary treasure.

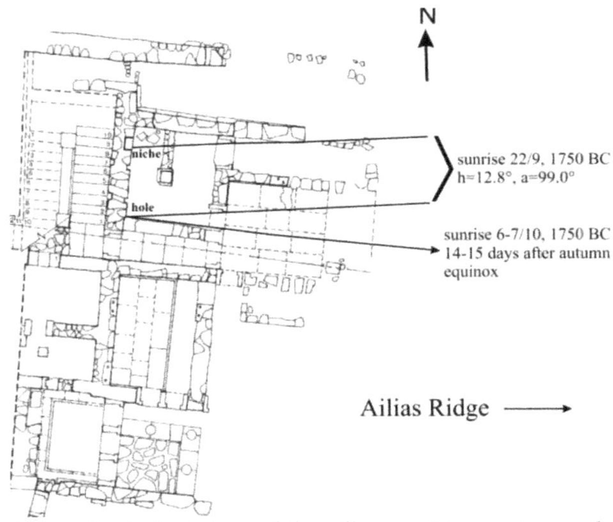

Figure 11. Orientations of the pillar crypt to sunrise on the autumn equinox. Orientations by Göran Henriksson

The position of the house on the slope in relation to the Ailias ridge opposite is responsible for the relationship of sunrise on the equinoxes to the pillar crypt, just as the alignment of the central palace sanctuary to the ridge is responsible for the phenomena to be observed there. The Sun appears at the intersection of the Ailias ridge and the northern door frame of the southern door (Figure 12), just as it did at the central palace sanctuary (Figure 6).

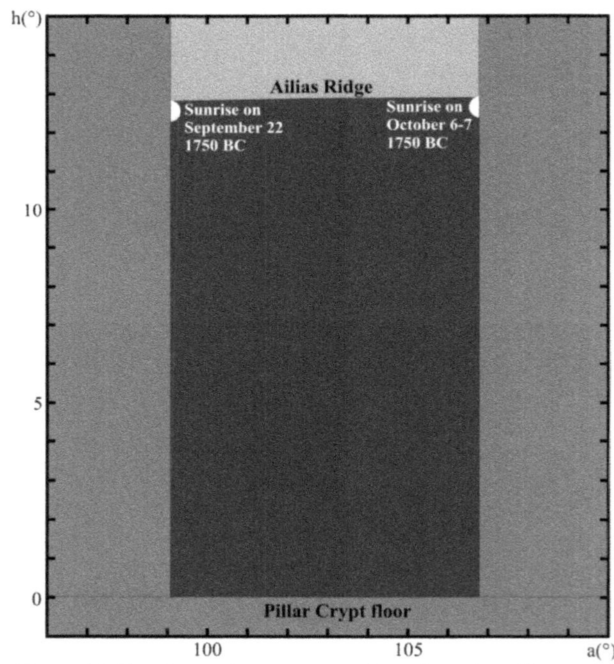

Figure 12. Sunrise at the autumn equinox from the pillar crypt. Reconstruction by Göran Henriksson.

Fourteen or fifteen days later it appeared at the southern side of the door opening. The results of the orientation of the south-east house are thus somewhat different than in the central palace sanctuary. If the new crescent Moon, for example, appeared on the equinox then the full Moon appeared on the 14th or 15th day. This could have been due to the different symbolic meanings of the days of the calendar or another possibility of calibrating the calendar in bad weather.

Discussion

The fundamental basis for good relations between humans and the divine world was honouring the gods in the proper ways. Paramount among these were the annually recurring monthly and seasonal festivals devoted to the gods. It was therefore essential that the calendar be correct, with the months occurring in their proper seasons. Learned members of elite political and religious groups with control over the calendar would have inspired deep awe. In turn their elite status may have been significantly supported by their knowledge of the heavens. We have in the area of Knossos three prominent buildings with different functions connected to the Minoan calendar, the palace as residence of the political elite, the sanctuary on Mount Juktas and the south-east house both presided over by priests. Despite the paucity of written documents these three monuments give us insight into the existence of an ancient and advanced astronomical knowledge shared by the highest elite of Minoan political and religious spheres, knowledge used to regulate the calendar. We can surmise the different calendar functions as follows.

At Knossos there was the installation of a reflecting bowl in the central palace sanctuary perhaps as early as the late Early Bronze Age whereby the morning of the equinox was clearly marked by the reflection cast by the Sun's rays as they fell in the liquid-filled bowl exactly at sunrise. The reflection was, and still is, cast on the western wall every morning for 11 days after the autumn equinox and then does not appear until 11 days before the spring equinox. In a consecutive series of three years, the rays of the Sun appear on the morning of the equinox after 365 days. However, on the fourth year the rays do not appear after 365 days but, as we all know today, after 366 days. This method of using reflections was not only a way of clearly and dramatically marking the moment of the equinoxes, but it also marked the cycles of the Sun with respect to the equinoxes and indicated when a day should be added to form cycles of whole days for the solar year (Figure 9). The knowledge of the lunisolar calendar probably also meant that the political elite knew of the 8- and 19-year lunar cycles. There is a tradition of the 8-year cycle as the period of rule for King Minos (Blomberg and Henriksson, 1996). The political leaders probably had ceremonial roles in connection with the beginning of the year, the beginning of cycles and the time for festivals.

At the peak sanctuary on Mount Juktas, the most impressive and probably the oldest one in Crete, sunrise on the equinoxes and on the 11th day after the autumn equinox were well marked from the altar (Figure 10) and also the ramp. This, together with evidence from three other sites, means that the Minoans had a lunisolar calendar beginning in connection with the autumn equinox by the end of the Early Bronze Age (before c. 2000 BC). The sanctuary, as the place for the observation and teaching of celestial phenomena, would most likely have been a focal point of important cultic activity.

At the south-east house the variety of relationships to the Sun from the pillar crypt may indicate the working base for the priests, where observations were recorded and contemplated, a theoretical framework constructed over time, and models for the calendar devised with methods of calibration. The priests working here and on the peak may have had a deeply esoteric role in mapping out the larger cycles for the motion of the Moon and the Sun.

Our results show that 10 of the 12 solar months were pointed out by orientations at 13 of the 15 Minoan buildings which we have so far evaluated (Figure 13).

Sites	Months
Petsophas, Phaistos, Knossos, Mt Juktas, Vathypetro	first (autumn equinox)
Malia, Vathypetro	second
Modi	third
Chamaizi, Vathypetro	fourth (winter solstice)
Modi	fifth
Malia, Vathypetro	sixth
Petsophas, Phaistos, Knossos, Mt Juktas, Vathypetro	seventh (spring equinox)
Gournia	eighth
Gonies, Petsophas, Pyrgos	tenth (summer solstice)
Gournia	twelfth

Figure 13. Solar months marked by orientations.

The publications of the buildings are given below in the references. Thus we have good indications that the Minoans used not only a lunar calendar but also a solar calendar and had methods for calibrating both of them. We do not have such clear evidence for any neighbouring culture.

Acknowledgment

We would like to thank the Greek Archaeological Service for permission to study the sites in our project and also Ch. Kritzas and Alexandra Karetsou, Ephors at Herakleion, for helping us in our work there. For efforts on our behalf we have been indebted to the late Berit Wells, former director, and also to Bodil Nordström, secretary, at the Swedish Institute in Athens. For permission to publish material for our figures, we are grateful to J.W. Myers, E.E. Myers, and G. Cadogan (Figure 1), S. Hood and W. Taylor (Figures 2 and 4), and J. Driessen (Figure 5).

References

Blomberg, M. and Henriksson, G. 1996. Minos Enneoros: Archaeoastronomical light on the priestly role of the king in Crete. In P. Hellström and B. Alroth (eds), Religion and Power in the ancient Greek world. Proceedings of the Uppsala Symposium 1993 (= BOREAS. Uppsala Studies in Ancient Mediterranean and Near Eastern Civilizations, 24), 27-39. Uppsala, Uppsala University Press.

Blomberg, M. and Henriksson, G. 2003. The Minoan peak sanctuary on Pyrgos and its context. In M. Blomberg, P. E. Blomberg and G. Henriksson (eds), Calendars, symbols, and orientations: legacies of astronomy in culture. Proceedings of the 9th meeting of the European Society for Astronomy in Culture, 27-30 August 2001. Uppsala Astronomical

Observatory Report 59,127-134. Uppsala, Uppsala University Press.

Blomberg, M. and Henriksson, G. 2005a. Orientations of the late Bronze Age villa complex at Vathypetro in Crete. Mediterranean Archaeology and Archaeometry 5, 51-61.

Blomberg, M. and Henriksson, G. 2005b. Some problems in Minoan archaeoastronomy. In M. Köiva, I. Pustylnik, and L. Vesik (eds), Cosmic Catastrophes: a collection of articles. Proceedings of the European Society for Astronomy in Culture (SEAC) 2002, 15-22. Tartu.

Blomberg, M. and Henriksson, G. 2007a. The function of the Minoan oval house at Chamaizi. In Emília Pásztor (ed.), Archaeoastronomy in Archaeology and Ethnography, Papers from the annual meeting of SEAC (European society for Astronomy in Culture) held in Kecskemét in Hungary in 2004, British Archaeological Reports International Series 1647, 15-18.

Blomberg, M. and Henriksson, G. 2007b. Orientations of the Minoan palace at Phaistos in Crete. In I. Liritzis (ed.), Ancient watching of cosmic space and observation of astronomical phenomena, Rhodes, Greece. Proceedings of the international conference on archaeoastronomy, SEAC 14. Mediterranean Archaeology and Archaeometry 6, 185-192.

Blomberg, M., Henriksson, G. and Papathanassiou, M. 2002. The calendaric relationship between the Minoan peak sanctuary on Juktas and the palace at Knossos. In T. M. Potyomkina and V. N. Obridko (eds), Astronomy of ancient Societies, Proceedings of the Conference 'Astronomy of Ancient Civilization' of the European Society for Astronomy (SEAC) associated with the Joint European and National Astronomical Meeting (JENAM)) Moscow May 23-27, 2000, 81-92. Moscow, Nauka.

Blomberg, P. E. 2006. A new interpretation of the figurines from Petsophas and Traostalos. In E. Tabakaki and A. Kaloutsakis (eds), Proceedings of the 9th Cretological congress, Elounda (Crete) 4 (1-7 October 2001), 333-347. Heraklion, Typokreta Kazanakis ABEE.

Catling, H. W. 1973-1974. Archaeology in Greece, 1973-1974. Archaeological Reports, 34-37

Davis, E. N. 1995. Art and politics in the Aegean: the missing ruler. P. Rehak (ed.), The role of the ruler in the prehistoric Aegean, Aegaeum 11, 11-19. Liege, Université de Liège.

Driessen, J. 1990. An early destruction in the Mycenaean palace at Knossos: new interpretation of the excavation field-notes of the South-East area of the West Wing, vol. 2. Leuven, Katholieke Universiteit Leuven.

Evans, A. J. 1901-1902. The Palace of Knossos. The Annual of the British School at Athens 8, 1-124.

Evans, A. J. 1902-1903. The Palace of Knossos. The Annual of the British School at Athens 9, 1-153.

Evans, A. J. 1921. The Palace of Minos at Knossos 1. London, Macmillan.

Guarducci, Margherita 1945. Note sul calendario cretese. Epigraphica 7, 72-87.

Hallager, E. 1987. A "harvest festival room" in the Minoan palaces? An architectural study of the pillar crypt area at Knossos. In R. Hägg and N. Marinatos (eds), The Function of the Minoan palaces, Proceedings of the Fourth International Symposium at the Swedish Institute in Athens, 10-16 June, 1984, Skrifter utgivna av Svenska institutet i Athen 32, 169-177. Göteborg, Åström.

Henriksson, G. and Blomberg, M. 1996. Evidence for Minoan astronomical observations from the peak sanctuaries on Petsophas and Traostalos. Opuscula Atheniensia 21, 99-114.

Henriksson, G. and Blomberg, M. 2008 Orientations at the Minoan peak sanctuary on Philioremos near Gonies (Maleviziou), Crete. In J. Vaiškūnas (ed.), Astronomy and Cosmology in Folk Traditions and Cultural Heritage, Archaeologia BALTICA 10, 125-130. Klaipėda, Klaipėda University Press.

Henriksson, G. and Blomberg, M. 2009. Solar Orientations of Bronze Age Shrines in Gournia, Crete. In Cosmology Across Cultures, Proceedings of a workshop held at Parque de las Ciencias, Granada, Spain 8-12 September 2008. Astronomical Society of the Pacific Conference Series 409, 430-435.

Henriksson, G. and Blomberg, M. in press. Archaeo-astronomical investigation of the peak sanctuary on Modi (Siteia). In Proceedings of the 10th Cretological congress, Chania (Crete), 1-8 October 2006.

Hood, S. and Taylor, W. 1981. The Bronze Age Palace at Knossos. London, The British School at Athens.

Manning, S. 1995. The Absolute Chronology of the Aegean Early Bronze Age: archaeology, radiocarbon, and history. Monographs in Mediterranean Archaeology 1. Sheffield, Sheffield Academic Press.

Myers, J. W., Myers, E. E. and Cadogan, G. 1992. An Aerial Atlas of Ancient Crete. London, Thames and Hudson.

PRINCE P. A. PUTYATIN WAS THE FORERUNNER OF RUSSIAN ARCHAEOASTRONOMY

N. DMITRIEVA

Abstract: One of the first researchers of Russian archaeoastronomy was Prince P. A. Putyatin (1837-1919), a member of the Tsar's family. He was an anthropologist, an ethnographer, an archaeologist, a collector, and an astronomer. Based on the first analysis of a certain artefact of the Neolithic epoch he concluded that ancient people already had observed the sky and developed some kind of astronomy. Putyatin moreover engaged himself in studying the history of astronomy.

There is a very interesting but little-known fact in the history of Russian archaeoastronomy. The first discovery in this area was made in 19th century (1885) by a lineal descendant of the Tsar's family in Russia (Rurik dynasty). The name of this discoverer was Prince Pavel Arsen'evich Putyatin (Figure 1).

Figure 1: Prince Pavel Arsen'evich Putyatin.

The Prince Pavel Arsen'evich Putyatin was born on the 4th of October in 1837. He died on the 6th of January in 1919 in Petrograd. Prince Pavel Arsen'evich Putyatin, 32nd *knyaz* (generation) from Rurik, belongs to the Russian princely family of the Drutsky branch. The forefather of the Putyatins was Prince Ivan Semenovich Mladshiy Drutskoj who had the nickname 'Putyata' and lived in first half of the 15th century. He was also known as the fellow-fighter Svidrigailov who signed the treaty of Lithuania as sword-bearer.

He was an anthropologist, an ethnographer, an archaeologist and a collector. Besides that he was also a good astronomer.

He was the first Russian archaeologist who began to cooperate intensively with high-ranking specialists in European archaeology and went abroad to take part in the work of European congresses and conferences from 1889 to 1905. He studied in detail the works of the leading ethnologists of Europe such as J. Lubbock (with whom he corresponded), C. Layel, D. Evans, A. Katrfazh, E. Larte, G. de Mortilje, E. Kartaljak, L. Kapitan and E. V. Rivjer (he was personally acquainted with M. S. Sautuola, who was the pioneer of cave art in France, the first Palaeolithic interments and the persevering supporter of 'rehabilitation' of Altamira). That is why he always kept track of up-to-date discussions and trends in Stone Age archaeology.

The talent of Putyatin as a collector is surprising. In his study room he kept a lot of remarkable things which could draw attention of the researcher. For example, a valuable collection of paintings, prints, sketches, water colours and architectural drawings made by Russian and foreign masters; a collection of photographic, oleographic, lithographic and other pictures; a huge archaeological collection that included findings from the Palaeolithic age to the Middle Ages; a numismatic collection; a collection of medallions and counters; a collection of rare busts; and at last a unique palaeographic collection of different manuscripts, old published books, rolls, columns and the papers of other forms; as well as autographs and facsimiles of well-known persons, from the Tsars Alexei Mikhailovich and Peter the Great to the author A. S. Pushkin. All these cultural treasures, being historically connected with the Valdai region where they existed and were collected, were opened for researchers.

During archaeological excavations of the late Stone Age site in Bologoe (the north of the European part of Russia), the Prince found a stone sharpener in the form of a trapeze with sockets on its surface. In the pattern of these sockets Putyatin saw the image of the constellation Ursa Major and some other subpolar constellations (Boötes, Draco and Canes Venatici) (Figure 2). He showed non-standard thinking, an inclination to innovation and the ability to go outside the limits of dogmatic directions, as well as the ability to transform a finding that was ordinary at first sight into an object of special importance. He was thus able to distinguish the hidden information in it and make a conclusion unexpected for nineteenth-century archaeology: people of the Stone Age not only had knowledge of the Earth but also knowledge of the sky. Moreover, they grouped figures of constellations from the stars that proved that an ancient person had rather a high

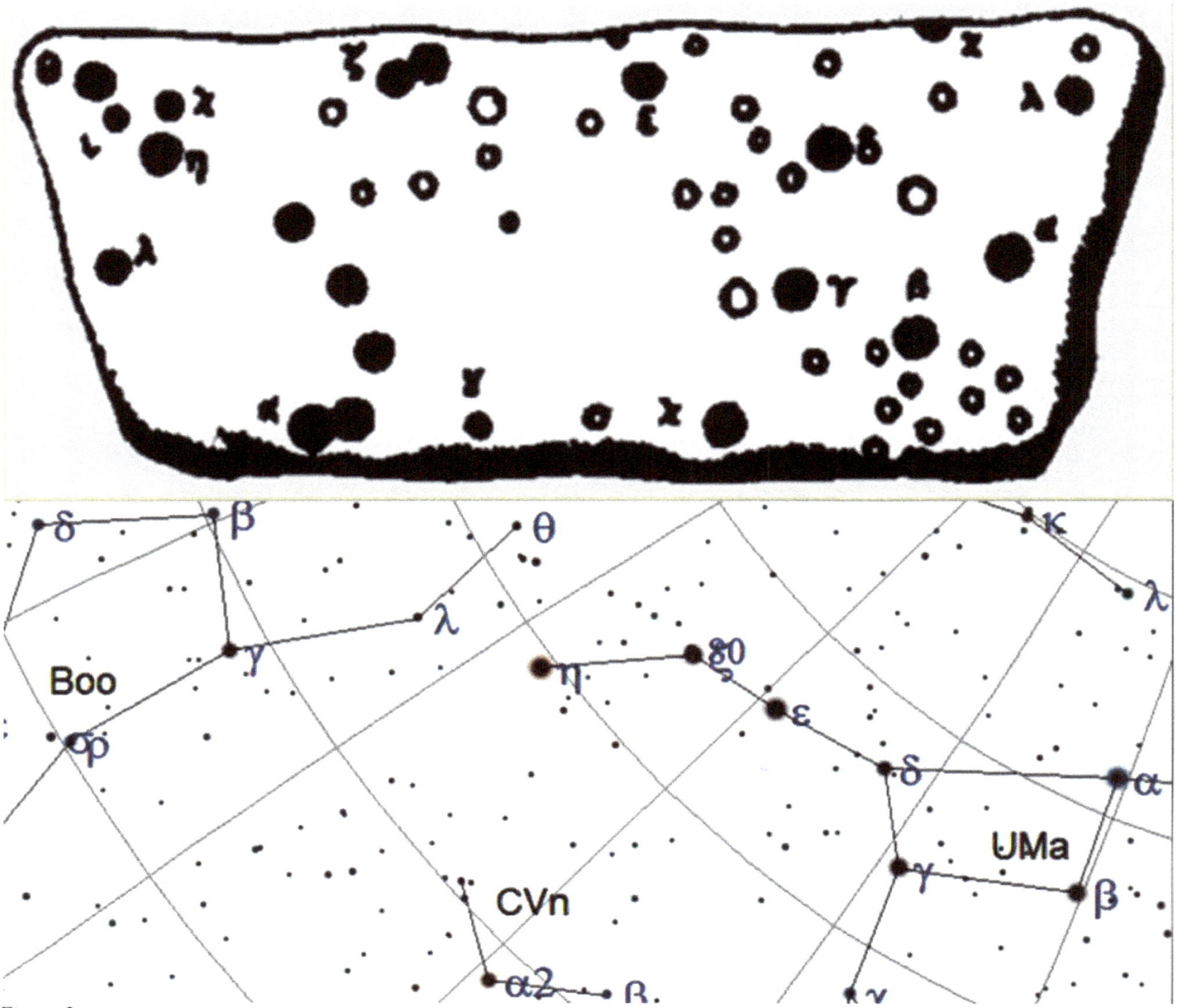

Figure 2.

level of culture. It also became evidence of the first foundations in natural-science knowledge.

P. A. Putyatin declared the discovery of a certain document of an astronomical category that testified to the interest of the creators of Eastern Europe Stone Age culture in the sky. The point is that if we accept the accuracy of his estimation of the essential contents of the 'pattern' engraved on 'the stone sharpener', we will have to accept the idea that the Stone Age ancestor perceived not only the Earth as his 'monastery', but he also looked at the open sky and not only did he notice the points, the stars, but he also began to group them by marking out the constellations. His attention was drawn to the North Star, around which all stars of the North revolve during a night. It was evidence of the fact that an ancient person thought not only of a terrestrial life but he also began thinking of something else. To Putyatin, it was proof of the complexity of the ancestor's worldview, as well as his intelligence and his acquiring of a certain volume of natural scientific knowledge.

After this discovery, in his articles and reports Putyatin began to talk about the necessity of a new sphere of archaeology that is 'archaeological astronomy' (or according to his other version—space astronomy). For example, during a session of the Archaeological Society he made a report, 'From the Area of Astronomical Archeology, The Image of the Constellation Ursa Major on a Sharpener of the Stone Age in Russia'. Unfortunately, his ideas about the rudiments of astronomical knowledge of the Stone Age people were not supported by archaeologists. These ideas were regarded as absurd and were not published in any archaeological resources. Putyatin collected and processed historical sources in ten languages in the fields of archaeology, observational astronomy, astrology and mythology.

Prince Putyatin corresponded with the well-known populariser of astronomy, Camille Flammarion. He consulted Flammarion about the astronomical sense of his finding:

Mister the Director!

Your investigation of changes of constellation outlines through the stars movements and your researches on ancient astronomy induce me to offer for your consideration the facsimile of the astronomical drawing made on a slate sharpener that I found in my manor of Bologoe estate near the lake of the same name in a Neolithic layer. This ancient object of the Stone Age the epoch of which is difficult to define, as it can be easily seen from the drawing enclosed full-scale, has a characteristic outline of Septentrion (seven stars), together with other stars, which it seems to me, belong to the same Ursa Major and [the next] constellations such as Canes Venatici, the Bootes and the Draco. The star Zeta of the Ursa Major is marked as double. The star Alpha, the Heart of Charles II of Canes Venatici constellations has the same image. These stars as well as the Star Alpha of the Ursa Major are represented in larger size than others. There are three big stars between Alpha and Delta, between Delta and Epsilon and near the star Zeta. These stars cannot be distinguished in the sky just with the eyes. There are also other smaller stars. Didn't these stars change their size since prehistoric time? Is the configuration of seven stars of the Ursa Major constellation is abnormality of the prehistoric worker or it corresponds to the valid figure which is going back to the old ages? Line AB can mean the horizon. Doesn't line CD meant a pole direction during an epoch of that time when stars Lambda and kappa of the Draco were polar?

Were other stars noted on a sharpener were ever visible just with the eyes? The knowledge of the sky might be practised in Russia Protoscythians.

Flammarion published this letter as well as the answer, called 'Archeologie Astronomique' ('Astronomical Archeology') in the magazine *L'Astronomie* (1885, Vol. 4, 48) in which he made a critical analysis of the image of the constellation Ursa Major on a stone sharpener:

> the Previous message is of great interest, and we thank prince Putyatin that he addressed it to us. It really sets many problems.
>
> First of all, the form of constellations changes with the lapse of time. If the drawing of the Ursa Major had been made 50 000 years earlier, 40 000 years or only 25 000 it would differ a lot from our so that we could define approximate time for it. The own movements with the help of which these seven stars were enliven have mainly the action that from century to century raises a star Ita to continuation of almost direct and slightly changed lines (delta, epsilon, zeta) and to that to remove it from other six to physical system of which it however does not belong and makes with them one group only optically. But in the drawing of a sharpener one can notice that the star Ita position does not differ so much from the present position <...>. And from this point of view the astronomy cannot define the exact time of this drawing. The Bootes even at the present time could be drawn similar to this image... .

One can conclude that the image of constellations on the plate from Bologoe was made during an epoch which is not far from the present time and thus it means that it is right to date this plate back to Neolithic times. At the beginning of the third millennium BC, the North Pole was really defined by the tails of Draco (line CD on P. A. Putyatin's scheme is focused on that direction, see Figure 2).

Further, Flammarion wrote:

> Anyway, it is the document of a rare interest (a unique document within our awareness until now) and thus we cannot thank enough those researchers, who search for traces of the last centuries and make them known for us, because being known these findings bring the contribution to the general course of human nature study.

Probably this very letter of P.A. Putyatin gave an initial impulse to Flammarion's future searches for astronomical knowledge sources in primitive cultures. He considered the history of astronomy to be the main area in studying the laws of society development:

> Without The History of Astronomy we cannot estimate anything neither in the History of Mankind nor in the History of the Universe.
>
> The primitive antiquity of astronomy, an origin of celestial sphere and constellations, views of ancient people on the World structure - all this scientific panorama represents an immense sight which reveal the whole soul and life of mankind with its power and powerlessness, with feverish curiosity and languor, with eternal persistent desire to reach, learn and rule over everything become visible. (Flammarion 1885).

Putyatin's knowledge of astronomy and research into the knowledge of ancient and medieval people about the sky is really unprecedented for the end of 19th century. The prince showed his 'daring' in his vision of an 'astronomical aspect' in the 'pattern' of a slate plate from Bologoe Station; it showed the ability of a creative person to perceive the facts out of the delusions of traditional archaeological dogmatics and his conviction in the fact that 'the primitive life was based on the universal principles which should be taken into consideration first of all'.

Besides the astronomical interpretation of this archaeological artefact, the Prince carried out unique research into ancient and medieval peoples' knowledge of the sky and gathered them in an integrated table, 'The Key Alchemical of the Book of XVI Century fra Basilii Valentini, Adapted to Astronomical, Natural-Historical and Other Concepts', which measured 35 x 100 cm.

The table contained 7 lines of stars (the Sun, the Moon, Mars, Mercury, Jupiter, Venus and Saturn) and 22 columns including their characteristics: signs, days of week—including Chaldean, Jewish, Egyptian, 'East Indian' (Sanskrit), Latin, Romanian, Italian, French, Swedish, German and English—metals, stones, sciences, colours, good and bad qualities, months of the conception and the birth of a person et al., all given according to the days of the week.

It won't be an exaggeration to say that Prince Putyatin was the forerunner of the Russian archaeoastronomy.

> The country which is able to become stronger making progress with the help of science and at the same time to support the workers from the rough environment is great. I think that there is much use from individual, unobscure, well-known scientists from different counties who are out of the pedantry and who for the sake of science share the knowledge for the common good of mankind.
>
> Prince P. A. Putyatin, 1886

References

Flammarion, Camille. 1885. Archeologie Astronomique. L'Astronomie 4, 48-50.

Putyatin, P. A. 1886. From the Area of Astronomical Archeology. The Image of Constellation of the Ursa Major on a Sharpener of the Stone Period. The Manuscript. It is stored in the archive of the Institute of History of Material Culture, f. 3, No. 557 (in Russian).

Putyatin, P. A. 1886. Iz oblasti astronomicheskoy archeologii. Izobrazhenie sozvezdiya Bol'shoy Medveditsu na tochilke kamennogo perioda. Rukopis'. g. Khranitsya v arkhive Instituta istorii material'noy kul'tury, f. 3, No. 557.

Mel'nikov, V. L. 2002. The Forerunner of the Russian Astroarcheology the Prince P. A. Putyatin and the Roerichs // History and Culture of East Asia, 196-202. Novosibirsk (in Russian).

Mel'nikov V.L. 2002. Predtecha rossiyskoy astroarkheologii knyaz' P. A. Putyatin i sem'ya Rerikhov// Istoriya i kul'tura vostoka Azii, 196-202. Novosibirsk (in Russian).

Larichev, V. E. 2009. Astroarcheology: Through Thorns To Stars! The beginning of «unpopular scientific tradition» formation // Astroarcheology is the natural-scientific tool for comprehension of protosciences and astral religions of hieratic priesthood of ancient cultures in Khakassia. The collection of scientific articles, 7-35. Krasnoyarsk, Gorod. (In Russian).

Larichev V. E. 2009. Astroarkheologiya: "skvoz' terni – k zvezdam!". Nachalo stanovleniya "nepopulyarnoy nauchnoy tradizii"// Astroarkheologiya – estestvenno-nauchnyy instrument poznaniya protonauk i astral'nukh religiy zhrechestva drevnikh kul'tur Khakassii. Sbornik nauchnukh statyay, 7-35. Krasnoyarsk, Gorod.

Fig 1. Source
http://www.lomonosov.org/article/jetapy_formirovaniya_kollekcii_muzeya_knyazya_putyatina.htm

Cultures in Europe, Asia, Oceania and Africa

THE CELESTIAL ENGINE AT THE HEART OF TRADITIONAL HAWAIIAN CULTURE

W. BRUCE MASSE

Abstract: The systematic observation and interpretation of the celestial heavens was of paramount importance to the fabric of pre-European contact Hawaiian culture. Hawaiian oral tradition is imbedded in royal chiefly genealogies extending back through time more than 95 generations. Because of the unusual richness of Hawaiian genealogical oral history, which includes observational details of spectacular transient celestial events (e.g., comets, supernovae, meteor storms) historically recorded in Asia and elsewhere in the world along with the occurrence of total solar eclipses whose characteristics can be reconstructed through astronomy software programs, we have a unique 1500-year chronological window on the role of traditional astronomy in Hawaiian culture. These data are augmented by the study of obscure classificatory lists of stars and planets associated with 19th century Hawaiian royal chiefs. Equally remarkable is the genealogical record of the Kanalu astronomical priesthood, serialized in 1900-1901 by a Hawaiian language newspaper. Together these data demonstrate that the celestial heavens permeated all aspects of traditional Hawaiian culture and played a key role in the conduct of war and other critical aspects of religion and chiefly power.
Keywords: traditional astronomy, myth, mana, transient celestial events

Introduction

Ever since the European discovery of the Hawaiian archipelago by Captain James Cook in 1778 and his subsequent quasi-deification and violent death on the shores of Hawai'i Island in 1779, Hawai'i and its native peoples have exuded mystery and power in the eyes of Western civilisation and captured our imagination. Royal chiefs were resplendent in their red and yellow feathered helmets and capes (Figure 1), but their persona was filled with such a sanctified presence that the mere touch of their shadow upon a commoner would spell immediate death for the unfortunate individual.

Figure 1. Hawaiian royal chief drawn by artist Jacques Arago in 1819 during the world voyage of Louis de Freycinet.

Temples were filled with mysterious tall fierce-looking wooden images (Figure 2). The overall *kapu* (*tapu*, taboo) system itself spelled out a strict set of rituals and behaviour because royal chiefs were considered partly divine. The four major gods (Kanaloa, Ku, Lono and Kane) demanded continual obsequious behaviours from all people including royal chiefs. And a host of demigods protected individual islands while spirits of all forms protected individuals and animated the natural world.

Figure 2. The Hale o Keawe temple on Hawai'i Island as observed by missionary William Ellis in 1823.

Adding to this sense of mystery and power, the Hawaiian Islands, at the extreme northern tip of the vast Polynesian triangle, are distinctive in their geographic remoteness from other island groups and continental land masses, a situation shared with Easter Island (Rapa Nui) and New Zealand at the extreme eastern and southern tips of this triangle. Such remoteness emphasizes the skill by which the Polynesian colonists and later voyagers sailed their canoes, in the case of Hawai'i more than 2000km over the open ocean, thus implying a sophisticated knowledge of the celestial heavens as a necessary aid to their long-distance navigation. Hawai'i Island is also unique in having the world's most active volcano, Kilauea, whose current eruption has been ongoing since 1983. Not unexpectedly, the Hawaiian demigoddess of volcanism, Pele, ranked equally with the four major gods in terms of her power and influence on Hawaiian peoples. And Hawai'i was unique in terms of European contact population size (perhaps over 400,000), and in terms of social and political stratification, perhaps becoming what archaeologists refer to as an archaic state (Kirch, 2010).

Hawai'i likewise is blessed in terms of the volume and quality of oral traditions and historical documents that have survived to the present. These materials, although largely underutilized to date, have allowed for the reconstruction of many aspects of pre-European contact Hawaiian society to a degree not possible throughout the rest of Polynesia and Oceania. Scholarly research interest has focused on the invaluable observations by 19th century native Hawaiian historians such as David Malo, Kepelino (Zephyrin), John Papa I'i, and especially Samuel Manaiakalani Kamakau, who either themselves participated in royal chiefly society prior to the overthrow of the *kapu* system following the 1819 death of King Kamehameha I, or had direct access to many priests and chiefs who had served under the *kapu* system. Kamakau's writings are part of a rich body of oral traditions preserved in native Hawaiian language newspapers during the 19th and early 20th centuries, much of which remains to be translated into English.

Of critical importance for our study of Hawaiian traditional astronomy is the recent publication by Malcolm Chun of the first complete English translation of the published portion of the extraordinary but esoterically obscure genealogy of the Kanalu priesthood of Hawaiian astronomer priests (Figure 3). This genealogy was initially recorded in 1900-1901 in the newspaper Ka Nupepa Kuokoa by Benjamin K. Namakaokeahi (2004), the self-proclaimed last known Kanalu priest.

Figure 3. Kanalu astronomy priesthood genealogy recorded by Benjamin Namakaokeahi (2004).

Kanalu was one of the most powerful priestly orders in Hawai'i at European contact (Malo, 1951; I'i 1959, 39), with Hewahewa, the high priest of Kamehameha I, being the head of the order when the *kapu* system was abolished. Namakaokeahi's book provides an unprecedented look at the history of a Polynesian priestly order, with most surviving genealogies being for the ancestral lineages of royal chiefs rather than the lineage of high priests. Ongoing analysis of this genealogy (Masse et al., 2015, in preparation), indicates an unbroken set of traditions that begin with a deadly tsunami in the Hawaiian archipelago genealogically dating to the 6th century AD (matching a major tsunami event geologically documented along the coast of Kauai'i and radiocarbon dated to the period of 440-670 AD—see Burney, 2002) through the early 15th century AD. The tsunami was stated as having been such a devastating event that one of the primary functions of the Kanalu order was that of the controlled repopulation of the Hawaiian Islands, which was maintained through strict training and through prophecies and their predicted events. Regrettably, for a currently unknown reason only the first 42 generations were published in Ka Nupepa Kuokoa, with the most recent 23 generations (c. 1400-1900 AD) apparently being lost to posterity. While surprising that esoteric knowledge of the priestly Kanalu order could survive largely intact for four generations (80 years) after the overthrow of the *kapu* system, analysis of the traditional astronomy data contained in the book (which among other observations includes identifiable known historic transient celestial events) demonstrates that the information in the genealogy is authentic and chronologically accurate, as will be discussed below.

While these sources provide significant details regarding Hawaiian traditional astronomy, the fact that Hawaiian myths and legends are embedded in royal chiefly genealogies is the most singular and invaluable aspect of Hawaiian traditional knowledge systems, particularly for our understanding of Hawaiian traditional astronomy. It turns out that many Hawaiian myths about the supernatural exploits of gods, demigods and culture heroes demonstrably describe the observation of remarkable historic transient celestial events such as total solar eclipses, the passage of unusual long-tailed great comets, supernovae, and great meteor storms (Masse et al., 1991; Masse, 1995; Masse et al., 2015, in preparation).

That these celestial events occur in myths attached to the reigns of named chiefs in Hawaiian royal genealogies provides the unique opportunity to study myth, traditional astronomy and other key aspects of Hawaiian history and knowledge systems using an absolute chronology extending back in time more than 1500 years. One result of such study is the finding that observed processes and events in the celestial heavens permeated all aspects of traditional Hawaiian culture and played a key role in the conduct of war and other critical aspects of religion and chiefly power. These celestial processes and events were the engine that drove Hawaiian culture.

Royal Chiefly Genealogies, *Mana*, Myth and Transient Celestial Events

Polynesians participated in a form of ancestor worship in which one of the most important criteria for being a royal chief was that of having a genealogy that demonstrated unbroken lineage with important earlier chiefs and culture heroes. Each royal chief owned their own genealogy and associated naming and genital chants (to help foster procreative abilities) that not only extolled their own virtues but also listed in epic form the adventures and exploits of their ancestors. The more a living chief could demonstrate connectivity with famous ancestors, the stronger their claim for being of the highest royal blood. In this regard, commoners were strictly forbidden to have their own genealogies. Missionary William Ellis (1967, 418), who lived for several years in Tahiti and other portions of the Society Islands of central Polynesia, noted that Hawaiian royal genealogies seemingly better preserved the chronological accuracy of their generation lists—at least 30 and perhaps as many as 100 generations—than was the case elsewhere in Polynesia.

The power of genealogical chants was not simply a demonstrated unbroken lineage succession but rather the accumulated *mana* associated with each named ancestral chief. The Polynesian concept of *mana* has been much debated by scholars in terms of its nature and meaning for traditional Polynesian society (e.g., Handy, 1927; Shore, 1989). For our purpose in discussing Hawaiian traditional astronomy, we turn to standard Hawaiian dictionary usage. In 1865 (Andrews 1974, 382), *mana* was defined as supernatural power, such as was supposed and believed to be an attribute of the gods. Edward Tregear's (1969, 203) 1891 Maori-Polynesian comparative dictionary further makes the distinction that something or someone possessing *mana* means having qualities that ordinary persons or things do not have.

Stokes (1930) noted that the mere recitation of a royal genealogy would transfer the accumulated *mana* of all named ancestors to the living chief. It is clear that the actual linguistic structural carriers of *mana* in the royal chiefly genealogies were the embedded supernatural-filled legends or myths associated with the named ancestral chiefs. This situation is evident in the famous Kumulipo creation myth (Beckwith, 1972), a 2000-line genealogical chant created to honour the birth of 18th century Hawai'i Island chief Lonoikamakahiki, recited in full in 1779 in a ceremony to honour Captain James Cook as being an incarnation of the god Lono. More than 60 lines of Kumulipo text are devoted to the supernatural exploits of ancestral demigod chief Mauiakalana.

Because the stories associated with many ancestral chiefs and super heroes were so widely known, simply invoking the names of famous ancestors was enough to conjure up the appropriate *mana* of that individual, a cognitive pattern referred to as the 'silence principle' (Barber and Barber, 2004; Masse et al., 2007). This explains why so few of the preserved genealogies (e.g., McKinzie 1983, 1986) actually list the deeds for each chief and yet there are large numbers of isolated myths linked to genealogical chiefs (e.g., Fornander, 1916-1920; Kamakau, 1991; Kalakaua, 1972; and Beckwith, 1970). Many of these now detached myths were undoubtedly once part of formal genealogical and naming or genital chants. An example of the silence principle is the statement by native Hawaiian historian David Malo (1951, 246): 'The traditions that have come to me of Maui are false, lies; and I repeat no falsehoods'. Thus while Malo includes Mauiakalana in his listing of chiefly genealogies, he includes none of the stories associated with the famous demigod chief (discussed below).

Fortunately for our understanding of myth, *mana*, and Hawaiian traditional astronomy, many myths have retained their original connection to a named genealogical chief so as to place the story into existing genealogies. This relative chronology of myth can be demonstrably matched with known and dated natural events to provide an absolute chronological anchor for Hawaiian myths and their associated genealogies. For example, during his tour of Hawai'i Island in 1823, missionary William Ellis was told stories of specific lava flows being created by the volcano goddess, Pele, during battles with demigods and chiefs during the reigns of named genealogical chiefs (Figure 4). Several of these same lava flows have been recently radiocarbon dated by geologists from the Hawaii Volcanoes Observatory (e.g., Holcomb, 1987).

Figure 4. Pele battles Kamapua'a, the half-pig half-human demigod who appeared during the mid-13th century reign of Oahu chief Olopana (as imagined by artist Dietrich Varez).

By comparing the radiocarbon dates with heuristic dates for the storied eruption events derived by assigning a value of approximately 22 years to the lengths of each chiefly generation prior to Kamehameha I (see Stokes, 1933), a reasonable match between the radiocarbon dates and the chronologically relative genealogical dates can be demonstrated for the eruptive events (Masse et al., 1991; Swanson, 2008). This strongly suggests that the Pele myths have captured real observations about the eruptive behaviour of Kilaeuea Volcano, and have correctly placed the myths and their associated observations within the relative chronology of the chiefly genealogies. Thus Hawaiian volcano myths record real history.

This model of mythology reflects the growing awareness that natural events are part of the fabric of mythology (Vitaliano, 1973; Piccardi and Masse, 2007; Masse and Masse, 2007; Cashman and Giordano, 2008):

> Myth is a structured narrative...derived from oral transmission, and typically created or assembled and perpetuated by knowledge specialists who use supernatural elements and images in order to categorize and explain observed natural phenomena and events that are of perceived vital importance or of special relevance to the social order and well-being of a given culture (Masse et al. 2007, 17).

Figure 5. Examples of transient celestial and meteorological phenomena encoded in Hawaiian mythology.

The fact that many myths were attached to naming chants and birthing genealogies meant that priests were required to memorize and chant them during periodic special moments within at least the lifetime of their chief (e.g., New Year festivals, temple rituals, the visits of other royal chiefs). Not only were the stories often repetitively chanted, but they also were depicted in hula dances and other performance media thus ensuring a surprising level of accuracy in their oral transmission during the course of many generations (Masse et al., 2007; Masse et al., 2015, in preparation).

This natural history model for myth becomes even more robust by considering that the *mana* being captured in royal birthing stories and naming chants associated with Hawaiian chiefly genealogies would also have included other evocative natural phenomena and events in addition to volcanic eruptions (Masse, 1995). Perusal of various Hawaiian myths indicates that many myths contained storyline details associated with naked eye processes and events in the celestial heavens (Figure 5).

Such matches can be demonstrated by correlation with the recorded historic observation of spectacular comets, meteor storms, supernovae, auroras and other observed celestial and meteorological events witnessed during the past 2000 years in Asia, the Near East, the Mediterranean and Europe (Masse, 1995; Masse and Tuggle, 1998; Masse et al., 2015, in preparation), along with accurate reconstructions of total solar eclipse paths and behaviour (Masse and Espenak, 2006; Masse et al., 2015, in preparation). These matches can be made because many spectacular transient celestial events have historically recorded unique signatures that are identifiable in the richly detailed Hawaiian myth storylines. Among many examples of genealogically-fixed celestial myths are the following:

(1) A supernatural owl befriended an Oahu Island commoner (Kapoi) who inadvertently collected owl eggs for breakfast, but returned them when asked by the owl. Kapoi built a shrine to honour the owl, which infuriated the chief of Oahu, Kakuhihewa, who has prohibited such construction during the period of his own religious observances. Kakuhihewa ordered the sunrise sacrifice of Kapoi at a temple on Waikiki Beach. A flock of owls covered and darkened the rising sun, frightening Kakuhihewa who immediately released Kapoi and made him a chief. This myth (Kamakau 1991, 23) captures the main observational details of a unique sunrise total eclipse on 10 April 1679 (during the nesting season for native Hawaiian owls) whose extremely narrow eclipse path was centred on Waikiki Beach.

(2) The mid-14th century Hawaiian sorceress, Wa'ahia, set herself on fire and changed into clouds that ascended into the heavens at twilight, fighting each other like gamecocks and dashing around like pigs. The vivid descriptions in this story (Malo 1951, 251-254) exquisitely capture the behaviour of historically observed auroral substorms (Akasofu, 2007; see also Figure 5) including that '...the clouds took on a singular appearance, some were white, some glistening, some green, yellow, red, black, blue black, black and glistening; and the sky sparkled and flashed with light'. Because there are several independent versions of the Wa'ahia myth, this provides the unusual opportunity to explore aspects of the process of Hawaiian myth-making, including that of the retelling

of the story under the influence of late 19th century Victorian scholarship in which most of the supernatural attributes are stripped from the myth (e.g., Kalakaua, 1972).

(3) The half-pig, half-man demigod Kamapua'a, attempted to seduce the volcano goddess, Pele, in her pit at Kilauea volcano (Figure 4). My translation of a portion of the naming/genital chant for Kamapua'a well captures his cometary nature (Fornander 1919, Vol. V, Part II, 316-319): 'Resembling precious flashing lightning; anointed flashings; the eyes of the gods appearing to fly in the heavens; bent flashings—persistent flashings; breathing flashings--dashing sun; suspended glittering outburst; suspended royal chief threatening the heavens...the grand stranger with dazzling eyes; Strange pig-shaped child of dust, the pig-shaped light in the heavens...' This chant describes the great comet of 1264 AD, one of the most remarkable of all historically recorded comets of the past 2000 years, with multiple tails that changed size and character several times over several months of naked eye visibility (Kronk, 1999).

(4) Missionary William Ellis (1967, Vol. 2, 415) collected the following story during his stay in the Society Islands:

> It is related that Maui, one of their ancient priests or chiefs, was building a marae, or temple, which it was necessary to finish before the close of the day; but, perceiving the sun was declining, and that it was likely to sink before the work was finished, he seized the sun by his rays, bound them with a cord to the marae, or an adjacent tree, and then prosecuted his work till the marae was completed, the sun remaining stationary during the whole period. I refrain from all comment on this singular tradition; which was almost universally received in the islands.

This story encodes a unique sunset total solar eclipse whose path encompassed the entirety of the Samoan Islands on 4 August 761. This story was subsequently spread throughout Polynesia during a wave of voyaging throughout eastern Polynesian (including Hawai'i) around 800-1200 AD. The fame of Mauikalana was considerably enhanced by a series of major comets and meteor storms between 759-765, including an apparition of Comet Halley simultaneously visible with another comet for several weeks in May and June of 760.

The Sociology of Hawaiian Stars

The myths and chronologically associated celestial events described above help to explain how spectacular celestial and meteorological events were converted into the supernatural-laden myth storylines that became powerful *mana* for their associated chiefs. Also apparent is the fact that the celestial heavens were not simply considered just the backdrop to these storied events, but rather were viewed as a separate world, operating parallel to but distinct from that of human-occupied Earth.

Western science and civilisation has largely lost this perspective in our current views of the origin and purpose of constellations (Motz and Nathanson, 1988). These views typically suggest that constellations were simply a convenient shorthand to enable astrologers and incipient astronomers to map the movements of the planets and other interesting aspects of the fixed celestial heavens.

Polynesian cosmology, however, makes its clear that even the fixed celestial heavens were animate and creative (Makemson, 1941; Johnson and Mahelona, 1975). In a Tahitian cosmological myth collected in 1818 (Henry 1928, 359), the shooting stars were the first born of the primeval creator deities, followed in succession by the Moon, the Sun, comets and finally by the constellations, planets and Milky Way. The meticulous construction of these births and their locations in the sky largely mirrors the initial cosmological creations of the Hawaiian Kumulipo, other than the fact that the Kumulipo begins with creatures of the sea and earth (e.g., Johnson, 1981), while the Tahitian chant populates the sky, which was considered a living realm just like the Earth.

The sky was the celestial firmament, a vast ocean-like space whose stars, constellations, Sun, Moon and planets formed a powerful deified geography, with comets, meteors, supernovae, auroras and other phenomena being active spirits and demigods (and sometimes interlopers) that interacted with and challenged the deities of the celestial sphere—*mana* in its most sanctified and supernatural form. This animate view of the sky extends to the manner in which it was viewed by the Kanalu order of astronomer-priests (see also Makemson, 1938, 1939). The Kanalu genealogy contains a classification of the stars into 92 individual stars organized into seven categories (Figure 6). Only two of these categories deal with navigation and the monthly calendar. Two of the categories are associated with religious functions including prophecy (revelation) and stars for specific gods. The final three classes—chiefly stars, priestly stars and commoner stars—mirror Hawaiian society itself.

STAR CATEGORY
15 Chiefly Stars
15 Commoner (People's) Stars
15 Revelation Stars
15 Priestly Stars
15 Land (Navigator) Stars
3 Stars for the Gods
14 Month Stars

Figure 6. The Kanalu Classification of Stars.

These 92 stars are not perceived of as being randomly scattered in the night sky. Rather, their observable and attributive qualities (size/magnitude, colour, position) allow for their categorization by the astronomer priests, including into male and female roles and their placement into spatial relationships that are appropriate to their defined roles.

Comets and the Iconography of Deity

To fully appreciate the role of the celestial heavens in daily Hawaiian society, we need only look at how the celestial heavens are portrayed in relation to the major Hawaiian deities. Valeri (1985, Figure 6) has aptly shown that the four principal gods (Ku, Lono, Kanaloa, Kane) were provided a series of specific attributes by which they were recognized and hierarchized. These include colour, direction, days of the month and periods of the day, natural and inorganic phenomena, plants animals, seasons and functions. Kirch (2004) and Ruggles (2007) take this one step farther by suggesting that Hawaiian temples can be sorted with respect to the four major gods by the nature of their orientations and other attributes.

While these models of temple construction likely capture significant aspects of these important deities and their associated attributes, the celestial manifestations of the four deities are even more profoundly evident in terms of pure iconography. For example, when missionary William Ellis (1979, 75) conducted his tour of Hawai'i Island in 1823, he notes that:

> The natives were very desirous to shew us the place where the [wooden] image of Kairi the war-god stood, and told us that frequently in the evening he used to be seen flying about in the neighbourhood, in the form of a luminous substance like a flame, or like the tail of a comet.

Approximately 100 wooden images of the Hawaiian war god, Ku ('Kairi' or Kaili is short for Ku-kaili-moku) have survived from the overthrow of the *kapu* system despite concerted attempts by missionaries to destroy them so as to encourage conversion to Christianity. As indicated in the top two images of Figure 7, the facial features and hairstyle of the depicted image of Ku closely match the details of a telescopic drawing of the nucleus and coma of Comet 1861 J1 (Kronk, 2003) made by Warren de la Rue (see Sagan and Druyan 1985, 175).

Although the 1861 comet was viewed and drawn by looking through a telescope, there were several comets during the 18th centuries that approached close enough to the Earth that the nucleus and coma might have appeared to the naked eye much like that of the 1861 drawing by de la Rue. Comet 1770 L1, which holds the record for the closest confirmed passage by Earth (Kronk, 1999) is an excellent candidate for the naked eye observations leading to the carving of the Figure 7 Ku image.

Equally remarkable is the fact that of the hundred or so surviving wooden images of Ku, approximately half exhibit the downflowing stylized hair referred to as Hairstyle 1 in Figure 7, while the other half exhibits the elaborate 'headdress' depicted in Figure 7 as Hairstyle 2 (see Valeri 1985, Plates 8 and 9). Given such a sample, it is reasonable to assume that the two hairstyles are representative of the pre-perihelion (style 1) and post-perihelion (style 2) orbital phases of an observed comet perihelion passage around the sun, which should appear in relatively equal numbers to observers on Earth. This relationship is depicted in the lower left image of Figure 7, in which the solar wind blows the tail of the comet behind the nucleus during the pre-perihelion passage toward the Sun whereas the solar wind is strong enough to blow the tail of the comet ahead of the nucleus (thus appearing like a headdress) as the comet moves back out through the Solar System.

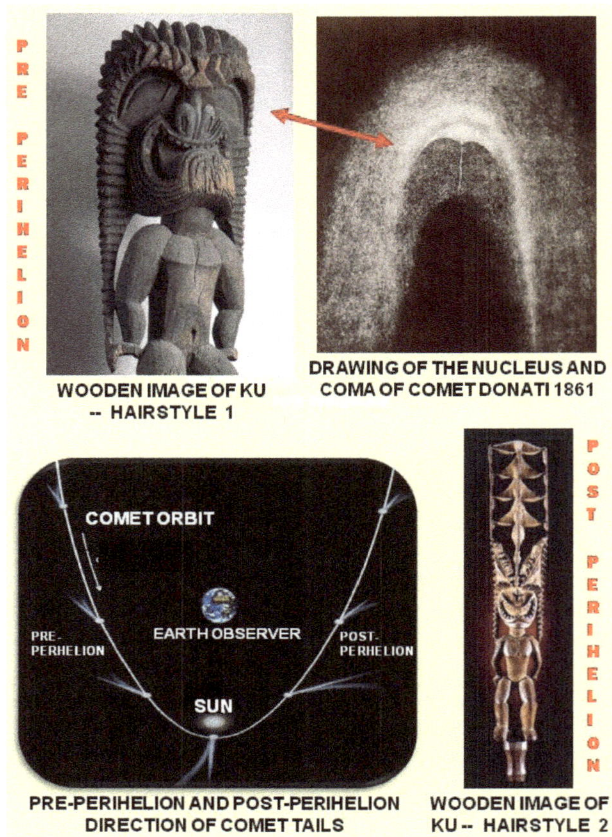

Figure 7. Comparisons of the wooden images of the war god Ku with aspects of observable comet attributes.

Acknowledgements:

I thank my colleagues Rubellite Kawena Johnson and H. David Tuggle for many years of productive dialog regarding Hawaiian traditional astronomy and myth. Fred Espenak of NASA Goddard kindly calculated the eclipses discussed in this paper. I also thank the organizers of the 2010 SEAC meetings for hosting and moderating an excellent conference, and for the lively comment and debate of the meeting participants.

References

Akasofu, S-I. 2007. Exploring the Secrets of the Aurora, 2nd edition. New York, Springer.

Andrews, L. 1974. A Dictionary of the Hawaiian Language. Rutland, VT., Charles E. Tuttle (1865).

Barber, E. W. and Barber, P. T. 2004. When They Severed Earth From Sky: How the Human Mind Shapes Myth. Oxford, Princeton University Press.

Beckwith, M. 1970 Hawaiian Mythology. Honolulu, University of Hawaii Press.

Beckwith, M. W. 1972. The Kumulipo: A Hawaiian Creation Chant. Honolulu, University of Hawaii Press.

Burney, D. A. 2002. Late Quaternary chronology and stratigraphy of twelve sites on Kuai'i. Radiocarbon 44, 13-44.

Cashman, K. V. and Giordano, G. (eds) 2008. Volcanoes and human history. Special publication of the Journal of Volcanology and Geothermal Research 176, 325-438.

Ellis, W. 1967. Polynesian Researches. London, Dawsons of Pall Mall (1829).

Ellis, W. 1979. Journal of William Ellis: Narrative of a Tour of Hawaii, or Owhyhee; with Remarks on the History, Traditions, Manners, Customs, and Language of the Inhabitants of the Sandwich Islands. Rutland, Vermont, Charles E. Tuttle (1825).

Fornander, A. 1916-1920. Hawaiian Antiquities and Folk-lore. Honolulu, Memoirs of the Bernice P. Bishop Museum 4-6.

Handy, E. S. C. 1927. Polynesian Religion. Honolulu, Bernice P. Bishop Museum Bulletin 34.

Henry, T. 1928. Ancient Tahiti. Honolulu, Bernice P. Bishop Museum Bulletin 48.

Holcomb, R. T. 1987. Eruptive history and long-term behavior of Kilauea Volcano. In R. W. Decker, T. L. Wright and P. H. Stauffer (eds), Volcanism in Hawaii, 261-350. Washington, D.C., U.S. Geological Survey Professional Paper 1350.

I'i, J. P. 1959. Fragments of Hawaiian History. Honolulu, Bishop Museum Press

Johnson, R. K. and Mahelona, J. K. 1975. Na Inoa Hoku: A catalogue of Hawaiian and Pacific Star Names. Honolulu, Topgallant Publishing.

Johnson, R. K. 1981. Kumulipo: Hawaiian Hymn of Creation, Vol. 1. Honolulu, Topgallant Publishing.

Kalakau, D. 1972. The Legends and Myths of Hawaii. Rutland, VT., Charles E. Tuttle (1888).

Kamakau, S. M. 1991. Tales and Traditions of the People of Old. Honolulu, Bishop Museum Press.

Kirch, P. V. 2004. Temple sites in Kahikinui, Maui, Hawaiian Islands: Their orientations decoded. Antiquity 78, 102-114.

Kirch, P. V. 2010. How Chiefs Became Kings: Divine Kingship and the Rise of Archaic States in Ancient Hawai'i. Berkeley, University of California Press.

Kronk, G. F. 1999. Cometography: A Catalog of Comets, Vol. 1, Ancient-1799. New York, Cambridge University Press.

Kronk, G. F. 2003. Cometography: A Catalog of Comets, Vol. 2, 1800-1899. New York, Cambridge University Press.

Makemson, M. W. 1938. Hawaiian astronomical concepts I. American Anthropologist 40, 3790-383.

Makemson, M. W. 1939. Hawaiian astronomical concepts II. American Anthropologist 41, 589-596.

Makemson, M. W. 1941. The Morning Star Rises: An Account of Polynesian Astronomy. New Haven, Yale University Press.

Malo, D. 1951. Hawaiian Antiquities. Honolulu, Bishop Museum Press (1898).

Masse, W. B. 1995. The celestial basis of civilization. Vistas in Astronomy 39, 463-477.

Masse, W. B., Carter, L. A. and Somers, G. F. 1991. Waha'ula heiau: The symbolic and regional context of Hawai'i Island's 'Red Mouth' temple. Asian Perspectives 30, 19-56.

Masse, W. B. and H. D. Tuggle. 1998. The date of Hawaiian colonization. In C.M. Stevenson, G. Lee and F.J. Morin (eds), Easter Island in the Pacific Context South Seas Symposium: Proceedings of the Fourth International Conference on Easter Island and East Polynesia, 229-235. Los Osos, Bearsville and Cloud Mountain Presses.

Masse, W. B. and Espenak, F. 2006. Sky as environment: Solar eclipses and Hohokam culture change. In D. E. Doyel and J. S. Dean (eds), Environmental Change and Human Adaptation in the Ancient Southwest, 228-280. Salt Lake City, University of Utah Press.

Masse, W. B., Barber, E. W., Piccardi, L. and Barber, P. T. 2007. Exploring the nature of myth and its role in science. In L. Piccardi and W. B. Masse (eds), Myth and Geology, 9-28. London, Geological Society of London Special Publication 273.

Masse, W. B. and Masse, M. J. 2007. Myth and catastrophic reality: Using myth to identify cosmic impacts and massive Plinian eruptions in Holocen South America. In L. Piccardi and W. B. Masse (eds), Myth and Geology, 177-202. London, Geological Society of London Special Publication 273.

Masse, W. B., Johnson, R. K. and Tuggle, H. D. 2015. Islands in the Sky: Traditional Astronomy and the Role of Celestial Phenomena in Hawaiian Myth, Language,

Religion and Chiefly Power. Honolulu, University of Hawaii Press. (In preparation; under contract).

McKinzie, E. K. 1983. Hawaiian Genealogies Extracted From Hawaiian Language Newspapers, Vol. 1. Honolulu, University of Hawaiian Press.

McKinzie, E. K. 1986. Hawaiian Genealogies Extracted From Hawaiian Language Newspapers, Vol. 2. Honolulu, University of Hawaiian Press.

Motz, L. and Nathanson, C. 1988. The Constellations: An Enthusiast's Guide to the Night Sky. New York, Doubleday.

Namakaokeahi, B. K. 2004. The History of Kanalu: Mo'oku'auhau 'Elua, translated and edited by M. N. Chun. Honolulu, First People's Productions (1900-1901).

Piccardi, L. and Masse, W. B. (eds) 2007. Myth and Geology. London, Geological Society of London Special Publication 273.

Ruggles, C. 2007. Cosmology, calendar, and temple orientations in ancient Hawai'i. In C. Ruggles and G. Urton (eds), Skywatching in the Ancient World: New Perspectives in Cultural Astronomy, 287-329. Boulder, University Press of Colorado.

Shore, B. 1989. Mana and Tapu. In A. Howard and R. Borofsky (eds), Developments in Polynesian Ethnology, 137-173. Honolulu, University of Hawaii Press.

Stokes, J. F. G. 1930. An evaluation of early genealogies used for Polynesian history. Journal of the Polynesian Society 39, 1-42.

Stokes, J. F. G. 1933. New bases for Hawaiian chronology. Hawaiian Historical Society 41st Annual Report, 23-65.

Swanson, D. A. 2008. Hawaiian oral tradition describes 400 years of volcanic activity at Kilauea. Journal of Volcanology and Geothermal Research 176, 427-431.

Tregear, E. 1969. The Maori-Polynesian Comparative Dictionary. Oosterhout, N. B., Anthropological Publications (1891).

Valeri, V. 1985. Kingship and Sacrifice: Ritual and Society in Ancient Hawaii. Chicago, University of Chicago Press.

Vitaliano, D. B. 1973. Legends of the Earth. Bloomington, Indiana University Press.

COSMIC POWER: THEMES OF ASTRONOMY AND POWER WITHIN THE FILM *COSMIC AFRICA*

JARITA C. HOLBROOK

Abstract: *Cosmic Africa* (Rogers and Rubin, 2002) is a South African produced film with the financial backing of Cosmos Studies (Druyan). Rodney Thebe Medupe, astrophysicist, and sometimes cultural astronomer, is the central focus of the film. The film captures the struggle of the main character as he seeks to reconcile his African heritage with his chosen profession of astronomy. Though it feels like a documentary it is a film. Themes of astronomy and power are found throughout the film; for example, there are several juxtapositions of big science astronomy with indigenous astronomy. On another level, there is the choice of filming locations: each place had some indigenous astronomy connection. Thus, the people featured at each location gained power through appearing in the film because of their proximity to sources of indigenous astronomy knowledge. A third level of the theme of astronomy and power is found within Medupe himself: he is African, yet an astronomer at the same time. Thus, he is both an insider and an outsider, he alternately enjoys an easy camaraderie with the people at each location but when he is at the telescope the power differential emerges. Finally, the film shows the power that astronomy has to take people who have been ignored by the power centres of the world and elevate them into the realm of people of international interest.

Keywords: Ethnoastronomy, Astronomer, Africa, Indigenous Knowledge, Filmmaking

The Film *Cosmic Africa*

Cosmic Africa is a film that focuses on the living traditions and practices in Africa that are connected to the sky. The film is a production of Cosmos Studios (Druyan)—a legacy of Carl Sagan. The central character is Rodney Thebe Medupe, a South African astrophysicist whose research focuses on stellar oscillations. Dr Medupe was among the first black South Africans to receive a doctorate in astrophysics in South Africa in 2002. The filmmakers present the film as a search undertaken by Dr Medupe to discover the connection between Africans and the sky, focusing on indigenous astronomy.

Figure 1: Dr Rodney Thebe Medupe with Dr J. C. Holbrook.

The film locations include South Africa, Namibia, Mali, and Egypt. Though Dr Medupe is Tswana, Tswana astronomy is not explored in any great detail in the film. There is only the brief enactment of the tale of how the Milky Way was formed within the context of a tale that his grandparents told him. Rather, Dr Medupe is shown at the South African Astronomical Observatory at Sutherland doing telescope observations; this reinforces the idea that he is an African removed from his heritage.

Cosmic Africa is beautifully filmed using real shots of the night sky rather than placing the sky in afterwards. Though it feels like a documentary it is a film because, although it is based on a real person who appears in the film, parts were scripted, filming sites were chosen ahead of time, and there is a professional narrator who narrates the entire film. The astronomy content of the film is within the context of each culture. The filmmakers did not shy away from the mystical and religious dimensions of astronomy knowledge. For example, they include a ritual sacrifice to bring rain, a healing trance dance and an elaborate divination practice. Thus, the film does a good job presenting African indigenous astronomy in its cultural context and the associated cultural practices. Because the film is such a rich resource, it is possible to analyse the context for themes of 'astronomy and power'.

Astronomy and Power

There are five themes of astronomy and power that are explored but these are not the only ones present in the film. The themes presented can be considered simply a starting point for a deeper analysis of astronomy and power.

1. WHO HAS THE RIGHT TO CREATE ASTRONOMY KNOWLEDGE?

Rodney Thebe Medupe had to struggle against the perception that Africans are not intelligent enough to study science in order to become an astrophysicist. Although he has been successful, the bias remains that Africans are incapable of doing science. *Cosmic Africa*, through using Dr Medupe, sets up the main questions of the film: are there Africans who have an astronomy tradition and are any of the African astronomy traditions still alive today? The narrator of the film, who is not Dr Medupe, but is meant to represent him, says in the introductory sequence 'While my science has uplifted me, it has also removed me from my heritage'. This statement gets to the heart of the issue: even Dr Medupe does not equate being an African with being an astrophysicist.

The question, 'who has the right to create astronomy knowledge?', is echoed throughout the film starting with the existence of Dr Medupe and continuing as he travels from place to place in Africa identifying individuals who have knowledge of African indigenous astronomy. The question is ever-present because those people he finds look vastly different from astronomers and their practices related to indigenous astronomy are different from those of astronomers.

2. Harnessing Cosmic Power for Personal Transformation (Rappenglück, 2010)
Returning to the narrator's statement 'While my science has uplifted me, it has also removed me from my heritage', Dr Medupe has personally transformed himself through becoming an astronomer. Thus, he has 'harnessed cosmic power' by learning the science of astronomy and producing original research and unique scientific results in astronomy. He has become a producer of astronomy knowledge. At the same time, he is aware that this transformation is important and in the film he extends this path to others.

While visiting the Dogon in Mali, he shows the children sunspots and the narrator states, 'these children are the same age that I was when I built my first telescope'. Though he is not teaching them how to make their own telescopes, he is using a telescope to show the sunspots.

At the end of the film, Dr Medupe is shown back in South Africa where he is lecturing to a class of African students and teaching them how to build their own telescopes. Thus, he is seen extending the pathway of learning astronomy (harnessing cosmic power) for personal transformation to this group of children, too.

3. Astronomy, Technology, and Power
Telescopes appear throughout *Cosmic Africa* and they convey astronomy in a way that nothing else does in the film. The telescope is *the tool* of astronomers. They perhaps mark the difference between scientifically trained astronomers and everyone else in that only astronomers have access to the biggest telescopes on Earth and those in space.

Telescopes demonstrate the connection between astronomy, technology and power. Astronomy is driven by technological advances in telescope and detector design. Observatories, which are collections of telescopes situated in the dominant position of mountaintops, are symbols of power. Each telescope represents hundreds of years of innovation and thousands of dollars of investment. The maintenance and daily functioning of observatories require a great deal of money as well. Thus, observatories and telescopes are reflective of intellectual and financial achievement for their nations and their populace.

In *Cosmic Africa*, Dr Medupe as he is today is presented against the background of the South African Astronomical Observatory in Sutherland, South Africa. He introduces himself while standing in the open dome of one of the telescopes several stories above the ground. He is shown in the computer control room of the telescope as he discusses how he wants to find Africans that still know their indigenous astronomy.

The sequences in Namibia, Mali and Egypt are dominated by nature and Africans rather than the high tech world of astronomy. However, each of these sequences is broken up by Dr Medupe using his telescope or his computer. Dr Medupe uses his portable telescope to perform being an astronomer. The telescope marks him as something different from his African hosts and at one point he acknowledges that 'they might think I am odd for knowing [about a partial solar eclipse]'. Thus, he is conscious of being an outsider but it is when he sets up his telescope that the differences are the most obvious.

4. Celestial Knowledge as a Manifestation of Cosmic Order, Natural Order (Rappenglück, 2010)
Dr Medupe holds a bachelor's degree in physics and a doctorate in astrophysics. As part of that training, he learned celestial mechanics and the physical processes underlying our understanding of the universe. Thus, he has a deep academic understanding of cosmic order.

While in Namibia, Dr Medupe uses his telescope and camera to record the partial solar eclipse. In order to alleviate the fears of his hosts, Dr Medupe explains the mechanics of solar eclipses demonstrating his celestial knowledge.

5. Interactive Powers of Nature reflective of Indigenous Worldview of Nature (Rappenglück, 2010)
Dr Medupe practices the worldview associated with astronomy culture and academic culture: technology is a tool for learning about nature as a means to understanding the cosmic order. In contrast, from what can be understood in the film, the worldview of the people in Namibia and Mali is different. The worldview of these people is one in which technology can influence nature. An example of this is the sequence of Dr Medupe photographing the partial solar eclipse. His Namibian hosts draw the conclusion that 'the telescope is eating the sun'. This prompts Dr Medupe to explain the mechanics of solar eclipse as mentioned already.

In Mali, the blacksmith is said to be the sorcerer of the Sun. He makes things of iron using smelting, firing, hammer and anvil (technology). However, as normal as it is to see a blacksmith at work, in the film he is creating a cloud hook—a device for capturing clouds to bring rain. It is a magical device that demonstrates the worldview of being able to use technology to influence nature.

Conclusions
The film *Cosmic Africa* contains several themes of astronomy and power. Dr Medupe as the centre point of the film is the focus of most of these themes because he always brings the astronomy point-of-view to the indigenous settings explored in the film. His scientific view is attached to and demonstrated by his telescope. The telescope is the symbol of astronomy power. It marks Dr Medupe as different from other Africans, but it also shows that he has the right to create astronomy knowledge. He demonstrates his expertise, his understanding of the natural world/cosmic order, at various times during the film using his telescope.

On the other hand, the indigenous Africans who have astronomy knowledge have a different worldview than that of astronomers. Their actions in the film suggest that their worldview is one of interacting with nature in order to influence nature, oftentimes to gain a desired outcome such as rain. Nature is not removed from human actions in general and the sky is included in this. Their view of technology is that it can influence nature and technology can be made for the purpose of influencing nature, such as with the cloud hook of the Dogon.

What are the realms of influence of astronomy and power found in the film *Cosmic Africa* (see 'Standpoint Theory' in Harding, 2004 and references therein)? Astronomy and power intersect with the political realm because Dr Medupe was raised in apartheid South Africa. There was the very real possibility that he would not be able to complete a doctorate in astrophysics. Astronomy and power is personal for Dr Medupe because it allowed for his personal transformation into an astrophysicist and the distance from his African heritage. Astronomy and power is societal for Dr Medupe because he now has a societal role as an astrophysicist and educator. Finally, astronomy and power is global for Dr Medupe as he circulates through the international astronomy community and through his participation in the film *Cosmic Africa*.

There is another thing to consider: the power of astronomy. Creating the film *Cosmic Africa* transformed certain groups of people from obscure figures to people of international interest because they appeared in the film. This is linked to both their indigenous astronomy knowledge and practices and their proximity to places connected to indigenous astronomy.

Tension [false dichotomies, necessary film devices]

[Astronomer]	[African]
Astronomy	Indigenous Astronomy
Celestial Realm Independent of actions of human	Actions of Humans can influence celestial realm
Technology to understand nature	Technology to influence nature
International Social Status for mastering Astronomy	International Social Status for mastering indigenous astronomy – only possible through the film, Cosmic Africa

Figure 2: Table of the artificial tension found in the film between Astronomer and African.

Returning to the film's tension between astronomy and African, Figure 2 summarizes the differences found between these two. The dichotomy established by the filmmakers between being an African and an astronomer is a way of having tension, which makes the film interesting. Thus in some ways it is artificial although perhaps something that Dr Medupe did struggle with but not on such grand a scale.

The tension can be reframed as being between 'astronomy' and 'indigenous astronomy'. The table captures the differences between the two worldviews and the powerful social status differences between astronomers, Dr Medupe and the indigenous astronomy knowledge holders.

In conclusion, *Cosmic Africa* is a resource for studying themes of astronomy and power. It is a good educational tool for teaching about contemporary cultural astronomy theories. The film is available for purchase at http://www.africanfilmny.org/dvd/dvd.html, the website for the New York African Film Festival.

Figure 3. Poster for the film Cosmic Africa.

Acknowledgements:

Special thanks to Dr R. Thebe Medupe for letting me pick apart this wonderful film.

References

Druyan, A. Cosmos Studios (Online). Available at: http://www.carlsagan.com/index_cosmos.htm [Accessed 13 July 2010].

Harding, S. (ed.) 2004. The feminist standpoint theory reader: Intellectual and political controversies. New York, Psychology Press.

Rappenglück, M. 2010. Keeper of Time and Guardians of Space – Some Basic Concepts of Astronomy and Power SEAC 2010 Conference: Astronomy and Power: How Worlds are Structured. Gilching.

Rogers, A., Rubin, C., Foster, C. and Foster, D. 2002. Cosmic Africa (film). Capetown, Cosmos Studios.

THE CONCEPT OF POWER AND COSMOLOGY: MANIPULATION OF COSMOLOGY BY SPIRITUALISTS OR NATIVE DOCTORS (*DIBIA*), A CASE STUDY OF THE IGBO SOCIETY OF NIGERIA

BARTH CHUKWUEZI

Abstract: In parts of Africa, and indeed among the Igbo of southeastern Nigeria, a group of renowned medicine men and spiritualists achieved tremendous fame and wisdom because of their manipulation and knowledge of cosmic forces. Basden (1966 [1921]) has tried to show that some Igbo medicine men were great diviners who wielded a lot of influence and helped to solve the intricate relationships of social existence and adaptation to a seemingly difficult cultural environment. They were able to give meaning to existence and to order the society in terms of meeting both spiritual and material existentialism. While they were not kings or top rulers, they were quite important in consummating the various rituals of succession to high office and ameliorating the problems of societal existence, and for this they were highly revered.
This paper will attempt to discuss how Igbo spiritualists and native doctors, through cosmic vision knowledge, have been able to acquire tremendous influence and power within Igbo society.
Keywords: cosmology; power; spiritualists; Nigeria; Igbo

Introduction

In various societies the knowledge of the cosmos has been a major factor for the acquisition of various powers and influence. In ancient Egypt and other ancient civilizations, the knowledge of the cosmos informed some of the ventures into what people may refer as mystic science and acquisition of esoteric powers. In all cultures human beings have tried to harness the powers of the cosmos to achieve various objectives, and people who are versed in the knowledge of these powers are highly regarded and at times referred to as 'wise men'. Astral travel, astrology and horoscopes all have a bearing on cosmic forces and the way people were able to manipulate the cosmic forces. The ancient civilizations of Egypt, Mesopotamia, China, Mesoamerica and India—and indeed various other civilizations—had great spiritualists who anchored a great deal of their mystic knowledge and fame through the manipulation of cosmic forces.

The Igbo spiritualists and native doctors were the soul of the society, for they ordered the meaning of existence. They were the custodians and interpreters of the knowledge of human existence. In a society where the religious inclination to well-being and good social existence is dictated by gods and spirits propelled by the knowledge of the cosmos, those who have the exclusive knowledge of these are highly revered and respected. They command great influence and power within the society.

In fact, most of the known spiritualists were known traditional leaders and titleholders who commanded much respect in the society. The spiritualists derived their influence, power and prestige through the manipulation of cosmic vision, which gave them various forms of social, religious and political power within the society.

In traditional Igbo society, spiritualists and native doctors were aware of the cosmic forces which they manipulated in various forms. The spiritualists were the custodians of the native gods and deities such as *Ala* (earth goddess), *Amadioha* (god of thunder), *Anyanwu* (the sun god), *Igwe Kala* (sky god) and numerous shrines and deities which informed Igbo cosmic worldview.

Concept of Spiritualists / Native Doctors

The Native Doctors

In Igboland the native doctors are usually those who acquired mystical power and knowledge on the various religious rituals in Igboland, and they are quite respected and feared by the people. The native doctors are diviners who struggle to anchor the religious problems within Igbo society.

The priests of various shrines in Igbo society are also revered; for example, the priests of ritual cults like the secret society Ekpe, found among the cross river Igbo, are quite powerful. They are imbued with magical powers. The priests are quite powerful through the use of manipulative power from the knowledge of the universe.

At this juncture it may be necessary to explain that there are two types of ritualists in Igboland. There are those who possess ritual mystical powers called spiritualists. They are usually diviners or shrine priests who look after various deities in Igboland. They may not be involved in healing and curing of the sick. However, their mystical powers are sought after and they play various roles in the community including directing the religious affairs of the community when necessary. At times they may offer curative or healing diagnosis. This class of people includes the diviners, shrine priest and priest-kings. This group also performs various rituals for the cohesiveness and orderliness of the society.

There are also the spiritualists who are called native doctors or (*dibia*) and this group performs curative and healing functions within the community. They are known for their curative power, which may involve the use of traditional herbs, and their spiritual powers. However it is quite necessary to point out that this dichotomy is not watertight, as one could be in both categories. In order words, one could be a shrine priest as well as a native doctor who could cure and heal the sick including divination. In this paper both may be used interchangeably while explaining the influence of their power in the society.

Concept of Cosmology and Power

This has to do with the understanding of the universe and the heavenly bodies, including the stars, Moon, planets, solar system and the Earth. The cosmos for the Igbo includes the heavenly forces enumerated above and other forces such as the deities and spirit forces within and under the Earth. It also relates to the dead ancestors hovering over the affairs of men. These forces work in various ways to influence man. They shape the people's ideas and actions including their worldview. The spiritualists harness these forces and through their manipulation are able to acquire immense power and prestige within the society.

The age of enlightenment and development of Western modern science tends to be rooted in scientific empiricism, which sees the existence of any form of knowledge outside the realm of Western science as non-existent and therefore a non-reality. This has denied various forms of social reality evident in the various activities of supernatural powers or invisible powers that may not be investigated empirically.

Modern scholarship and scholars are beginning to probe into the various cultural configurations reflected in megalithic structures like the Pyramids and other amazing structural feats left behind by the ancients. They are equally accepting of the fact that the ancients possessed more than physical knowledge and must have been imbued with great spiritual intuition. Some even refer to them as ancient astronomers. In one television documentary by the *Discovery Channel*, they documented how the ancients practiced astronomy and space travel through spiritual intuition. They attributed such mystic knowledge to the shamans or spiritualists.

Power and Cosmos in Igbo Society

Igbo society is found in the southeastern part of Nigeria. They are largely a patrilinial society although a few incidents of matriliny are observed. They are influenced by the knowledge of the cosmos in their various social structures and social organisations. Chukwuezi (2007) has tried to explain how social structures and social relationships are influenced through the knowledge of the cosmos such as the Moon, stars, solar system and the spirits and deities that are found in Igboland. They have bearing on the activities of the people. Through the knowledge of the cosmos the various belief structures, such as the worship of deities and spirit beings from above and from underground spirit forces, are propitiated. Chukwuezi (2008), in another forum, has explained how the spirits and deities such as *Ala* (an earth goddess), *Amadioha* (god of thunder) and others were worshipped through the influence and knowledge of the cosmos. The focus of this paper is directed to how this knowledge of the cosmos exudes power and prestige in Igbo society.

The Igbo society believes very much in the celestial forces which influence most of the various activities in Igbo society. The Igbo people believe in heavenly forces, especially the divine beings in heaven, the gods. They believe that the abode of the god of creation (*Chukwu*) is in heaven and that from there he visits the earth at will. The stars, moon and the planets are the creator's handwork. There is a popular Igbo belief that the creation of humanity was through a divine spirit from the heavens referred to as god the creator (*Chineke*) in the name of Eri. Eri is regarded in Igboland as a culture god who came from the skies or heaven to create the world. Eri had mystical power which caused the earth, then full of water, to form land which he dwelt in to create man and woman. He then became the first man to give the Igbo their food crop yam as well as the market days. He was credited as being the god who organised the various cultural institutions of the people such as priest-king and other cultural activities that form some of Igbo ritual activities. The Eri god man is credited with magical powers and he instituted the divine kingship and priests in parts of Igboland. Onwuejeogwu (1981) captures this ideology when he stated:

> Owing to the profound influence that Nri exerted on the development of some Igbo cultures. It is almost impossible to isolate what derives from Nri culture from what derives from other Igbo cultures. But it is generally accepted by traditional Igbo people that the distinctive features of Nri culture are the Nri title complex, which is epitomized in a hierarchy of titles, such as the Ichi, Ozo or Nri titles, and the Nri religious concepts of chukwu, Alusi, Nmuo, Ifejioku and the ideology of abomination.... the Nri title complex and / or their religious ideology are found all over Igboland in one form or another, probably as a result of centuries of dissemination by the Nri people.

The descendents of Eri are called UmuNri. However, in UmuNri the representative of Eri is called the Eze Nri; he has tremendous power and is known as the spiritual head of the Igbo people. The priestly cult of Eri is quite powerful. The Nri have groups of ritualists and medicine men who cleanse Igboland of pollution and other evil forces. Through the priests, the kings of the Nri were able to hold sway over parts of Igboland without any organised military force. Rather, the power of the Eze Eri depended on the purely divine ritual of kingship instituted by Eri the heavenly mystical figure. In furtherance to the powerful nature of Eri, Ifesie says,

> It is primevally true that the Igbo worldview is heavily anchored in Eri religion, tradition and culture. God sent Eri to Igbo world for a mission that was diversified in nature and character. The first of the assignments from Chi–ukwu to Eri was to make food out of his children (Eri's). Eri's acceptance of Chi-ukwu's moral and ethical message gained for him religious character with accompanying powers sequel to it. As he was obedient to Chi-ukwu, the Igbo people whom he supplied yams and coco-yams paid homage to him. Eri, the Eze-Nri and his descendants took up the function of seeing the new crops increased all over Igboland by making yam medicine for the people who came for it and they in turn paid sacrificial tithes of goats, fowls, chalk (Nzu) and pots with which they took the medicine to their homes. It was through this religious and agrarian function of

Nri-Eri priest that the cult of Ifejioku became widespread throughout Igboland.

It might be necessary to explain that the predominant influence of Nri over Igbo cultural formation has been challenged in terms of its ubiquitousness and period of introduction (Nwaezeigwe 2007). Nonetheless it is important to point out that the general belief in the roles of gods and cosmology are noted generally among the Igbo. The story tends to express the affairs of the heavenly gods among the Igbo which informs their worldview. It may not be seen as historical exactitude, as always is the problem with myths and legends.

As a follow up to the priestly Eze Nri role in Igboland there are also powerful ritualists from Nri called Umunwanshi. The Umunwanshi are quite powerful through the knowledge of celestial rituals. They were believed to be imbued with heavenly power from the heavenly culture god Eri, according to Acholonu-Olumba (2005). They alone had the power to set up shrines of worship. They were the year counters, time recorders and living archives of the history of Igboland:

> The Nhi year-counters... are the ones that do the Igo Ojo (Invocation of Word of Power), for which reason they were called ojo-mvanchi (Oru dialect for mvanchi: Eri dwarf). They were very small. Even though of great Age, they looked like little children, whether men or women...
>
> And they all had the Ozo title (which means they bore the Eri facial scarification marks as the emblem of their sacred identity), and wore the ozo string. People would lift them up and place them on a stool and from there they would rattle off history... We were told (by people from Nri) that it was an occasional occurrence, the birth of a midget within the Nhi (Nri) clan. It is like a calling and it falls once in a while within a family. When such a child is born, the parents know immediately that (it) bears the (divine) calling of a priest, so they perform rites and release him to the spirits, who will then lead him from place to place. He keeps the spirits of the ancestors alive by his memory through which he recounts the years, the births and deaths of great men, the great happenings in Igbo land, and also by the power, which is given him to bless and curse and to foretell the future... They are very special children, blessed by the gods and given to man on special mission... (Acholonu 1989, 119-121), (Acholonu-Olumba 2005, 397).

There is also the Earth goddess who is said to be the most powerful female cult or deity in Igbo culture and whose power emanated from god above. She is referred to as the mother god with divine attributes.

The Earth goddess has priests who minister to the needs of the Earth goddess. The priests of the earth goddess are quite powerful and they dictate the various ritual rites that involve the veneration of the earth goddess for bounteous harvests and the cleansing of some taboos and pollution in the land. They also see to the need of ministering to the people for the proper adjustment in the society. The priests of the Earth goddess in parts of Igboland organise the Mbari ceremony which is popularly known in parts of southern Igbo of Owerri; it is essentially done to appease and venerate the Earth goddess. The priests of the Earth goddess are very influential and they are highly revered and obeyed. As the goddess is quite revered, the priests are equally held in high esteem and they are imbued with immense power. They can pronounce spells for any miscreant working against the welfare of the land or community. The various ceremonies linked to the welfare of the community, such as marking the beginning and end of the year cycles, are traced to them.

The Mbari art and ceremony is a reflection of the cosmos, as various gods and deities that influence the life of humans are identified in the art forms. It is a celebration of the cosmos because various rituals of cosmic cycle are re-enacted in the ceremonies and art forms. The priests of the Earth goddess stand in the centre of all this due to their enduring knowledge, power and status.

There was a cult in Arochukwu (Ibin Ukpabi), which was named after the god Chukwu, meant for providing solutions to various problems of existence. It became a very fearful cult and people from various parts of Igboland were travelling to Arochukwu to consult the oracle. The cult deity, Long Juju, has an elaborate labyrinth of underground steps and caves where one can find the god Chukwu inside the cave. Within the underground journey one ascends to the level where he consults the god. In fact the Arochukwu people were called 'Umuchukwu', son of god. People believe that they could consult god in the cave. One could liken this to the Egyptian pyramid where the Pharaohs sojourned underground through the pyramid's labyrinth in their journey to meet god and be transported to the next life of immortality. The priests were very powerful and they were highly revered in Igboland. The shrine has an intricate labyrinth that led to the delta part of the sea. In the course of time, the cult became a tool for slavery and it was destroyed by the British circa 1902.

Some Igbo spiritualists posses immense power which may even render them invisible to opponents as well as allowing them to control some actions without physically being there, a form of remote control. There are instances from various accounts in the course of my research told by informants where people could disappear into the thin air only to resurface again somewhere else. There is a popular story of a powerful spiritualist / native doctor who could transform into a ball of fire as he disappeared only to reappear somewhere else later. This is usually done to make sure he meets no one waiting on the way to harm him after performing very powerful ritual for his clients.

In many ancient societies including the Igbo there are spiritualists who display some amazing feats which beat the imagination of modern science. In Igbo society the influence of the spiritualists / native doctors were quite renowned and Umeh has captured these feats. Umeh (1997) has argued that native doctors were next to god in

Igbo society through their manipulation of various forces which include cosmic vision. Umeh (1997) has discussed how native doctors and spiritualists performed various amazing tasks through the manipulation of cosmic forces. They were able to accomplish incredible tasks such as cloning into animals and disappearing which made them invisible to enemies. He even mentioned that native doctors were able to confer and consult the spirit of the dead to solve persisting human problems and put things aright where there is some tension and confusion in the society.

The Igbo spiritualists could induce rainfall through the knowledge and manipulation of the sky. They know the mood of the cloud and how to bring the clouds together to cause rainfall. Rainmakers are well known in Igbo society. It is quite evident that even today in Igboland, when there is a major celebration in various parts of Igbo land, the rainmakers will be consulted and appeased for them to withhold back the rain otherwise the ceremony will be a flop. They have the power to cause rain to fall at a particular place. In the pre-colonial period this art was quite preponderant and was widely evident.

There are some spiritualists or native doctors who could induce lightning and thunder to destroy any evil persons. Some of them are credited with the power of inducing lightning through the manipulation of god of thunder (*Amadioha*) to strike miscreants. The priest shrines of *Amadioha* are credited with such power and they are highly feared and respected. In some indeterminate cases the priests could be consulted and they would use their mystical power to get at the truth of the matter by threatening to invoke the wrath of the god of thunder, which produced a quick response. In this way they were able to adjudicate cases and restored normality within the society. People believe that the god of thunder comes from heaven and it is instant justice.

On issue of succession to high office, the spiritualists enact the various rituals that bestow high office including kingship. The nature of the cosmic and mystic power relationship is seen very clearly in the enactment of rituals for the succession to high office in parts of Igbo society. The Eze Nri succession to high office supports this point according to (Nwaezeigwe 2007). The ritualists who perform these acts are specially designated and they are quite powerful and they are also seers or diviners. In the first instance, there appears to be a generally accepted, or so to say, popular method of selecting a new king. It is the belief that the selection of a candidate was not based on such criteria as personality, wealth or title. The selection was believed to be undertaken by virtue of supernatural forces manifested in the frequent falling down of the walls of the house of the would-be Ezi-Nri during the dry season. (Nwaezeigwe 2007, 166) Commenting further,

> Noted that the ritual rite of the succession was meant to transform the priest-king from mere mortal to a supernatural being. From the above stage, the candidate then proceeded to the next, which involved the ritual death, burial and resurrection. Although, in describing the general significance of this coronation ceremony, Jeffrey had asserted that it had the two-fold objective of transforming a man into a god, and of recreating the world, one could only but agree with the former, at least at the present stage. And that is that the rites of ritual death, burial and resurrection were meant to explain the cosmic transformation of the candidate from man to spirit-being. (Nwaezeigwe 2007, 168)

These ritual ceremonies are quite significant in Igboland and it conferred on the initiated much power and status. The Eze-Nri after the ceremony is seen as priest king, a personification of the god. They command great magical powers surrounded by various taboos. Those who perform these rituals and have the knowledge—and not only the initiated—are highly revered for their immense power.

The immense power of the spiritualists and native doctors helped the traditional Igbo society to have a relatively reduced rate of crime and to work towards justice and honesty. They could point out a miscreant without having to be at the scene of the crime. Due to their immense power that these spiritualists possess and the ability of the people to believe in the efficacy of their powers, the society tends to be more orderly and honest. This is the opposite of what obtains in our modern society where criminals ravage with impunity because they might not be caught and they remain anonymous.

The spiritualists and native doctors, due to their knowledge of cosmos, have acquired much power and respect within the society. They are the titleholders, the opinion holders, people of wealth and status. At times they are even feared and people dare not play around with them. They remain a special class of people set apart.

Conclusion

The paper has tried to discuss various manifestations of power through the knowledge of the cosmos in Igbo society. The cosmos informed various activities of the Igbo Spiritualists / Native Doctors which are reflected in their various form of activities. The spiritualists were able to solve the issue of existentialism and other issues that affect the welfare of the society. The spiritualists / native doctors, through their mystical feats, were able to garner a lot of power and prestige within the society. The spiritualists performed incredible feats that have defied Western thought and scientific empiricism. There is a need to begin to reshape our understanding of knowledge in such a way that we can begin to understand other spiritual forces that influence social realities. In this way we can start to decode ancient civilizations and then learn from them to improve humanity.

References

Acholonu-Olumba, C., Prabhakar, A. and Nneka, C. Egbuna. 2005. The Gram Code of African Adam: 450,000 Years of Africa's Lost Civilizations. Abuja, CARC Publications.

Acholonu, Catherine Obianuju. 1989. The Igbo roots of Olaudah Equiano: an anthropological research. Owerri, AFA Publications.

Basden G. T. 1921. Among the Ibos of South East Nigeria: An Account of the Curious & Interesting Habits, Customs, & Beliefs of a Little Known African People by One who Has for Many Years Lived Amongst Them on Close & Intimate Terms. London, Frank Cass & Co. Ltd. (this edition 1966).

Chukwuezi, B. 2007.The Relationship between Social Structure and Cosmology in Igbo Society. Paper presented at the Cultural Astronomy Conference, University of Cape Coast, Ghana.

Chukwuezi, B. 2008. Interpretation of Ethnographical Artefacts (A case study of the Igbo Society of Nigeria). Published in the Astronomical Society of the Pacific Conference Series 409, Cosmology Across Cultures.

Chukwuezi, B. 2009. The Anthropological Method of Studying Cultural Astronomy. Paper presented at the Cultural Astronomy Conference at the University of Cape Coast, Ghana.

Cole, H. M. 1982. Mbari: Art and Life Among the Owerri Igbo. Indiana, Indiana University Press.

Nwaezeigwe, Nwankwo T. 2007. The Igbo and their Nri Neighbours: a Study in the Politics of Igbo Culture and Origins. Enugu, Snaap Press.

Onwuejeogwu M. A. 1981. Nri Kingdom and Civilization: A dawn of Igbo Civilization. Benin, Ethiope Publishing.

Umeh J. A. 1997. After God is Dibia: Igbo cosmology, healing, divination & sacred science in Nigeria, Vol. 1. London, Karnak House Publishers.

Umeh J. A. 2000. After God is Dibia: Igbo cosmology, healing, divination & sacred science in Nigeria, Vol. 2. London , Karnak House Publishers.

SOME ASPECTS OF EUROPEAN MOON MYTHOLOGY

MARE KÕIVA AND ANDRES KUPERJANOV

Abstract: The article overviews myths in Estonian and European traditions about the Moon and the nature of moon-spots. Moon-creation myths have reached us mostly thanks to oral tradition and publications. The best-known myths about the origin of the moon spots concern people who have ended up there either as a result of their own actions or at the discretion of the Moon, or describe how human activity has caused the spots. The supernatural power of the Moon is also recorded in incantations which reflect the belief that the Moon has power over the thriving and failure of everything from wealth to health to farming, as well as the magic nature of moonlight itself.
Keywords: Estonian and European astral lore, moon myths, astral charms

European astral mythology and astral lore consist of different layers. There are features common to all of Europe as well as regional variation and unique aspects. Taking the Estonian example, the geographically next-door neighbours in the south have very different astronyms, except for the Moon and the Milky Way. However, Latvians and Lithuanians (Baltic peoples) and Estonians (Finno-Ugric people) share the same explanations about Moon spots and beliefs connected with Moon phases.

The orbital and visible movement of the Moon is very complicated. In the winter season, the full Moon is up high and in the summer the Moon is low—and its visible height changes dramatically from 12° to 4° during a period of 18.6 years. When the Moon is low, close to the horizon, it seems reddish in color and optical illusion makes it look big and move fast. When the Moon is more than 20 degrees above the horizon, visual azimuth evaluation by human eye is challenging and people usually say it is 'shining'.

The face of the Moon that people see also depends on the observer's latitude: rising and setting, the Moon is tilted relative to the azimuth. Obviously, therefore, European peoples see different figures on the Moon and you would expect that to reflect also in their astral myths.

In this paper, we are analysing myths and charms related to the Moon, giving a comparison of the main motifs. Our study is based on the corpus of Estonian astral lore (myths, beliefs and narratives from the 17th to 21st century) and the corpus of Estonian charms compiled at the Department of Folkloristics in the Estonian Literary Museum, and on the manuscript card files of Baltic Finnic peoples' (Livonian, Vepsian and Votian) folklore, all from the Estonian Folklore Archives.

The corpus of astral lore consists of unequal, diffuse texts. The Estonian corpus contains myths (etiological or cognitive), belief and personal experience narratives of different genres and multifunctional short texts: common beliefs, sayings, meteorological knowledge and different omens. All together the astral lore and charm corpora form a mixture of sacred texts and utilitarian heritage.

We also use parallels referenced from books and research papers on the field, especially from the studies of A. Gura on Slavic peoples (Gura, 2010), by Y. Berezkin on wider Eurasian sources (Berezkin, 2010), and by M. Kõiva (2011) on Estonian charms.

We used methods of comparative folkloristics to combine astronomical knowledge with folklore texts in order to find similarities. Folklore methods are effective for finding main motifs and for mapping the motifs.

Myths About the Moon

There are several Estonian myths about the creation of the Moon that have survived to the present day thanks to folklore publications. One is in the Creation Song (Est. *Loomislaul*), where the Moon emerges from the egg of the world bird. The bird throws its hatched eggs all around, so that they become the Sun, the Moon, the stars and the land with creatures living upon it (Loorits, 1949).

We find the Moon in the role of a luckless suitor in the epic song of Star Bride (Est. *Tähemõrsja*), where he is faulted for his inconsistency: waning, waxing or being totally out of sight. In other mythical songs, the Moon is a son of a God (Sun), he is usually an old man, but also mentioned as the Sun's sister. The Moon and other celestial bodies are depicted as living creatures—Heavenly Beings (Est. *taevalised*). Since the Moon, like other celestial bodies, was sacred, it could not be pointed at with the finger. Also, one was not to perform impermissible actions in sight of the Moon, i.e., in the moonlight: steal, have sex, hit somebody.

According to folk belief, a moon's life is short, only four weeks: old moons are behind the world's border, and there are related funny beliefs, for example that Chinese women use cast-off moons as fireplace pokers. There are stories that there are 13 moons, one of which is stolen. In belief accounts, the Moon is also a mentioned as a window to hell.

Why the Moon has Spots

There are quite a few explanations about the origin and nature of the spots on the Moon. Many people believe the spots form a face with a nose, eyes and a mouth. Many believe the spots are connected with the Moon's ability to take people up on its own—we see people who for various reasons have ended up on the Moon. Other explanations attribute the spots to a range of human actions.

There Are People On The Moon

The Moon which is like Earth: people live there and the darker spots are their cabbage fields (e.g., corpus of astral lore, ERA II 129, 382 (45) < Rapla).

Some say the spots are a man on the moon cutting trees, a motif also known in the wider Finnic-Ugric area (Harva 1948, 159).

Water carrier on the moon

One major belief cycle about the spots on the Moon says that we see the figure of a person fetching water and holding water pails in her or his hands.

Versions of this story are known in all Europe, but in the Baltic-Finnic region female water carriers predominate. There is also a version of the story where the main protagonist is male (man, lad) or two children. This story is less widespread or absent in Southern Europe, where the girl in the Moon is known, however.

The distribution of the 'water carrier on the moon' story in Europe coincides approximately with the distribution of Balts, Slavs and Baltic-Finnic nations: Lithuanians (Kerbelite 2001, 70; Laurinkene 2002, 365; Vaiškunas 2006, 158), Letts (Pogodin 1895, 440), Estonians (Kuperjanov 2003, 72), Votians (Ariste 1974, 5; Ernits and Ernits 1978, 579), Bulgarians and Ukrainians (Gura, 2010).

An early printed version of the story (Jakobson, 1867) was spread through Estonian school textbooks and tells of an innocent maiden from Vaskjala who gets onto the Moon. The Moon takes pity on her and takes her up to the sky where we can see her on moonlit nights.

There are three main motifs relating to the female water-carrier on the moon in Estonia:

1. The moon takes someone up out of pity.

The most popular version tells of an orphan (or woman) exhausted from a hard life and work, who, when carrying water from the spring or brook, asks the Moon to take her up to him:

> Moon, dearest,
> Take me to be your back-whisker
> And head-washer.
> Corpus of charms, H, Wiedemann 3, 134

> Moon, call me,
> Dawn star, help me,
> Whisk leaf, take me away.
> Corpus of charms, H, Wiedemann 1, 203

The Moon pities her and takes her up to the sky where we can see her on moonlit nights.

How a girl carrying buckets was raised to the moon where she can still be seen was recorded from a Votian informant (Ernits and Ernits, 2011).

It should be noted here that the formula the maiden uses to address the Moon has often been recorded without the accompanying story and was used independently as an incantation asking the Moon for a boon.

2. The moon takes someone up for insolence

A woman carries water buckets and asks the Moon for help. When the Moon does not help, she abuses the Moon and the Moon snatches her.

> In the old times once it was that a woman went to fetch water. She had shoulder poles and buckets on it. It was late evening and the Moon was shining. She looked up and was in a bad mood because she needs to carry water so late: 'Why are you lazing about, take up the poles, try carrying them, see if it's easy!' The Moon promptly pulled her up together with the buckets, right up there: 'Now you have it good and easy here!' For this reason there is a human face in the Moon. Corpus of astral lore, ERA II 77, 281/2 (74) < Hageri

3. Punishment for offending the Moon by breaking a law or taboo in its light

The crime is breaking a law or taboo in the moonlight. It can either be theft (a globally known motif), wife killing (in South Estonia), an act of love (in West and North Estonia), or insulting the Moon (in Lutsi, Estonia, a woman calls her butt whiter than the moon). There are no gender differences—both a man or a woman can be taken up to the sky.

An intriguing taboo found among belief motifs: anyone in the sauna late on a Saturday evening (or generally late) is taken to the Moon. To wit, many North-Eurasian people share the command to leave the sauna late at night for the sauna spirits, namely Christ, Mary or God, who either live there or arrive at the dusk of day.

> I've heard that Christ went to sauna and is holding a whisk in his hand. Corpus of astral lore, ERA II 163, 261 (39) < Setu

All three Estonian main motifs have widespread parallels. Among the Samis—the Sun takes the girl to give her in marriage to his son and throws her onto the Moon with her pails (Charnoluski, 1930). Other, eastward parallels are found among the Komi-Zyrians, Komi-Permiaks, Udmurts, Chuvash, Mari, Bashkirs, Volga Tatars, Samoyeds, Selkups, Khanti, Russians and Ukrainians, etc.; only a few versions are found in Archangelsk and other places close to ethnic groups of Finnic Ugric peoples (Gura, 2010). Yuri Berezkin (2010) finds that some peoples of Asia and America have the same motif: the Kazakh, Kirgiz, Kets, Khakas, Evenki, Nanai, Lamut, Nikhv, Buryat, Mongolian, Japanese, Ainus, Chinese and Tlingit, as well as some Palaeoasiatic peoples, the Chukchis, Koryaks and Kamchadals.

Mythical And Christian Protagonists On The Moon

It is interesting to remark that predominantly Lutheran Estonians, Finns and Latvians have few astral myths with Christian or Biblical protagonists, especially when compared to Orthodox Slavic peoples (Gura, 2010). However, many Europeans do indeed see Biblical characters, saints and historical figures on the Moon.

These beliefs are mostly spread in European Slavic areas. Adam and Eve, the icon of Matthew the Evangelist, the head of John the Baptist, Saint George frightened by lightning or playing musical instruments, a chained Satan crucified by God, the witch Tvardovski from Krakow, the Czars Ivan the Terrible and Peter the Great and the piebald head of King Marko's horse: this is only a short list based on research by A. Gura.

One of the few such beliefs held by Estonians is that the spots are the result of a quarrel or argument between the two brothers Cain and Abel (cf., the corpus of astral lore, ERA II 206, 100 (22) < Mustjala).

Other mythical protagonists as moon-spots include the legend about a man, wolf and bull on the Moon. This is probably influenced by myths about Ursa Major (a poor farmer with his wagon and horse, and a wolf attacking his horse, taken to heaven) and maybe adaptation and transformation from similar folklore (e.g., the corpus of astral lore, ERA II 77, 572 (86) < Nissi).

In Lithuania it is also believed that in the Moon we can see the legendary Lithuanian wizard Tvardauskas, whom the devil took to hell but lost on the Moon (Vaiškunas 2006, 176, citing Balys 1951, 9-11).

Moonspots As A Result Of Human Actions

Another very widely spread motif, in addition to the water carrier, is the story about tarring the Moon. Moonlight is too bright for those wanting to carry out an act of malice or sin, therefore the perpetrators decide to tar the Moon. The evildoer gets stuck in the tar or is pulled (often together with the tubs of tar) up to the Moon, where we can still see him.

The following two examples describe the crime and moon-tarring motifs:

> One man had killed his wife. The moon was watching and laughing—'I see you!' The man got angry, took a pot of tar and a brush. He climbed up on to the Moon and started to tar it. His hand was shaking. In some places there was a lot of tar, in others very little. That's why the Moon is spotted. Corpus of astral lore, ERA II 115, 644 (3) < Urvaste

> The boy took buckets of tar and went to tar the Moon so it would not throw its light when he goes stealing. But he got stuck to the tar. Since then there is a Moonboy on the Moon, his hands spread, holding the buckets of tar. Corpus of astral lore, RKM II 195, 688/9 (9) < Iisaku

A Votian writing indicates that the shadows of three thieves are seen on the surface of the moon. Three men went onto the moon to cover it with tar: they wanted to go thieving but there was moonlight. So they decided to tar the moon but got stuck there (Posti 1980, 203).

The Moon Itself Has A Human Face

The full moon has also been compared to the human face. The motif is commonly known in the wider European area. One example from the Votic area:

> As if eyes and nose and like that. As if the face of a man. (Ernits and Ernits, 2011).

Charms Addressing the Moon

Moon charms come from the oldest layer of incantations, and in Estonia written records of incantations about the Moon are dated to the 17th century. The corpus of moon charms consists of nearly a hundred distinct texts. The great majority of moon charms are short formulae accompanied by uncomplicated rites.

Common beliefs about Moon phases (waning—old, waxing—new, full—stable, no Moon—dangerous period) are used for organizing work, making predictions and healing and charming. The phase of the Moon is a very common time anchor in incantations. For warding off disease, the old Moon was best suitable; to promote economy, heal broken bones, sprains, guarantee subsistence, a favourable court ruling and plant growth, the new Moon was addressed.

Charming According To Moon Phases

One of the older formulae is greeting the Moon (Est. *kuu teretamine*). The incantation is similar to greetings addressing other natural forces and fairies (e.g., sauna and sauna spirits, earth spirits, wind, snakes). The incantation had to be chanted when you saw the new crescent Moon in the sky and the purpose was to gain a) youthfulness, b) health and recuperation, c) money, or d) food. Usually, people presented one wish at a time, making moon charms a single-episode formulae. Moon incantations were performed by adults, but in the 20th century the repertoire spread also to children.

> 'Hello, Moon, you [will grow] old, I [will grow] young!' Said my grandmother in 1918, when we were holding hands and walking in moonlight home from the sauna. Corpus of charms, RKM II 200, 186 (109) < Lüganuse

In stereotypical moon incantation verses, the worse half is wished to the Moon, the better half to the charmer. In the following sample text, the Moon grows a butt end of a bread loaf while the self or human a slice of bread:

> ... You grow on a butt-end of bread,
> I stretch on a slice [of bread].
> Corpus of charms, H III 12, 703 (10) < Hanila
>
> or
>
> Hello, hello, dear moon!
> You heavy as iron,
> And me light as gold.

Corpus of charms, RKM II 368, 417 (10) < Maarja-Magdaleena

or

you the size of a flea,
me the weight of a bull
Corpus of charms, ERA II 260, 561/2 (167) < Mustjala.

Many traditional song types in southern Estonia have longer and more archaic versions of the same. Similarly, south Estonian moon incantations are longer in their addressing. They often use alliteration to combine several extremely short wishes into one whole. South Estonia also sees more petitions for sufficient bread. Some formulae add to wishes for youth the words for disease and sleepiness transfer, typical of healing incantations:

Sleep to the swine,
Laziness to the sheep,
Illness to the rich [Germans]
But to me good health.
Corpus of charms, KKI, WS < Setu

One such health formula found all over Estonia is, for example, 'Your eye to grow dim, my eye to be clear' (Corpus of charms, KKI 45, 300 (23) < Türi) as well as the health transfer formula 'Health to me, disease to you' (Corpus of charms, E,StK 22, 98 (11) < Põlva).

The whole Europe charmed the new Moon for money. In Estonia, these customs and formulae are used in South Estonia: money was jingled in the pocket or the pocket was given a slap, asking the new Moon for subsistence:

You [will] eat gold,
I [will] get gold.
Corpus of charms, A 2367 (1) < Saarde

The formulae wish the sayer the lightness or shine of gold, while the Moon is offered, in magic transactions, the weight of iron and the colour of rust.

Money was jingled in the pocket – this attracts more money – and the moon was greeted:
'Golden coins to bag,
Silver money to pockets.'
Corpus of charms, RKM II 212, 221 (16c) < Põlva

Nearly the same logical structure applies to the next charm.

If you are holding something when you see the new moon, you need to say:
'Luck to finder, harm to sufferer!'
Corpus of charms, RKM II 195, 689 (12) < Iisaku

M. Arukask reports on the Votians: 'When the new moon appears, they petitioned god asking food for the family and crossed themselves three times' (Arukask, 1998).

From Lithuanian tradition we can find the same greeting of the new moon:

Young Moon,
King of the sky,
You light the Earth and clean it,
Clean my sinful body too.
(Masalskiene, 1990, cited via Vaiškunas, 2006)

Bugs and parasites could be 'given' to the Moon.

If there were roaches in the house, one person had to go out on the new moon Thursday and say: 'New moon, take my bosses [bloodsuckers] for yourself!'
Corpus of charms, H II 9, 586 (26) < Haljala

Addressing The Moon In Healing Charms

The Moon is addressed in short charms warding against disease and often disease is transferred to the Moon; some healing charms are used even today.

Small growths could be reattributed to the Moon—warts, boils and bone deformities—as well as skin ailments and some contagious diseases. Such incantations are common in Scandinavia (Norway, Denmark and Sweden), the British Isles (Ireland, Scotland and England), in central European traditions (German, Austrian, French and Hungarian), among Western and Southern Slavs (Bulgaria, Macedonia, Serbia, Slovakia, Czech and Poland), and eastern Slavs (Russia, Ukraine and Belorussia). The Moon is shown a wart and offered:

See what I have and you don't have. Take it!
Corpus of charms, RKM II 159, 81/2 (5) < Rakvere

or

Salt that has been in contact with the wart is thrown towards the moon and the incantation is said: 'Get, eat peas!' Corpus of charms, ERA II 250, 542 (15) < Tapa

If you see someone riding in moonlight, you have to say: 'And the wart makes three!' Corpus of charms, H II 54, 815 (47) < Põltsamaa

To make an illness recede, the old or full Moon was addressed. The final formula 'Go like the old moon disappearing from the sky!' is found in many healing words, repeated three or nine times. This was common practice in the early 20th century, and there are still people using the short charms and customs in Estonia today.

The Healing Properties Of Moonlight

Moonlight was an important magical instrument, detrimental to disease and promotional to growth. Moonlight was used to perform magic symbolic rites for washing off the disease. For example, the most common method to heal chicken pox was whisking in moonlight—the patient was in the moonlight, the charmer in the shadows, shaded from moonlight (no healthy person or body parts could remain in the moonlight), and performing the whisking with a specially made sauna whisk (of birch or juniper). Sprains and warts were also healed by folk doctors by symbolically palming moonlight and 'washing' the disease hotbed with circular motions. Sometimes the patient was told to wash hands with moonlight or to imitate handwashing.

Conclusions

The most popular Moon myths are very widely known indeed, the distribution areas stretching across the world on similar latitudes. The most common motifs of our Moon spot stories are related to the special power the Moon has to take people up at its discretion: out of pity or as punishment for giving offence to the Moon or for performing immoral or forbidden acts.

The man on the Moon has spread in oral heritage as well as in literary fiction and both strata—oral and literary—have influenced each other. In children's books, the Moon is often a kind Moon-man. Although the Estonian language lacks gender and the Moon has no gender, contextual hints let us conclude that the Moon is a he. We can see allusions to this also in old epic songs (the daughter of day marries the Moon, they are sitting together on the edge of the world). In the 19th century and more recently, visual media has played its part: postcards, book illustrations, etc., depicting the Moon as male.

It is interesting how often Moon myths found their way into 19th century primers and school textbooks where they were chosen as one example of national mythology. Water carrier on the moon obviously fitted well because of its human(itarian) message—there is always hope in a difficult situation. The story also carries key concepts of the ethnic identity—pointing out the difficult situation and backbreaking workload of the poor (Estonians were lower class labourers). The humour and didactic messages found in other versions of the tale (Moon tarrer) made it an indispensable school myth.

The attributes of the phases of the Moon—the waxing Moon supporting undertakings and the waning Moon having a dampening effect—are universally recognised. Moon charms are widely known in Europe and the power of the Moon and moonlight was and is often used in healing. The simplicity of the charming text and rituals is the reason why some have stayed in active use until the present day.

Acknowledgements:

The research was supported by the project SF0030181s08 and ESF grant 8137.

References

Ariste, P. 1974. Vadja muinasjutte ja muistendeid [Votian Fairy Tales and Legends]. Töid eesti filoloogia alalt 4, 3-34.

Arukask, M. 1998. Vaateid vadja rahvausundile. [About Votic Folk Belief]. Available at: http://www.folklore.ee/rl/folkte/sugri/vadja/mytol/mytindex.htm (accessed 21 July 2015).

Berezkin, Y. 2010. The Pleiades as Openings, The Milky Way as the Path of Birds, and The Girl On the Moon: cultural links across Northern Eurasia folklore. Electronic Journal of Folklore, 44, 7-32.

Ernits, E. and Ernits, T. 2011. Vadja rahva astronoomiast 20. sajandil [Votian Folk Astronomy in the 20th Century]. In M. Kõiva (ed.), Paar sammukest 35. Tartu, Scholarly Press of ELM.

Charnoluski, V. 1930. Materialy po bytu loparei [Materials on the Everyday Life of the Lapps]. Leningrad, Izdatelstvo Russkogo Geograficheskogo Obschestva.

Gura, A. 2010. On the Methods of Constructing a Mythological Text: Slavic Folk Beliefs regarding the Spots of the Moon. Folklore: Electronic Journal of Folklore 44, 35-50.

Harva, U. 1948. Suomalaisten muinaisusko [Finnish Ancient Religion]. Helsinki, Werner Söderström Osakeyhtiö.

Jakobson, C. R. 1867. Kooli Lugemise Raamat [School Reading Book]. I. Tartu, Laakmann.

Kerbelyte, B. 2001. Tipy narodnyh skazanii [Types of Folk Narratives]. St. Petersburg, Evropeiski Dom.

Kõiva, M. 2011. Eesti loitsud [Estonian incantations]. Tallinn, Pegasus.

Kuperjanov, A. 2003. Eesti taevas. Uskumusi ja tõlgendusi [Estonian Sky. Beliefs and Interpretations]. Tartu, Eesti Folkloori Instituut.

Laurinkienė, N. 2002. Predstavlenia o mesiatse i interpretatsia vidimyh na nem piaten v baltiiskoi mifologii. Balto-slavianskie issledovania [Balto-Slavic Studies] 15, 360-385. Moscow, Indrik.

Loorits, O. 1949–1957. Grundzüge des estnischen Volksglaubens. I–III. Skrifter Utgivna av Kungl. Gustav Adolfs Akademien för Folklivsforskning 18, 1-3. Lund, Carl Blom.

Pogodin, A. 1895. Kosmicheskie legendy baltiiskih narodov [Cosmic legends of Baltic peoples]. Zhivaia starina 5, 3-4, 428-448.

Posti, L. 1980. Vatjan kielen Kukkosin murteen sanakirja [Dictionary of Kukkosi dialect of Votic language]. Lexica Societatis Fenno-Ugricae 19. Helsinki, Suomalais-Ugrilainen Seura.

Vaiškūnas, J. 2006. The Moon in Lithuanian folk tradition. Folklore: Electronic Journal of Folklore 32, 157-184.

STARS OF POWER – ASTRONOMICAL OBJECTS ON ANCIENT PRINCELY INSIGNIA

MAX E. LIPPITSCH AND SONJA DRAXLER

Abstract: Homer's detailed description of the shield of Achilles is well known. Can this description be taken for real, or is it only the expression of exuberant poetic fantasy? Why is it not Medusa's head like in the Aegis or another mythological monster deterring the enemy?
Sumerian cylinder seals of the 3rd millennium BC already used a star as a determinative for divinity. It can be assumed that powerful meteorological events had led to the association of the sky (and the objects therein) with deities. From prehistoric times there is evidence for deification of rulers. The highest conceivable power was to command the heavenly bodies. Therefore the stars became obligatory symbols of power, from Egyptian royal tombs to the Stars and Stripes of today.
Do we have any archaeological proof that shields with heavenly bodies existed in that time? Shields from the Late Bronze Age are rare in prehistoric collections. Most of them are rather simply designed, either with a smooth surface with only a few concentric circles as the ornamentation, or with numerous close-set circles of small bosses. Only a few have a more complex structure. The most impressive example we have with a shield and armour is from c. 1200 BC, unearthed some years ago near Graz in Austria. It is adorned with suns and several similar symbols (moons?) together with two sun-barques. It is far from the artistic wonders of Homer, but nevertheless it gives impressive proof that also at that time the celestial bodies were favoured symbols of power and leadership.
Keywords: bronze-age, shield, armour, astronomical objects.

1. Introduction

Most experts take it for granted that in early times astronomy was an important factor of power. However, the reason for this well-accepted notion is far from being evident. Our early ancestors had other things to fear – large predators, other strong animals, natural catastrophes – but certainly not the stars. Even the Sun, the heavenly body exerting the strongest physical influence on our planet, usually is not harmful to us in a vigorous way. Therefore we should have no reasons to fear the stars – and no reason to ascribe them special powers.

On the other hand, we have numerous unquestionable hints from literature and architecture that in every culture star symbols have been connected with power – and even today military persons show their rank by displaying astronomical symbols on their collars and epaulettes. It is worthwhile to dedicate some attention to this question.

2. Literary texts and archaeological evidence

The best-known example is the shield of Achilles described in Homer's Iliad. Hephaestus, the heavenly blacksmith, on the request of Achilles' mother, 'the silver-footed Thetis', forges armour and a shield for the hero. Homer devotes no less than 128 verses to the description of the shield. Its rich pictorial elements contain mainly everyday scenes, but the centre is filled with astronomical objects (Figure 1):

> Therein he wrought the earth,
> therein the heavens, therein the sea,
> and the unwearied sun, and the moon at the full,
> and therein all the constellations wherewith heaven is crowned –
> the Pleiades, and the Hyades and the mighty Orion
> and the Bear, that men call also the Wain, that
> circleth ever in her place, and watcheth Orion,
> and alone hath no part in the baths of Ocean.
> (Homer, Iliad, XVIII 482-489; transl. Murray, 1924)

The surface of the shield obviously was structured in concentric rings (reconstruction Willcock, 1976). The whole iconographic program of the shield is rather peaceful, with cattle and sheep and dancing youth, with ploughing and reaping and vintage. Only one of the rings shows, and there merely in one half, a city at war, and even these verses are focussed not on violent actions but mainly on the appearance of the gods on the battle field. The shield does not display rude force, as would have done the Gorgon's head or a furious lion. What, then, is the meaning of astronomical objects in this context?

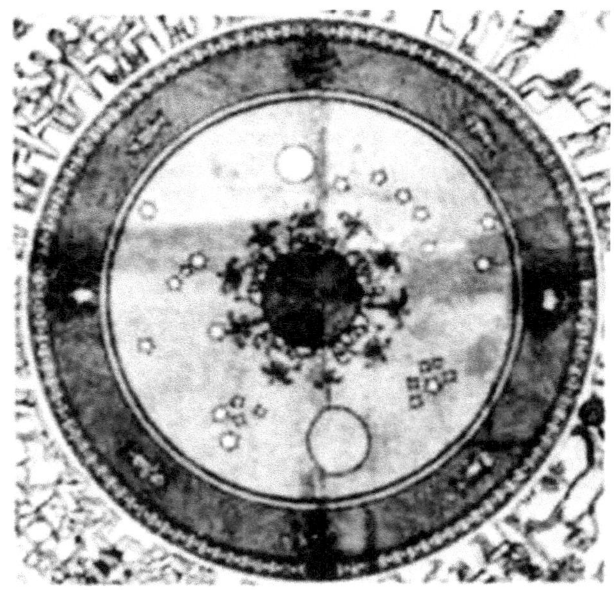

Figure 1. Centre of the shield of Achilles with astronomical objects (from http://www.mlahanas.de/Greeks/Mythology/).

Prior to the relationship between powerful warriors and the stars we already find the heavenly bodies connected to even more powerful beings: the gods. Clear evidence reaches back at least to the 3rd millennium BC. How this connection emerged is not quite clear, but maybe it emerged from the perception of weather. Sunshine and rain, thunder and lightening come from the sky, therefore this region is the origin of supernatural power. The stars are in the sky, therefore they are the sources of this power. A cylinder seal (Akkadian period, c. 2330-2150 BC) from the Oriental Institute at the University of Chicago (OIM A27903) shows the goddess Inanna, her foot on a lion, clearly demonstrating that she has power over even the strongest creatures. Above her head a star is signifying her divine status. In Sumerian script an eight-pointed star is the determiner for deities in general. This signification remained until Neo-Assyrian times (1st millennium BC), and was adopted by other cultures.

It is clear that rulers always were eager to emphasize their intimate connection with the gods, and with time they tended to accroach divine insignia for them. A nice example is a stele of the Akkadian king Naram-Sin (c. 2200 BC), on display in the Louvre in Paris. It shows the king defeating Satuni, king of Lullubi. Here the stars are not above a god but over the head of the king: the kings are becoming gods, and the stars symbolize their power, obtained directly (and legitimately) from the gods.

3. Bronze-age shields

Coming back to the time of Achilles, we have to ask whether shields like the one described ever existed during the Bronze Age. A number of shields of that period have been excavated, but the results are somewhat disappointing. If we stick to about 1200 BC as the date for the Trojan War (whether historical or not) we get nil return when looking for an adornment in any way comparable to Homer's description. A shield from the Villanova culture in Italy (c. 800 BC), and one from Tepe Hasanlu, Iran (c. 900 BC), both much younger than Troy, display the ring structure as suggested by Homer, but the ornaments are merely abstract symbols. Only the feature at the very centre in both shields may be interpreted as a sun.

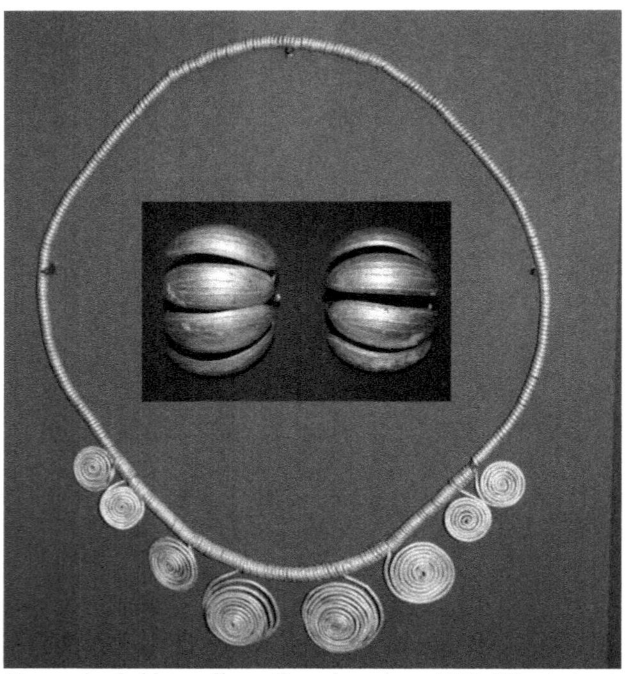

Figure 2. Gold jewellery. Graz hoard, c. 1200 BC. Author's photo.

The closest similarity to Homer's description is found on a shield from Toprakkale in Turkey. A central sun is surrounded by three rings bearing animal figures, mostly identifiable as lions and bulls, thus suggesting an astronomical connotation. The date of production, however, is c. 650 BC, thus probably younger than Homer's text. Even if we assume that people from Urartu had not read the Iliad, the shield represents the design of the time and not that of the 12th century BC.

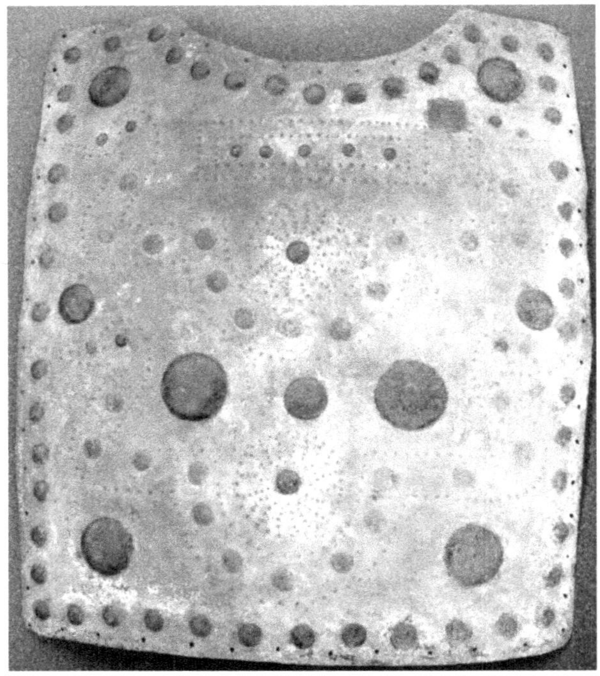

Figure 3. Pectoral armour. Graz hoard, c. 1100 BC. Author's photo.

Figure 4. Bronze shield, diameter c. 50 cm. Graz hoard, c. 1200 BC. Author's photo.

In 1998 there appeared a remarkable Bronze Age hoard at an antique market in Graz, Austria. It consisted of 3000 objects, of which there were 2700 pieces of gold weighing 4.2kg (Figure 2). It had been excavated by a metal detectorist on a hill near Graz. It was offered to the Prehistoric Museum Deutschlandsberg, but for financial reasons only a few pieces could be purchased and the rest was sold to a collector in the United States. The largest pieces were pectoral armour (Figure 3) and a shield (Figure 4). The shield has been dated to c. 1200 BC, the armour is probably a little later.

Both pieces are decorated with patterns of embossed points forming larger objects. On the shield as well as on the armour, structures are shown where rays emerge from a central bump. On the armour the rays are radial with a slight curvature, on the shield they are straight but inclined to the radial direction by about 35°, so that they seem to emerge not from the centre but tangentially to a circle smaller than the central bump. The differences in style most likely characterize their different ages.

On the shield there are 10 more round objects, consisting each of a central bump and three circles of embossed dots around it. They are significantly smaller than the sun symbol.

In addition, the shield shows two rectangular structures. Each of them has a small bump in the centre, surrounded by a row of small dots, then a row of larger dots and again a row of small dots, forming a rectangle each. In the following these structures will be referred to as 'boxes'. They are resting in curved bands ending in duck heads (Figure 5). Similar duck-headed bands appear on the armour, but here the round radiating structures are resting on them.

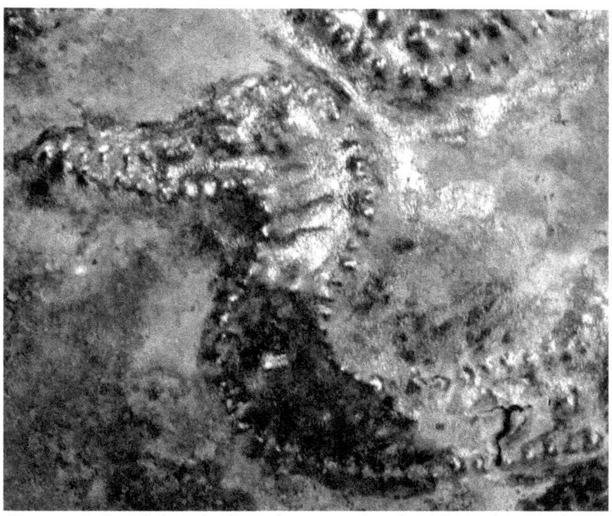

Figure 5. Duck head of sun boat on the shield. Author's photo.

The interpretation of these ornamentations must be speculative. From comparison with other artefacts from the same age it seems reasonable, however, to identify the central structure on the shield as a sun symbol. The smaller circular structures could be interpreted either as suns or moons. Their number, 10, reminds us of the ancient Roman calendar that named ten months only, leaving the winter lunations without appellation due to their insignificance for rural activities. If this interpretation is right, one could conclude that the two boxes might be the winter months. The curved bands then could mean sun-boats, carrying the Sun through the dark time of the year. This assignment is corroborated by the comparison with other findings from that time. Boats with bird heads on the bow and stern are well-known from the Sea People (as in the relief of Ramses III in Medinet Habu, c. 1180 BC). Duck-headed boats bearing radiating circles, resembling closely those on the armour, are known from Hallstatt age findings from Denmark, northern Germany and Slovakia and from the Villanovan in Italy (all about 1200-1000 BC; Wachsmann, 1996). The boxes appearing on the shield in place of the radiant circles are uncommon, but analogies could be assumed with Egyptian illustrations. In the Amduat the sun god Re is depicted travelling in his barque, and the divine figure is always enclosed in a shrine resembling our boxes. This analogy is strengthened by the relationship of Re with water fowl: One of his manifestations (besides a man and a ram) was as a goose. Sa Re is a sun god known at least from the Middle Kingdom, and the hieroglyph for *Sa* was a goose. (The god Sa Re 𓅭 ☉ should not be confused with the royal title *Sa Re*, Son of Re. The word *Sa* had the meaning 'goose' and 'son' as well). A nice example for this triad is found in the Amun Re Stele from Abydos (c. 700 BC, Figure 6) in the Egyptian Museum in Berlin (Inv. no. 7295).

Figure 6. Amun Re Stele from Abydos (c. 700 BC). Egyptian Museum Berlin (Inv. no. 7295) (in the public domain).

If these speculations are valid, we could conclude that the shield displays an astronomical content, with connections to the calendar. The central radiant circle then is the Sun. The smaller circles are months, either representing the Moon, or, more probable, the Sun in a given month – larger in summer, smaller in spring and autumn, and travelling through the underworld in winter.

On the armour we have nine bumps, two radiant circles born by duck-headed boats, and an elongate rectangular box with five small embossed dots. Thus we have twelve structures, representing twelve lunations, and five extra dots. Again as a speculation, this could be read as 12 months with 30 days and 5 extra days, thus agreeing with the Egyptian civil calendar.

The speculation could be drawn even further, counting the dots of the various structures on the shield and the armour and comparing the numbers to those of astronomical importance. Sixty and 120 points in the inner and outer margin of the central sun remind us of the hexagesimal system. Fifty-three rays of this sun roughly correspond to the number of weeks in a year (regardless of serious doubts whether weeks were observed at that time). Along the duck-headed boats there are 366 dots, approaching the number of days in a year ... but all these conjectures cannot be regarded as sufficiently verified and remain pure speculations, albeit fascinating ones.

If we accept this astronomical interpretation, the shield and the armour once more give evidence that astronomical objects were displayed on princely insignia during the Bronze Age. However, despite the bellicose purpose of these objects, the design lacked any allusion to force and violence. The objects prove the power of their possessor not by showing terrifying horror, but by knowledge mastering the world. Knowing the stars means to command them and to dominate the world. So the stars were the symbols of legitimate governance, not of brute militant force. The scenes on Achilles' shield therefore express that the hero (at least in the imagination of his mother) is not primarily a merciless warrior but the rightful conqueror of Troy.

It is interesting to see how things change with time. While Augustus in his famous statue of Primaporta displays gods and defeated enemies on his armour, Titus already has winged gryphons (Museo Archeologico Nazionale in Naples, Inv. 6059). During the Middle Ages we have a completely different situation. Some examples of coats of arms in the Holy Roman Empire are shown in Figure 7 and of Great Britain in Figure 8. Most of them show lions or eagles, only a negligible minority has heavenly bodies. Maybe this reflects the more violent culture of the Middle Ages.

Nowadays, about one third of all national flags show heavenly bodies. Nepal has a sun boat, in Kazakhstan the Sun is born by a bird, and the flag of the most influential country on earth bears 50 stars.

Let's hope that this fact reflects a society drawing their power far more from knowledge than from violence.

4. Conclusion

From the evolution of princely insignia over time it can be concluded that the display of astronomical objects at the most early stages is connected with the heavenly powers of storm and thunder. When rulers tried to express their good relationships to the gods, they used heavenly bodies to do so. Knowing the course of the stars provided prestige and esteem. There may be a direct relationship between the use of astronomical objects as symbols of power and the degree of humanity in a society.

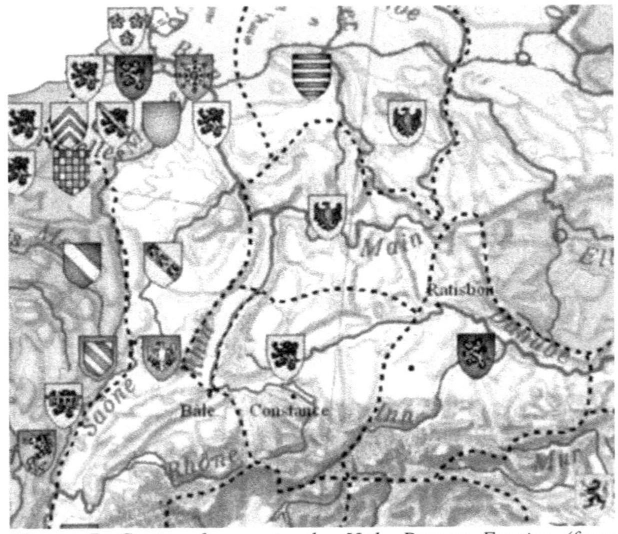

Figure 7. Coats of arms in the Holy Roman Empire (from http://www.earlyblazon.com/).

Figure 8. Coats of arms in medieval Great Britain (from http://www.earlyblazon.com/).

Acknowledgements:
Support from Burgmuseum Archeonorico Deutschlandsberg and its Director Mr Anton Steffan is gratefully acknowledged.

References

Homer. The Iliad. 1924. With an English Translation by A. T. Murray, Ph.D. in two volumes. Cambridge, MA., Harvard University Press; London, William Heinemann, Ltd..

Wachsmann, S. 1996. Bird-Head Devices on Mediterranean Ships. In H. Tzalas (ed.), TrOpIS IV (IVth International Symposium on Ship Construction in Antiquity, Athens, August, 28-31, 1991). Athens, Hellenic Institute for the Preservation of Nautical Tradition, 539-572.

Willcock, M. M. 1976. A Companion to the Iliad. Chicago, University of Chicago Press.

Cultures of North America, Mesoamerica and South America

ASTRONOMY AND POWER IN MESOAMERICA

IVAN ŠPRAJC

Abstract: The political relevance of celestial imagery and astronomically derived concepts in ancient Mesoamerica is evidenced in a wide variety of sources. Prehispanic codices often contain astronomical tables used by a priesthood intimately connected with the governing class, and early colonial documents allude to the astronomical expertise of particular Contact-period rulers. The iconography of sculpted monuments and decorative elements associated with elite residences and administrative buildings discloses a close relationship between rulers and certain celestial bodies, whose orderly behavior was believed to be responsible for a proper course of cyclical changes in the natural environment. Mythological narratives and, particularly, the Maya hieroglyphic texts and the accompanying iconography reveal that the kings personified not only the most important deities but also their celestial avatars. And finally, the architectural orientations, largely intended to pinpoint seasonal astronomical phenomena whose relevance can be understood in terms of their concomitance with agriculturally important moments of the tropical year, are most prominently and consistently incorporated in monumental buildings of civic and ceremonial cores of ancient cities throughout Mesoamerica.
Keywords: Mesoamerica, astronomy, astrology, written sources, architectural orientations, political ideology.

Introduction

In order to understand the role of astronomy in the mechanisms of power of ancient societies, we should recall, in the first place, its practical uses, which can account for both its antiquity and its importance as attested in early civilizations (see Van der Waerden 1974, 1; White, 1959). The celestial poles and directions in which the objects in the sky rise and set provide basic references for orientation in space, whereas the cyclical motions of celestial bodies allow orientation in time. The problem of the measurement of time has been defined as 'peculiarly one of astronomy' (Woolard and Clemence 1966, 326); moreover, various cyclical changes observable in the sky coincide with seasonal changes in the natural environment, but since the periodicity of celestial events is much more stable and exact, the observation of these regularities allowed ancient societies to predict annual changes in their environments and to regulate their activities in time. Therefore, the need for astronomical observations increased notably with the origin of agriculture. Since this form of subsistence requires an orderly scheduling of labors in the yearly cycle, astronomical knowledge offered adaptive advantages to the societies possessing better specialists in this field and, consequently, acquired great importance in early states: making subsistence strategies more efficient, it contributed to the legitimation of the power of the ruling class (see Aveni and Hartung 1986, 56; Broda, 1982; Iwaniszewski, 1989; Reyman, 1975; Šprajc 1996, 20-22).

This paper summarizes the available information revealing how the knowledge and beliefs derived from the observation of the sky were used for political purposes in ancient Mesoamerica. The term refers to the culturally defined geographical area corresponding to central and southern parts of modern Mexico and the northern part of Central America, where civilizations with a number of common cultural traits flourished from the 2nd millennium BC until the Spanish conquest in the early 16th century AD. The history of Mesoamerica is traditionally divided into three main periods or evolutionary stages. The earliest complex societies appeared during the Preclassic period (c. 2000 BC-250 AD) along the southern part of the Mexican Gulf coast, in central Mexico and in the Maya area in the Mesoamerican southeast. The greatest splendor, particularly notable in fine arts, architectural achievements and writing systems, was reached during the Classic (c. 250-900 AD), whereas the Postclassic period (c. 900-1519 AD) was characterized by intensified migrations, pronounced militarization and, particularly among the Maya, by increased political fragmentation.

Astronomy, priesthood and government

It is rather clear that celestial observations, closely related with calendrical matters, were systematically performed by the priestly class. Indirect evidence can be found in prehispanic manuscripts or codices, which contain a number of astronomical tables accompanied by figures and religious texts. Since literacy was not a commonly shared skill but rather a privilege of the elite (Brown, 1991), we can infer that the most sophisticated astronomical knowledge was possessed by the upper social class. It is also obvious that the codices, considering their largely religious contents, were produced and employed mainly by the priestly segment of the elite.

On the other hand, as noted by León-Portilla (1986, 13), the following early post-Conquest source from central Mexico, which presents a discussion between the first Christian missionaries and native dignitaries, clearly situates the Nahua astronomers among those who were in charge of the worship of gods, ritual performances and the study of painted books, as well as the counting of days and years:

> Y, he aquí, señores nuestros, están los que aún son nuestros guías, ellos nos llevan a cuestas, nos gobiernan, en relación al servicio de los que son nuestros dioses [...] los sacerdotes ofrendadores [...] Sabios de la palabra [...] los que miran, los que se afanan con el curso y el proceder ordenado del cielo, cómo se divide la noche [...] los que tienen a su cargo las pinturas. Ellos nos llevan, nos guían, dicen el camino. Los que ordenan cómo cae el año, cómo siguen su camino la cuenta de los destinos y los días, y cada una de las veintenas, [...] (Sahagún 1986, 139-141).

In the Codex Mendoza, painted in central Mexico soon after the Conquest, two priests are depicted, one observing the sky and the other one beating on a drum to announce the time of night (Aveni 2001, 19-20). The meaning of *Ah Kin*, 'he of the Sun', which was one of the titles used for Maya priests (Tozzer 1941, 27), also suggests that astronomical observations were among the most important tasks of the clergy.

Even if the study of celestial phenomena must have been an official duty of religious specialists closely connected with the ruling elite, there is evidence indicating that the rulers themselves were directly concerned with astronomical matters. It is well known that king Nezahualpilli, who reigned shortly before the Conquest in Texcoco, a polity lying east of its powerful ally, the Aztec capital of Tenochtitlan, was not only very much interested in promoting astronomical knowledge, but was himself a keen observer of the sky. The Spanish writer Fray Juan de Torquemada describes an enclosure he had on the roof of his palace, 'where he went by night attended by his astrologers to contemplate the heavens and the stars' (Aveni 2001, 17). And when the Aztec emperor Motecuhzoma saw a comet, which his astrologers and soothsayers neither had observed nor were capable of explaining its significance, he sent for Nezahualpilli, whose astronomical and astrological expertise is reflected in the account given by another Spanish chronicler, Fray Diego Durán (1995, vol. I, 535):

> El rey de Tezcuco le respondió: "por cierto, señor; grande a sido el descuido de tus vasallos los astrólogos y agoreros y adivinos, pues siendo ya tan vieja y tan antigua esa señal en el cielo, me digas agora eso como de cosa nueva [...] as de saber que ya a muchos dias que apareció en el cielo esa estrella con ese resplandor, la qual sale de oriente y se acaba en derecho de México y deste reyno todo [...]

Since the king of Texcoco goes on to explain the misfortunes presaged by the comet, which eventually materialized in the Spanish Conquest, the negligence he reprimanded was obviously a serious fault, not only of the careless priests, who were severely punished (Durán 1995, vol. I, 534), but also of Motecuhzoma himself: according to the chronicler Tezozomoc, the emperor had the obligation to get up at midnight to observe four constellations in certain parts of the sky (Reyes 1979, 37).

Paradoxically, while Motecuhzoma's image, as depicted in different reports, is certainly not one of a king profoundly conscious of his astronomical duties, the responsibilities that the Mesoamerican rulers had in this respect are illustrated by yet another piece of evidence related with the same monarch. Fray Toribio de Benavente o Motolinía (1971, 51), a Spanish friar who arrived in Mexico soon after the Conquest, writes that the Aztec feast of Tlacaxipehualiztli 'took place when the sun stood in the middle of [the Temple of] Huitzilopochtli, which was at the equinox, and because it was a little out of line, [King] Moctezuma wished to pull it down and set it right' (Aveni 2001, 236-238). This comment, as well as the map of Tenochtitlan attributed to Cortés, where the face of the Sun is shown between the twin sanctuaries of the Templo Mayor (Aveni 2001, Figure 84), suggest that the main temple in the Aztec capital was aligned to sunrise or sunset on a certain date. While we know that Mesoamerican civic and ceremonial buildings were largely orientated on astronomical grounds (see below), the fact that Motolinía's statement was corroborated specifically by the orientation measured on the remains of the Templo Mayor of Tenochtitlan (Šprajc 2000b, S24-26) confirms that Motecuhzoma's concern was, indeed, of an astronomical nature.

Since the knowledge of celestial events is useful for solving practical problems, particularly in agricultural societies, this must have been a very important reason for the Mesoamerican rulers' interests in astronomy, but it was not the only one. In view of the parallelism observed between the movement of celestial bodies and the alternation of seasonal changes in the natural environment, and because the intervals at which astronomical phenomena recur are much more constant and precise than those separating other cyclical natural events, the sky was considered, since time immemorial, to be the image of divine perfection and supreme order to which human and earthly order was subordinated. The ideas establishing causal links between the behavior of celestial bodies and whatever was occurring on the Earth resulted in a wide variety of myths and beliefs found all over the world, commonly labeled astrology. To a greater or lesser extent, daily life was, within all social classes, influenced by astrological concepts, but it was, again, the priesthood that had the major responsibilities. The contents of Mesoamerican codices are largely of an astrological nature, suggesting that the priests observed the sky not only with the purpose of determining suitable moments for activities of wider social significance, such as agricultural works, but also in order to monitor the events believed to cause harm to humanity. Eclipses and certain moments of the synodic cycle of Venus, for example, are known to have been particularly ominous, and this fact explains the presence of Venus and eclipse tables in prehispanic codices (see Aveni 2001, 173-196; Milbrath 1999, 111-117, 163-186): if the moments of potentially dangerous phenomena could be predicted, the appropriate ritual acts intended to prevent their negative influences could be performed on time. The astrological beliefs of this kind must have been maintained and even reinforced by the ruling elite, because they provided an additional instrument of domination: the ceremonies required by diverse circumstances and carried out by those in charge of the state cult must have been viewed no less indispensable than, say, agricultural scheduling.

But the leaders of Mesoamerican states were not only interested in monitoring the motion of celestial bodies; they also personified them.

Celestial identities of Mesoamerican rulers

Just as in many other archaic civilizations, in Mesoamerica the rulers acted as impersonators of the most important deities (see Frazer, 1922; López Austin, 1973), and since the latter very commonly had celestial avatars, the rulers were also associated with the most important heavenly bodies.

Early colonial written sources from central Mexico reveal quite eloquently the sovereign's god-like nature (Reyes García 1979, 35-37). A sixteenth-century Spanish historian mentions the oaths that the newly elected

Mexica *tlatoani* had to make during the accession ceremony carried out on top of the temple of the patron god Huitzilopochtli:

> El rey se levantaba entonces, echaba de aquel incienso en las brasas, y con gran mesura y reverencia sahumaba a Vitcilopuchtli y se sentaba. Llegaba luego el gran sacerdote y le tomaba juramento de palabra, y le conjuraba que tendría la religión de sus dioses, [...] que haría andar al sol con su claridad, llover a las nubes, correr a los ríos y a la tierra producir todo género de mantenimientos. Estas y otras cosas imposibles prometía y juraba el nuevo rey. (López de Gómara 1987, 434-435)

Motolinía (1971, 283), describing the same ceremony, mentions a similar exhortation directed to the new ruler: 'habeis de velar mucho en hacer andar al sol é á la tierra'. If the Aztec emperor was believed to be capable of controlling such important natural phenomena, he must have been viewed as an incarnation of a powerful deity. Both the circumstances of the ritual of investiture and other data suggest that he represented Huitzilopochtli (see Broda 1976, 39-40), the god of both warfare and the Sun.

We know that the lords of a number of Mesoamerican kingdoms commonly acted as impersonators of supreme deities, such as Quetzalcoatl in central Mexico and Itzamna in the Maya area (see Šprajc 1996, 76-106), whose names are often included in their royal titles (Garza, 2002; Houston and Stuart, 1996; López Austin, 1973; Martin and Grube, 2000; Nicholson, 2001; Sellen, 2002; Sharp, 1981, 16-17; Thiemer-Sachse, 1992). Both gods had their origins in an early Mesoamerican reptilian deity which evolved into a feathered serpent, a common representation of Quetzalcoatl, on the one hand, and into a two-headed dragon, which normally represented Itzamna, on the other. A close relationship between Quetzalcoatl and Itzamna is reflected in a number of attributes they shared: both were creators, presiding over water and fertility, incorporating an aspect of cultural hero (inventor of arts, calendar and writing) and having a special connection with rulers; and both were essentially sky deities (Šprajc 1996, 76-106). Since the bicephalic monster representing Itzamna is often depicted with a body in the shape of the so-called celestial band composed of astronomical signs, including the Sun, Moon and Venus glyphs, it is significant that the sky bands frequently decorate the images of Maya rulers, particularly the ceremonial bars they often hold in their hands. In agreement with this is the fact that Maya royal titles most frequently incorporated the names or appellatives of the gods that were, in some way, related with celestial phenomena (García Barrios 2007, 16-17; Grube 2002, 349).

While both Itzamna and Quetzalcoatl (as well as Kukulcan, the Postclassic Maya version of Quetzalcoatl) were associated with the sky in general, they also had a more specific relationship with the planet Venus (Milbrath 1999, 177-186; Šprajc 1996, 76-106). It thus does not come as a surprise that also the rulers commonly claimed a special connection with this celestial body. Possibly the best known example of it is found in the story about Topiltzin Quetzalcoatl, lord of Tula, which was the capital of the most powerful Early Postclassic state in central Mexico. In the famous myth–which may well be based on some historical facts, but conflates different individuals carrying the highest god's name into a single character–the legendary Toltec king, upon being forced to abandon the throne and leave the city, is transformed into the planet Venus (Nicholson, 2001). Even more explicit is the evidence from the Maya area. In a hieroglyphic text from Quirigua, Guatemala, a king of that polity appears to be declaring a familial relationship with Venus (Closs, 1994). In some of the numerous Maya reliefs representing royal persons, a Venus symbol is placed in the ruler's headdress. In a palace at Tonina, Chiapas, Mexico, one of the sites that symbolize the decline of the Classic Maya in the late first millennium AD, a throne was found decorated with a giant Venus glyph elaborated in stucco (Šprajc 1996, 102-106, Figure 3.10, Plate 7). It has also been argued that the planet was linked to the founders of some Maya lineages (Milbrath 1999, 196-197). At Uxmal, Yucatán, Mexico, a lord called Chac ruled around 900 AD; the decoration of the façade of his residential building, traditionally known as the Governor's Palace, incorporates various Venus-related iconographic elements, including hundreds of Venus glyphs placed in the cheeks of the masks of the rain god Chac (Šprajc, 1993a; 1993b; 1996). While we know of a number of other Maya rulers whose titles included the name of Chac (García Barrios, 2007; Martin and Grube, 2000; Rivera Dorado and Amador Naranjo 1994, 35-36; Sharp 1981, 16-17), the Mesoamerican rain gods, in general, were associated with Venus (Milbrath 1999, 197-209). The conceptual relationship of the planet with rain deities, as well as with supreme gods who, like Quetzalcoatl and Itzamna, held absolute power over water and fertility and actually integrated minor deities, is but one aspect of the broader Mesoamerican rain, maize and fertility symbolism in which Venus, particularly its evening manifestation, had a prominent role. As argued elsewhere, this conceptual complex must have been based on the agriculturally significant seasonality of the easily observable extremes of the evening star: the northerly extremes are always reached between April and June, which means that they announce, or coincide with, the onset of seasonal rains necessary for planting, while the southerly extremes, occurring from October to December, mark the end of the wet season and the time of harvest (Šprajc, 1993a; 1993b; 1996).

The associations with Venus claimed by the Maya rulers are also evident in the contents of monumental hieroglyphic inscriptions, in which the records of dynastically important events, such as enthronements, or captures and battles with royal protagonists, frequently include symbols referring to Venus. An analysis of the dates of these events has revealed a statistically significant correlation with characteristic positions of Venus, particularly with the morning and evening star's first and last appearances (Aveni and Hotaling, 1994; Milbrath 1999, 187-196). The fact that most of these dates fall within the periods of visibility of the evening star agrees with a prevalent importance of the latter aspect of Venus, indicated by other data and mentioned above. For some rulers an astrologically correct timing of crucial events in their lives seems to have been particularly important. Such was the case of the famous Pacal, who ruled in the 7th century AD in the powerful city now known as Palenque, in the Mexican state of Chiapas: the dates of his birth (probably contrived) and accession correspond to the first appearances of the evening and morning star, respectively (Milbrath 1999, 193).

The Mesoamerican rulers identified themselves with various deities and, consequently, with different celestial objects. A number of Maya kings had the name of the Sun god Kinich in their titles, and they are often represented with solar attributes (Baudez, 1985; Colas, 2003; Garza, 2002; Houston and Stuart, 1996; Martin and Grube, 2000; Milbrath 1999, 83-87). A king of Palenque even had a special relationship with the planet Jupiter (Lounsbury, 1989). It is quite likely that significant positions of bright planets were relevant for timing politically important events (Aveni and Hotaling, 1994; Milbrath 1999, 218-248).

Clearly, by claiming a relationship with the most prominent heavenly bodies, the Mesoamerican rulers reaffirmed their divine nature and, thereby, additionally sanctioned their power.

Architectural alignments and political ideology

A substantial amount of information on the uses of astronomy in ancient Mesoamerica is materially preserved in architectural vestiges. Systematic archaeoastronomical studies have revealed that many important buildings were orientated to astronomical phenomena observable on the horizon, particularly to sunrises and sunsets on certain dates (Aveni, 2001; 2003; Aveni and Hartung, 1986; 2000; Galindo, 1994; Šprajc, 2001a; 2001b; Tichy, 1991). While the earliest orientations refer to solstitial sunrises and sunsets–probably because the solstices, marked by easily perceptible extremes of the Sun's movement along the horizon, must have been the most elementary references for orientation in time–in later times more complicated orientation principles began to prevail. Recent studies in central Mexico and the Maya area have shown that the alignments allowed the use of observational calendars composed of calendrically significant and, therefore, easily manageable intervals: the intervals separating the sunrise and sunset dates recorded by orientations at a particular site tend to be multiples of 13 or 20 days, i.e., the basic periods of the Mesoamerican calendrical system.

The correspondence between the most frequently recorded dates and the crucial moments of the cultivation cycle suggests that the observational schemes, reconstructed for a number of sites, served for predicting important seasonal changes and for an efficient scheduling of the corresponding agricultural and associated ritual activities (Aveni and Hartung, 1986; Aveni et al., 2003; Šprajc, 2001b; 2008; Šprajc et al., 2009). It should be recalled that the Mesoamerican calendrical year of 365 days, due to the lack of intercalations, did not maintain a perpetual concordance with the tropical year of 365.2422 days; direct astronomical observations were, therefore, always necessary. The orientations, marking critical and canonized moments of the year of the seasons, not only allowed their determination by means of direct observations; if the observational schemes were composed of elementary periods of the formal calendrical system, it was relatively easy to *anticipate* the relevant dates (which was important because cloudy weather may have impeded direct observations on these dates), knowing the structure of a particular observational calendar and the mechanics of the formal one. Particularly important for these purposes must have been the 260-day count, in which the cycles of 13 and 20 days were intermeshing, so that every date had a name composed of a number from 1 to 13 and a sign in the series of 20. Given the structure of this calendrical count, the sunrises and sunsets separated by 13-day intervals and their multiples occurred on the dates with the same numeral, while the events separated by periods of 20 days and their multiples fell on the dates having the same sign (Šprajc, 2001b).

Even if the observational function of architectural orientations indicates their relationship with practical needs, the alignments cannot be understood in purely utilitarian terms. It is well known, for example, that the Mesoamerican architectural orientations, in spite of exceptions found in certain periods and areas, tend to be skewed clockwise from cardinal directions (Aveni 2001, 233; Aveni and Hartung 1986, 10; Šprajc, 2001a; 2001b; Tichy, 1991), which means that the orientations referring to the Sun and exhibiting this skew recorded sunrises in autumn and winter and sunsets in spring and summer. This peculiarity, considering it cannot be compellingly explained in practical or observational terms, must have been based on the symbolism related to the world directions: the south-of-east/north-of-west skew of orientations implies that the dates recorded on the eastern and western horizon fell mostly in the dry and wet seasons, respectively, and this is precisely what the Mesoamericans most probably intended to achieve. There is evidence indicating that the dry season was conceptually related to the eastern and the rainy season to the western side or part of the universe. The symbolism and directional associations of the Sun, Moon and Venus are particularly revealing: the Sun, presiding over the east, was related with heat, fire and drought, whereas the Moon and Venus, primarily its evening manifestation, were associated with the west and, on the other hand, with water, maize and fertility. A large amount of data supporting this conclusion, as well as the observational facts that may have accounted for these concepts, have been extensively discussed elsewhere (Šprajc, 1993a; 1993b; 1996; 2004a). It is thus highly likely that one of the most pervasive features of architectural and urban planning in Mesoamerica had a cosmological rationale.

This conclusion is in agreement with opinions expressed formerly by different researchers. Several authors have argued that the ancient Maya architecture and urbanism reflects cosmological concepts and directional symbolism, and that such principles of site planning were used by Maya rulers to express and reinforce their status within the political order they controlled (e.g., Ashmore, 1989; 1991; Ashmore and Sabloff, 2002; 2003; Rivera Dorado 2001, 113-140). Indeed, since the repeatedly occurring directions are most consistently incorporated in the monumental architecture of civic and ceremonial urban cores, i.e., in the buildings whose construction was evidently commissioned by the political elite, it is clear that both practical uses of astronomical observations and the ideas about the structure and functioning of the universe formed a very important part of ideology of the ruling class. The astronomically orientated buildings are largely temples; however, considering what Carrasco (1985, 72) labeled 'institutional fusion of economical, political and ceremonial activities' in Mesoamerica, the design and layout of structures whose functions were mainly or even exclusively religious undoubtedly reflects the ideology of power.

On the other hand, the astronomical alignments are also incorporated in a number of constructions whose characteristics indicate quite clearly that they were built primarily for residential or administrative, rather than religious, purposes. In some cases, Teotihuacan being the most prominent example, the whole urban core is laid out in conformity with the astronomical orientations of the main temples (see Šprajc, 2000a; 2001b). Although in such cases the orientations of most buildings were not based on direct observations and were, therefore, neither precise nor observationally functional, they certainly indicate the importance, in social and political terms, of incorporating an astronomically-based alignment in secular structures.

Even more indicative are examples of monumental but essentially non-religious buildings that were, *per se*, orientated on astronomical grounds. Such is the case of the Acropolis of Xochicalco, a Late Classic site in central Mexico: the orientation of this high status residential compound, which occupies the most elevated part of the site, exhibits the orientation pertaining to one of the most widespread alignment families in Mesoamerica (Šprajc 2001b, 258-275). While there are other similar examples in central Mexico, the astronomical orientations of structures that appear to have had predominantly secular functions seem to be even more common in the Maya area, where an additional and possibly related peculiarity can be observed: even the buildings whose formal characteristics suggest they were not intended primarily or exclusively for performing religious ceremonies, but rather served as elite residences or for activities associated with government and public events, very frequently exhibit a plethora of decorative elements with patently religious significance. The presence of such iconography combined with the astronomical orientations of these buildings can be interpreted as reflecting the aforementioned merging of religious and secular authority, as well as the outstanding importance of astronomically-derived concepts in the Maya worldview, in general, and political ideology, in particular.

One illustrative example is the already mentioned Palace of the Governor at Uxmal, which is orientated to the northerly extremes of Venus as evening star (for a different interpretation see Aveni 2001: 283-288). Even if few similar cases are known in Mesoamerica, the idea that this building was intentionally orientated to Venus has strong iconographic support. More than 350 Venus glyphs are still visible in the decoration of the Palace's façade. Since the northerly extremes of the evening star always occur from April to June, heralding the beginning of the rainy season, it is particularly significant that Venus glyphs are placed in the cheeks of the masks of the rain god Chac. Furthermore, the Chac masks are arranged in groups of five, whereas eight stylized double-headed serpents are set in the decoration above the central doorway, and the numeral eight in Maya dot-bar notation appears above the eyes of the Chac masks at both northern corners of the Palace; five and eight are, obviously, Venus numbers *par excellence*, considering the commensurability, well known to the Maya, of five synodic periods of the planet and eight calendar years (Šprajc 1993a, 45-47; 1996, 170-178).

The orientation of the Governor's Palace cannot be regarded as particularly useful in practical terms. Venus extremes, since they are not annual phenomena and do not occur constantly on precisely the same dates of the tropical year, are rather unsuitable for accurate measurement of time. However, since the orientation to these phenomena is incorporated into what is probably the finest building of Uxmal, it must have had enormous symbolic significance. Indeed, the Governor's Palace was built around 900 AD by the ruler named Chac, under whose reign Uxmal reached its greatest splendor (Kowalski, 1987). Aside from the very name of the ruler, the faces of the god Chac decorating his residence and having Venus signs suggest that this personage, in accordance with the practice mentioned in the previous section and common among the Maya kings, pretended to be an incarnation of the rain deity as well as of his celestial manifestation. If Venus, whenever it was visible as evening star reaching its northernmost position, was believed to bring about the rainy season, then the House of the Governor can be viewed as a monumental materialization of the direction that must have been sanctified, because it marked the phenomena whose timely occurrences, conditioning annual climatic changes that were crucial for a successful agricultural cycle, were vital for subsistence. We can also imagine that lord Chac, by orientating his residence to the relevant position of the rain god's celestial avatar whose power he assumed or shared, displayed in a singular way his divine identity and, consequently, his kingly responsibility for a proper development of natural cycles and for maintaining the ideal cosmic order, which guaranteed the survival of his subjects.

While buildings resembling the Governor's Palace of Uxmal, in spite of being laden with religious iconography, disclose rather clearly their primary purpose, it is often difficult or impossible to reliably ascertain a structure's functionality exclusively on the basis of its architectural characteristics. Probably the most illustrative examples can be found in the Maya Late Classic Río Bec style architecture, named after an archaeological site in the central lowlands in the Yucatan peninsula and characterized by buildings whose evidently residential nature, amalgamated with an overload of religious symbolism embedded in their profuse decoration, indicates their dual function and suggests, as Andrews and Gendrop (1991, 15) observe, that 'the normally separate functions of "Church" and "State" were combined into a single entity and assumed by a single person (or family or clan) as a way of legitimizing their divine authority, and hence their political control'. The term 'temple-palace', commonly used in Mesoamerican archaeology and reflecting the ambiguity of information relevant for determining the function of the buildings to which it is applied, can thus be accounted for, to a considerable extent, by the very complexity inherent in the structure of government in prehispanic societies. The orientations of various Maya buildings of this kind in the Río Bec area have been shown to belong to the alignment groups commonly found in Mesoamerica and are arguably based on astronomical considerations (Šprajc, 2004b).

The astronomical alignments, just like other types of evidence, clearly show that practical uses of astronomy were embedded in ritual and intimately related to the social organization, religion and political ideology of Mesoamerican societies. The apparently immutable and perfect order observed in the sky, obviously superior to the one reigning on the earth, must have been the primary source of deification of heavenly bodies, whose cyclic behavior thus was not viewed as being simply correlated with seasonal transformations in the natural environment but rather as provoking them. If we assume, therefore, that timely occurrences of these changes were believed to be conditioned by the arrival of celestial bodies, such as the Sun and Venus, to specific points on the horizon, the architectural alignments reproducing directions to these phenomena may well have been intended to secure, in accordance with principles of magic, their regular sequence. The astronomical orientations of important civic and ceremonial buildings suggest that this was a major concern of the rulers, who as men-gods pretended to be responsible for the proper functioning of the universe (see Šprajc 2005, 211-212).

Conclusions

Mesoamerican astronomy had a complex social role. According to a number of sources, the most sophisticated astronomical knowledge was possessed by the priestly class closely associated with the ruling elite. However, both prehispanic vestiges and early colonial documents suggest that also the rulers themselves were deeply concerned with what was occurring in the sky.

Astronomical interests of those holding political power come as no surprise. In both Mesoamerican and other ancient civilizations, whose subsistence was based on intensive agriculture, the ability to predict important seasonal changes in the natural environment was of paramount importance; in the absence of a calendar accurately reproducing seasonal cycles, reliable predictions could only be based on astronomical observations performed by specialists familiar with cyclical celestial phenomena and their concomitance with annual climatic variations. Considering that a more efficient distribution of activities in the agricultural cycle increased productivity and secured survival to a larger population, the astronomers-priests' professional skills must have been vital for a successful economy and smooth functioning of the existing social and political system.

However, the study of celestial regularities, rather than being a strictly scientific or technical endeavor isolated from the rest of ideas and activities and aimed at solving practical problems only, was deeply intertwined with religious beliefs and ceremonial activities. The celestial order, apparently invariable and perfect, came to be considered superior to the terrestrial and human order, giving rise to the myths and beliefs according to which the events on the earth depended on the phenomena observed in the sky. With the origin and development of social stratification, such beliefs were modified and incorporated into the ideology that was elaborated, declared and imposed by the ruling elite, with the purpose of sanctioning and maintaining the existent social order. The rulers were responsible for performing rituals that guaranteed a proper development of natural cycles and the preservation of the ideal cosmic order. Advances in astronomical knowledge made the achievement of these objectives more effective, as they allowed the most appropriate moments for every ceremonial act to be determined with greater precision. Moreover, reliable predictions of celestial events and the corresponding astrological auguries contributed to the legitimation of power, justifying the privileges enjoyed by the rulers and their collaborators dedicated to priesthood, astronomy and calendar.

References

Andrews, G. F., and Gendrop, P. 1991. Río Bec Tower Complexes: Forms and Functions. Cuadernos de Arquitectura Mesoamericana 12, 9-21.

Ashmore, W. 1989. Construction and cosmology: politics and ideology in Lowland Maya settlement patterns. In W. F. Hanks and D. S. Rice (eds), Word and Image in Maya Culture: Explorations in Language, Writing, and Representation, 272-286. Salt Lake City, University of Utah Press.

Ashmore, W. 1991. Site-planning principles and concepts of directionality among the ancient Maya. Latin American Antiquity 2, no. 3, 199-226.

Ashmore, W., and Sabloff, J. A. 2002. Spatial orders in Maya civic plans. Latin American Antiquity 13, no. 2, 201-215.

Ashmore, W., and Sabloff, J. A. 2003. Interpreting ancient Maya civic plans: reply to Smith. Latin American Antiquity 14, no. 2, 229-236.

Aveni, A. F. 2001. Skywatchers: A Revised and Updated Version of Skywatchers of Ancient Mexico. Austin, University of Texas Press.

Aveni, A. F. 2003. Archaeoastronomy in the ancient Americas. Journal of Archaeological Research 11, no. 2, 149-191.

Aveni, A. F., Dowd, A. S., and Vining, B. 2003. Maya calendar reform? Evidence from orientations of specialized architectural assemblages. Latin American Antiquity 14, 159-178.

Aveni, A., and Hartung, H. 1986. Maya city planning and the calendar. Transactions of the American Philosophical Society 76, no. 7, Philadelphia, American Philosophical Society.

Aveni, A., and Hartung, H. 2000. Water, mountain, sky: the evolution of site orientations in southeastern Mesoamerica. In E. Quiñones Keber (ed.), In Chalchihuitl in Quetzalli: Precious Greenstone Precious Quetzal Feather: Mesoamerican Studies in Honor of Doris Heyden, 55-65. Lancaster, Labyrinthos.

Aveni, A. F., and Hotaling, L. D. 1994. Monumental inscriptions and the observational basis of Maya planetary astronomy. Archaeoastronomy 19 (Journal for the History of Astronomy, Supplement 25), S21-S54.

Baudez, C. F. 1985. The sun kings at Copan and Quirigua. In V. M. Fields (ed.), Fifth Palenque Round Table, 1983, 29-37. San Francisco, The Pre-Columbian Art Research Institute.

Broda, J. 1976. Los estamentos en el ceremonial mexica. In P. Carrasco, J. Broda et al., Estratificación social en la Mesoamérica prehispánica, 37-66. México, Centro de Investigaciones Superiores, Instituto Nacional de Antropología e Historia.

Broda, J. 1982. Astronomy, cosmovisión, and ideology in pre-Hispanic Mesoamerica. In A. F. Aveni and G. Urton (eds), Ethnoastronomy and Archaeoastronomy in the American Tropics, Annals of the New York Academy of Sciences 385, no. 1, 81-110.

Brown, C. H. 1991. Hieroglyphic literacy in ancient Mayaland: inferences from linguistic data. Current Anthropology 32, no. 4, 489-496.

Carrasco, P. 1985. La economía del México prehispánico. In P. Carrasco and J. Broda (eds), Economía política e ideología en el México prehispánico, 4th ed., 15-76. México, Editorial Nueva Imagen – Centro de Investigaciones Superiores del Instituto Nacional de Antropología e Historia (1st edition 1978).

Closs, M. P. 1994. A glyph for Venus as evening star. In V. M. Fields (ed.), Seventh Palenque Round Table, 1989, 229-236. San Francisco, The Pre-Columbian Art Research Institute.

Colas, P. R. 2003. K'inich and king: naming self and person among Classic Maya rulers. Ancient Mesoamerica 14, no. 2, 269-283.

Durán, D. 1995. Historia de las Indias de Nueva España e islas de tierra firme, 2 vols. México, Consejo Nacional para la Cultura y Las Artes.

Frazer, J. G. 1922. The golden bough: A study in magic and religion. London, Macmillan.

Galindo Trejo, J. 1994. Arqueoastronomía en la América antigua. México, CONACYT.

García Barrios, A. 2007. El dios Chaahk en el nombre de los gobernantes mayas. Estudios de Cultura Maya 29, 15-29.

Garza, M. de la, 2002. El puesto del gobernante en el cosmos y sus ritos de poder. Estudios de Cultura Maya 22, 247-259.

Grube, N. 2002. Onomástica de los gobernantes mayas. In V. Tiesler Blos, R. Cobos and M. Greene Robertson (eds), La organización social entre los mayas: Memoria de la Tercera Mesa Redonda de Palenque 2, 321-353. México, Instituto Nacional de Antropología e Historia - Universidad Autónoma de Yucatán.

Houston, S. and Stuart, D. 1996. Of gods, glyphs and kings: divinity and rulership among the Classic Maya. Antiquity 70, no. 268, 289-312.

Iwaniszewski, S. 1989. Exploring some anthropological theoretical foundations for archaeoastronomy. In A. F. Aveni (ed.), World Archaeoastronomy, 27-37. Cambridge: Cambridge University Press.

Kowalski, J. K. 1987. The House of the Governor: A Maya palace of Uxmal, Yucatan, Mexico. Norman – London, University of Oklahoma Press.

León-Portilla, M. 1986. Astronomía y cultura en Mesoamérica. In M. A. Moreno Corral (ed.), Historia de la astronomía en México, 11-16. México, Fondo de Cultura Económica.

López Austin, A. 1973. Hombre-dios: Religión y política en el mundo náhuatl. México, Universidad Nacional Autónoma de México.

López de Gómara, F.. 1987. La conquista de México, ed. J. L. de Rojas. Madrid, Historia 16 – Información y Revistas.

Lounsbury, F. G. 1989. A Palenque king and the planet Jupiter. In A. F. Aveni (ed.), World Archaeoastronomy, 246-259. Cambridge, Cambridge University Press.

Martin, S., and Grube, N. 2000. Chronicle of the Maya Kings and Queens: Deciphering the Dynasties of the Ancient Maya. London, Thames & Hudson.

Milbrath, S. 1999. Star Gods of the Maya: Astronomy in Art, Folklore, and Calendars. Austin, University of Texas Press.

Motolinía, T. de Benavente o, 1971. Memoriales o libro de las cosas de la Nueva España y de los naturales de ella, ed. E. O'Gorman. México, Universidad Nacional Autónoma de México, Instituto de Investigaciones Históricas.

Nicholson, H. B. 2001. Topiltzin Quetzalcoatl: The Once and Future Lord of the Toltecs. Boulder, University Press of Colorado.

Reyes García, L. 1979. La visión cosmológica y la organización del imperio mexica. In B. Dalhgren (ed.), Mesoamérica: Homenaje al doctor Paul Kirchhoff, 34-40. México: Instituto Nacional de Antropología e Historia.

Reyman, J. E. 1975. The nature and nurture of archaeoastronomical studies. In A. F. Aveni (ed.), Archaeoastronomy in Pre-Columbian America, 205-215. Austin, University of Texas Press.

Rivera Dorado, M. 2001. La ciudad maya: Un escenario sagrado. Madrid, Editorial Complutense.

Rivera Dorado, M., and Amador Naranjo, A. 1994. Más opiniones sobre el dios Chak. Revista Española de Antropologia Americana 24, 25-46.

Sahagún, B. de, 1986. Coloquios y doctrina cristiana: con que los doce frailes de San Francisco, enviados por el Papa Adriano VI y por el emperador Carlos V, convirtieron a los indios de la Nueva España, 1st edition, ed. M. León-Portilla. México, Universidad Nacional Autónoma de México – Fundación de Investigaciones Sociales.

Sellen, A. T. 2002. Storm-god impersonators from ancient Oaxaca. Ancient Mesoamerica 13, no. 1, 3-19.

Sharp, R. 1981. Chacs and Chiefs: The Iconology of Mosaic Stone Sculpture in Pre-Conquest Yucatan, Mexico. Studies in Pre-Columbian Art & Archaeology 24. Washington, Dumbarton Oaks, Trustees for Harvard University.

Šprajc, I. 1993a. The Venus-rain-maize complex in the Mesoamerican world view: part I. Journal for the History of Astronomy 24 (Parts 1/2), 17-70.

Šprajc, I. 1993b. The Venus-rain-maize complex in the Mesoamerican world view: part II. Archaeoastronomy 18 (Journal for the History of Astronomy Supplement 24), S27-S53.

Šprajc, I. 1996. La estrella de Quetzalcóatl: El planeta Venus en Mesoamérica. México, Editorial Diana.

Šprajc, I. 2000a. Astronomical alignments at Teotihuacan, Mexico. Latin American Antiquity 11, no. 4, 403-415.

Šprajc, I. 2000b. Astronomical alignments at the Templo Mayor of Tenochtitlan, Mexico. Archaeoastronomy 25 (Journal for the History of Astronomy Supplement 31), S11-S40.

Šprajc, I. 2001a. La astronomía. In L. Manzanilla and L. López Luján (eds), Historia antigua de México, vol. IV: Aspectos fundamentales de la tradición cultural mesoamericana, 273-313. México, Universidad Nacional Autónoma de México - Instituto Nacional de Antropología e Historia - Porrúa.

Šprajc, I. 2001b. Orientaciones astronómicas en la arquitectura prehispánica del centro de México. México, Instituto Nacional de Antropología e Historia (Colección Científica 427).

Šprajc, I. 2004a. The south-of-east skew of Mesoamerican architectural orientations: astronomy and directional symbolism. In M. Boccas, J. Broda and G. Pereira (eds), Etno y arqueo-astronomía en las Américas, 161–176. Memorias del Simposio ARQ-13 del 51° Congreso Internacional de Americanistas, Santiago de Chile.

Šprajc, I. 2004b. Astronomical alignments in Río Bec architecture. Archaeoastronomy: The Journal of Astronomy in Culture 18, 98-107.

Šprajc, I. 1996. La estrella de Quetzalcóatl: El planeta Venus en Mesoamérica. México, Editorial Diana.

Šprajc, I. 2005. More on Mesoamerican cosmology and city plans. Latin American Antiquity 16, no. 2, 209–216.

Šprajc, I. 2008. Alineamientos astronómicos en la arquitectura. In I. Šprajc (ed.), Reconocimiento arqueológico en el sureste del estado de Campeche, México: 1996-2005, British Archaeological Reports International Series 1742 (Paris Monographs in American Archaeology 19), 233–242. Oxford: BAR Publishing.

Šprajc, I., Morales-Aguilar, C., and Hansen, R. D. 2009. Early Maya astronomy and urban planning at El Mirador, Peten, Guatemala. Anthropological Notebooks 15, no. 3, 79-101.

Thiemer-Sachse, U. 1992. Der Dualismus von weltlicher Herrschaft und Priesterschaft im alten Mexiko. Ethnographisch-Archäologische Zeitschrift 33, 499-525.

Tichy, F. 1991. Die geordnete Welt indianischer Völker. Das Mexiko-Projekt der Deutschen Forschungsgemeinschaft 21, Stuttgart, Franz Steiner Verlag.

Tozzer, A. M. 1941. Landa's Relación de las Cosas de Yucatan: A Translation. Papers of the Peabody Museum of American Archaeology and Ethnology, Harvard University, Vol. 18, Cambridge (reprinted in 1966).

Van der Waerden, B. L. 1974. Science Awakening II: The Birth of Astronomy. Leyden, Noordhoff International – New York, Oxford University Press.

White, L. A. 1959. The Evolution of Culture: The Development of Civilization to the Fall of Rome. New York, McGraw-Hill.

Woolard, E. W., and Clemence, G. M. 1966. Spherical Astronomy. New York, Academic Press.

POSSIBLE MESOAMERICAN NAKED-EYE OBSERVATION OF SUNSPOTS – I: EVIDENCE FROM THE TIKAL BALL COURT MARKER

RICHARD R. ZITO

Abstract: Many cultures of antiquity have made naked-eye observations of large sunspot displays. The Chinese, Japanese, Koreans and Greeks have all made such observations. Pictorial evidence on dated Mesoamerican monuments suggests that similar observations may have been made in the New World as well. Mesoamerican observations may help fill in gaps in the observational records and corroborate data. A long continuous record of large sunspot displays is essential for understanding the Sun, sunspot cycles and Earth's climatology.
Keywords: Maya, Astronomy, Sunspots, Solar Cycle, Climate Change

1. Background

The Sun has occupied a position of central importance, sometimes quite literally, in the cosmology of many cultures. Our own modern heliocentric solar system is no exception to this rule. This interest in the Sun is natural, since most cultures have correctly believed that life and the habitability of the Earth depends largely on the Sun's radiation. Any change in the opacity of the Earth's atmosphere (in the visible, IR or UV), the angle of attack of solar radiation on the Earth's surface, or the solar irradiance has profound implications for the Earth's climate. Currently, the solar constant has a mean value of 0.140 w/cm^2. However, sunspot formation can increase the solar constant by a few tenths of a percent (Fröhlich et al. 1991, 11), accompanying solar flares can increase the Sun's X-ray output from 10 to 100,000 times for wavelengths between 200Å and 2Å respectively (Johnson 1965, 95), and charged particle output (solar wind) can increase by a factor of 40 from sunspot minima to maxima (Johnson 1965, 95). The increase in ionizing radiation greatly increases the density and depth of the Earth's ionosphere, and may affect atmospheric layers below in a complex way.

Eddy (1977, 80) has shown that an absence of sunspots correlates with a cold climate. This conclusion is based on comparison of climate records with accurate sunspot counts made since Galileo first used the telescope for solar observation in 1610. During the Maunder Sunspot Minimum (1645-1710) the Earth, or at least Europe and China, became considerably colder. This can be seen in the Flemish paintings of the era. In Madrid's Thyssian-Bornemisza Museum, art works by Ruysdael (*A Winter Landscape with Skaters and Sleigh Before a Town*, 1660-1670), Neer (*Winter Landscape with Skaters on a Frozen Waterway*, 1650-1655), Grimmer (*A Winter Landscape with Village and Skaters on a Frozen River, Hunters in the Foreground*, 1525-1590), and others, attest to a much colder climate (Arnaldo 2004, 87, 103, 113). A less well known sunspot minimum, the Dalton Minimum, occurred between about 1800 and 1830. Again, temperatures dropped in both Europe and North America. And again, the change was recorded by artists of the time: Foschi (*Winter Landscape in the Apennines*, and the author's favorite for the period, *Winter Landscape in the Apennines with a Ruined Temple*, both circa 1805) (Andrade and Guerrero 1992, 398) and Thomas Birch (*Philadelphia Winter Landscape*, 1830-1845) (Arnaldo 2004, 237).

Before 1610, only the largest sunspots, or sunspot groups, had been observed by ancient astronomers via the unaided eye. However, these direct observational records are discontinuous. It would be of primary scientific interest if the gaps in these records could be completely, or even partially, filled. Large sunspot groups can appear during maxima in the sunspot cycle, therefore, identification of these outstanding naked-eye sunspot displays would help extend the solar activity record back into pre-telescopic times. A longer sunspot history may allow identification of longer Fourier components of the sunspot cycle. Furthermore, more correlations can then be made between sunspot activity and climate indicators (such as dendrochronology). In the current debate over man-made climate change, an understanding of naturally caused climate changes is of paramount importance. The political will of nations to curb greenhouse gasses may depend on the outcome of such studies.

Possible sunspot observation in Mesoamerica may provide some of the missing data, and may also corroborate observations made in other cultures. There are some dated monuments in Mesoamerica that show the Sun God with a spotted body; usually three or four spots on the cheeks (Milbrath 1999, 87). Cheek spots have been variously interpreted as an artistic motif possibly representing death spots or jaguar spots (Milbrath 1999, 87). Both of these theories have their difficulties. The Sun promotes the growth of maize, and is associated with life. It is the night that is associated with death in Mesoamerican culture. Furthermore, the jaguar is also associated with night. However, there is another possible explanation for cheek spots. It may be that, like so many other cultures, Mesoamericans simply made naked-eye observations of outstanding sunspot displays. Indeed, the astronomical 'observatory' at Tikal, Guatemala is well situated and designed for such observations. However, before the information on Mesoamerican monuments can be used to fill in missing sunspot maxima, it must be tested by identifying observations that overlap with those of other cultures, and by comparison with tree ring records and carbon-14 data. This report, the first in a series, is concerned with establishing a link between the information found on the ball-court marker at Tikal (having an equivalent Gregorian date of 378 AD) and the sunspot cycle.

2. Brief Description of the Marker and its Glyphs

The marker and its site have been described in detail by Fialko (1988, 117). It appears that the marker was used as a kind of dynastic public relations device, and was left in place for a period of 45 years. Text on the marker contained an equivalent Gregorian date of 378 AD (Fialko 1988, 117) and probably marked a major shift in political power at Tikal. Distinctive features of Sun God glyphs on this device include three facial spots near the eye and a *Kin* (Sun) sign joined to the head. The Sun sign contains four points on its interior that represent east, west, zenith and nadir. The *Kin* is a kind of 'autograph' for this deity (Fialko 1988, 117).

3. Physics of Naked–Eye Sun Spot Observation

The best time for observing sunspots is when the Sun is on a misty horizon (Murdin et al. 1979, 159). These are precisely the conditions found at Tikal, Guatemala. The North Acropolis functioned as a kind of observatory. An observer stood on temple 5D-33 and was able to position the rising Sun over three diminutive early classic structures (not numbered and not excavated) of which only the tops are visible. These three structures lie on a north-south line and are spaced by a center to center distance of 13m and 17m respectively (Coe 1967, map). The northern most structure marks the summer solstice sunrise point, the center structure marks the equinox, and the southern most structure marks the winter solstice. The distance from 5D-33 to the summer solstice structure is about 49m east (Coe 1967, 42). Since the rainy season lasts from May to October at Tikal, plenty of atmospheric water is available for attenuation. When the Sun is high in the sky, intensity reductions can be produced by reflection from a basin of water, preferably with a little dye in it. When the Sun is at zenith, the ratio of the reflected power (P_r) to the incident power (P_i) from pure water (index of refraction n = 1.33) is given by (Lorrain and Corson 1970, 518)

$$|P_r / P_i| = [(1-n)/(1+ n)] = (-0.33/2.33)^2 = 0.02 . \qquad (1)$$

Therefore, only 2% of the incident power is reflected back to the eye. With this reduction, and a little haze (but not too much, or sunspots will be obscured), it just might be possible to tolerate (with some eye damage) naked-eye observation.

4. Historical Data

Large sunspots are naked-eye objects which, during multiple spot displays, can cover as much as 1% of the solar disk (Murdin et al. 1979, 159). However, coverage in excess of 0.05% of the central part of the solar disk allows naked-eye observation of sunspots (Bray and Loughhead 1964, 229). Asian observers employed several methods to reduce the Sun's brilliance (all unsafe) for thousands of years. In China, the fine dust of the Gobi desert can remain suspended in the atmosphere by wind turbulence. Mei scattering then reduces the Sun's visible intensity to levels tolerable to the human eye. Japanese observers used smoked glass to make naked eye sunspot observations. Asian observations from China, Japan and Korea cover a period from 28 BC (Julian calendar) to 1604 AD (Gregorian calendar) (Clark and Stephenson 1978, 387). However, there are only four observations prior to 268 AD, and there are some large gaps, especially in the 5th, 7th and 15th centuries AD. Fortunately, the interval between 268 AD and 400 AD is rich with observations and brackets 378 AD nicely. In fact, an outstanding display was observed in China on 10 January 375 AD. Chin-shu writes, 'Within the Sun there was a black spot as large as a hen's egg'. Displays were also recorded on 6 April 374 AD, and on both 28 January (approx.) and 26 December of 373 AD. Prior to 373 AD, no sunspots had been observed since 29 March 370 AD. After 375 AD, the next sunspot observation came on 2 April 388 AD. All these observations were made by Chin-shu in China.

In the west, sunspots were first observed by the Greek philosopher Anaxagoras in 467 BC. Outstanding sunspot displays are often accompanied by magnificent aurora, and this case was no exception. Book 3 of Livy's *The History of Rome from Its Foundation* (Sélincourt 1960, 199) records outstanding auroral activity for the years 464-463 BC, when 'lights blazed in the sky'. And, again in 462-461 BC, 'fires blazed in the sky'. Apparently the Sun was very active from 467-461 BC; a period of 6 years, or about half an average sunspot cycle. Again, Livy (Yardley, 2006) records exceptional auroras in 218-217 BC (Book 21) when 'glowing ship-like figures had appeared in the sky', in 206 BC (Book 28) when 'daylight appeared during the night', and in 204 BC (Book 29) when 'there had been moments of daylight during the night'. These exceptional displays at the latitude of Rome (41.54^0) were carefully recorded by Roman historians since they were considered 'omens' of terrible things to come. Pliny the Elder (23-79 AD) (Healy 1991, 22) also mentions auroras in his *Natural History*. Pliny describes a time 'during the Antonine War' (44 BC) 'when there was continuous twilight (at night) for almost a whole year'. In another possible auroral reference (86 BC) Pliny describes a 'blazing shield that traversed the sky from west to east'. The excellent year-by-year history of Ammianus Marcellinus (Yonge 1911, 576) survives and covers the years of 353-378 AD (including the omens). Marcellinus was a staff officer in the *prefectores domestici* (imperial bodyguard) and he writes about events of which he had direct knowledge, making him a primary source of information. In autumn of the year 375 AD (the same year as Chin-shu's observations), while campaigning along the Danube River with the emperor Valentinian near Aquincum (Budapest) at a latitude of about 47.5^0, Marcellinus writes (Book 31, Chapter 1, Section 2) that 'lurid sunrises made the mornings dark' and that 'birds of night constantly uttered querulous and mournful cries'. According to Webster's Dictionary, the word 'lurid' means 'ghostly pale in appearance, causing horror, shinning with the red glow of fire'. This is actually a fairly accurate translation of the original Latin text (White, 2009). The actual word used by Marcellinus was *hebio*. This unusual Latin word is only used in one other place in his history (Book 19, Chapter 6, Section 7), where it is used to describe the night of the new Moon. That is, the Moon occulted by the Earth. In fact, the modern English word 'occult' comes from this phenomenon, and the connotation is definitely spectral. Bright auroras normally show in colors of green, blue and blood red. The latter would definitely qualify as lurid, and in Marcellinus' recollection would have portended Rome's disastrous defeat at Adrianople (Edirne, Thrace) in 378 AD. But, *hebio* has another meaning, viz. 'sluggish and slow' (Lewis and Short, 1879). From Budapest, the near point to an expanded auroral oval would lie to the northeast (Davis 1992, 25). Therefore, the light of a pre-dawn aurora would have seemed like an extended sunrise to an ancient observer unfamiliar with the phenomenon. So, Marcellinus' use of the word *hebio* has a double meaning. Finally, it should be noted that auroral displays are most common in the autumn and spring, and may persist from dusk till dawn, but are brightest at local midnight, and may emit as much light as a full Moon (Davis 1992, 35, 64). Such illumination levels are known to induce bird calls in anticipation of Sun rise. Given Marcellinus' location, the year and season of his observations, animal reactions, and his use of the word *hebio*, it is likely that his somewhat ambiguous description of the omens was actually an auroral sighting. This would be a confirmation of Chinese sunspot observations.

5. Tree Ring Data

In the American southwest, the annual rain fall is very systematically tied to the sunspot cycle. In fact, every sunspot cycle minimum is also a year of *El Niño* rains. Of course, there may also be extra *El Niño* years that are not sunspot minima. The contrapositive of the rule above is also true, i.e., the absence of *El Niño* rains implies the absence of a sunspot minimum (or the presence of higher than average sunspot activity). *El Niño* rains produce exceptional tree ring growth, while dry, non *El Niño*, years are associated with the narrowest tree rings. In the desert mountains of the United States, only the Rocky Mountain Bristle Cone Pine (*Pinus longaeva*) lives long enough to test Mesoamerican dates of maximum sunspot activity; in this case, 378 AD. Two sites were chosen as representative of the southwestern climate of the United States. Figure 1 shows that the tree ring width minimum is consistent with higher than average sunspot activity between the years of about 376 AD to 378 AD. It should also be noted that, for *Pinus longaeva* from the White Mountains of California (north of Death Valley) in the United States, the effect of prior growth on the width of any particular growth ring is small (Fritts 1976, 379). Therefore, tree ring data for this species and location is particularly reliable.

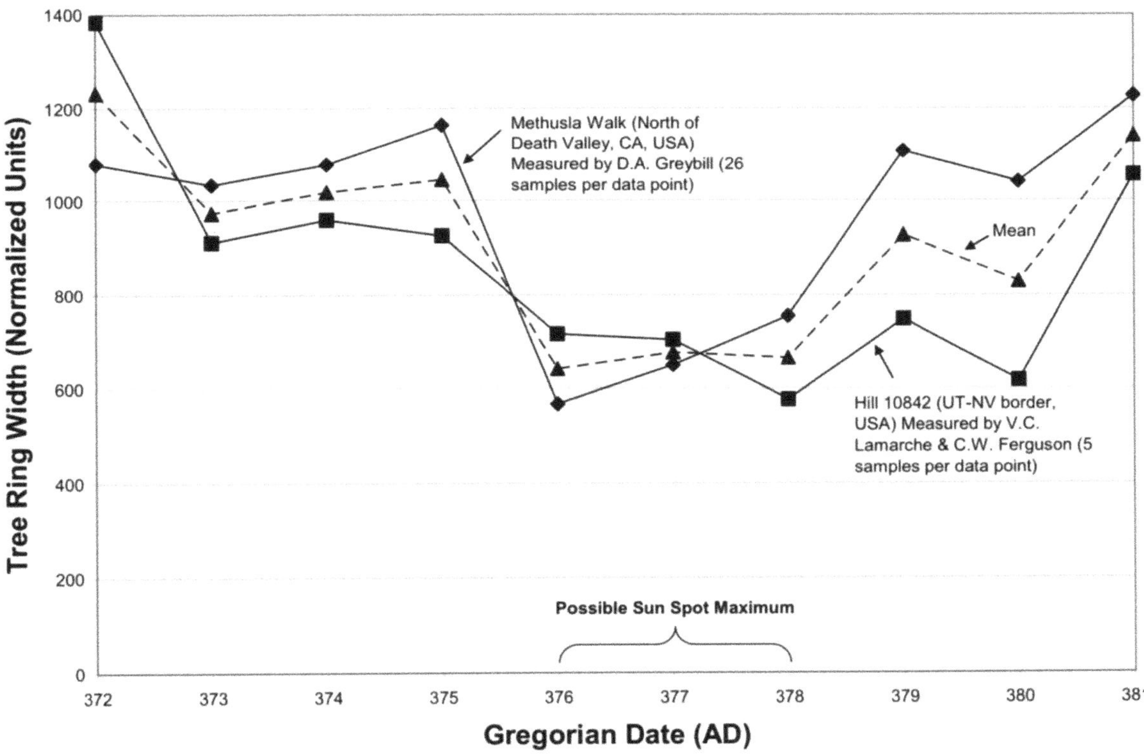

Figure 1. Tree ring data for the June to August growth of Pinus Longaeva *(Rocky Mountain Bristle Cone Pine). Tree ring widths are from the International Tree Ring Data Bank (www.ncdc.noaa.gov/paleo/treering.html). Widths are in 'normalized' units such that 1000 represents average growth. A 'width' less than 1000 indicates drought conditions, and above 1000 indicates wet conditions. Climate accounts for 63% of the growth variance (Fritts, 1976). The time scale covers one sunspot cycle. A sunspot maximum probably occurred somewhere in the 376 AD to 378 AD time interval.*

6. Carbon-14 Data

Carbon-14 is produced in the upper atmosphere by the impact of high-energy galactic cosmic ray produced neutrons on nitrogen-14. This atmospheric carbon-14 is eventually taken up by plants and can be extracted from tree rings in a year-by-year fashion. After correcting for the beta decay of carbon-14 with time, the original atmospheric level of radiocarbon can be estimated (Damon and Sonett, 1991). The production of ^{14}C is almost constant, but small variations exist due to changes in the number of high-energy neutrons that are produced in the upper atmosphere. Neutron production is modulated by the sunspot dependent solar wind and changes in the Earth's dipole moment. Carbon-14 production decreases when the sunspot count is a maximum, and vice versa (Jokipii 1991, 205). After removing the trend line due to terrestrial dipole moment variations in the anomalous ^{14}C production data, a minimum is found near 375 AD (±3 years). Therefore, ^{14}C supports the conjecture that this time neighborhood was a period of high sunspot activity.

7. Conclusion

All the data taken together seems to indicate that use of the sunspot motif in the Sun God image on the ball court marker at Tikal was a response to prior sunspot activity that probably peaked on 375 or 376 AD. Therefore, the dates on monuments showing a spotted Sun God cannot be taken as *exact* dates of sunspot activity. However, spots cannot necessarily be taken as arbitrarily used artistic devices either. Given the Mayan use of astronomical events as portents of the future, the sunspot events of 375 AD may have been considered, in retrospect, a portent of the political change that was to occur during the *entrada* of 378 AD, with the coming of the ambassador Fire Born (Smoking Frog), and the political dominance of *Teotihuacan*. In a sense, perhaps the Maya did interpret sunspots as death spots; the 'death' of one dynasty, and the beginning of another. Ironically, and also in retrospect, Marcellinus in Europe considered auroras associated with the same sunspot event to be a portent as well, but in this case, of military disaster.

Many more artifacts will have to be examined before a final judgment can be made on the connection between the spotted Sun God motif in Mayan iconography and the existence of physical sunspots. And each artifact will have to be evaluated on a case by case basis. A useful artifact will have to pass four tests.

1. It will have to show a spotted (or unspotted) Sun God.
2. It will have to be dated.
3. There will have to be corroborating direct observations from another culture.
4. There will have to be corroborating physical evidence (tree ring data, climate data, flood levels, aurora, etc.).

References

Andrade, J., and Guerrero, M. 1992. Old Masters, Vol. 1. Thyssen-Bornemisza Museum. Madrid, Fundacion Coleccion Thyssen-Bornemisza.

Arnaldo, J. (ed.). 2004, Carmen Thyssen-Bornemisza Collection, Vol. 1. Madrid, Fundación Colección Thyssen-Bornemisza.

Bray, R. J., and Loughhead, R. E. 1964. Sunspots. London, Chapman and Hall.

Clark, H. D., and Stephenson, F. R. 1978. An interpretation of the pre-telescopic sunspot records from the orient. Quarterly Journal of the Royal Astronomical Society 19, 387-410.

Coe, W. R. 1967. Tikal, 41-50. Map of Central Tikal, Philadelphia, University of Pennsylvania.

Damon, P., and Sonett, C.P. 1991, Solar and terrestrial components of the atmospheric ^{14}C variation spectrum. In C. P. Sonett, M. S. Giampapa and M. S. Matthews (eds), The Sun In Time, 360-388. Tucson, University of Arizona Press.

Davis, N. T. 1992. The Aurora Watcher's Handbook., Fairbanks, University of Alaska Press.

Eddy, J. A. 1977. The case of the missing sunspots. Scientific American 236 no. 5 (May 1977), 80-92.

Fialko, V. C. 1988. El marcador de juego de pelota de *Tikal*: nuevas referencias epigraficás para el periodo clásico temprano. Mesoamérica 15, 117-135.

Fritts, H. C. 1976. Tree Rings and Climate. Caldwell NJ, Blackburn.

Fröhlich, C. et al. 1991, Solar irradiance variability from modern measurements, In C .P. Sonett, M. S. Giampapa and M. S. Matthews (eds), The Sun In Time. Tucson, AZ, University of Arizona Press.

Healy, J. F. (trans.). 1991. Pliny the Elder: Natural History. London, Penguin.

Johnson, F. S. 1965. Satellite Environment Handbook. Stanford CA, Stanford University Press.

Jokipii, J. R. 1991. Variations of the cosmic-ray flux with time, in C. P. Sonett, M. S. Giampapa and M. S. Matthews (eds), The Sun In Time, 205-220. Tucson, University of Arizona Press.

Lewis, C., and Short, C. 1879. Latin Dictionary. Oxford, Oxford University Press.

Lorrain, P., and Corson, D. R. 1970. Electromagnetic Fields and Waves, 2nd edition. San Francisco, Freeman.

Milbrath, S. 1999. Star Gods of the Maya. Austin TX, University of Texas Press.

Murdin, P., Allen, D., and Malin, D. 1979. Catalogue of the Universe. New York, Crown.

Sélincourt, A. de. 1960. Livy: The Early History of Rome, London, Penguin.

White, C. 2009. Private Communications. Tucson, University of Arizona, Classics Department.

Yardley, J. C. 2006. Livy: Hannibal's War. Oxford, University of Oxford.

Yonge, C. D. 1911. The Roman History of Ammianus Marcellinus. London, Bell and Sons.

CALENDRIC-ASTRONOMICAL ORIENTATION AS AN EXPRESSION OF POWER IN MESOAMERICA

JESÚS GALINDO TREJO

Abstract: In this paper we present a brief description of how Mesoamericans mainly orientated architectural structures where not only astronomical issues were involved but also aspects related to the calendar. This cultural practice is emphasized as a clear expression of political power that helped the sovereign to maintain his prestige and authority.
Keywords: Mesoamerica, calendar, architectural alignments

1. Introduction

Over the millennia the various Mesoamerican cultures built great buildings and planned splendid cities. The choice of where their settlements were founded had to do with criteria regarding the conditions of survival, such as the proximity to food and water supplies. They also took into account situations of security against possible attacks by enemy groups. However, ideological aspects related to the surrounding landscape, especially with the day and night sky, were also considered. Thus the spatial coincidence of the solar disk with a feature of the landscape on a particular date could be observed from a given building. In most ancient cultures important moments such as solstices and equinoxes were chosen mainly to achieve an architectural alignment with the solar disk on the horizon. The practice of observing the sky has accompanied man since his origins. Thanks to his innate intelligence, man quickly accumulated sufficient knowledge about the functioning of the sky and he was able to use it for a variety of situations. As this is a highly specialized knowledge not accessible to common people, the ruling class soon took charge of preserving, transmitting and promoting the development of such knowledge. A particularly important use of that knowledge was the development of organizational systems to carry out efficiently every activity in the community. Calendars were generated with certain characteristics, according to the observational aspects of the sky. Since a calendar was a quantitative model to describe the periods of observation of a heavenly body, the careful tracking of time achieved such prominence that the elite recognized the desirability of using it as an ideological means to strengthen its social control and to maintain the prevailing status quo.

The cultural area of Mesoamerica includes not less than three thousand years of social evolution, achieving a refined level of development in many areas of knowledge. Astronomy, architecture and urbanism played a key role in defining Mesoamerican societies. In particular, they draw attention to the fact that the Mesoamerican calendrical system was valid for several millennia. This calendar spanned various periods and exhibited varying levels of sophistication. It was even considered as a work of the gods who gave it to mankind for the proper functioning of their society. The study of the orientation of Mesoamerican architectural structures has established in recent years that the importance of the calendar was also expressed in the choice of solar orientations, a fact that is only explicable on the basis of the characteristics that define this calendar. Canonical orientations of large buildings and urban outlines were defined. These architectural features followed divine precepts because they signalled the proper passing of time.

In recent decades persistent archaeoastronomical work has been performed in many Mesoamerican cities to determine the orientation of their most important architectural structures and their urban layouts.

When Paul Kirchhoff (1943) defined Mesoamerica by numerous cultural characteristics over several millennia extending into large geographic regions, the vast amount of archaeoastronomical research available today had not yet been carried out (Aveni, 2001; Galindo Trejo, 1994 and 2003; Šprajc, 2001 and 2008). Therefore, recent archaeoastronomical research may provide new cultural characteristics to be considered in the definition of Mesoamerica.

The Mesoamerican practice of orienting architectural structures: an exaltation of political power

An excellent example of the use of pictorial art to exalt power can be recognized in the Mayan city of Bonampak. There the ruler Chan Muan commissioned one of the most expressive and realistic murals of Mesoamerica to make known his deeds on Earth. The Building of the Paintings has three rooms fully painted with different scenes in which the ruler is depicted in various situations. The vault of every room was painted with representations of the so-called Sky Monster. Thus, the ruler is presented in front of his defeated enemies in various ceremonies, battles and sumptuous parties with orchestras. On their murals the Mayas wrote dates that correspond to the eighth century AD. Painted on the vault of the central room are several square frames with star glyphs and what are believed to be constellations. In the early hours of the date written on the walls of the room, one region of the sky, which corresponds to the constellations of Taurus and Orion, extended along the room´s vault (Figure 1). These star configurations allowed an identification of the frames. The turtle is Orion, the herd of wild boars is the Pleiades, the person with the wand is identified with the red star Aldebaran and located on the horn of Taurus, and Mars is represented by another personage who holds what could be a mirror. For this date, at the beginning of the night, the Milky Way was aligned to the axis of symmetry of the Building of the Paintings. About nine hours later, the Milky Way was placed along the facade of the building, which coincided with the position of the Sky Monster. This is clearly a clever use of architecture, mural painting and astronomy to frame the deeds of the Mayan ruler and to achieve an effect of admiration, prestige and legitimation before his society. In this practice the priest-astronomer played an important role in maintaining social stability and in strengthening the power of the ruler.

Figure 1. Mural Painting from the central room of the Building of the Paintings in Bonampak. Possible Mayan constellations are identified from the position of sky objects on the date written inside the room. Artwork drawn by Francisco Villaseñor.

As a result of the intensive work of many scholars, one can assert that the Mesoamericans developed a very peculiar way of choosing the orientation of architectural structures. Like other cultures of the ancient world, the Mesoamericans occasionally orientated buildings towards the rising or setting sun on days of solstices, equinoxes and the solar zenith passage. However, most of the major Mesoamerican temples are oriented so that the Sun is aligned to these structures on certain days which themselves have no important astronomical meaning. These dates of solar alignment can be explained by some properties of the Mesoamerican calendrical system. These dates essentially allow the solar year to be split into counts of days expressible by numbers that define some characteristics of the two Mesoamerican calendrical counts, namely, the 365-day solar count, the 260-day ritual one, and the relationship between both. As is well known, both counts are strongly coupled, since they both start simultaneously, just after 52 periods of 365 days, and end at the same time. The ritual count in the meantime will have completed exactly 73 cycles. Therefore, it is possible to obtain a Mesoamerican calendrical equation: $52 \times 365 = 73 \times 260$. Likely this system of time tracking originated in Olmec culture, which has been considered to be the mother culture of Mesoamerica from about 1500 BC. On the other hand, it has been proven that Mayan astronomer-priests recorded in the Dresden Codex the synodic period of Venus of 584 days and various observation stations, such as its appearance as the Morning and Evening Star, and the non-observation intervals of this planet. Curiously, the number 73, representing one fifth of the solar year of 365 days, can be correlated with that period through accumulating 8 times 73.

The interesting thing about the orientation of architectural structures in Mesoamerica is that the most emblematic buildings in many cities show a solar alignment at dawn or sunset on dates of no astronomical significance. However, these dates are separated from the solstices by a number of days expressible by 52, 260, 73 and 65. The number 65 can be understood from ethnohistorical sources from the Zapotecan culture in Oaxaca in the 16th century which indicate that the 260-day ritual count was divided into 4 parts of 65 days and that this number of days was identified with one of the most important deities in the Zapotecan pantheon, Cocijo (Córdova 1886, 202). It is interesting to emphasize that the dates associated with the number 65 lead to a balanced embedding of the ritual count into the solar count taking the solstices as pivots for the counting of days. Note that the calendar in Mesoamerica reached the highest level of relevance and was regarded as the work of the gods. This practice of orientating buildings, which was in force for several millennia, can be considered as another cultural manifestation that should be included in the definition of Mesoamerica. In this paper, we present three families of calendric-astronomical orientation identified so far in Mesoamerica with numerous examples of each.

A remarkable example of the Mesoamerican situation in this regard is the Pyramid of the Sun in Teotihuacan (Figure 2). Being the largest building at the site, its orientation defines the urban outline inasmuch as its symmetry axis is perpendicular to the so-called Avenue of the Dead. The dates of solar alignment of this great pyramid are at sunset on 29 April and 13 August (Galindo Trejo, 1994; Šprajc, 2001). Clearly these dates do not correspond to any astronomical event of particular importance. However, in the Mesoamerican cultural context we are faced with a choice of orientation of great ritual significance. Indeed, these dates divide the solar year at a rate defined by certain properties of the Mesoamerican calendrical system. If we start counting the days from the first alignment on 29 April, on every day after this moment the solar disk will set increasingly to the north (right of the observer) until 52 days after that day, when the summer solstice will arrive. From this astronomical event, it will be another 52 days until the date of the second alignment in the year comes around on 13 August. Continuing the count of days, the sunsets after the previous date will mark positions of the solar disk increasingly to the south (left of the observer). Thus every day the sun will move in the direction of its southern extreme, which is attained on the day of the winter solstice. Then sunsets slowly indicate subsequent positions farther north, until sunset number 260 which will happen on 29 April of the following year. It is clear that the dates of solar alignment are tacitly suggesting a division of the solar year in the 260/104 ratio, taking into account a solstice as pivot.

It is noteworthy that this pattern of dividing the year is repeated considering the alignment of the pyramid towards the rising sun. In fact, the dates of solar alignment of the Pyramid of the Sun are in the early hours of 12 February and 29 October. Similarly, from the second alignment in the year, 29 October, 52 days must pass until the day of the winter solstice. After another 52 days the solar alignment

Figure 2. The orientation of the Pyramid of the Sun in Teotihuacan is a striking example of one of the families of calendric-astronomical orientation. Its urban layout is defined by dividing the solar year according to a ratio given by the characteristics of the Mesoamerican calendar.

of the pyramid will return on 12 February of the following year. Counting this alignment, sunrise number 260 will be on the next 29 October. One remarkable thing about this calendric-astronomical orientation is that throughout Mesoamerica and over the course of many centuries it appears in countless architectural structures. Here are some examples: the Pyramid of the Five-Stories in Edzná, House E of the Palace of Palenque, the Main Temple of Tula, the Palace of Tomb 105 at Monte Albán, the Pyramid of the Feathered Serpents of Xochicalco, the Temple of the Sun at Malinalco in relation to a cut in the opposite mountain, Structure A from Tamuín in the Huaxtecan region, the Upper Temple of Jaguars on the ball court at Chichén Itzá and Structure I of Calakmul, among others (Galindo Trejo, 1990, 2001, 2003 and 2008; Malmström, 1997; Šprajc, 2001 and 2008).

Another orientation family in Mesoamerica can be illustrated with an example which at the same time shows how some modern cities in Mexico still retain in their urban outline the original orientation set in Prehispanic times in concordance with calendric-astronomical principles. This is the case of the Templo Mayor at Tenochtitlan. This great pyramid, as the center of the Mexica worldview, represented perhaps the most important symbol of power and prestige of one of the largest cities in Postclassic Mesoamerica. The upper shrine of this temple was occupied by sanctuaries dedicated to the gods Tlaloc and Huitzilopochtli. With a double staircase, this imposing pyramid was orientated towards the west and reached a height similar to the existing towers of the Catedral Metropolitana. Around this pyramid, the city spread in four directions, according to the vertices of the building. The front of the Templo Mayor is aligned to the sunset on 9 April and 2 September (Galindo Trejo 1994, 166). Without apparent astronomical significance, these dates again indicate a unusual division of the solar year. From the first alignment in the year, 9 April, each day the Sun will set gradually further on the northern side of the western horizon (i.e., right of the observer) so that after 73 days the day of the summer solstice will arrive. Then another 73 days will lead us back to the date of the second alignment, 2 September. Following the registration of the next sunsets, one can note that the sun will be moving towards the south; it will come to its southernmost position on the day of the winter solstice and slowly will undertake its return journey, so that on the following 9 April it will complete its yearly course. It is remarkable that, from 2 September to 9 April, 219 days will elapse, i.e., 3 times 73 days. Note moreover that 5 times 73 days completes a solar year of 365 days. We recognize again the desire to emphasize the importance of the number 73 as a fundamental part of the Mesoamerican calendrical equation. This kind of division of the solar year is also given to the east. Indeed, the Templo Mayor of Tenochtitlan is aligned to the Sun in the early morning of 4 March and 9 October. As stated earlier, another feature that shows the importance of the calendrical number 73 lies in the fact that it can easily help to calibrate the synodic period of Venus of 584 days through the successive observation of solar alignments on the dates indicated above. This comes from the identity 584=8x73. For example, once we have observed the sunset on 2 September, we have to wait for the sunset of 9 April of the second year so that exactly 584 days will have elapsed. According to Chronist Sahagún (1979, 161), in Tenochtitlan there was a temple, called Ilhuicatitlan, dedicated precisely to the worship and the observation of Venus. Note that the correlation gathered by this chronicler indicates that the beginning of the Mexica year is 12 February. Therefore, the day of the solar alignment of the Temple Mayor at dawn on 4 March corresponds precisely to the start of the second *veintena* called Tlacaxipehualiztli (Sahagún 1979, 77-78).

Other examples of this orientation family are: the Pyramid of the Niches at El Tajín (Figure 3), the Great Pyramid of Xochitecatl, in its upper bodies from the Epiclassic period, the Olmec Pyramid of La Venta, the Great Solar Mask from the East Court at Copan, the Upper Building of Tomb 103 at Monte Albán, the Calendrical Temple of Tlatelolco, the solar painting of the Hall of the Frescoes in Mayapan, in lateral illumination, Structure 38 at Dzibilchaltún, the Great Arch known as Structure CA-9a at Oxkintok, and the Building E1 of Tehuacalco in the state of Guerrero, among others. (Galindo Trejo, 1994, 2003, 2004, 2008, 2010 and 2011; Šprajc, 2001; Casares Contreras, 2001 and 2002; Martz de la Vega, 2010).

The third family of orientations is represented abundantly in the Zapotecan region. The axis of symmetry of a building called the Jeweled Temple or Teotihuacan Embassy (Figure 4) on the northern platform of Monte Albán points to sunrise on 25 February and 17 October (Galindo Trejo, 2008). Without particular astronomical significance, both dates are located 65 days before and after the day of the winter solstice. As explained previously, according to ethnohistorical sources from the 16th century, the Prehispanic Zapotecs divided the 260-day ritual count into four periods of 65 days. They called this period of 65 days Cocijo and it was worshiped intensely because they asserted that this god was the origin of all things. Moreover, documents seized in the 17th century by the Spanish Inquisition recorded that in the Zapotecan Sierra the year began on 25 February (Franch 1966, 131). Inside Building P at Monte Albán there is a zenithal observatory whose geometry was configured so that, besides recording the passage of the sun through the zenith, the extreme incidence of sunlight into the enclosure´s floor indicates two particular dates: 17 April and 25 August (Morante López, 1995). These dates are also located 65 days before and after the day of the summer solstice. Therefore, we note that the Prehispanic Zapotecs embedded the ritual count into the solar year in a balanced way, since the two pairs of dates mentioned above are basically separated by 52 days. Another striking example of this Zapotecan practice is the group of El Arroyo at Mitla. On the lintel of the northern room of this group there was painted a mural of obvious astronomical significance: a solar disk, 'tied up' by two men holding ropes that flank it, seemingly descending from heaven. The scene is framed by what appear to be two architectural structures. For both pairs of dates described above, in the morning and afternoon, the mural was illuminated for a few seconds by lateral sunbeams when the sun was on the east and west horizons, aligned along the wall of the lintel (Galindo Trejo, 2008). On the other hand, recent archaeoastronomical studies in the north-central region of Mesoamerica report the presence of the family of orientations defined from the period of 65 days in various Epiclassic cities (Granados Saucedo, 2008 and 2011). Such a situation suggests that this family of orientations probably also extends, like the other two described before, throughout Mesoamerica.

Given the great importance that the calendar in Mesoamerican culture reached, it is obvious that such a way of choosing the orientation of important architectural structures is fervent and explicit praise to the creator gods of the calendar. Although there are still archaeoastronomical analyses which seek purely astronomical explanations for Mesoamerican alignments, it is increasingly clear that Mesoamericans mainly privileged calendric-cultural aspects over astronomical ones (equinoxes and solstices). One approach in this last sense for the Templo Mayor has been recently presented by Anthony Aveni (2006).

The analysis of a large number of important architectural structures orientated to sunrise or sunset points to certain days which suggest that perhaps the Mesoamericans structured a general pattern of orientation considering counts of days given by multiples of 13 days. In fact, two of the families of orientation considered here correspond to 4 and 5 times 13 days before and after the solstices. Moreover, starting from a solstice, 365 days divided by 13 leads to 28 groups of dates with a residue of only one day. So far we have identified numerous buildings that show a solar alignment for some of those dates so obtained. Of course, it is necessary to investigate more cases of buildings erected during the entire Prehispanic period and throughout all Mesoamerica before presenting a comprehensive proposal to explain this peculiar calendric-astronomical orientation.

This raises the possibility that the ruler who decided the orientation of each Mesoamerican temple according to that pattern certainly obtained the blessing of the gods because he could show that his material work was erected according to the divine canon. Thus, the ruler could appear before his people as fully endorsed by sacred favour and at the same time obtain a strong legitimation of the political and religious power he exercised over society. Certainly, this calendric-astronomical allegory associated with a temple constitutes only part of the complex symbolic discourse, complemented by the ruler with the additional choice of other artistic elements such as sculpture, mural painting and even the ceremonies that were held inside the temple. In order to unravel this discourse it is necessary to combine efforts from different areas of knowledge not only astronomical but also from the Humanities. Further archaeoastronomical research into unexplored archaeological sites in Mesoamerica may clarify the possible existence of other families of orientation, their origin and geographical distribution.

Figure 3. The Pyramid of the Niches at El Tajín probably had as many niches as days in the year, but its calendrical vocation may be recognized with certainty from corroborating that it is aligned to the rising Sun on the dates that divide the year into multiples of 73 days.

Figure 4. The Jeweled Temple or the Embassy from Teotihuacan at Monte Albán shows a solar alignment at dawn on 25 February and 17 October. Both dates are located just one Cocijo, i.e., 65 days, before and after the day of the winter solstice. This count of days identified one of the most important deities in the Zapotecan pantheon.

Conclusions

The calendrical system reached such great prominence for the Mesoamerican idiosyncracy that calendrical numbers were sometimes used both to establish architectural proportions of plazas and buildings and to define the ideal number of architectural elements (e.g., the number of stairs or bodies of a pyramid). In addition, some of the dates associated with the described solar alignments were used to fix transcendental moments in the calendar such as the beginning of the solar year or the starting point of the tracking of time. Thus, every aspect related to the calendar became canonical and therefore consistent with the principles dictated by the gods.

This brief study shows that the architectural practice of orientation in Mesoamerica followed its own course, somewhat different from other cultures of the ancient world, by observing the sky and using it as a reference to express the importance of certain canonical dates. The characteristics of such dates manifest the exceptional transcendence of the system of tracking of time for the ideology of Prehispanic peoples. However, we should note that this practice accounted for only a small part of the ritual and symbolic discourses that express all buildings dedicated to religious cult. Of course, a calendric-astronomical orientation awarded to the building an intangible value of the greatest solemnity and in this manner it was placed in complete harmony with the calendrical principles established by the sacred will. In addition to this, mural painting, sculpture, architectural elements and the religious ceremonies that were held inside the building would complete the complex message of fervour and political power that the ruling elite offered to generous deities who would favour, with all kinds of benefits, not only the aristocracy but also the people.

The ruling class of Mesoamerican society, by means of a thorough knowledge of the calendrical system, was able to justify and consolidate its position of political power before a people who recognized the action of the gods through the earthly works of their leaders.

Acknowledgements:

I am thankful to David Wood for his English corrections and comments on this paper.

References

Aveni, Anthony F. 2001. Skywatchers: A Revised and Updated Version of Skywatchers of Ancient Mexico. University of Texas Press, Austin.

Aveni, Anthony F. 2006. The Templo Mayor in Sacred Space. In Arqueología e historia del Centro de México: Homenaje a Eduardo Matos Moctezuma, coordinators Leonardo López Luján, David Carrasco and Lourdes Cue. Instituto Nacional de Antropología e Historia, México.

Casares Contreras, Orlando. 2001. The Astronomical Orientation of Structure 38: the Presence of Astrocalendric Alignments in Dzibilchaltún. The Precolumbian Art Research Institute Journal 2, no.3 (Summer), 6-9.

Casares Contreras, Orlando. 2002. Un estudio arqueoastronómico en Oxkintok. BA thesis, FCA, Merida, Universidad Autónoma de Yucatán.

Córdova, fray Juan de. 1886. Arte del idioma zapoteco. Imprenta del Gobierno en la Escuela de Artes, Morelia, Michoacán, México.

Franch, José Alcina. 1966. Calendarios zapotecos prehispánicos según documentos de los siglos XVI y XVII. Estudios de cultura náhuatl 6, 119-133.

Galindo Trejo, Jesús. 1990. Solar Observations in Ancient Mexico: Malinalco. Archaeoastronomy 21, no. 15, 17-36.

Galindo Trejo, Jesús. 1994. Arqueoastronomía en la América Antigua. Coleccion La Ciencia y la Tecnologia en la Historia. Madrid, Equipo Sirius.

Galindo Trejo, Jesús. 2001. Transfiguración Sagrada de Visiones Celestes: Alineación Astronómica de Estructuras Arquitectónicas en Cuatro Sitios Mayas. In Beatriz de la Fuente (ed.), Pintura Mural Prehispánica en México, Universidad Nacional Autónoma de México 2, Tomo IV, 294-310.

Galindo Trejo, Jesús. 2003. La Astronomía Prehispánica en México in Lajas Celestes: Astronomía e Historia en Chapultepec. Mexico City, Museo Nacional de Historia, Instituto Nacional de Antropología e Historia, 15-77.

Galindo Trejo, Jesús. 2004. Orientación Calendárico-astronómica en Teotihuacan y en El Tajín: alternativas de una misma tradición mesoamericana In Memorias de la Segunda Mesa Redonda de Teotihuacan, Universidad Nacional Autónoma de México Instituto-Nacional de Antropología e Historia, 369-391.

Galindo Trejo, Jesús. 2008. Calendario y orientación astronómica: una práctica ancestral en la Oaxaca Prehispánica. In Beatriz de la Fuente (ed.), La pintura mural prehispánica en México, vol. III: Oaxaca, t. III: Estudios, 295-345. Mexico City, Instituto de Investigaciones Estéticas, Universidad Nacional Autónoma de México.

Galindo Trejo, Jesús. 2010. El tránsito de Venus por el disco del Sol de 2012. Arqueología Mexicana 17, no. 103, Instituto de Investigaciones Estéticas, 49-51.

Galindo Trejo, Jesús. 2011. Orientación calendárico-astronómica en el Preclásico: el caso de La Venta en El Legado Astronómico. Universidad Nacional Autónoma de México, Instituto de Astronomía, Mexico, 154-167.

Granados Saucedo, Francisco. 2008. Observaciones astronómicas en el Centro Norte de México. Los casos de El Cerrito, Querétaro y Cañada de la Virgen, San Miguel de Allende, Guanajuato, en Tiempo y Región. Estudios Históricos y Sociales 2, 137-176. Universidad Autónoma de Querétaro, Instituto Nacional de Antropología e Historia., México.

Granados Saucedo, Francisco. 2011. El Culto a la Montaña in El Centro Norte de México y sus implicaciones calendárico-astronómicas. Los casos de El Cerrito, El Barrio de la Cruz, El Rosario y La Trinidad, en Querétaro; y Huamango, Estado de

México. PhD. dissertation, FFL, Universidad Nacional Autónoma de México, Mexico City.

Kirchhoff, Paul. 1943. Mesoamérica. Acta Americana 1, 92-207.

Malmström, Vincent. 1997. Cycles of the sun, mysteries of the moon: the calendar in Mesoamerican civilization. Austin, University of Texas Press.

Martz de la Vega, Hans. 2010. Los alineamientos y el paisaje en el sitio arqueológico de Tehuacalco, región Centro de Guerrero. BA thesis, La Escuela Nacional de Antropología e Historia, Mexico.

Morante López, Rubén. 1995. Los Observatorios Subterráneos. La Palabra y el Hombre 94, 35-71.

Sahagún, fray Bernardino de. 1979. Historia General de las Cosas de Nueva España. Mexico City, Editorial Porrúa.

Šprajc, Ivan. 2001. Orientaciones Astronómicas en la Arquitectura Prehispánica del Centro de Mexico. Instituto Nacional de Antropología e Historia, Mexico.

Šprajc, Ivan. 2008. Alineamientos astronómicos en la arquitectura in Reconocimiento Arqueológico en el sureste del estado de Campeche, México: 1996-2005. British Archaeological Reports International Series 1742 (Paris Monographs in American Archaeology 19). Oxford, BAR Publishing.

POSSIBLE MESOAMERICAN NAKED-EYE OBSERVATION OF SUNSPOTS – II: EVIDENCE FROM THE CODICES

RICHARD R. ZITO

Abstract: In the first paper of this series a suggestion was made that Mesoamericans may have observed sunspots, as did many other prominent cultures of antiquity. Furthermore, it was postulated that active years of the sunspot cycle were followed by Sun God images on various artifacts possessing facial or body spots. This second paper will test the sunspot postulate and its converse further. Evidence from the Codices spanning a time interval of about 200 years suggests that facial/body spots appear and disappear in the iconography as sunspot activity increases and decreases, respectively, immediately prior to the time of document creation.
Keywords: Maya, Astronomy, Sunspots, Solar Cycle, Climate Change

1. Background

In order to establish an 'if and only if' relationship between the occurrence of the Sun God facial/body spot motif and the presence of physical sunspots, it is necessary to show that active sunspot years eventually result in images of a spotted Sun God *and* that a lack of spotted images implies low sunspot activity. In the first paper of this series four requirements were laid out for testing this hypothesis. First, the artistic motif must be present (Sun God with or without spots). Second, the artifact upon which the image occurs must be dated or dateable. Third, solar activity levels must be verifiable by direct observations from some ancient culture. And, finally, observations must be substantiated by scientific data (i.e., tree ring analysis, Nilometer measurements, auroral activity, carbon 14 measurements, etc.). It is difficult to find artifacts that satisfy all these criteria, so most can be quickly eliminated. The Mesoamerican codices (both precolonial and early colonial) provide a written record that tracks sunspot activity via Sun God iconography.

2. The Codices

2.1 Tudela Codex

The Tudela Codex, named after José Tudela de la Orden, is an early post Cortéz Aztec document that is part of the permanent collection of the Museum of the Americas in Madrid, Spain. After the conquest of Mexico in 1521 AD newly produced artifacts could be reliably dated. The creation date of pages 11-125 of the Tudela Codex was about 1540 AD. The remaining ten pages (only four of which survive) were joined to the codex after 1554 AD. These four pages, called the 'Libro Pintado Europeo' (the 'Book of European Pictures') are identified by their renaissance artistic style. On the right side of page 97 is a red-orange Sun God (*Tonatiuh*), and he is without facial spots. Facing him on the left is one of the black celestial demons of darkness. The demon is in possession of a collection of stars. The Sun is also in possession of such a collection, but now each star is enclosed by a red-orange ring, indicating an understanding that the stars are still present during the daylight hours but are hidden by the light of the Sun; a very sophisticated idea! The Sun God has a semi-circle on his back containing rays. The semi-circle is the outer edge of the Aztec calendar with the Sun in the center and rays on the edge (Taube 1993, 43). It should also be noticed that the background of the image is heavily glossed in faded Spanish script. In the lower left hand quadrant the words 'el sol' ('the Sun') can be discerned. This further reinforces the identification of the protagonists.

2.2 Vaticanus A Codex

A similar representation for the Sun can be seen on pages 11 and 19 of the early colonial (1556-1562 AD) Vaticanus A Codex (Codex Rios) of the Tolteca-Chichimeca people of the Tehuacan Valley in modern Puebla and Oaxaca (Pohl, 2009). European chroniclers have labeled this image *Touatioh*, and the descriptive Italian text below these images also describes them as '*il sole*'. Again, the Sun God is without spots.

2.3 Borgia Codex

The Aztec Borgia Codex contains numerous planting almanacs. These almanacs show scenes of maize plants, rain, and pests attacking maze plants. Hernández and Bricker (2004, 279) suggest that accompanying dates mark the beginning of the planting season in central Mexico. The 260 day calendar dates would correspond to the traditional onset of the planting season for the years 1467 to 1519 AD (Hernández and Bricker 2004, 283). The almanacs seem to be historical, rather than prognostic, in nature since there is reference to drought conditions that actually occurred and are described in early colonial Mexican records. Since the last almanac in the codex is associated with the Gregorian date of 1519, and since the conquest of Mexico began in 1519, this preconquest codex must have been created in 1519 AD. What is important about the Borgia Codex is that the Sun God *Tonatiuh* on plate 71 (Díaz and Rodgers 1993, XXX, 7) is without facial or body spots. The God's identification is certain since his red image lies over a solar disk with rays.

2.4 Madrid Codex

The Mayan late post classic (1250-1521 AD) Madrid Codex also contains numerous planting almanacs that have been dated to the period from the mid 1440's to the early 1480's (Hernández and Bricker 2004, 279), as well as two remarkable astronomical almanacs on pages 32 and 33 of the Codex. Again, these astronomical almanacs seem historical rather than predictive and record characteristic astronomical events within a 260 day sacred calendar cycle (Vail and Aveni 2004, 17). There must have been many such astronomical compilations prior to the massive destruction of Mesoamerican books (Zito 2008, 277).

The 260 day almanac on page 32 shows God Z below a sky band. Hanging from the sky band is a lunar eclipse sign on the left and a solar eclipse sign on the right. Between these eclipse signs hangs a snake, which in Mayan astronomy is associated with the modern constellation of Sagittarius on the ecliptic. In fact, a partial solar eclipse was visible from northern Yucatan on 6 April 1503 AD, about half an hour before sunset (Hernández and Bricker 2004, 300). Curiously, Sagittarius was near the meridian just before sunrise on the day of the eclipse. Hence, a serpent hangs from the middle of the sky band on page 32 (Hernández and Bricker 2004, 303).

The 260 day almanac on page 33 shows the rain god Chac under a sky band. Also, the rain is pouring under the sky band, signifying rainy season (April or May to late November). On the left end of the sky band hangs a solar eclipse sign. Under the opposite (right) end of the sky band a frog can be seen (corresponding to the modern constellation of Cancer). This almanac has been dated to 1483 AD. An annular solar eclipse (i.e., the Moon's shadow did not reach the Earth's surface) occurred on 16 April 1483, and Cancer was near the meridian at sunset (Hernández and Bricker 2004, 299). However, the author thinks it more likely that the almanac on page 33 refers to the great solar eclipse of 8 August 1498 AD, which was well into the rainy season. This eclipse was total in Central Mexico just before sunset (Aveni 2001, 26). On the day of the eclipse, Cancer was rising in the eastern sky before dawn. Hence, the glyphs on opposite sides of the sky band. Furthermore, the close spatial proximity of the two almanacs is now matched by a close temporal proximity. The two almanacs are also stylistically similar since both God Z and Chac are solid black figures. This may be a reference to the occultation, or 'death' of the solar disk. And, the shields they each carry on their arm, fringed by rays, are probably to cover the Sun so that only the corona shows. For now, the almanacs on pages 32 and 33 seem to be the latest in the codex.

Cautiously taking the date of the almanac on page 32 (i.e., 1503 AD) as the approximate creation date of the codex seems reasonable, since there is little point in having a significant time gap between the last almanac and the creation date of the codex. In any case, the late post classic period came to an abrupt end in the 1519-1521 time period.

In agreement with the other codices of similar vintage, the Sun God on page 66 of the Madrid codex is without spots (Vail and Bricker 2004, 181). His identification is certain. He is seen being 'eaten' by a serpent; a reference to the Mesoamerican belief that a solar eclipse is due to the Sun being eaten by a prodigious serpent. In the text above, the solar eclipse glyph can be seen. The Sun God's eye is made of a *Kin* or sky band variant, and the god holds a *Kin* (Sun) glyph in his hand. As a matter of reference, *kin* (*k'ēn*) means Sun, day, or time in Yucatec Mayan. The same glyph can also be read as *kan* (*k'ăn*) meaning yellow. On top of the hand-held kin glyph is a *yax* glyph. The two symbols together refer to the month *yaxkin*. A late post classic eclipse occurred during *yaxkin* on 31 December 1451 (Vail and Bricker 2004, 185). So, the glyphs in the Sun God's hand have a double meaning. Our scribe has a bit of a sense of humor, as evidenced by the slightly upturned corners of the Sun God's mouth! Not the facial expression one would expect when being devoured by a serpent-monster, and not one of John Pohl's usual 'snarling' Mayan gods (Pohl 2004, 375). Another interesting bearded, red, Sun God image appears on page 68. His face and body are free of spots but, in the author's opinion, his identification is less certain than the god on page 66.

2.5 Dresden Codex

The creation date of the Mayan Dresden Codex has recently been determined relative to the creation date of the Madrid Codex (Aveni 2004, 167). As a compilation of astronomical observations, the Madrid codex covers about 500 years of data. Fortunately, one of the astronomical almanacs pertaining to the solar eclipse of 30 April 906 AD is shared by both the Madrid and Dresden Codices (D.38b-41b and M.10a-13a). A calendar adjustment was made in the Madrid Codex to accommodate the change in seasonal events such as the onset of the nourishing rains. The size of the adjustment indicates that the creation dates of the two documents are separated by 131 years. The situation is complex and involves changes in almanac structure, epoch, and the difference between the vague year (365 days) and the tropical year (365.24 days). However, it is easy to understand how calendrical changes might be needed over 131 years when one considers that the leap year correction alone would amount to 33 days. That is enough of an error to break the connection between an almanac and seasonal events. If the creation date of the Madrid Codex is about 1503, then the creation date for the Dresden Codex would be about 1372 AD. Knorozov's page 22 of the Dresden Codex shows a Sun God with three facial (cheek) spots; two large and one small (Villacorta and Calderón 1930, 44). The identification of the Sun God is certain since he has a *Kin* (Sun) sign 'tattooed' on his back!

3. Ancient Asian Observations

Chinese, Japanese and Korean observations of unusually large solar outbreaks, visible to the naked eye, cover a period from 28 BC (Julian calendar) to 1604 AD (Gregorian calendar). During the 77 year period between the years 1278 AD and 1356 AD, there was not a single naked eye sunspot observation. Then, starting in 1356 solar activity began to rise rapidly. From 1356 to 1387 an astonishing 33 observations were made. Some of these were duplicate observations from the three Asian cultures. Others were the same sunspot group observed a month later after solar rotation. In any case, the estimated creation date of the Dresden Codex (1372) falls within this active period, as might be expected. In fact, there were five observations in the year 1372 alone! After 1387 solar activity decreased, and there were no other sightings for the remainder of the 14th century. Only three large sunspot displays have been recorded for both the 15th and 16th centuries in the years 1402, 1520 and 1603 (Clark and Stephenson 1978, 387). None of the proposed creation dates for the Madrid, Borgia, Tudela or Vaticanus A Codices lies shortly after one of these events, so it is not surprising to find that sunspots are not reflected in the iconography of these documents. Of course, without any other sources of information, it is impossible to tell if the paucity of records during certain centuries is due to a real physical decline of sunspots or a lack of interest in making such observations.

4. Carbon-14 Data

The production of Carbon-14 in the upper atmosphere by cosmic rays is modulated by sunspot activity. This isotope eventually ends up in the biosphere were it can be extracted from tree rings, thereby establishing a relationship between the Gregorian date and ^{14}C levels. The ^{14}C levels, in turn, imply the level of solar sunspot activity. In general, a high sunspot count implies a low level of ^{14}C production, due to shielding of the Earth from productive energetic galactic cosmic rays by the enriched solar wind, and vice versa. The method is not exact enough to reveal the detailed structure of the sunspot cycle; however, general trends can be established. Figure 1 shows the radiocarbon anomaly (i.e., the difference between the observed radiocarbon level and the mean cosmic ray ^{14}C production level without solar modulation) as a function of the Gregorian date from 950 AD to 1950 AD, a span of 1000 years (Damon and Sonett, 1991). The year 1372 AD lies near the radiocarbon minimum between the Wolf and Spörer Epochs; a time of high solar activity. The 15th and 16th centuries lie within the Spörer Epoch, an era of high ^{14}C production and, therefore, low sunspot count. These measurements tend to reinforce Korean observations, and indicate that the lack of data in these latter two centuries is due to a physical lack of sunspots.

5. Tree Ring and Rain Fall Data

There are a few locations on Earth where the weather pattern is strongly coupled to the sunspot cycle. One such location is the southwestern United States where several rules apply concerning the sunspot cycle and El Niño rains. These can be summarized as follows:

1) every sunspot cycle minimum is also a year of El Niño rains;
2) there may be extra El Niño years that are not sunspot Minima; and
3) the absence of El Niño rains implies increased Sun spot activity.

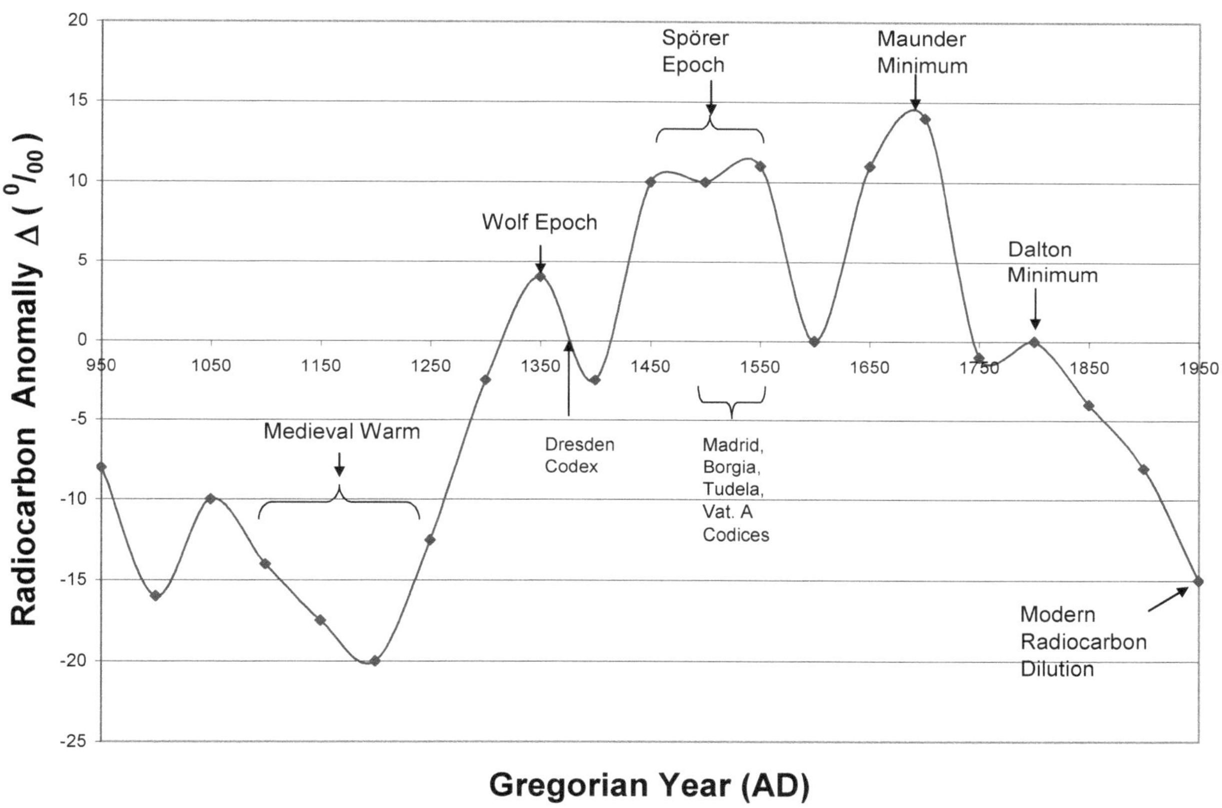

Figure 1. The radiocarbon (carbon-14) anomaly as a function of time. The anomaly (Δ) is measured in parts per thousand (⁰/₀₀) from the mean level. In modern times, the combustion of fossil fuels has diluted the natural abundance of C-14, resulting in an artificially negative value of Δ. The Dalton and Maunder sunspot minima have been documented by telescopic observations, and correlate with maxima in the cosmic ray production of Carbon-14. The Spörer Epoch also corresponds to a sunspot minimum, with the single outstanding exception of 1520 occurring at the depression in the center of the epoch. However, during the Wolf Epoch several large sunspots were recorded for the 31 year period from 4 April 1356 AD to 15 April 1387 AD, but none for 78 years before, or 15 years after, this anomalous period. The Medieval Warm corresponds to a period low Carbon-14 production, high sunspot count, and warm weather. Even Greenland was habitable!

El Niño rains produce rapid tree growth and wide tree rings in the pine trees of the American southwest. Drought years result in narrow rings. Three sites were used to sample climatic conditions in the southwest. These were Methuselah Walk (MW) in the White Mountains of California (north of Death Valley), Hill 10842 in Nevada near the Utah border, and Charleston Peak (C. Pk) Nevada, near Las Vegas. Pinus longeva (PILO), commonly known as the Bristlecone Pine, was used as the model plant for these sunspot investigations for several reasons. First, tree ring growth for PILO is primarily dependent on conditions just prior to the growth season. By contrast, the ring growth of other species tends to average moisture conditions over several years, making data interpretation very difficult. Second, PILO records are the longest continuous tree ring records in North America, and go back 8000 years. Third, all the data for the southwestern United States was collected by only three investigators. Therefore, sampling errors due to individual technique tend to be reduced. All data at MW was collected by D. A. Graybill (NOAA, 2009) and involved 49 to 53 samples per year for the 1503-1562 period and 41-42 samples per year for the 1356-1387 period. All data at Hill 10842 was collected by V. C. Lamarche and C. W. Ferguson (NOAA, 2009) and involved 12 samples per year for the 1503-1562 period and 11 samples per year from 1356-1387. Finally, all data from C. Pk was, again, collected by D. A. Graybill and involved 38-41 samples per year from the 1503-1562 period and 29-30 samples per year from 1356-1387.

Examination of rings from PILO show that the period from 1356 to 1387 was a time of mild drought in the American southwest, the hardest hit area being the White Mountains of California where annual growth was reduced by 9% over this period. The data imply an active Sun, in agreement with Asian observations. The situation for the years 1503 to 1562 is also straightforward. The only naked eye Asian observation occurred in the year 1520. This record correlates well with the severe droughts in the years 1518, 1519, 1522 and 1523, when the average annual ring width, over the three sites surveyed, was reduced by 20-30%. The remainder of the tree ring record from 1503-1562 is generally consistent with a quiet Sun. However, tree ring data from 1540-1550 may support a weak solar maximum with a few telescopic sunspots.

6. The Nilometer

Another location where the sunspot cycle is closely linked to climate is the Nile Valley. Secular (slow) variations of Nile river levels directly correlate with solar variations in the medieval to modern epoch. The annual high and low water levels of the Nile have been measured since pharaonic times at several locations. The 'new' Nilometer at Roda Island in Cairo, better known as 'Al Miqyas' ('The Measurer') or 'Miqyas al-Nil,' was built in 715 AD and is essentially an enormous 'dip stick' for the Nile River. A graduated column extends into a cistern that is connected to the Nile by three tunnels. Measurements at this location extend throughout the Islamic period from 622 AD (AH 1) to the present century (1922). From 622 AD to about 1470 AD the record is fairly complete. The year 1372 closely corresponds to a high water maximum for the entire data series, implying an active Sun. After 1471 Nilometer data is discontinuous, and there is an unfortunate paucity of data for the 16th century. However, all is not lost. Kondrashov et al. (2005, L10702) have made a Fourier reconstruction of the missing data. After correcting for changes in the units of length, siltation and calendrics, a time sequence was constructed that represents the difference between high and low water marks normalized by the standard deviation of the data. The time period from 1450 AD to 1550 AD shows a particularly small amplitude, with a slight increase about 1525 AD. This is consistent with a period of generally low solar activity, containing a temporary increase about the time the Korean observation of 1520 was made.

7. Conclusion

The creation dates of the Dresden, Madrid, Borgia, Tudela and Vaticanus A Codices cover 1372, 1503, 1519, 1540 and 1556-1562 respectively, a period of almost 200 years. Only the estimated creation date of the Dresden Codex is embedded in a period of high sunspot activity that could be observed with the naked eye. The Borgia creation date of 1519 is close to, *but before*, the sunspot event of 1520. And, 1519 is probably a hard upper limit on the creation date of this pre-colonial document due to the conquest. It is remarkable that the Dresden Codex is the only document of the five that shows a spotted Sun God. This correlation implies that Mesoamericans were observing sunspots, and that their iconographic art reflected the general trends in sunspot activity. For the 453 years from 1278 to 1731, there are only 33 years for which such a correlation could be achieved; keeping a fixed time interval of 131 years between the creation dates of the Dresden and Madrid codices. The probability of that happening by random chance is only 33/453, or 7.3%, if only Mayan documents are considered (i.e., the Madrid and Dresden codices), and less than 5.7% if all Mesoamerican documents are considered. It would be a most useful check on the sunspot hypothesis if the Madrid Codex could be carbon dated; taking care to avoid the 'European patches' on pages 1 and 56 (Bricker 2004, 34; Chuchiak 2004, 58). Only a few milligrams of fig bark paper from an unmarked portion of the codex are required for accelerator dating. It should be possible to distinguish 1400 AD from 1500 AD material; sufficient to prove or disprove the hypothesis. By implication, the creation date of the Dresden Codex would be established as well. Furthermore, current notions on historical almanacs versus predictive almanacs would be verified or rejected. Finally, a carbon date would verify if current research is correctly matching 260 day calendar cycles to Gregorian dates.

References

Aveni, A. F. 2001. Skywatchers. Austin, Univ. of Texas Press.

Aveni, A. F. 2004. Intervallic structure and cognate almanacs in the Madrid and Dresden codices. In G. Vail and A. Aveni (eds), The Madrid Codex, 147-170. Boulder, University Press of Colorado.

Bricker, H. M. 2004. The paper patch on page 56 of the Madrid Codex. In G. Vail and A. Aveni (eds), The Madrid Codex, 33-56. Boulder, University Press of Colorado.

Chuchiak, J. F. 2004. Papal Bulls, Extirpators, and the Madrid Codex: The Content and Probable Provenience of the M. 56 Patch. In G. Vail and A. Aveni (eds), The Madrid Codex, 57-88. Boulder, University Press of Colorado.

Clark, D. H., and Stephenson, F. R. 1978. An interpretation of pre-telescopic sunspot records from the orient. Quarterly Journal of the Royal Astronomical Society 19, 387-410.

Damon, P., and Sonett, C. P. 1991. Solar and terrestrial components of the atmospheric C-14 variation spectrum. In C.P. Sonett, M.S. Giampapa, and M.S. Matthews (eds), The Sun in Time, 360-388. Tucson, University of Arizona.

Díaz, G., and Rodgers, A. 1993. The Codex Borgia, XXX, 7. New York, Dover.

Hernández, C., and Bricker, V.R. 2004. The inauguration of planting in the Borgia and Madrid codices. In G. Vail and A. Aveni (eds), The Madrid Codex, 277-320. Boulder, University Press of Colorado.

Kondrashov, D., Feliks, Y., and Ghil, M. 2005. Oscillatory modes of extended Nile River records (AD 622-1922). Geophysical Research Letters 32, no. 10, L10702.

NOAA 2009. Available at: www.ncdc.noaa.gov/paleo/treering.html (Accessed 13 May 2009).

Pohl, J. M. D. 2004. Screenfold manuscripts of Highland Mexico and their possible influence on Codex Madrid: a summary. in G. Vail and A. Aveni (eds), The Madrid Codex, 367-413. Boulder, University Press of Colorado.

Pohl, J. 2009. A Colonial Era Decipherment of Codex Rios. Available at: http://www.famsi.org/research/pohl/jpcodices/rios/ (Accessed 16 February 2009).

Taube, K. 1993. Aztec and Maya Myths. Austin TX, British Museum Press and University of Texas Press.

Vail, G., and Aveni, A. 2004. Research methodologies and new approaches to interpreting the Madrid Codex. In G. Vail and A. Aveni (eds), The Madrid Codex, 1-30. Boulder, University Press of Colorado.

Vail, G., and Bricker, V. R. 2004. *Haab* dates in the Madrid Codex. In G. Vail and A. Aveni (eds), The Madrid Codex, 171-214. Boulder, University Press of Colorado.

Villacorta, C. A., and Calderón, J. A. V. 1930. The Dresden Codex: drawings of the pages and commentary in Spanish. Walnut Creek CA, Aegean Park Press.

Zito, R. R. 2008. A comparison of two independent models of the solar system: Bode's Law and the 260 day Mesoamerican calendar. In J. A. Rubiño-Martin, J. A. Belmonte, F. Prada, and A. Alberdi (eds), Cosmology Across Cultures ASPCS Vol. 409, 277-282. San Francisco, ASP.

POWER, DANGER AND LIMINALITY: MOON, STARS AND WOMEN AMONG THE TOBA OF WESTERN FORMOSA (GRAN CHACO, ARGENTINA)

CECILIA PAULA GÓMEZ

Abstract: Among the Western Tobas of the province of Formosa, women will hardly talk to strangers about certain topics, including those related to the celestial sphere. Their knowledge about this topic is usually important although the Tobas tend to transmit it through the men. However, a process of change has been noticed after many years working in the field, when women were asked about the Moon in one-to-one conversation. When talking about the Moon with the Tobas, everything seemed to point to women and they felt free to express fully their knowledge about it. The Moon is thought to be a masculine entity in charge of 'making women'. This idea is directly associated with women's first menstrual period. Thus, on the representational level the Moon is clearly responsible for women's first menstrual period. Several representations and taboos are linked with this event. From this point of view certain characteristics associated with women are analysed, especially those which express women's power, feminine danger and liminality since females are regarded as entities that mythically came from the sky.
Key words: Toba, Women, Moon, Menstruation, Sky, Chaco

Among the indigenous societies that currently live in the Gran Chaco the Tobas have been labelled as 'central chaqueños' or 'typical chaqueños' (Braunstein 2008, 20-21). Among the 'typical' Chaco ethnic groups we find several peoples belonging to different linguistic stocks: Mataco-Maká (Wichí, Nivaclé, Chorote, Maká), Guaycurú (Toba, Pilagá, Mocoví, Mbayá-Caduveo) and Maskoi (Lengua, Sanapaná, Angaité, Toba-lengua, Guaná-lengua). By the same criteria the 'atypical' groups would be the 'northern' ones, which belong to the Zamuco linguistic stock (Ayoreo, Chamacoco), and the 'western', who speak related but diverse languages of Tupí-Guaraní origin (Chiriguano, Chané, Tapiete, Simba, Guaraní) (Braunstein, 2008).

The groups that constitute the Toba ethnic chain occupy the watersheds of the Pilcomayo and the Bermejo rivers, and together they compose one of the three biggest ethnic groups which can be distinguished among the indigenous peoples of the Gran Chaco. Their subsistence is traditionally based on hunting and gathering techniques, fishing and basic agriculture—in fact, in archaeological terms, they could be regarded as 'complex hunter-gatherers'. The Tobas that settled in the west of the province of Formosa (Argentina) specifically reside in the Bermejo Department, approximately eighty kilometres from the city of Ingeniero Juárez. Their settlements are varied, with Vaca Perdida, La Rinconada and El Churchal being the most populated. There are other smaller settlements such as La Mocha, Seis de Abril, Tres Yuchanes, Pozo Ramón, Isla García, La Madrugada, Pozo Charata, El Breal, La Bolsa, Laguna Cansino, Quebrachito and Barrio Méndez, among others. There are also an important number of seasonal settlements (Córdoba 2008a, 133; Arenas 2003, 41). According to a survey carried out in 1999 by the Civil Association *Comlaje'pi naleua*, almost 2000 people lived in these settlements. During 1989 these communities had access to the property of a territory of 35,000 acres, which at least includes part of their traditional hunter and gatherer circuits (Gordillo 2005, 204; Córdoba 2008a, 135).

The women and the transmission of knowledge about the sky

Before analysing the ethnoastronomical data it is necessary to take into account the difficulties in talking about the sky with those who are not 'the elders'. Together with other factors, this fact tells us how a Toba woman should behave according to social requirements. There is a constant when you talk to the young Tobas about phenomena such as the sky (*pi'yem*), the sun (*a'hewa*), the stars (*waqachi'ñi*), the Moon (*a'woGoyk*), or the night (*'piyak*). They always start by saying they don't know anything about it, that it would be better to ask 'the elders': 'I am new, therefore I don't know'.

'The new' is a category that includes young men and also those who are already 40 years old but are still considered 'new', and who are thus not allowed to talk in public while someone older can. These are 'the elders' and they are permitted to talk about any of the topics related to the social knowledge transmitted by previous generations. It could be considered that what is related to the collective memory of society, what Tobas regard as 'tradition' and is considered important to be remembered, must only be discussed with 'the elders'. Even though men who are considered 'new' show an initial ignorance during an interview; after a brief pause they may start to 'remember' and name some stars and asterisms and even myths about them. It should be clearly stated that, on the one hand, memory is a collective practice, and on the other hand authority of 'the elders' is still cardinal: they are the official authorized spokesmen of the whole society, whose knowledge and wisdom are transmitted orally from generation to generation.

It was definitely harder to work on the topic of the sky with the women than with the men from places such as Vaca Perdida and Rinconada. Traditionally women used to talk about stars and asterisms through the men in their family; one can easily deduce this by noticing their gestures and interactions when talking about the sky within their family group. Even if the conversation took place within any other environment, when the man was alone feminine wisdom was constantly cited.

Tobas usually make reference to the way they used to look at the sky together with their wives, mothers or sisters. In these situations the woman is the one who recalls the most about the sky, and about how 'the elders' thought about it. Therefore it can be said that the collective memory is allocated according to age, and quite especially the specific kind of knowledge related to the sky. However, there are only a few 'elders' who are implicitly authorized to acknowledge these facts, which are probably shared by many other people. Women have a wide knowledge of the sky but they hardly pass it on outside of their family group: their husbands will do it, and especially 'the elders' who are well regarded and authorized by society.

However, the distant attitude that women used to keep concerning these topics has slowly changed over time and these kinds of conversations have started to be more common, especially when talking about the Moon. On these occasions certain women, particularly those who could manage to speak Spanish, showed a definite eagerness to talk, as if they had the social permission to do so. The Moon and its implications seemed to belong to the semantic and symbolic realm of femininity. On the other hand, talking about the Moon with men did not cause any inconvenience whatsoever.

So, to start with the analysis it is important to consider that there are only a few women who can openly talk about the topic. Nevertheless, some testimonies were obtained and they clearly show that the sky is a subject close to the women's point of view. Consequently, the significant relationship between the celestial sphere and the social representation of femininity has been documented, taking into account not only 'the elders' opinion but also the general women's viewpoint.

Relationship between women and the sky

Now we will analyse the meanings attached by Toba cosmology to the symbolic association between women and the sky. We will try to show how this association is related to the intrinsic danger of the female figure and power a woman latently owns, which should be moderated by social means in order to avoid personal, familial and collective misfortunes. Therefore, we will describe the symbolism and praxis in the relationship between the feminine universe and the mythical character portrayed by the Moon. On the narrative level the Moon is a masculine entity which conducts the mythological conformation of the Toba women. Bearing this in mind, a series of social mechanisms that contribute to the cultural conformation of women can be described. The analysis of the symbolic framework concerning these significant relationships shows that, from the Toba point of view, women are dangerous entities which display strongly marked characteristics of liminality. This leads to the fact that Toba society builds up a kind of femininity which is obedient to social requirements: insubordination and rule breaking in everyday life can trigger certain individual or collective maladjustments which can unchain cosmic disorders.

According to the famous ethnographer Alfred Métraux, the Tobas believe that the Moon dies and that it is a masculine entity:

> 'When Moon wanes, people say that he is dying or that he is going hunting. When Moon reappears, they salute him and ask, "How are you? Have you had a good journey?" Men and women rejoice because they soon will be able to dance at night. According to his shape Moon changes his position. This is because Moon is a tired man who cannot lie too long on the same side' (Métraux 1946, 19).

This statement does not contradict what is told by contemporary Toba. Even today, the Moon is culturally present at many different levels among women both in symbolic and practical aspects. The lunar imaginary is used to measure diverse physiological and temporal cycles. The phases of the Moon reveal a particular development though there is not an exact term to define each of them; however, from a general point of view the lunar phases refer globally to the vital cycle of any living entity. For example, when the Moon is considered "young" it can be *called a'woGoyk dalaichi* (tender moon) or *ka'leka, kat'ñwk* (very young). It is labelled this way, approximately, when it is in its first phase cycle. When the Moon is 'mature enough', approximately in its second phase cycle, it can be called a*'woGoyk dai'ñi* (hard moon). When it is found in its old age, approximately in its third phase cycle, the Toba call it *a'woGoyk yagai'ki* (old moon), and when the Moon 'dies', approximately in its fourth phase cycle, it is called *a'woGoyk yi'lew* (dead moon).

The Toba explanation about the Moon and his apparent movement in the sky clearly refers to a group of organic ideas which evoke youth, maturity and death of a living agent. It is a social agent that disappears and reappears; therefore 'the elders' explain that the new Moon is the son of the former one, etc. At the same time the term *a'woGoyk* (moon) indicates the amount of time which the Moon takes to travel from west to east until he disappears and finally reappears in the west: that is the reason why, in many lexical glossaries, the translation for *a'woGoyk* is either 'moon' or 'month'.

Rafael Karsten's 1912 ethnography of the Bolivian Toba also suggests that they thought that menstruation was caused or influenced by the Moon (Karsten 1923, 26). Once again, the Toba of western Formosa still think that the relationship between the Moon and female sexuality seems to be quite evident. Métraux also referred to this link in his classic studies about Toba and Pilagá mythology: he indicated that 'when a girl's liver exudes blood (i.e., when she menstruates), some people say that she has had intercourse with the Moon'. The myth about sexual relationships between the Moon and women who have had their first period can still be heard among the Toba. The Moon is still thought of as a masculine entity which is responsible for the mythological conformation of Toba women: in the words of a Toba, 'Moon is the first man of every woman'. When their first period appears mothers still say to their young daughters that 'Moon is to blame'. Therefore, in this context mothers utter things such as 'Your first man is the Moon' or 'It is as if Moon had raped my daughter'. As we have already said, according to Toba oral tradition the Moon is responsible for the mythical formation of femininity. However, it has been claimed that among the eastern Tobas the first menstruation of women is due to Moon's gaze and not to sexual intercourse (Tola 2001, 77).

We can also mention the fact that the lunar phase in which the menstrual period appears for the first time will be the same as the following months pass by: women say 'that it is my Moon'. The position of the Moon in the sky and the shape it forms will then 'register' the menstrual cycle of each woman. Among the Toba, therefore, recalling the moons in which women have their period is considered of great significance. Nowadays some women use the western, Gregorian calendar, but in general they do know in which lunar phase they will have their period. On the other side, if a young woman's expected menstrual period does not start she begins to 'count the moons' as she is likely to be pregnant.

Femininity is shaped daily among the Toba (Córdoba 2008b). The first menstruation which the Moon causes marks a clear conceptual distinction though it is not just an isolated event at a social level. The young woman who starts to menstruate is also socially recognized as a marriageable spouse. This is precisely the moment when young women are finally given what is generally known as 'the studies' by adult women and the elder ones. Even though there are traditional rituals requirements which no longer apply, young women who have their first menstrual period are sometimes set aside from social ordinary life. Nowadays extreme measures, such as covering the girls' heads, may not be applied; but there are still some taboos which are kept, whose disruption can motivate doubts, misgivings and cultural concerns. In other words, during their menstruation women seem to reveal their intrinsic qualities in an exacerbated way (Métraux 1937, 190-192).

It is interesting to mention the taboos that still exist around menstruating women, which can suppress the danger and power that each woman latently owns. When women have their menstrual period, their danger, power and liminality can suddenly emerge: in fact it can easily bring misfortune to herself and even to her family and kin. That is precisely why they try not to break any social requirement. Let's exemplify what was already stated. When a woman is having her period she cannot approach any river or well, since the *wo'sak* (rainbow) would smell her and individual or collective problems could arise. The *wo'sak* (rainbow) is a snake shaped entity which prowls about near the water and is quite able to move both in the sky and on land. Also named *na'naik 'poleo* (lit. 'big snake'), the rainbow serpent pursues women who break social taboos and menaces the whole community; storms and floods can be unchained and even the whole village could disappear. As regards the woman herself, the *wo'sak* may penetrate her to settle in her womb, finally provoking her death.

An old man who has just passed away told his own experience concerning these beliefs. Some girls had gone with him and other boys to the river. One of the girls was around 12 years old and she did not mention she was having her period. They were all bathing when suddenly it became cloudy and it started to rain so hard that the waters rose. Some 'bugs' (amphibians, fish) also appeared. If the old man hadn't realized what was happening the *wo'sak* would have penetrated the girl by her vagina. If this had happened she would have been the one to blame, because she had not given a proper warning. The old man went on saying they had been saved thanks to *pioGo'nak* (shaman), who had the courage to face the *wo'sak*. After telling the story, he mentioned these episodes are currently disappearing.

This man was one of the first health workers in the town. He was educated in a school conducted by Anglican missionaries and he was highly respected both in Vaca Perdida and in the nearest villages. This narration was not told in Spanish, and it was his son-in-law, who lived with him in the same house, who translated it into Spanish. The old man was speaking in one of the patios in the house, which was shared by many parental kin. As the story went along, his grand-daughters, who were around fourteen and sixteen years old, approached him and were kneading while staring at him and listening attentively. He also stared back at them and started to tell the story directly to the girls, who were quite absorbed in it. It all ended up in laughter. This was really interesting: the look, the gestures and the body language signalled the grandfather was talking to me, but his look was clearly diverted when his granddaughters arrived, clearly addressing the *wo'sak* story to them. However, at the end he mentioned that these events are less common nowadays.

Another prohibition which women should respect while menstruating has to do with the food they are not allowed to eat, such as fish, meat or honey. If they do eat any of these misfortunes would unchain and hurt their family; for instance, the husband would not be able to hunt or to fish with success—a woman explained that then the wife is to blame since this is 'the significance of menstruation'. So, when each woman sees 'her moon' she knows she will have her period and that she will have to follow social regulations to hold the power that month after month remind her of her heavenly origin.

The woman and her arrival from the sky

Toba femininity can be analytically linked in a social, practical and narrative fashion with the apparent movements of the Moon and the sky. This idea is strictly related to the mythical origin of women. As in other indigenous cosmologies of the Gran Chaco (cf. Karsten 1923,104-106; Lehmann-Nitsche 1924, 284-285; Métraux 1946, 102-107; Cordeu 1969/70, 143-148; Palavecino 1969/70, 185-186; Miller 1977, 323-325; Wilbert and Simoneau 1982a, 67-77, 1982b, 101-143, 1985, 29-38, 1989, 130-155; Perez Diez 1983, 134-169; Idoyaga Molina 1986, 165-174; Barúa y Dasso 1999, 253-257; Tola 2005: 59-79), the mythological origin of the Toba women is purely stellar. In fact, women who stayed at the sky are seen as the current stars, and even today the stars are regarded as female beings. Women are regarded as social agents which came from the sky *(pi'yem)* to the earth *(a'lewa)* to feed themselves and who stayed here due to the first mythical men/animals. In fact, the trickster (the fox *-wayaka'lachigi-*) made it possible: at the beginning women had 'a second mouth'' a vagina with teeth, and it was the fox who broke these teeth, therefore allowing primordial men/animals to have wives. Regarding this fact, it is important to remember that the first women did not stay here at will: they were trapped while they tried flee back to the sky.

It is noticeable that in a symbolic plane the character played by Moon replicates the mythic actuation of the trickster, giving men the possibility of having wives and in a general sense making human reproduction possible. Taking this as a starting point, the following may be stated: women are entities whose origin is mythically in the sky. They are not allowed to stay on earth at will; they must be humanized, 'domesticated'; and only the actions by the trickster enable the first mythical men/animals to have sexual intercourse with females, thus originating their family and descendants. Nowadays it is the Moon who, as well as the fox, lets men have sexual intercourse with females, making them their wives and originating families.

Conclusion

According to the extended Chaco myth, the fact that the fox breaks 'the second mouth' of the pre-social women symbolizes the 'domestication' (Palmer, 2005) of the great powers due the female stellar origin. Power, danger and liminality are marked characteristics of the cultural representation of Toba femininity and they seem to be exacerbated during the menstrual period. That is why women must fulfil social requirements especially during these periods, so that their latent powers are not unleashed, bringing misfortunes to their kin. Lets recall here that the sky is regarded as an extremely cold place which is populated by perilous beings, except for those who have some special powers, such as *pioGo'nak* (shamans). To sum up, the sky and everything related to it generates experiences and situations which are extremely dangerous to human beings.

By regulating their bleeding, Toba women can channel their organic danger. They can also impose limits to the power which is intrinsic to them, and which could lead to personal, family and community misfortunes if left boundless. Most of the interviewed women affirm that they comply with traditional rules which are still respected and not easy to break; this has to do with a social symbolic operation which generates regulations, norms and values that provides a practical orientation for everyday praxis. By taking care of herself the woman takes care of everybody else, and this seems to be a cardinal point regarding the cultural association between the Moon and the female vital cycles.

The symbolic meanings associated with the stars (*waqachi'ñi*) and the Moon (*a'woGoyk*) thus express in an encoded way the social duties and requirements related to femininity. This symbolic framework is more than merely representational, and it is closely linked to the collective behaviour regulation and to the way in which women proceed daily. As terrestrial entities, Toba women need to accomplish the social stereotype of femininity by running a proper family and fulfilling a set of culturally assigned moral values. The women's femininity is contained in the private realm rather than the public sphere. That is why women will not talk to strangers about the sky. As it was said earlier, in general female roles are structurally central especially among domestic units where they hold authority. Therefore, a share of the specific knowledge women have got about the sky and specifically about the Moon (*a'woGoyk*) is a crucial part of 'the studies' that when they age they will pass to the new generations.

Acknowledgments:

I am grateful to Dr Diego Villar for his critical insight about several drafts of this paper.

References

Arenas, P. 2003. Etnografía y Alimentación entre los Toba-Ñachilamoleek y wichí-Lhuku'tas del Chaco Central (Argentina). Buenos Aires, Dunken.

Barúa, G. and M. C. Dasso. 1999. El papel femenino en la hostilidad wichí. In M. Califano and M. C. Dasso (eds.), El chamán wichí, 251-297. Buenos Aires, Ciudad Argentina.

Braunstein, J. 2008. Muchos caciques y pocos indios. Conceptos y categorías del liderazgo indígena chaqueño. In J. Braunstein and N. Meichtry (eds.), Liderazgo, representatividad y control social en el Gran Chaco, 3-31. Corrientes, Editorial Universitaria de la Universidad Nacional del Nordeste.

Cordeu, E. 1969/70. Aproximación al horizonte mítico de los Toba.. Runa 12 nos. 1-2, 67-173.

Córdoba, L. 2008a. Liderazgo, grupos locales y organización sociopolítica entre los Toba del Oeste formoseño. In J. Braunstein y N. Meichtry (eds.), Liderazgo, representatividad y control social en el Gran Chaco, 135-140. Corrientes, Editorial Universitaria de la Universidad Nacional del Nordeste.

Córdoba, L. 2008b. ¿Existe la iniciación?: procesos de construcción social de la femineidad entre los Toba del oeste formoseño. Acta Americana 16, no. 2, 61-83.

Gómez, C. 2008. Venus and the Star Woman. Archaeologia Baltica 10, 25-28.

Gómez, M. 2006. Representaciones y prácticas en torno a la menstruación y menarca entre mujeres Tobas: entre la salud de las mujeres y la construcción social del género femenino. Papeles de Trabajo 14, 9-51.

Gordillo, G. (ed) 2005. Nosotros vamos a estar acá para siempre: historias Tobas. Buenos Aires, Editorial Biblos.

Idoyaga Molina, A. 1976/77. Aproximación hermenéutica a las nociones de concepción, gravidez y alumbramiento entre los pilagá del Chaco Central. Scripta Ethnologica 4, 78-98.

Idoyaga Molina, A. 1986. Estudio fenomenológico del mito de origen de las mujeres entre les Pilagá (Chaco Central). Anales de arqueología y etnología 41-42, 165-174.

Karsten, R. 1923. The Toba Indians of the Bolivian Gran Chaco. Helsingfors, Åbo Akademi.

Lehmann-Nitsche, R. 1924. Mitología Sudamericana X. La astronomía de los Tobas. Revista del Museo de La Plata 27, 267-285.

Lévi-Strauss, C. 1986. El sexo de los astros. In Antropología estructural II, 203-211. Madrid, Siglo XXI Editores.

Lévi-Strauss, C. 1997. Mitológicas III. El origen de las maneras en la mesa. México, Siglo XXI Editores.

Métraux A. 1937. Etudes d'ethnographie toba-pilagá (Gran Chaco). Anthropos 32, 171-194, 378-401.

Métraux A. 1946. Myths of the Toba and Pilagá Indians of the Gran Chaco, Vol. 40. Philadelphia, American Folklore Society.

Miller, E. 1977. Simbolismo, conceptos de poder y cambio cultural de los tobas del Chaco argentino. In E. Hermitte (ed.), Procesos de articulación social, 305-338. Buenos Aires, Amorrortu.

Nordenskiöld, E. □1912□ 2002. La vida de los indios. El Gran Chaco (Sudamérica). La Paz, Editado por Apoyo Para el Campesino-Indígena Boliviano.

Palavecino, E. 1969/70. Mitos de los indios Tobas. Runa 12, nos. 1-2, 177-197.

Palmer, J. 2005. La buena voluntad wichí. Una espiritualidad indígena. Formosa y Salta, Asociación para la Promoción de la Cultura y el Desarrollo y otros.

Perez Diez, A. 1983. A tale of the mataco about the origin of women. Latin American Indian Literatures. 7, no. 2, 134-169.

Tebboth, T. 1943. Diccionario toba. Revista del Instituto de Antropología de la Universidad Nacional del Tucumán 3, no. 2.

Tola, F. 2001. La luna en el imaginario masculino y femenino de los Tobas (qom) orientales del Chaco argentino. Ava 3, 75-88.

Tola, F. 2005. Sociabilidad y mito: hombres, mujeres y animales desde la perspectiva toba. Latin American Indian Literatures Journal 21, no. 1, 59-79.

Wilbert, J. and K. Simoneau (eds.) 1982a. Folk Literature of the Mataco Indians. Los Angeles, University of California, Latin American Center Publications (UCLA Latin American Studies, 52).

Wilbert, J. and K. Simoneau (eds.) 1982b. Folk Literature of the Toba Indians. Volume One. Los Angeles, University of California, Latin American Center Publications (UCLA Latin American Studies, 54).

Wilbert, J. and K. Simoneau (eds.) 1985. Folk Literature of the Chorote Indians. Los Angeles, University of California, Latin American Center Publications (UCLA Latin American Studies, 60).

Wilbert, J. and K. Simoneau (eds.) 1989. Folk Literature of the Toba Indians. Volume 2. Los Angeles, University of California. American Center Publications (UCLA Latin American Studies, 68).

A TOPOLOGY OF POWER: SKY AND SOCIAL SPACE IN THE ARGENTINEAN CHACO

ALEJANDRO MARTÍN LÓPEZ

Abstract: This paper will discuss the ways in which social relations build the sky among several aboriginal groups of the Argentinean Chaco, creating a space that can be described in topological terms. Shaped by power relations between humans and non-humans, the sky of the Mocoví, Toba and Wichí is a real map of the symbolic struggles for identity and leadership among these groups. To perceive the 'texture' of the celestial space means to be able to account for the types of beings that inhabit it. We discuss the use of Corbin's notion of 'Mundus Imaginalis' and Jung's concept of 'active imagination' to deal with the flexible cosmologies of these aboriginal people. To do this we reformulate these ideas by postulating that the active imagination and the archetypal images are not ahistorical categories but are socially constructed. And the process of this construction mainly involves the bodies, the 'habitus' and the 'logics of practice'. in the sense of Bourdieu. These reflections are based in a long field experience. Since 1998, we have been working on a research project that involves anthropological fieldwork and ethnohistorical research in order to understand the astronomy of these groups, focusing on Mocoví ethnoastronomy.

Keywords: Topology; Chaco; ethnoastronomy; habitus, Power relations, Mundus Imaginalis

Mocoví, Toba and Wichí

Originally hunters and gatherers, the Mocoví, Toba and Wichí are three of the most relevant aboriginal groups in the Argentinean Chaco. The Mocoví and the Toba belong to the Guaycurú linguistic group and the Wichí to the Mataco-Mataguayo linguistic group. These aboriginal people underwent several changes during colonial times (e.g,. some of them adopted the horse); however, they retained a significant important degree of autonomy until the 20th century, when Argentine national society carried out a systematic campaign of expansion within their traditional territory. Nowadays, a population of 15,000 Mocoví, 70,000 Toba and 40,000 Wichí survive by working as rural labourers and by developing a variable amount of hunting and gathering.

Roberto Lehmann-Nitsche's works represent the first research into these groups' astronomy (Lehmann-Nitsche, 1923; Lehmann-Nitsche, 1924-25; Lehmann-Nitsche, 1924). Since the 20th century some important scholars have been working on the cosmology and astronomy of the Toba people (Wright, 2008; Cordeu, 1969-70; Braunstein, 1995; Ruiz Moras, 2000; Miller, 1975; Métraux, 1935), the Pilagá and Toba-pilagá people (Idoyaga Molina, 1989; Gómez, 2006), the Wichí people (Dasso, 1989; Barúa, 2001; Braunstein, 1989), and the Mocoví people (Giménez Benítez et al., 2004; Giménez Benítez et al., 2006; López and Giménez Benítez, 2008; López, 2009).

Ethno-territory

The celestial space should be addressed in the context of the study of the ethno-territory among these groups. The concept of ethno-territory refers to a 'socially constructed spatiality, essential to the collective identity' (Toledo Llancaqueo 2005, 17, author's translation). These ethno-territories are the basis on which indigenous movements and the power relations that flow through them are articulated. Among many South American indigenous groups, the ethno-territories acquire a distinctly relational feature based on a conception of space linked to the human body, a body with appetites or desires that establish relations with other bodies (Echeverri 2005, 238). The ethno-territory is, therefore, an arena of disputes and power relations between groups in unequal conditions. There you can see struggles for material and symbolic resources, including the links with the sacred. The celestial space, conceived as the region that is the source of wealth and power, is transformed by the groups studied in the centre of their ontological idea of territory.

Social topology

Therefore, we intend to draw a topology of the cosmos of these groups. As the term suggests, it means to delineate a qualitative description of the 'geometry' of a space that is characterized by its ability to be 'deformed' or 'distorted' continuously. The first description of one of these cosmologies with a topological view is that of Pablo Wright on the Toba's cosmos (Wright, 2008). In the present work our aim is not only to extend this topological description to other related groups, but also to think about these topologies in terms of 'social' topologies and use them as analytical tools.

The cosmos of these groups is a space tensioned by the actors who occupy it and their linkages. In the cosmos constructed by the ethno-territoriality of the Mocoví, Tobas and Wichí, space is a real social field in Bourdieu's terms (Bourdieu, 2005; Bourdieu, 1997a): a field whose 'geometry' is largely shaped by the capital and trajectories of those who inhabit it, human and non-human. In the social play of spatial relationships in these groups, one of the dominant forms of capital is power. So our topology of the cosmos among the Mocoví, Tobas and Wichí will not only be a mapping of their ethno-territories as conceived and lived by them but also a cartography of power and power relations from their perspective. To feel the texture of these territories is to perceive the roughness, the tensions generated by these links. It is thus a quite mobile and changing 'topography', but not a capricious one.

Cosmology and power conceptions

In order to pursue this endeavour, we will discuss the idea of power. Among these groups, power and how to manage it—how to get it, increase it and keep it, how to relate to those who have more than one—are fundamental questions of cosmology.

Agreeing with Cordeu (1998), we can refer to two main styles of worldview in terms of conceptions of power. On the one hand, we have an 'active' world-view style, organized around the idea of gaps and differences, and for which the bonds between people are generally hierarchical and asymmetrical. On the other hand, there is a 'speculative' style focusing on ideas about harmony and agreement aimed at the integration of beings and energies. The first of these styles is associated with a dynamic conception of power, in which this is seen as a 'non-ordinary energy' characterized by its activity, its efficacy and its link to fertility. This kind of conception is the dominant one in the groups that we study. This is power as a capacity to act, as potentiality, as a force that has someone or something, that makes it capable of being fertile, rich and abundant. Any activity that requires some talent needs that its agent has power, and not just any power but the specific power for that task. And that power is derived from a specific power-being, or beings, associated with the concerned task, skill, or space region.

In this sense, the fundamental notion that structures life experience in these groups is the notion of 'meeting' with an 'other' that has plenty of power. These meetings, organized according to the scale of power of the actors involved, are the units of experience in this space-time continuum. The cosmography of these groups contains the memory of relevant events, focusing on 'resources', 'powers' and 'dangers'. And memory is always exegesis. In these kind of meetings, the presence of the powerful being transforms the qualitative nature of the local time-space. During the meeting this time-space has properties linked with the scale of power of this powerful being, typically features of the original times.

Body and habitus

Due to the fact that we are not only referring to native theories about the nature of celestial space, but also the implicit ideas about it, person, body, language and habitus play a main role. In modern anthropology, the body is seen as a key to understand the connections between person, cosmology, territory and society (Seeger et al., 1979; Echeverri, 2005).

In the groups we study, the body is seen as a 'condition of possibility of the person and vehicle of relations' (Tola 2009, 46). It is a body always in transformation, with fuzzy and changing boundaries (Wright, 2008). A body crossed by fluids entering and leaving. A porous and multiple body, long interpenetrated by other bodies. And this exchange of fluids is a particular record of intersubjective relations. The body is not a border but a space of mediation (Tola 2009, 46). Tola (2009, 47) suggests the concept of an 'embodied person' that includes human and non-human beings with different and changing 'body regimes'. The human bodies are in permanent contact with bodies, objects and fluids from celestial beings. In this sense, the meeting with a powerful being can affect the human body. For example, we can consider meteorites. Contact with meteorites can originate rain, and only the shamans can manipulate these objects without danger. Also, certain fungi are another 'body regime' of metallic meteorites, and people use these fungi to treat injuries, especially those caused by iron weapons (Giménez Benítez et al., 2004).

Powerful celestial beings can appear on Earth in different 'body regimes': some strange exemplar of an animal or vegetal specie associated with this being or also a white man. The color or brightness of the body is a manifestation of power. The interpretation of the nature of power of different parts of the sky are linked with this idea of brightness = power. And the nature of power of some regions is connected with the presence of certain kind of beings there.

The topology of the celestial space is formed by power and relations between the beings that inhabit the sky. And this topology is transmitted to the 'practical conscience' through the constitution of habitus. In terms of Bourdieu (1997b), habitus refers to a generative principle that functions as a kind of matrix of principles of perception, appreciation and action. The social agents tend to operate from a set of knowledge, preferences and feelings, shaped by past experiences, which have become a more or less stable principle of action. The habitus is a 'practical sense' or 'sense of the game' that lets players guide their actions and conscious perception of the situation and develop appropriate responses to it, as a sort of 'smell' that suggests how to anticipate future developments from the present situation. But, these 'logics of practice' are coherent only in general terms and their incompleteness is an important part of their nature.

An example of this is how the structure of language conveys a certain apprehension of space and time. In particular, the Toba and Mocoví languages have structures that incorporate the notion of 'meeting' as a basic structure of experience. These languages use explicit forms to indicate the degree of confidence that the speaker attributes to the information provided by his statement. The key criterion for this is if the speaker is a direct witness or not of the events described (evidentiality). These linguistic structures therefore give great importance to direct contact with the powerful beings that shape the world as sources of information about it. Thus the shamans and elders are privileged sources of knowledge in cosmology. And cosmology is a sort of 'know how' in the context of social relationships on a cosmic scale.

In a similar way, these languages give a central role to the concept of 'path' in the sense of a general structure to organize knowledge. Narrative strategies and classification procedures are based in this metaphor. For example, the construction of personal biographies and mythical stories are like series of wanderings. In this sense the path is like a chain punctuated by marks. And these marks are important encounters with 'others'. In the sky, the Milky Way is seen through this model (López and Giménez Benítez, 2008).

Flexible topology and metaphors

Habitus, mental images, metaphors and conceptions of the body and person have a central importance in the analysis of the cosmological discourses from these groups. Not taking into account these concepts has been the origin of some misunderstandings in academic research.

The existence of almost three levels in the cosmos of these people is well known (Wright, 2008; Miller, 1975; Cordeu, 1969-70; Idoyaga Molina, 1989). But these levels are not 'layers' in the sense of a rigid system of flat strata. First at all, the important fact is the existence of differentiated domains and relations between them. These relations are expressed many times by the idea of how the layers are ordered. The existence of an order and the social value of this order, not the exact number or sequence of layers, is the important thing (this point is also referred to for the Tobas by Wright (2008, 197). In the same sense, the flat nature of these layers is not an important matter. For this reason the term 'plane' is not a good choice. The same person can draw these domains in a vertical arrangement or in a horizontal one (Wright 2008, 145-149). These cosmological models are relational and situational constructions, as Wright shows in the case of the Toba (Wright 2008, 151 and 171). And the search for a unique world representation is an ethnocentric mistake (Wright 2008, 150). Axes like up-down, west-east and day-night are part of an essentially fluid and dynamic worldview (Wright 2008, 174-175).

We will think of the spatial images and metaphors of these groups in terms of contextual world-models, and not as if they were 'candid pictures'. For example, the world tree that connects these layers is much more than a 'naive mystical tree'. It is a true artery of the world where resources and beings travel up and down, a fundamental communication to life on earth. These layers and the tree that connects them are, as stated above, flexible and changing images. On the one hand, the layers can change their shape from one communication situation to another. On the other hand, this universe is interconnected by a tangled network of tunnels, in which the world tree is the most important. Associated with lagoons, swirls of dust, and still completely invisible and only detectable by shamans, these tunnels create a very dynamic and porous universe (López and Giménez Benítez, 2008).

As mentioned, the non-human powerful beings, which guard the vital resources, make up domains of authority and influence in space. Due to their presence in these areas, space takes a 'texture' that refers to the time of origins. The relations between these beings and humans are fundamental in the understanding of the economy of power that shapes the space of heaven in these communities. In this sense, the Milky Way area is at the present time an area restricted to the shamans. And the space near the Sun and the Moon can be inaccessible even for the most powerful of them (Sánchez 2009, 391). This inaccessibility has an important social effect, because if you have a meeting with the Sun or Moon, they can give you the management of metals and fire. The white people have these secrets, and for this reason they can oppress the aboriginal people.

Shamans as cosmologist

Shamans, men full of power, are the human equivalent of the non-human masters. Their ability to 'see' and 'travel' through the different layers of the world makes them the main characters in the contest for leadership and the definition of identity. As mentioned, the sky is not an open region, not everyone can access it. This unique cosmic experience of the shamans makes them the most important traditional cosmologists in these groups.

The topology of the celestial space among these groups, being fundamentally shaped by the scales of power of the beings that inhabit it and their relationships, becomes essential to the experience of inhabiting. In this sense, perceiving the 'texture' of the celestial space means to be able to account the types of beings that inhabit it. Thus, the topology turns into a mapping of these groups' social space, which involves the bounds between humans and non-humans. And again it is not mainly a theoretical understanding but a corporeal experience based on everyday habitus.

In this sense, the idea of *pacto* (pact) as a mechanism for managing unequal power relations is transformed into another nucleus of organization of experience and relationships. The pact is a medium to 'domesticate' the exuberance and excess of the powerful beings and obtain from them some power. The human needs an attitude of 'respect' and the accomplishment of some restrictions in food, circulation, etc. To make a pact, the human in question also needs some amount of power. For this reason, traditionally, only the shamans can make pacts with the beings that rule the world structure.

New cosmologist

Between 1940 and 1960 an important series of changes took place in the Chaco. The growing presence of Evangelical churches, schools, medical facilities, government agencies and NGOs became alternative spaces that provided resources and ways to construct prestige and social networks both inside and outside the community. Thereby, young teachers, teachers' assistants, nurses, pastors and politicians positioned themselves as new leaders, and for this reason, new cosmologists. The topology of celestial space as it is transmitted by the Christian scriptures and scholarly books, especially by the drawings that illustrate them, have created 'areas' of interaction and symbolic resignification (Wright, 2008; Ceriani Cernadas Citro, 2005). Now, even the radio and television play an active role in the new experience of the celestial space in these groups.

This access of literate young people to cosmologic production generates conflicts with the elders who value the traditional ways and context of cosmological knowledge. The new cosmological production is also flexible and situated. But new truth criteria, ways of legitimation and some preoccupation for 'consistency' has appeared.

Mundus Imaginalis

We consider that it is important to incorporate in the study of the social construction of space what might be called, according to Henry Corbin, 'Mundus imaginalis' or 'visionary space' (Corbin, 1996). The way in which the territory is conceived among the communities we studied it is not only based on the evidence of the senses in the waking state but also on three other types of experiences: dreams, altered states of consciousness and 'active imagination'. The role of shamanic experiences and dreams in the conceptualizations of space that we studied, and the ideas about the constitution of the person and experience in these groups, makes Corbin's suggestions particularly relevant. Moreover, the connections between the astronomical and the night and between the night and the dreams creates important links between astronomy and the dream world, providing to the celestial space many features of the oniric reality. This oniric condition of the topology of the cosmos is not only visible in its fuzzy and variable character. Another important feature of this cosmic topology is related to the visionary space that Corbin studied. Corbin (1996, 125) says that, in the Mundus Imaginalis, the internal acts of beings project visible forms. In fact, each conscious being generates it own visionary space. These forms are real ones but at the same time are ways of being or attributes of the conscious being in question. In this sense, the visionary space of a conscious being is its own celestial body. Certainly, the cosmologies that we study are not considered by Corbin. But his analysis and phenomenological descriptions has a value that goes beyond the particular case studied by him. In particular they are well suited to address the experiences of our informants about how the presences of powerful beings give 'texture' to their spatial and temporal environment. Corbin (1996, 61 and 227) also mentions the connection between power or glory and the 'visionary' light (he calls it 'light-being'). In fact, in the tradition that he studied, bodies and souls are different states of this 'light-being'. He (Corbin, 1996, 60-63) remarks that many geographical, historical and calendrical descriptions are in fact 'visionary geography' and 'visionary history'. The reason is that, in the tradition he studied, all the important things, the fundamental things for the cosmos, occur in this visionary world. Corbin (1996, 51) prefers the word 'visionary' or 'imaginal' and not the word 'mythical'. This is because, for these people, this visionary dimension is not only a real aspect of the world, in fact it is the most real one. The access to this dimension is, at the present, restricted to who has certain special abilities. Corbins (1996, 51) calls 'active imagination' the mental operation that allows the visionary creativity. If we remember that Corbin and Jung were both members of the 'Eranos Circle', we can use the work of Jung to clarify the nature of the 'active imagination'. Jung called 'active imagination' a method that allows the establishment of an active dialogue between the unconscious and the waking state.

We think that this framework is very interesting in order to understand the nature of the cosmologies of these aboriginal people and their flexible topology. We can add to the Corbin-Jung conception the idea that the active imagination and the archetypal images are not ahistorical categories but are socially constructed. And the process of this construction mainly involves the habitus and the logics of practice. Thus, we can understand how social practices shape the topology of these cosmic spaces. And this topology is a sensorial representation of the social relations that structure these groups.

One of the most important practices in this process of social construction is the experience of the meeting with powerful beings. The young people learn by imitation to construct the socially accepted categories to make sense of the experiences of the loneliness of the woods or in the exaltation of the cult. People think of these experiences using categories, images and metaphors that refer to the idea of a meeting with powerful beings.

Another important point is the group of ideas about the body and the degree of reality that is given to oniric experiences. In dreams the person makes acts with consequences. These acts are real ones. For this reason the dream is viewed as a time when the celestial beings come to the earth and the human can interact with these entities (Wright 2008, 168-170).

This 'imaginal' character of many characteristics of this cosmos doesn't means that these are 'fantastic' or 'unreal' characteristics. The 'imaginal' is a dimension of the real, and beyond the metaphysical status of their phenomena, the psychological and sociological reality of this aspects of the cosmos is indubitable. In fact, many mental disorders in these groups are linked with unhappy meetings with powerful beings. The fear of these meetings and the strange phenomena connected with them are important factors of social control in these communities—something similar is reported for Mayan communities (Hermitte, 2004).

A socio-historical view of the construction of these visionary spaces is an important tool for understanding many cosmological systems. In each case it is important to reconstruct the specific mechanisms that construct, in practice, this imaginary. And it is important to pay attention to the development in time of these Mundus Imaginalis.

References

Barúa, G. 2001. Semillas de estrellas. Los nombres entre los wichí. Buenos Aires, Editorial Dunken.

Bourdieu, P. 1997a. Meditaciones pascalianas. España, Anagrama.

Bourdieu, P. 1997b. Razones prácticas. Sobre la teoría de la acción. Barcelona, Editorial Anagrama.

Bourdieu, P. 2005. Las Reglas del Arte. Génesis y estructura del campo literario. Barcelona, Anagrama.

Braunstein, J. 1989. Astronomía mataca. Revisión crítica de la Mitología Sudamericana V de Roberto Lehmann-Nitsche. Scripta Etnológica, 43-60.

Braunstein, J. 1995. Las figuras de hilo del Gran Chaco III. Figuras de hilo de los pilagá y los toba-pilagá (1ª parte). Hacia una carta étnica del Gran Chaco 6, 139-150.

Ceriani Cernadas, C. and Citro, S. 2005. El movimiento del evangelio entre los tobas del Chaco argentino. Una revisión histórica y etnográfica. In Guerrero Jiménez, B.

(ed.) De indio a hermano. Pentecostalismo indígena en América Latina, 111-170. Iquique, Chile, Ediciones Campus, Universidad Arturo Prat y Ediciones El Jote Errante.

Corbin, H. 1996. Cuerpo espiritual y tierra celeste: Del Irán mazdeísta al irán chiíta. España, Siruela.

Cordeu, E. 1969-70. Aproximación al horizonte mítico de los tobas. Runa. 12, no. 1-2, 67-173.

Cordeu, E. J. 1998. Los ropájes de lo insólito. Dualidad de las cosmovisiones e ideas del `poder´ en tres mitos tomaraxo. Scripta Ethnologica 20, 7-29.

Dasso, C. 1989. Las transformaciones del sol y las estrellas en la cosmovisión de los mataco (grupo tachóhnai). Scripta Etnológica 9, 29-42.

Echeverri, J. Á. 2005. Territory as Body and Territory as Nature: Intercultural Dialogue?, in A. Surrallés and P. García Hierro (eds), The Land Within: Indigenous Territory and the Perception of Environment, 234-250. Copenhagen, International Work Group for Indigenous Affairs.

Giménez Benítez, S., López, A. M. and Granada, A. 2004. Suerte, riqueza y poder. Fragmentos meteóricos y la presencia de lo celeste entre los mocovíes del Chaco, in J.B. M. Boccas and G. Pereira (eds), Etno y Arqueo-Astronomía en las Américas. Memorias del Simposio ARQ-13: Etno y Arqueoastronomía en las Américas, 51° Congreso Internacional de Americanistas, Santiago de Chile.

Giménez Benítez, S., López, A. M. and Granada, A. 2006. Sun and moon as marks of time-space among Mocovies from the Argentinian Chaco. Archaeoastronomy. The Journal of Astronomy in Culture 20, 54-69.

Gómez, C. P. 2006. Algunas representaciones Toba-Pilagá sobre las constelaciones celestes. 26th Congreso de Geohistoria Regional. Resistencia, Chaco. IGI, CONICET.

Hermitte, M. E. 2004. Poder sobrenatural y control social en un pueblo maya contemporáneo. Buenos Aires, Antropofagia.

Idoyaga Molina, A. 1989. Astronomía Pilaga. Scripta Ethnológica. 17-28.

Lehmann-Nitsche, R. 1923. La Astronomía de los tobas. Revista del Museo de La Plata. 27, Tercera serie,, 267-285.

Lehmann-Nitsche, R. 1924. La Astronomía de los Matacos. Revista del Museo de La Plata. 27, Tercera serie, 253-266.

Lehmann-Nitsche, R. 1924-25. La Astronomía de los Mocoví. Revista del Museo de La Plata. 28, Tercera serie, 66-79.

López, A. M. 2009. La Vírgen, el Árbol y la Serpiente. Cielos e Identidades en comunidades mocovíes del Chaco. PhD dissertation, Facultad de Filosofía y Letras, Universidad de Buenos Aires.

López, A. M. and Giménez Benítez, S. 2008. The Milky Way and its structuring functions in the worldview of the Mocoví of Gran Chaco. Archaeologia Baltica, 21-24.

Métraux, A. 1935. El universo y la naturaleza a través de las representaciones míticas de dos tribus salvajes de Argentina. Sur 5, 19-46.

Miller, E. S. 1975. Shamans, Power Symbols, and Change in Argentine Toba Culture. American Ethnologist. 2, no. 3, 477-496.

Ruiz Moras, E. 2000. El rumor de los astros: una aproximación a la etnoastronomía toba taksek (Chaco Central), in III Congreso Argentino de Americanistas, 1999, 263-289. Buenos Aires, Sociedad Argentina de Americanistas.

Sánchez, O. 2009. Da na'aqtaguec nam Qompi Toba mayi Ima'na Ita'adaic Chaco. Nam ÿa'axat som lquedoxonecpi na mayipi. Historias de los aborígenes tobas del Gran Chaco contadas por sus ancianos. Resistencia, Librería de la Paz.

Seeger, A., Da Matta, R. and Viveiros de Castro, E. 1979. A construçao da pessoa nas sociedades indígenas brasileiras. Boletim do Museu Nacional, 2-19.

Tola, F. 2009. Les conceptions du corps et de la personne dans un contexte amérindien. 'Je ne suis pas seul(ement) dans mon corps'. Indiens toba du Gran Chaco sud-américain. Paris, L'Harmattan.

Toledo Llancaqueo, V. 2005. Políticas indígenas y derechos territoriales en america latina 1990- 2004. ¿Las Fronteras Indígenas de la Globalización?, in G. d. T. s. M. Indígenas (ed.), Pueblos Indígenas y Democracia en América Latina, Seminario Quito, Julio 2004. Quito. CLACSO, Bs. As.

Wright, P. G. 2008. Ser-en-el-sueño: Crónicas de historia y vida toba. Buenos Aires, Editorial Biblos.

ANTIQUITY

AMONG THE CIRCLES: A GEOMETRICAL ANALYSIS OF THE TEATRO MARITTIMO IN VILLA ADRIANA

MARZIA MONACO, SILVIA GAUDENZI AND MARCELLO RANIERI

Abstract: Among the many architectural achievements of the Roman Emperor Hadrian (76-138 AD), those of Villa Adriana in Tivoli are the most representative. Circular geometries, frequent in Hadrian architecture, are well represented at Villa Adriana. The most attractive and charming is in the so-called 'Teatro Marittimo'. Its circular structure is emphasized by the annular water canal surrounding the 'isle', which is accepted as the very place where the Emperor used to stay when in Villa Adriana (Chiappetta, 2008). For the geometrical analysis of the structure, the plan of M. Ueblacker (Ueblacker, 1985) has been utilized: it emerged as the best among the available surveys. The plan was imported into the CAD (Computer Aided Drawing) application through a calibrated scanner; the analysis was conducted by drawing best-fitting circles on the water canal annulus as well as on the Porticus annulus. The result is that the water canal annulus is interpretable as a 41-58 (x2) CQC geometry (Ranieri, 2009) and the Porticus is consistent with an annulus capable of bearing a regular pentagon inscribed in it. The length-unit derived for both annuli equals the roman foot of 0.296m within few mm, although a shorter unit of 0.291m also appears possible.
Keywords: Roman architecture, Villa Adriana, ancient circular geometry, CAD analysis, Length-units.

Figure 1. The Teatro Marittimo in Hadrian's Villa near Tivoli, Rome.

Circular geometries of antiquity: premises

As reported for other architectonic structures of antiquity, (Ranieri, 2007, 2009) a relevant circular geometry scheme that frequently appears is that of an annulus, the dimensions of which are such that a square is inscribable in between the two circular borders. Generally speaking, the most relevant circular geometries of antiquity are annuli with inscribable regular polygons (Figure 2); namely

Circle-Triangle-Circle CTC (a triangle is inscribed)
Circle-Square-Circle CQC (a square is inscribed)
Circle-Pentagon-Circle CPC (a pentagon is inscribed)
Circle-Hexagon-Circle CEC (a hexagon is inscribed)

Each one of these is characterized by the ratio between the outer and inner diameters of the circles, Dext /Dint:

CTC: Dext /Dint = 2
CQC: Dext /Dint $\approx$ sqrt(2)
CPC: Dext /Dint $\approx$ sqrt(5) -1
CEC: Dext /Dint $\approx$ 2/sqrt(3)

The $\approx$ sign (almost equal to) accounts for the fact that in antiquity all lengths are supposed to be expressed by integer numbers of length-units, so that Dext and Dint are integers; the ratio Dext/Dint is consequently a rational number and, as such, will never be equal to the exact (irrational) value. However appropriate integer values for

Dext and Dint exist by which Dext/Dint will arrive close enough to the exact (irrational) value.

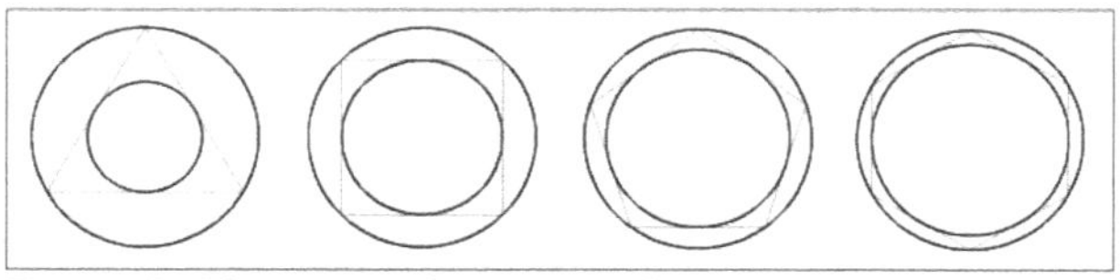

Figure 2. The most relevant circular structures of antiquity: annuli with inscribable regular polygons.

A - B	A - B
12-17	**56-79**
17-24	**57-81** (19-27)x3
19-27	**58-82** (29-41)x2
22-31	**60-85** (12-17)x5
24-34 (12-17)x2	**62-88** (31-44)x2
27-38	**63-89**
29-41	**65-92**
31-44	**66-93** (22-31)x3
34-48 (17-24)x2	**68-96** (17-24)x4
36-51 (12-17)x3	**70-99**
38-54 (19-27)x2	**72-102** (12-17)x6
39-55	**73-103**
41-58	**75-106**
43-61	**76-108** (19-27)x4
44-62 (22-31)x2	**77-109**
46-65	**78-110** (39-55)x2
48-68 (12-17)x4	**80-113**
51-72 (17-24)x3	**81-114** (27-38)x3
53-75	**82-116** (41-58)x2
54-76 (27-38)x2	**84-119**
55-78	**85-120** (17-24)x5

Figure 3. Couples of integer numbers suited to achieve CQC annuli (in bold for Dext < 120).

For CQC annuli, the available couples of integers that produce the proper geometry are shown in Figure 3. For the selected A-B couples (primaries and multiples) the difference of B/A from √2 is less than 0.5% and (for primaries only) A√2 is different from B by less than 0.25. For CPC annuli the available couples of integers that produce the proper geometry are shown in Figure 4. For the selected B-C couples (primaries and multiples) the difference of C/B from (√5-1) is less than 0.5% and (for primaries only) B(√5-1) is different from C by less than 0.25. CTC and CEC annuli are not present in the Teatro Marittimo.

Geometrical CAD Analyses and the choice of the most reliable plan

For any geometrical CAD analysis, a mandatory requirement is that it must be performed by using a reliable survey plan. There are many survey plans of Teatro Marittimo available in the literature of Villa Adriana (see Figure 5); after a thorough evaluation of their reliability, the Ueblacker plan (Ueblacker, 1985) has been chosen. The plan was scanned into the image editor (Adobe Photoshop) using a calibrated size 3A professional scanner and imported as a tiff file in the CAD application (Micrografx Designer).

B-C	B-C
13-16	**78-96** (13-16)x6
17-21	**80-99**
21-26	**81-100**
25-31	**84-104** (21-26)x4
26-32 (13-16)x2	**85-105** (17-21)x5
29-36	**86-106** (43-53)x2
30-37	**87-108** (29-36)x3
34-42 (17-21)x2	**88-109**
38-47	**89-110**
39-48 (13-16)X3	**90-111** (30-37)x3
42-52 (21-26)x2	**91-112** (13-16)x7
43-53	**92-114** (46-57)x2
46-57	**93-115**
47-58	**94-116** (47-58)x2
50-62 (25-31)x2	**97-120**
51-63 (17-21)x3	**98-121**
52-64 (13-16)x4	**100-124** (25-31)x4
55-68	**101-125**
56-69	**102-126** (17-21)x6
58-72 (29-36)x2	**104-128** (13-16)x8
59-73	**105-130** (21-26)x5
60-74 (30-37)x2	**106-131**
63-78 (21-26)x3	**108-134** (54-67)x2
64-79	**110-136** (55-68)x2
65-80 (13-16)x5	**111-137**
67-83	**112-138** (56-69)x2
68-84 (17-21)x4	**114-141** (38-47)x3
71-88	**115-142**
72-89	**116-144** (29-36)x4
73-90	**117-144** (13-16)x9
75-93 (25-31)x3	**118-146** (59-73)x2
76-94 (38-47)x2	**119-147** (17-21)x7
77-95	**120-148** (30-37)x4

Figure 4. Couples of integer numbers suited to achieve CPC annuli (in bold for Dint < 120).

AUTHOR	*YEAR*
Andrea Palladio	*1554*
Francesco Contini	*1668*
Giovanni Piranesi	*1781*
Pierre-Gérôme-Honoré Daumet	*1859*
p. Blondel	*1881*
Hermann Winnefeld	*1895*
Pierre Gusman	*1904*
Vincenzo Reina	*1906*
Giuseppe Lugli	*1932*
Herbert Bloch	*1937*
Mathias Ueblacker	*1985*
Jeffrey Jacobson	*1986*
Eugenia Salza Prina Ricotti	*1992*
William MacDonald - John Pinto	*1995*
Marina De Franceschini	*2006*
Benedetta Adembri - Giuseppina Cinque	*2006*
Federica Chiappetta	*2008*

Figure 5. The available plans for Villa Adriana/Teatro Marittimo.

The circular geometry of the Teatro Marittimo.

The major geometrical peculiarity of Teatro Marittimo is that it consists of three concentric circles forming two concentric annuli. The hypothesis that we make is that the two concentric annuli possess a CQC + CPC geometry, as illustrated in Figure 6.

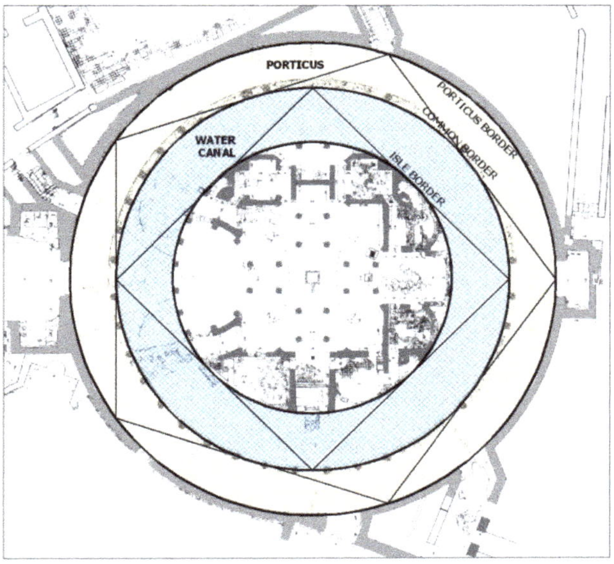

Figure 6. The CQC+CPC geometry of the Teatro Marittimo to be verified with CAD analysis.

First annulus: The water canal

In order to achieve the best fitting annulus of the water canal some preliminary checks have been necessary. Initially, by comparing the Ueblacker plan with other available plans and with on-site measurements by archaeologists working in Villa Adriana, the validity of the metric scale has been confirmed. This yielded the best possible and most reliable calibration in meters of the measurement device of the CAD application. To achieve the best fitting circles, a sufficient number of points were established on the two borders of the annulus (26 on the internal and 32 on the external). The averages of the lines' thickness at the points was evaluated (8cm ~ 0.3% of inner diameter and 7cm ~ 0.2% of outer diameter) and introduced as measurement uncertainties. The points were chosen in diametrically opposite positions, deriving 13 and 16 'diameters'. From the 'cloud' of 13+16 positions of the half-way points [xi,yi] (Figure 7 left) an average [x0,y0] position was calculated and used as the initial centre for the fitting procedure.

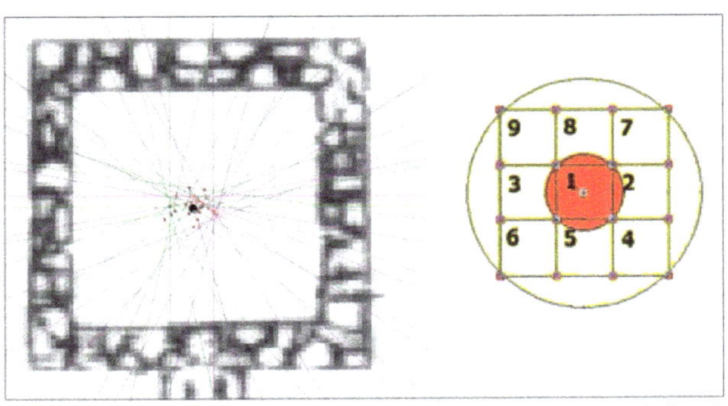

Figure 7. Best-fitting the water canal annulus: the search for the circle's centres.

Distances between the initial centre [x0,y0] and the selected points were successively measured, deriving 26 and 32 radial distances for the internal and external borders, respectively; the corresponding averages and standard deviations were computed. In order to allow a two degree of freedom variation for the centre position, (the third degree being on the radius), a grid of 9 (3×3) squares with side size close to 1/3 of the initially obtained values for the standard deviation (Figure 7, right) was created around [x0,y0]. The final distributions are shown in Figure 8, the best fitting values being:

Internal border $\langle DI \rangle \pm \sigma DI = (24.51 \pm 0.06)$m,
External border $\langle DE \rangle \pm \sigma DE = (34.38 \pm 0.08)$m,
or
Internal border DI-MIN = 24.45m; DI-MAX = 24.57m,
External border DE-MIN = 34.30m; DE-MAX = 34.46m.

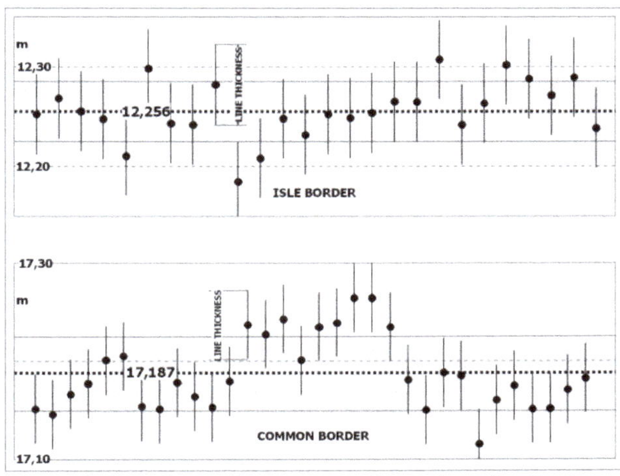

Figure 8. Best-fitting the water canal annulus: distribution of radii measurements.

DIAMETERS RATIO VERSUS √2.
$r = \langle DE \rangle / \langle DI \rangle = 1.403$
$\Delta r = \pm r\,(\sigma DE / \langle DE \rangle + \sigma DI / \langle DI \rangle) = \pm 0.007$
$r \pm \Delta r = 1.403 \pm 0.007$

That is, the diameters ratio r is within 1.396 and 1.409, its difference from √2 ranges from ~ -1.3% to ~ -0.4%. Such figures allows to claim that a square is inscribable in the water canal annulus.

DIAMETER VALUES VERSUS THE ROMAN FOOT VALUE.
In the reasonable hypothesis that the roman foot employed at the Teatro Marittimo is not far from the canonical value of 0.296m more than ~ 2%, the diameter values are to be compared with values ranging from prMIN = 0.290m to prMAX = 0.302m. The integer numbers of roman feet representing the obtained values for the diameters have been calculated: the results, reported in Figure 9, point out that the integers representing DI may range from 81 to 85 while those representing DE from 114 to 119.

DIAMETER	/PR$_{MIN}$ (0.290)	/PR$_{MAX}$ (0.302)
$D_{I\text{-}MIN}$ (24.45)	84 (84.3)	81 (81.0)
$D_{I\text{-}MAX}$ (24.57)	85 (84.7)	81 (81.4)
$D_{E\text{-}MIN}$ (34.30)	118 (118.3)	114 (113.6)
$D_{E\text{-}MAX}$ (34.46)	119 (118.8)	114 (114.1)

Figure 9. The range of possible integer numbers that can represent the best-fitted circles of the water canal annulus for roman foot values not more far than 2% from 0.296m.

DIAMETERS VALUES VERSUS CQC COUPLE OF INTEGERS

By comparing the values of Figure 9 with the CQC couples of integers in Figure 3, two couples only emerge as consistent: 82-116 and 84-119. The corresponding roman foot values are shown in Figure 10.

82-116			
D_I /82	FOOT	D_E /116	FOOT
$D_{I\text{-}MIN}$/82	0.298M	$D_{E\text{-}MIN}$/116	0.296M
$D_{I\text{-}MAX}$/82	0.300M	$D_{E\text{-}MAX}$/116	0.297M
84-119			
D_I /84	FOOT	D_E /119	FOOT
$D_{I\text{-}MIN}$/84	0.291M	$D_{E\text{-}MIN}$/119	0.288M
$D_{I\text{-}MAX}$/84	0.293M	$D_{E\text{-}MAX}$/119	0.290M

Figure 10. Possibly used values of the roman foot in achieving the CQC geometry for the water canal annulus.

For the 82-116 CQC, a roman foot value of 0.297m would establish for DI a measure of 82x0.297 = 24.354m with a difference of 15.6cm (0.6%) from <DI> = 24.51m, and for DE a measure of 116x0.297 = 34.452m with a difference of 7.2cm (0,2%) from <DE> = 34.38m.

For the 84-119 CQC, a roman foot value of 0.291m would establish for DI a measure of 84x0.291 = 24.444m with a difference of 6.6cm (0.3%) from <DI> = 24.51m, and for DE a measure of 119x0.291 = 34.629m with a difference of 24.9cm (0.7%) from <DE> = 34.38m. Both 82-116 and 84-119 CQC appear as possible at this stage of the analysis.

Second annulus: the Porticus

The Porticus (Figure 11) is the annular zone in between the outer border of the water canal and the internal side of the circular enclosing wall. Being the internal border of the Porticus in common with the external border of the water canal, the best-fitting procedure has been applied to the Porticus external border alone. A minor ambiguity exists on whether the points to be considered should be set at the level of the concrete or at the level of plaster (see Figure 11). The thickness of this latter is anyhow on the order of ~10cm on average, which corresponds to less than a small percentage of 0.5% on the diameter.

However, 22 points were established on both plaster and concrete borders of the annulus and the best fitting results were as follows:

at plaster:
DP = <DP> ± σDP = 42.700 ± 0.124m (0.3%)
or DP-MIN = 42.576m; DP-MAX = 42.824m
at concrete:
DC = <DC> ± σDC = 42.906 ± 0.115m (0.3%)
or DC-MIN = 42.791m; DC-MAX = 43.021m

DIAMETERS RATIO VERSUS √5-1.

<rP> = <DP>/<DE> =1.242
ΔrP = ± rP (σDP/<DP> + σDE/<DE>) = ± 0.006
rP ± ΔrP = 1.242 ± 0.006;
<rC> = <DC>/<DE> =1.248
ΔrC = ± rC (σDC/<DC> + σDE/<DP>) = ± 0.006
rC ± ΔrC = 1.248 ± 0.006

That is the ratio is within 1.236 and 1.254, the difference from √5-1 (=1.236) ranging from ~ 0.01% to ~ 1.4%. Such figures allows to claim that a pentagon is inscribable in the Porticus annulus.

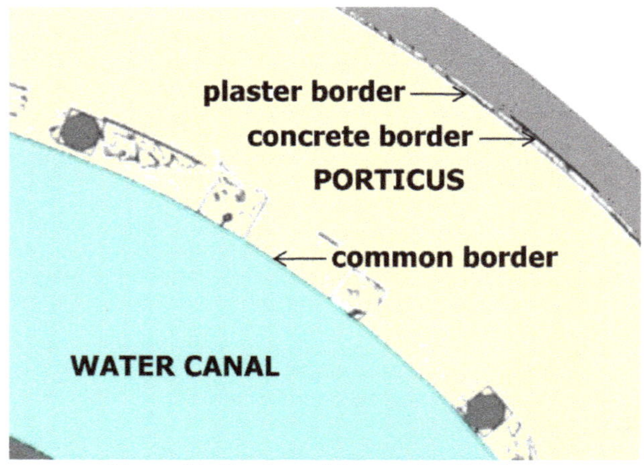

Figure 11. Borders of the annular Porticus.

DIAMETER VALUES VERSUS THE ROMAN FOOT VALUE.

In the same hypothesis that the roman foot employed at the Teatro Marittimo is not far from the canonical value of 0,296m more than ~ 2%, the diameter values are to be compared with values ranging from prMIN = 0,290m to prMAX = 0,302m. The possible integer numbers of roman feet that may represent the two diameters can be calculated, with the results shown in Figure 12. The integer representing DP ranges from 141 to 148 while those representing DC from 142 to 148. Aside from 141 for DP, the possible integers are the same, confirming that the ~10cm plaster do not appreciably change the results.

DIAMETER	/PR$_{MIN}$ (0.290)	/PR$_{MAX}$ (0.302)
$D_{P\text{-}MIN}$ (42.58)	147 (146.8)	141 (141.0)
$D_{P\text{-}MAX}$ (42.82)	148 (147.7)	142 (141.8)
$D_{C\text{-}MIN}$ (42.79)	148 (147.6)	142 (141.7)
$D_{C\text{-}MAX}$ (43.02)	148 (148.4)	143 (142.5)

Figure 12. The range of possible integer numbers that can represent the best-fitted circles of the Porticus annulus for roman foot values not more far than 2% from 0.296m.

DIAMETER VALUES VERSUS CPC COUPLES OF INTEGERS

By comparing the values of Figure 12 with the CPC couples of integers in Figure 4, several couples emerge: **114-141, 115-142, 116-144, 117-144, 118-146 119-147** and **120-148**. In front of the only two possible CQC couples for the water canal annulus (**82-116** and **84-119**) only two CPC couples are consistent: **116-144**, and **119-147.** Therefore, the CQC+CPC permitted configurations are **82-116-144** and **84-119-147**; it is worth noting that **82-116-144** is twice 41-58-72 while 84-119-147 is seven times 12-17-21.

The **82-116-144** (41-58-72 x 2) configuration implies a roman foot value of 0.297m while **84-119-147** (12-17-21 x 7) points to a roman foot value of 0.291m.

Conclusions

The analysis performed on the Ueblacker plan of the Teatro Marittimo has shown that the water canal that encircles the 'isle' is an annulus shaped as a CQC, which can be arrived at with the use of two possible numerical combinations being: **82-116** which yields for the implied length-unit a value close to 0.296m (known value of the Roman foot) within ≈ 1% on the internal border and ≈ 0,01% on the external one; **84-119** which yields for the implied length-unit a value close to 0.291m far from the known value of the Roman foot as much as ≈1.4% on the internal border and ≈ 2.4% on the external one. The larger ring that encircles the water canal is an annulus shaped as a CPC, which can be arrived at with the use of two possible numerical combinations being: **116-144** which yields for the implied length-unit a value close to 0.296m within ≈ 0.01% on the inner border and ≈ 0.1% on the external one; **119-147** which yields for the implied length-unit a value close to 0.291m far from the known value of the Roman foot as much as ≈ 2.4% on the internal border and ≈ 2.0% on the external one.

On these bases a preference should be assigned to the triplet of numbers **82-116-144**, it being the one that is consistent with the known value of the Roman foot of 0.296m (Figure 12).

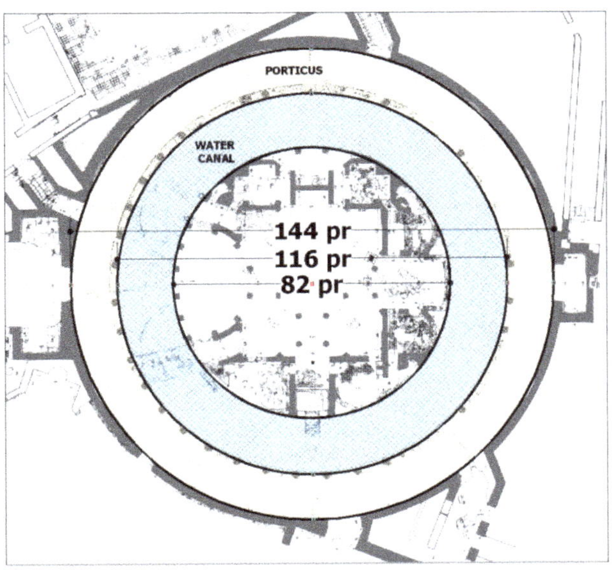

Figure 13. Geometrical and numerical solution for the Teatro Marittimo.

We are extending this type of study to other structures of the Villa Adriana, in particular to those with marked circularity. We hope to arrive at sound conclusions about the used value of the roman foot, the use of numbers and the geometrical forms.

References

Chiappetta, F. 2008. I percorsi antichi di Villa Adriana. Roma, Quasar.

Monaco, M. 2009. Indagine, tramite tecnica CAD, sui contenuti geometrici, numerici e metrici delle strutture circolari del Teatro Marittimo in Villa Adriana. Unpublished thesis, Università di Roma 'La Sapienza', Dipartimento di Fisica.

Ranieri, M. 2007. A geometrical analysis of Mesoamerican pre-Hispanic Architectures: triads, numbers, length units and the calendar. In Astronomy and cosmology in folk traditions and cultural heritage, Proceedings of the 6th SEAC Meeting 2007 and Oxford VIII International conference, Vol. 10, 183-194. Klaipėda, Archaeologia Baltica.

Ranieri, M. 2009. Beyond the Cosmological: Numerical Scenarios underneath Ancient Annular Architectural Structures. In Cosmology across Cultures, Proceedings of the 6th SEAC Meeting and IX Oxford International Conference, Granada, 8-12 Sept. 2008 Conference Series', Vol. 409, 209-244. San Francisco, Astronomical Society of The Pacific.

Ueblacker, M. 1985. Das Teatro Marittimo in der Villa Adriana bei Tivoli. Deutsches archäologisches Institut Rom – Sonderschriften 5. Mainz am Rhein, Verlag Philipps Von Zabern.

MEDIEVAL TIME IN ORIENT AND OCCIDENT

ASTRONOMY AND THE STATE: TIME, SPACE AND POWER IN THE FOUNDATION OF BAGHDAD

NICHOLAS CAMPION

Abstract: This paper will consider the application of the heavens to the organisation of the 'human polity modelled on the assumption of a close relationship between society on the one hand, and planetary and stellar patterns on the other. I will examine accounts of the astrological foundation of Baghdad within the tradition of celestial town planning, consider contradictions in the data for the foundation, and argue that the city was designed to connect heaven to earth and ensure peace, stability and political success by harmonising time and space. It is therefore a prime example of the use of astronomy in the service of political power.
Keywords: astronomy, power, cosmopolis, town planning, horoscope

Introduction

The inauguration of the city of Baghdad in July 762 AD (145 AH) marked a seminal moment in the legitimisation of Abassid power. The move of the capital city of the Islamic Empire from Damascus under Caliph al-Mansur was a huge undertaking, initiated in 758 AD and involving around 100,000 labourers and artisans (Wiet, 1971). It signified a break with the past and a new beginning which was conducted in conformity with appropriate astronomical alignments and the rules of astrological interpretation. It therefore provides a case study in the relationship between astronomy and political authority. My purpose is to analyse the horoscope, using literary sources, to establish whether the textual evidence adds to our understanding of the cultural application of astronomy in the eighth century.

The relationship between astronomy and power is often treated as one in which those people in positions of authority apply their astronomical knowledge in order to manipulate those whom they control. As Steve McCluskey (1993, 429) pointed out in relation to Mesoamerica, 'Astronomical observation and knowledge were... signs of sacred power and status'. Elaborating on this argument, and using the Pre-Columbian Mesoamerican application of the movements of Venus to regulate warfare, Clive Ruggles described the manner in which astronomical knowledge can function as 'a resource that can be brought to bear in support of the dominant ideology to reaffirm and reinforce the structures of a particular society' (Ruggles and Saunders 1993, 4). No doubt there is some truth in this perspective, and it resonates with Antonio Gramsci's (1971) theory of cultural hegemony, in which ideology rather than physical force serves as a tool of repression. In terms of the theory of space, Deyan Sudjic (2005) encompasses the hegemonic model (although he does not use the term) in his analysis of the phenomenon whereby powerful people construct monumental structures, including, of relevance to this paper, Saddam Hussein's reconstruction of Babylonian sites. Imposing architecture is used as an instrument of state power, to reinforce a cosmology in which the leader is interposed between the mass of the population and their ability to control their destinies.

The foundation of Baghdad is an archaeoastronomical issue, but Clive Ruggles has made an eloquent case for examining astrology in relation to archaeoastronomy. Speaking of academic interest in astrology, he argued, 'What interests these scientists [archaeologists and anthropologists, whose interest is human behaviour] is the fact that people throughout the ages have drawn direct connections between the appearance of the sky and events on earth, and that this forms an integral part of their understanding of the way the world works' (Ruggles 2005, 24). Again, in relation to megalithic astronomy in particular, and all pre-modern astronomy in general, he wrote,

Why should the archaeologist, and in particular the student of Neolithic and Bronze Age Wessex, be interested in the perception of astronomical phenomena? A general answer is that in many, if not virtually all, non-Western world-views celestial phenomena are not separated from terrestrial ones, but form part of an integrated whole with complex interconnections. The association may often be viewed as closer in nature to modern astrology than modern astronomy (Ruggles 1997, 204).

For the astronomers of the eighth-century Near East, the use of astrology in the sense that Ruggles defines it was reinforced by Claudius Ptolemy's insistence that the astronomers work is divided into two phases. The first was the measurement of celestial positions, the second the measurement of their effects (Ptolemy 1940, I.1). To understand archaeoastronomy in the Islamic world, it is insufficient to consult Ptolemy's astronomical work, the *Almagest*; his astrological text, the *Tetrabiblos*, is the essential text. Ptolemy also developed the notion of cosmography, defined by Short as the use of geographical and celestial coordinates to map the harmonious relationship between sky and earth so that order might be maintained in terrestrial society (Short 2001, 28).

Cosmic Building

The evidence suggests that traditions of cosmic building appear to be as old as settled human life and, as Colin Renfrew said of Stonehenge, the practice seems to represent a 'deliberate attempt to align the human society in question with the cosmos' (Renfrew 2008, 183). Such concepts are evident in the ancient Near East. For example, in first-millennium BC Persia, Persepolis, the ceremonial capital of the Achaemenid Empire of c. 550-330 BC, was conceived of as what we might call a cosmic city, a cosmopolis in Greek. Paul Wheatley (1971, 439) described the city as follows:

With its acres of buildings, with a reception hall open on all sides to symbolize the diffusion of divine authority to the four quarters, and its triple wall, itself symbolic but further strengthened by symbolic defensive signs and enormous supernatural figures standing guard before its gates, with its sacred groves in stone, its man-headed and lion-slaying bulls, sphinxes and paws uplifted in adoration before the Tree-of Life, throne room scenes, and all-pervading symbolic emblems, Persepolis constituted a magnificent demonstration of abundance, the contribution of the Persian people to the maintenance of harmony between the heavens and the earth.

To make better use of the heavens, key temples and palaces appear to have been aligned roughly to the north-east, facing the rays of the rising summer solstice sun. There are also accounts that Seleucia, capital of the Seleucid empire, one of the successor states to Alexander's empire, was founded on the basis of astrological advice. The story is possibly apocryphal but the second-century AD Alexandrian historian, Appian (II.9.58 in Cramer 1996, 11), emphasised the importance given to the correct foundation of cities by giving to the Babylonian astrologers of the time the following words: 'That which is fated, O king... neither man nor city can change, for there is a fate for cities, as well as for men!' Records of an astrological component in the consecration of Constantinople on 11 May 330 may also be apocryphal, as Pingree (1977) suggested, but also reinforce a prevailing idea that great cities should have great horoscopes.

The Astronomical Foundation of Baghdad

The best study of the foundation of Baghdad as a cosmopolis is Ibrahim Allawi's 'Some Evolutionary and Cosmological Aspects to Early Islamic Town Planning' (1988). Allawi drew attention to the astronomical and astrological features of Baghdad's foundation but, while he addressed the former, he did not possess the technical expertise to apply the latter.

Allawi's thesis was based on the following assumptions. First, town planning in the Islamic world was, in important instances, cosmological, and in this respect based on pre-Islamic examples (of which the most notable is Mecca, the Ka'aba being aligned with the spring equinox) and which have antecedents which may be traced back to Sumerian culture in the 3rd millennium BC. Second, there were two principal models of Islamic *cosmopoleis*; one was the city of Kufa, which had replaced Medina as the Islamic capital under Caliph Ali in 656, and the second was Baghdad, founded as a replacement for Damascus in 762, under Caliph al-Mansur. When Al Mansur decided to create his new capital he intended it to be, as Allawi (1988, 62) put it, 'the centre of a world empire of power and commerce, and the navel of the whole cosmos'. Allawi identified various cosmological-numerological templates for urban planning, such as a quadruple system analogous to the cardinal points, and a hexagonal system which he identified with the many septenary systems in the ancient word, including the seven planets. He argued that the design of Kufa had been essentially open and aligned, like Mecca, with the spring equinox. Baghdad, meanwhile, was essentially closed and, unusually, round, and envisaged as 'a grand cosmic astrolabe' (1988, 62). Allawi concluded that,

The design of the Round City of Baghdad is based on an astrological scheme. To attempt an explanation of its intricate conceptions would call for a technical treatment and some familiarity with Islamic astronomy and astrology, subjects still unfortunately unintelligible save to the initiated few and historians of science. Yet such a discussion is unavoidable if the discussion is to go beyond statement of hypothesis (1988, 62).

Again, Allawi reiterated a few lines later that the design of Baghdad 'cannot be separated from astrology' (1988, 62). However, he restricts his discussion to a kind of general astronomy/astrology in which the round city is divided into twelve sections and related to the great circles of the universe, such as the tropics and the equator, as well as to the sun's apogee and perigee (1988, 66-9). As an example, the sun covers a longer distance when at its perigee (when furthest from the earth) than at its apogee (when closest to the earth) and the area of the city corresponding to the Sun's perigee therefore, he claimed, contained more streets.

However, for a full picture of the astronomical considerations behind Baghdad's foundation, the horoscope needs to be analysed. In order to ensure the best possible moment for his new capital's inauguration, al Mansur employed a committee of astrologers which, according to the chroniclers al-Yaqubi (d. 897) and al-Biruni (973-1048), included Naubakht the Persian, of whom little is known, Masha'allah (c. 740-c. 815), whose work was to be influential both throughout the Islamic world and in medieval and Renaissance Europe, Umar ibn al-Farrukhan al-Tabari (known in Latin Europe as Omar of Tiberius), and al-Fazari, reputedly the first Arabic astrologer in the Islamic world (Pingree 1970, 104). Again, according to Allawi,

More relevant to our theme is the team of astronomers, astrologers and *muhandisun* to whom was entrusted the design of the city and the execution of the work. Translations of Ptolemy's *Almagest*, Indian astronomical treatises and Persian manuals were ordered by al Mansur (1988, 62).

The astrologers active in the region at the time were not necessarily Muslims (Pingree, 1997, 41-9), a significant point because they were unaffected by previous Islamic suspicion of non-Islamic learning. One, Kankah was Indian, Masha'allah was a Jew, while those who were Muslims, such as Abu Ma'shar, Omar of Tiberius, and Naubakht, tended to be Persians. Such astrologers formed part of an international community, exchanging information and texts between Greek, Syriac, Pahlavi and Sanskrit. The Greek texts which formed the basis for pre-Islamic Persian astrology included works by Dorotheus of Sidon, Vettius Valens and Hermes, dating to the first two centuries AD (Pingree 1997, 49). The last Greek astrological author known to have been consulted by the Arabic-speaking Persian astrologers of the 8th century, was Rhetorius (late 6th to early 7th century), who summarised earlier astrological works including Claudius Ptolemy's second-century AD *Tetrabiblos*, which I shall refer to in the original (Pingree 1989, 232-3). The first major astrological author of Islamic Persia was a Greek, Theophilus of Edessa, (c. 685-785), who concluded his career by working for Caliph al-Mahdi (775-85). I shall therefore consider texts from the Greek-speaking world, together with works in Arabic, which tend to be written immediately after the foundation of Baghdad, but which, as Pingree (1997) argued, are a summation of astrology prior to its foundation.

The Horoscope Data

The horoscope data was given by al-Yaqubi (Pingree 1970, 104) in the 9th century and al-Biruni (1879, 262) in the tenth-eleventh. However, there is a contradiction in the data. Al-Yaqubi listed the planetary positions on the basis of information from the astrologer al-Fazari. I recalculated the data using Solar Fire software, and the positions of the sun and moon indicate a time of 23.37 LMT on 30 July 762 and an ascendant (rising degree) of 20;45 Taurus (see Figure 1).

	1.Al-Fazari	2.Calc	3.Al-Biruni	4.Calc
Saturn	26;40 Ar. Ret.	1;10 Ta	26;40 Ar. Ret.	1;27 Taurus
Jupiter	Sag.	8;48 Sag	Sag.	8;47 Sag. Ret.
Mars	2;50 Ge.	6;31 Ge.	2;50 Ge.	6;55 Ge.
Sun	8;10 Leo	10;11 Leo	8; 10 Leo	10;47 Leo
Venus	29;0 Ge.	2;01 Can.	29; 0 Ge.	2;44 Can.
Mercury	25;7 Can.	26;59 Can Ret.	25; 7 Can.	26;54 Can. Ret.
Moon	19;10 Lib.	19;10 Lib.	19; 10 Lib	27;37 Lib
North Node		28;48 Cap	25; Cap	28;47 Cap
Ascendant		20;45 Ta.	Sag	6;09 Sag

Figure 1. The horoscope of Baghdad. 1. Planetary positions from al-Fazari, recorded by al-Yaqubi. 2. Planetary positions from al-Fazari recomputed for 23.37 LMT 30 July. 3. Planetary positions from al-Biruni's horoscope given by Sachau. 4. Planetary positions computed for the presumed ascendant of 6⁰ Sagittarius.

Al-Biruni reproduced what Pingree assumed was Naubakht's horoscope for the city, giving it an ascendant in Sagittarius. This, though, is inconsistent with al-Yaqubi's data, for Sagittarius was rising on the afternoon of 31 July. The horoscope from al-Biruni's manuscript reproduced by Allawi appears to contain no degree positions for either planets or ascendant (Adams, 2011). The printed chart diagram created for Sachau's English translation of al-Biruni reproduced al-Yaqubi's degree positions for all the planets except Jupiter (for which al-Yaqubi records no position), but not for the ascendant (Figure 2). A tradition then arose, the origin of which I have not been able to trace, that the ascendant was 6⁰ Sagittarius. Allawi assumed that this was correct, and it occurs elsewhere in the modern literature. James H. Holden uses this position and assumes a time and date of 14.40 LMT on 31 July 762, on the basis that, although 'No cuspal degrees are given... we can assume that the astrologers chose the time when Jupiter was in the ASC degree' (Holden, n.d.). The correct calculated time for an ascendant of 6⁰ Sagittarius is 14.28 LMT. The result of this uncertainty is confusion even in the date of the city's inauguration; Pingree assumed 30 July, but Allawi and Holden assume 31 July. This makes it difficult to retrospectively reconstruct the intentions of al-Mansur's committee of astrologers with any certainty. However, I shall examine what the horoscopes may indicate, referring to the al-Fazari data as AF, and the al-Biruni data as AB.

Al-Biruni himself did not analyse the horoscope and gave only the following general commentary:

Winterly air on sea, winds (Philippus and Metrodorus); beginning of the blowing of the Etesian winds (Egyptians). On this day 'Abu Ja'far Almansur began to build Baghdad, that party which is called *Mansur's town*, on the western side of the Tigris in the present

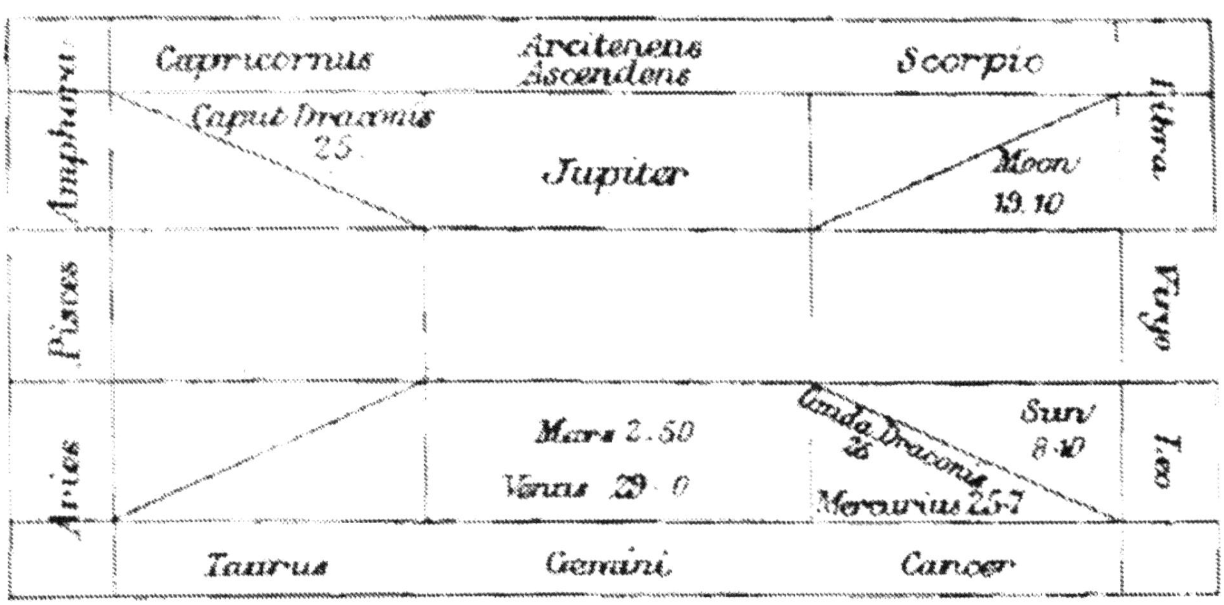

Figure 2. The printed horoscope chart from Sachau's 1879 English Translation of al-Biruni's Chronology of Ancient Nations.

Baghdad. This was A. Alexandri, 1074 [762 AD]. Astrologers are obliged to know dates like this, and must date from such an epoch by means of their knowledge of the Permutations, Terminations, Cycles, and Directions, until they find the horoscopes of those people who were born at those times. It was Naubakht who determined the time (for the commencement of building). The constellation which heaven showed at the time, and the stations of the planets which appeared on heaven, were such as are indicated in the following figure (1879, 262).

Note that he specified that a precise time for the inauguration was chosen, confirming that an ascending degree was specified, but that he records neither a time nor ascending degree.

Allawi's Analysis

In Allawi's view, the crucial determining fact for the horoscope for Baghdad was that the ascendant was set for 6^0 Sagittarius. This is equivalent to 246^0 of celestial longitude which Allawi believes was significant because it represents the terrestrial longitude of Baghdad if the mythical city of Kandakes, supposedly located in eastern China, is taken as 0^0. I make no comment on this argument. Aside from this, Allawi writes that 'it is not easy to comprehend the astrological scheme underlying the horoscope at a glance' (1988, 62). However, the astrological scheme might be understood by consulting the astrological handbooks of the time.

Analysis of the Horoscope

There are abundant textual sources for eight-century astrology, but few indications of the specific rules that might have been implemented in the foundation of a city. The astrologer Sahl ibn Bishr al-Israili (c. 786-845?), did include a section in his work 'On Elections' titled 'If you wished to elect so that you might build a house...' (Sahl 2008, 199). He recommended taking the relevant factors in the following order of importance: the Moon, the planet ruling the sign containing the Moon, the ascendant, the planet ruling the sign containing the ascendant, the lot of fortune and Mercury. He then argued that Mars should be avoided with the qualification that, if he is present, then he should be under the sway of a strong Venus. He concluded with a string of increasingly minor and more intricate rules. However, given the uncertainty of the horoscope data, I need to start with the obvious common factor in both horoscopes, planetary locations in the zodiac signs rather than the houses (or 'domiciles' from the Greek, which are based on the diurnal cycle): except for the moon, the planets move only fractionally through the zodiac signs in one day but pass through all twelve houses. Thus their sign positions in the AF and AB charts are the same, but their house positions radically different.

To begin with Sahl's criteria, the moon is in Libra, a location where, taking all factors into account, it is strong: in particular it is in the sign ruled by Venus, a fortunate planet (Ptolemy 1940, 1.5, 1.17). Ignoring the ascendant, for the moment, as it is the cause of uncertainty, the lot of Fortune is not specified, so I will not consider it. Mercury, next in the list, is in Cancer, the sign ruled by the Moon and so, indirectly also, drawn under Venus' benevolent sway (Ptolemy 1940, 1.17). Mars, a destructive planet (Ptolemy 1940, 1.5), meanwhile is not avoided, as Sahl advised, but is in a close opposition to Jupiter, a difficult alignment (Ptolemy 1940, 1.13), a point to which I shall return. Mars is, though, in Gemini, a sign ruled by Mercury (Ptolemy 1940, 1.17) and so, via Mercury and the Moon, drawn under Venus. Venus itself is in Gemini, ruled by Mercury, so it does seem as if the astrologers may have been attempting to create a system of interlocking relationships between the planets. Saturn, being in Aries, a signed ruled by Mars (Ptolemy 1940, 1.17), is therefore also drawn into this scheme. The sequence of relationships could be considered a form of reception, a means of bringing the planets into mutually supportive relationships (Masha'allah 2008a, 439).

Only two planets are not part of this system: Jupiter is in Sagittarius, a sign which it rules, and the Sun is in Leo, which it rules (Ptolemy 1940, 1.17). Both planets therefore possess a kind of autonomy, and my suspicion is that these are the two most important factors in the chart. The Sun is in its own sign of Leo, which was characterised by Hephaistio of Thebes (1994, I.1) as 'kingly': hence we can assume it represents al-Mansur himself. The Sun itself was described by Vettius Valens (1993, I.1) in the following terms:

the all-seeing Sun, then, being truly fire-like and the light of the mind, the organ of perception of the soul, is significant at birth for a kingly office, hegemony, mind, practical wisdom, outward form, motion, height of fortune, public registration, action, popular leadership, judgment, father, mastership, friendship, persons of high repute, the honors of images, statues and crowns of office, ach-priests of the fatherland.

At 10^0 Leo, the Sun had just entered the sign's second decan, for which Hephaistio of Thebes (1994, I.1) gave the following reading:

he who is brought forth upon the second decan will be a master, a king; he will be declared a plenipotentiary, and he will subject many and found cities, and many will pay tribute to him, and there will be knowledge of god in him, and the discovery of gold, and his mind will be inspired by him, and his honor equal to the gods.

There was a tradition in the classical world, expressed by Julius Firmicus Maternus in the 4th century, that the emperor was not subject to the stars (1975, II.XXX.4-5). However, if he is not subject to them, heaven at least indicates that al-Mansur's position was powerful.

Jupiter, meanwhile, was considered the most benevolent of the planets (Ptolemy 1940, I.5) and its benign role was strengthened due to its rulership of Sagittarius (Ptolemy 1940, I.17), the sign containing it. In the Babylonian tradition, Jupiter appears to have been benevolent for kingship: if the planet was present in an eclipse then it signified, in one text, 'well being for the king' (Hunger 1992, 535). Another recorded that, if Jupiter was bright, the king would prevail (Hunger 1992, 489). Also significant was that that the Sun and Jupiter are in a trine (separated by three zodiac signs), which brings them into a harmonious relationship (Ptolemy 1940, I.13,18). They are also both in their own signs within the fiery triangle (Aries-Leo-Sagittarius), which the Sun rules by day (which is relevant if the AB chart is correct) (Ptolemy 1940, I.17).

So far then, we have considered only general factors. The Sun was in Leo for an entire month and in a close angular relationship with Jupiter for a few days: we cannot say how close a relationship al-Mansur's astrologers required the planets to be in, but some quality about the days 30-31 July must have been significant.

We need to consider, though, the AB horoscope, with its Sagittarius ascendant (the ascendant came high on Sahl's list of criteria for choosing the time to erect a building), which, as I argued, is entirely inconsistent with the planetary (especially lunar) data given by al-Yaqubi. There is a clear problem. However, to place the ascendant in Sagittarius also places Jupiter, being in the same sign, close to the ascendant and in the first house. For Masha'allah, this could not be better: 'Jupiter, if he were in the Ascendant, signifies reverence and beauty and the honesty of faith and religion, of teaching also, and reason, and the end of those things in which is involved, will be towards salvation, and sovereignty will be handed to him in this' (Masha'allah 2008b, 507). From the classical literature, we find the following from the fourth-century astrologer Julius Firmicus Maternus:

if Jupiter is found exactly on the first angle, that is, on the ascendant, especially in those signs in which he rejoices or in his own house or terms or in his exaltation, he will make the native high-born, famous, always ruling great states, perhaps the first ten of the great states. He will also make him... virtuous, charming, benevolent, cheerful, rich, especially if Jupiter is in this house by day...' (1975, III.3.1).

Again, this is a powerfully benevolent indication for al-Mansur. Further, a Sagittarian ascendant would place the Sun in the ninth house which, in Masha'allah's view, signified 'divine culture faith and the fear of God the Highest, and the remembrance of angels' (2008b, p. 499). Going back to the classical tradition, Firmicus (1975, III.V.33) wrote that

Natives who have the Sun in the ninth house make images of the gods or worship divine images; or they decorate the gods, design their temples or write hymns and prizes of the gods. They make ostentatious display of holy songs and gain fame and honor from this activity.

He also wrote that 'If the Sun is on the ascendant by day, Jupiter on any angle... the native, although not a lord or king, will have power throughout the whole world' (1975, III.V.1). Now, the Sun was not on the ascendant, but Jupiter was on an angle (the ascendant), so his reading gives us something of the flavour of the pattern.

The AB chart, with its Sagittarian ascendant, is quite clearly so benevolent while being at odds with the data in the AF chart that it is clearly possible that it is rooted in a tradition which was already current when Al-Biruni wrote. We may point to two solutions to the problem.

1. The AB horoscope might have been retrospectively rectified and attributed to Naubakht, that is, a new time and date assumed on the basis that Sagittarius 'should' have been ascending when Baghdad was inaugurated, moving the data from 30 July, as chosen by al-Mansur's committee, to 31 July.

2.There was never any intention that the ascendant and the planetary data in the AB horoscope should correspond. This may be a heretical suggestion, but it was already accepted that the ascendant, as the starting point of a twelve-house system, need not be astronomically correct. For example, it was possible to use the Lot of Fortune, calculated on the basis of the Sun, Moon and astronomically-correct ascendant, as the basis of a twelve house system known as the athla (Manilius 1977, 3.43-159), with the Lot of Fortune functioning as a derived ascendant. Thus Allawi may be correct that the ascendant was 6^0 Sagittarius. However, it may not have been the case that al-Mansur's astrologers chose a corresponding time of 14.28 LMT 31 July, but that a derived, rather than astronomically-correct, house system based on 6^0 Sagittarius, was imposed on plants calculated for 23.37 LMT on 31 July. This is indeed a complicated solution, but not an impossible one.

We should also bear in mind a third option; that the Sagittarius-ascendant horoscope presented by al-Biruni is genuinely that cast by Naubakht, but that the planetary positions, especially that of the Moon, were miscalculated in the original. In that case, though, we have even more reason to consider that benevolent Sun-Jupiter relationship as crucial. There is, though, one further problem with both the AB and AF charts: Mars, a naturally malefic planet (Ptolemy 1940, 1.5), was in a close and inauspicious opposition to Jupiter. Worse, in the AB chart Mars was on the descending angle (in the seventh house), strengthening its destructive power rather than, as Sahl was to recommend, minimising it. This is what Masha'allah had to say about Mars in the seventh house (2008b, 510): 'business dealing and contention and sexual intercourse with injustice, and harm on account of this, and disarray in every matter, and contention and horribly [so], and grief because of them'. Firmicus (1975, III.4.17) had considered that the same position 'indicates the greatest evils and enormous perils'. He also suggested that 'Jupiter and Mars in opposition indicate an unstable life and continual loss of possessions. Hostility of friends is stirred up, and there is the greatest danger from riots and rash activities' (1975, VI.VXI.1). What seemed profoundly auspicious (Jupiter in Sagittarius in the first house trine Sun in Leo in the ninth house) now appears to carry equally profound dangers. A resolution to this conundrum has to be found.

One possible solution is suggested by the astrological allocation of Zoroastrian millennia to planets. According to Masha'allah, the tenth of the twelve one thousand year periods in the life span of the Persian world was ruled by Mars (Kennedy and Pingree 1971, 72-3). Pingree and Kennedy, noting that the births of both Christ and Mohammed occur in a Martian period, find this unsatisfactory and argue that the better option would have been the Sun (Kennedy and Pingree 1971, 74-5). However, there is no problem: all the evidence from medieval and Renaissance Europe, where Masha'allah's scheme was adopted with enthusiasm, indicates that the planets' rulerships of historical periods are not associated with their inherently benign or malign natures, which are disregarded for this purpose (Campion 1994, 354-9).

A good precedent was set by the emperor Augustus when he promoted the zodiac sign in which the Moon was located at his birth, Capricorn, for propaganda reasons. The Moon is weak in Capricorn—Masha'allah (2008c, 371) referred to the position as 'evil'—which is ruled by the planet Saturn, which was considered to be destructive (Ptolemy 1940, 1.5). It appears that the emperor was exploiting an alternate tradition in which Saturn, as the Greek Kronos, was ruler of the Golden Age (Campion 2009, 234). The strict rule which designated planets weak or strong depending on their zodiac sign was therefore trumped by a historical imperative. Thus, if we adopt a millenarian Zoroastrian scheme, rather than a horoscopic one, the threatening qualities of the Jupiter-Mars opposition may have been replaced by a neutral indication that the foundation of Baghdad conformed to a divine, historical plan. This, at any rate, is a suitable solution.

Conclusion

Clive Ruggles (2005, 23) wrote that,

From ancient Babylon to China, Mexico and Peru, from empire and city state to tribe, astronomical information was gathered and recorded and use by those interests lay as much in the spheres of status enforcement and political ideology as in predicting rainy seasons or planning agricultural schedules.

In the case of Baghdad we have a prime example of the encoding of astronomical data in space, through the structure of the city and through time in the moment chosen for its inauguration. That we have doubts over the data used for this moment does not change the principle. The city was intended to create a physical link between heaven and earth, to synchronise human actions with God's will as expressed in the unfolding of planetary patterns.

Although the horoscopic data for the foundation of Baghdad is inconsistent, and unresolved questions remain, a good argument can be made that it was designed to bring the power of the celestial spheres and of the divine plan of history to bear on the new city. The purpose may also have been to actively play a role in ensuring the city's success. As the pseudo-Ptolemaic Centiloquium (probably composed in eighth-century Persia) put it: 'One skilful in this Science may evade many effects of the Stars, when he knows their Natures, and diligently prepares himself to receive their effects' (Ptolemy 1679, 305). I would therefore like to suggest that we can identify different perspectives on the use of astronomy in relation to power: (1) to harmonise with the divine plan as expressed through celestial cycles, (2) to call down the blessing of God through such harmonisation, (3) to gain power through correct alignment with history and (4) to gain the knowledge necessary to actively manipulate fate through correct behaviour.

David Harvey (1989, 175) has drawn attention to the fundamental importance of time-space relations to social relations, but argues that, in conventional urban studies, time is prioritised. This is particularly clear in the case of Baghdad, in which its astronomical qualities were encoded in the time for its foundation. As Vincent Berdoulay (1989, 135) wrote, 'a place comes explicitly into being in the discourse of its inhabitants, and particularly in the rhetoric it promotes. Thus the geographers' discourse uses the same ways as the people who define their own place'. In these terms, astrology was the discourse used by the founders of Baghdad.

Bibliography

Adams, Ben. 2011. Personal communication 27 January 2011.

Al-Biruni. 1879. The Chronology of Ancient Nations, trans. C. E. Sachau from the Arabic text 'Athar al-Baqiyah' or 'Vestiges of the Past' (1879). London, W. H. Allen.

Allawi, Ibrahim. 1988. Some Evolutionary and Cosmological Aspects to Early Islamic Town Planning. In Margaret Bentley Sevcenko (ed.), Theories and Principles of Design in the Architecture of Islamic Societies. Cambridge, Massachusetts, Aga Khan Program for Islamic Architecture. Available at http://archnet.org/library/documents/one-document.jsp?document_id=3776 (Accessed 4 July 2010).

Berdoulay, Vincent. 1989. Place, Meaning and Discourse in French Language Geography. In John A. Agnew and James S. Duncan (eds.), The Power of Place. London, Unwin Hyman.

Campion, Nicholas. 1994. The Great Year: Astrology, Millenarianism and History in the Western Tradition. London, Penguin.

Campion, Nicholas. 2009. A History of Western Astrology, Vol. 1. The Ancient World. London, Continuum.

Cramer, Frederick H. 1996. Astrology in Roman Law and Politics. Chicago, Ares Publishers.

Gramsci, Antonio. 1971. Selections from the Prison Notebooks. London, Lawrence and Wishart.

Harvey, David. 1989. The Urban Experience. Baltimore, Maryland, Johns Hopkins University Press.

Hephaistio of Thebes. 1994. Apotelesmatics, 2 vols., trans. Robert Schmidt (1994). Berkeley Springs WV, Golden Hind Press.

James H. Holden. (no date). The Foundation Chart of Baghdad, (online). Available at http://cura.free.fr/xxv/25hold3.html (Accessed 28 Jan 2011).

Hunger, Hermann. 1992. Astrological Reports to Assyrian Kings. Helsinki, Helsinki University Press.

Kennedy, E. S. and Pingree, David. The Astrological History of Masha'allah. Cambridge, Massachusetts, Harvard University Press.

Manilius, Marcus. Astronomica, trans. G. P. Goold (1997). Cambridge, Massachusetts and London, Harvard University Press.

Masha'allah. 2008a. On Reception. In Sahl and Masha'allah, Works, translated by Benjamin Dykes, 437-97. Golden Valley, New Mexico, Cazimi Press.

Masha'allah. 2008b. The Planets in the Twelve Domiciles. In Sahl and Masha'allah, Works, translated by Benjamin Dykes, 499-513. Golden Valley, New Mexico, Cazimi Press.

Masha'allah. 2008c. Signification of the Planets in a Nativity. In Sahl and Masha'allah, Works, translated by Benjamin Dykes, 361-90. Golden Valley, New Mexico, Cazimi Press.

Maternus, Julius Firmicus. Mathesis. Translated as Ancient Astrology: Theory and Practice, Jean Rhys Bram (1975). Park Ridge, New Jersey, Noyes Press.

McCluskey, Stephen. 1993. Native American Cosmologies. In Norris Hetherington (ed.), Encyclopaedia of Cosmology. New York and London, Garland Publishing

Pingree, David. 1970. The Fragments of the Works of al-Fazari. Journal of Near Eastern Studies 29.2, 103-123.

Pingree, David. 1977. The Horoscope of Constantinople. Prismata, Festchrift fur Willy Hartner, 305-15. Wiesbaden: Franz Steiner Verlag.

Pingree, David. 1997. From Astral Omens to Astrology: From Babylon to Bikaner. Rome, Istituto Italiano per l'Africa et l'Oriente.

Pingree, David. 1989. Classical and Byzantine Astrology in Sassanian Persia. Dumbarton Oaks Papers 43, 227-239.

Ptolemy, Claudius. 1679. Centiloquium. In John Partridge, Mikropanastron, or an Astrological Vade Mecum, briefly Teaching the whole Art of Astrology - viz., Questions, Nativities, with all its parts, and the whole Doctrine of Elections never so comprised nor compiled before, &c., 305-321. William Bromwich, London.

Ptolemy, Claudius. Tetrabiblos, trans. F. E. Robbins (1940). Cambridge, Mass., London, Harvard University Press.

Renfrew, Colin. 2008. Prehistory: the making of the human mind. Oxford, Oxbow Books.

Ruggles, Clive. 1997. Astronomy and Stonehenge. In Barry Cunliffe and Colin Renfrew (eds.), Science and Stonehenge, 203-229. Oxford, Oxford University Press.

Ruggles, Clive. 2005. Ancient Astronomy: An Encyclopaedia of Cosmologies and Myth. Santa Barbara, California, ABC CLIO.

Ruggles, Clive and Saunders, Nicholas. 1993. The Study of Cultural Astronomy. In Clive Ruggles and Nicholas Saunders (eds.), Astronomies and Cultures, 1-31. Niwot, California, University of Colorado Press.

Sahl bin Bishr. 2008. On Elections. In Sahl and Masha'allah, Works, translated by Benjamin Dykes, 187-222. Golden Valley, New Mexico, Cazimi Press.

Short, John Rennie. 2001. Alternative Geographies: from Cosmography to Geography. In Philip P. Arnold and Ann Grodzins Gold (eds.), Sacred Landscapes and Cultural Politics, 27-34. Aldershot, Ashgate.

Sudjic, Deyan. 2005. The Edifice Complex: How the Rich and Powerful Shape the World. London, Allen Lane.

Valens, Vettius. 1993. The Anthology, Book I, trans. Robert Schmidt (1993). Berkeley Springs, Virginia, Golden Hind Press.

Wheatley, Paul. 1971. The Pivot of the Four Quarters: A Preliminary Enquiry into the Origin and Character of the Ancient Chinese City. Edinburgh, Edinburgh University Press.

Wiet, Gaston. 1971. Baghdad; metropolis of the Abbasid caliphate. Translated by Seymour Feiler. Norman, University of Oklahoma Press.

ASTRONOMY AND POLITICS: THREE CASE STUDIES ON THE SERVICE OF ASTROLOGY TO SOCIETY

S. MOHAMMAD MOZAFFARI

Abstract: This paper is based on the assumption that the 'egocentric perspective' of the human mind acted as an efficient psychological factor for evolving astrological doctrines. We adopt a psychological approach to studying the relation between historical-social events and the astrological doctrines of the Medieval period. The main focus of this study deals with the three celestial phenomena observed in the Middle East from the beginning of the 14th century to that of the 15th, whose astrologically interpretations or prognostications had political, social and historical effects. The examples have been selected from primary historical sources. They are Comet 1402 D1, the 1299 AD solar conjunction of Mars, and the unordinary conjunctions between Mars and Saturn. Considering them in various contexts, it is shown how those phenomena were the cause of anthropocentric historical events such as wars, or identified as the justifier of social or natural accidents like epidemics. The explanation of the role astrology played in such events in particular, and its impact upon societies in general, must be sought in the psychological effect caused by astrological ideas on the human mind and its relation with the physical world. We conclude that as the human egocentric mind was a main cause of the formation of astrology, astrology itself penetrated into the consciousness of the human mind and objectively appeared in the physical world, and therefore guided history in a specific direction. It is illustrated that this effect was so extended as to give birth to astrological history. In addition, having classified the four ways of interpreting celestial phenomena in the ancient-medieval period (meteorological/weather prognostication, natural philosophical, meteorological/atmospheric, astrological), we briefly explore the situation of astrological dogma in Islam and the methodological distinction of astrology from astronomy by Avicenna in that period.
Keywords: Astrology, Astrological History, Comet 1402 D1, Egocentrism, History of Astronomy, Planetary Conjunctions, Sociology of Astronomy, Maragha Observatory, Tamerlane, Ilkhanid Dynasty

> It is incredible how useful belief in astrology has been to humanity. If Kepler and Tycho Brahe made a living, it was because they sold to naïve kings predictions founded on the conjunctions of the stars. If these princes had not been so credulous, we should perhaps still believe that nature obeys caprice, and we should still wallow in ignorance. Henry Poincaré (1913, 296-7)

Introduction

To begin with, the Aristotelian background of the ancient–medieval intellectual system provided a stable, unchangeable and harmonic cosmos wherein any phenomenon had been known as *a priori*. Yet, a more ancient idea was the direct effect of celestial phenomena on terrestrial incidents, which echoes the 'egocentric' perspective that appears to be a principle characteristic of human mental activities throughout the everyday activities of humans (see Richardson et al. 2010, especially 309–311). The same perspective, as a striking example, was also responsible for the tendency of man to resist for ages every idea rejecting the 'geocentric' cosmos. One can simply perceive that primary simple notions, like the influence of the heavens on the Earth, emerged from the egocentric viewpoint of the human mind. Finally, they evolved through time and became embedded in astrological doctrines. Similar psychological considerations focusing on the interrelation of the human mind and the physical world may be adopted as a criterion to explain the continuous survival of astrological dogmas. In addition, they can also be employed for the sake of interpreting the practical usage of them in any social context. Looking at issue from such a perspective, the purpose of this study was to investigate three important cases in which the astrological interpretation or prediction of celestial phenomena played a considerable role influencing historical events. We believe that these examples are appropriate for showing how and to what extent astrological doctrines (as matters arising from the intrinsic part of the human mentality) could efficiently be engaged in social and historical mechanisms. Moreover, we are also motivated by the reason that they belong to the late Islamic period (the 10th to 15th centuries) when not only astrological opinions were treated as heresies, but also when a methodological distinction between astronomy and astrology had already been crystallized in its philosophical system (as evidenced later in this paper).

This paper tried to highlight several main issues. First, a short classification of the quasi-astrological and astrological ideas is introduced. Second, a passing allusion to the situation of astrology with regard to orthodox religious opinion in the period in question is reviewed. Third, the rest of the paper is devoted to three case studies which constitute the bulk of the present study. Finally, we will discuss the case studies in light of a psychological interpretation of the impact of astrological doctrines upon history.

It is worth noting that to carry out research on a historical topic in its own context, especially when it accompanies a psycho-sociological approach, we are likely influenced by the problem of anachronistic prejudice. For this reason, we have avoided the science/pseudo-science boundary which is prevalent in the modern astronomical history studies we refer to in this work.

What is 'astrological'?

To take into account historical matters, not all of the celestial phenomena that were assumed to have some effect on the terrestrial domain were interpreted or treated solely 'astrologically' up to the late Hellenistic period. To clarify this point, some main variants are classified briefly as following:

1. Some celestial 'arrangements' were used for weather prognostication (e.g., predicting the direction of winds or storms). Examples of such arrangements include the inclination of the line passing over centers of the Moon and the Sun from the ecliptic during an eclipse and the direction marked on the horizon by the lunar latitude-line,

by the moon's sickle or by the comets' tails. Therefore, these belong to 'meteorological' (in the modern sense of word) theory, not to astrological doctrines (Neugebauer 1975, 141-142).

2. Sometimes, the belief in the effect of the celestial phenomena on the terrestrial domain was rooted in Aristotelian 'Natural Philosophy'. In this light, the importance of the solar and lunar eclipses became apparent. Consequently, a great deal of research has been done in this realm to exactly predict the eclipses. The attempts and insights of such astronomer/astrologers resulted in establishing observational programs, geometrically modeling the luminaries' motions, and finally developing astronomy. Hipparchus (second century BC) and Ptolemy (second century AD) are the best-known examples. The first claimed that he had attained an accuracy of about ±2 digits (one digit = one-twelfth part of the luminaries' apparent diameters) in the prediction of the eclipse's magnitude (Neugebauer 1975, 667). The main motive for this interest can be attributed to the high-ranking role given to the Sun and the Moon in Aristotelian philosophy for their effects on the continual stream of 'generation and corruption' in the sublunary domain. For instance, the motion of the Sun in its declination was known as the 'principal cause' of generation and corruption (Aristotle, On Generation and Corruption, 336a35, in Tricot 1971, 141), or the Sun was a 'moving cause' for generating humankind (Aristotle, Physics, 194b13, in Apostle 1969, 29, 209). Accordingly, the parameters of every eclipse (e.g., duration, magnitude, phases, etc.) must be precisely calculated for the sake of confidently predicting its physical effects on human beings.

3. Another important point is the sublunary (atmospheric or 'meteorological', in the historical sense) phenomena such as comets, shooting stars and so on. Again, according to Aristotelian philosophical tradition the heavens were incorruptible. Hence, all instantaneous appearances, such as comets, had to be attributed to the terrestrial world. The comets, having inauspicious outcomes, were the symbol of the widespread changes in the normal course of the life. The comet 1402 D1 may be an appropriate instance to show the important political effect of a comet as a celestial phenomenon on human life (see Case 1).

4. The last group consists of 'astrologically' important aspects between the planets, i.e., conjunctions, oppositions and so on; these will be our main concern (see Cases 2 and 3).

A quick glimpse at the above four sorts of celestial-terrestrial interactions revealed noticeable differences among them. However, they manifested altogether and were mostly treated as an integrated body of astrology up to the late medieval period. It is essential to add that the two latter sorts were more the focus of attention. Also the second and the fourth used the astronomical tools for determining the ecliptical coordinates of the planets (e.g., those used in horoscopes) and caused mathematical procedures to be important for providing solutions for their problems. The first case lost its importance in the medieval period.

Astrology and Islam

Astrology was soon being spread to other communities and nations, each holding various and different 'ideologies' (in the anthropological sense). Since astrological doctrines gave rise to serious challenges to the fundamental tenets of every monotheistic religion, there was deep conflict over accepting the astrological prognostications in both Islam and Christianity (for Islam, see Saliba 1995, 53ff). On the one hand, the controversies between Islam and astrology began very early. The orthodoxy constitutes a main part of the core of the ideology of every community formed based on a specific religion. On the other hand, Islam, becoming a 'knowledge'-based intellectual system (see: Rosenthal, 2007), at least for the sake of itself, had to put astronomy in its own service. This made the Muslim jurists draw a virtual distinction between astronomy and astrology. For the same reasons, very soon similar controversies arose among philosophers who assumed themselves to be the polar point of the intellectual sphere of the day. Accordingly, the jurists and the philosophers began to write a good number of extensively severe critical treatises against astrology.

Nevertheless, for the first time, the Iranian philosopher Avicenna (980-1037 AD), in his influential work *Treatise on the classification of the rational science*, highlighted the 'methodological' difference between 'astronomy' and 'astrology': 'the first is 'verifiable', but the second, purely 'conjectural'. (Avicenna 1986, 88; Gutas 2003, 148). As a result, astrology, which was not known to be so differentiated from astronomy throughout the Greek and Hellenistic periods, now became distinguished from astronomy and needed to be redefined. Such a rigid religious-philosophical distinction, however, did not change anything in the realm of astronomy. Thus, astrology received a new definition and that methodological difference gives it a new title. It was no longer the *ᶜIlm*, 'Knowledge/Science', but *Ṣināᶜa*, 'Art' (our account here is considerably different from that of Saliba 1995, 66ff). In Islamic medieval astronomy, the astronomers who were working on 'observational astronomy' were called *Aṣḥāb al-irṣād* (lit., possessors of the observations); those who was interested in making geometrical planetary models ('physical astronomy') were called *Aṣḥāb al-hay'a* (lit., possessors of the configuration [of the planetary spheres]); those who were working on 'mathematical astronomy' were called *Aṣḥāb al-zījāt* (lit., possessors of the astronomical tables), and so on. The astrologers now were called *Aṣḥāb al-Ṣināᶜa*. These terms are distributed throughout the Islamic treatises; see *Ṣināᶜa* in the title of many Islamic treatises, e.g., that of Abū Muḥāmid Ghaznawī; *Aṣḥāb al-irṣād* in Shīrāzī 1, fol. 50r and Shīrāzī 2, fol. 38v; *Aṣḥāb al-zījāt* in al-Wābkanawī's *Zīj*, III, 5.4 or *Aṣḥāb al-Ṣināᶜa* in III, 7.1.) *Ṣināᶜa* was even applied sometimes as the name for the whole of 'astronomy' itself (e.g., al-Wābkanawī, III, 1.2.1 and III, 13.14).

That methodological difference was also transferred to the West and became so prevalent in Europe that the extensive references to the Latin synonyms for *Ṣināᶜa* /Art may be found in the treatises written during the late medieval period. As a result, in Christianity of the late medieval period, there is a ready solution for the problem. However, Gregory Chioniades (c. 1240 - c. 1320 AD) may yet be an amazing instance: he spent some years in Tabriz, Iran, studying the texts of astronomy and translated some of them into Greek (see Pingree, 1985). After returning to Byzantium, he had to write a treatise entitled *Profession in Faith*, apparently in defense of himself (see Westerink, 1980).

Astrology in the Service of Society

Our aim is to show how belief in the effect of celestial phenomena on 'anthropocentric historical events'—like wars, political decisions, etc.—can have various and specific kinds of functional application, or how it took the role of 'justifying' happenings that were out of humankind's power, e.g., floods, epidemics and so on.

In what follows, we present three historical cases related to Iranian history to show how some astronomical phenomena (astrologically interpreted) may have the considerable political or social implications. These historical cases are comet 1402 D1, the 1299 AD conjunction of Mars with the Sun, and the unordinary extensive conjunctions of Saturn with Mars at the turn of the 14th century AD. Interestingly, these phenomena, either directly or indirectly, are in relation with two periods of scientific activity at the Maragha Observatory (1259-1283 AD and 1283 - c. 1320 AD, respectively), which had been founded by the Ilkhanid dynasty of Iran (1265-1344 AD).

CASE 1: COMET 1402 D1

In general, two Persian sources refer to the observation of comet 1402 D1. One is Khāndmīr's *Ḥabīb al-Sīyar* (1954, written c. 1523-4 AD), and the other is an astrological work called *Latā'if al-Kalām fī Aḥkām al-Aᶜwām* by Sayyid Muḥammad the Astrologer (written after March 1421 AD in Lāhīdjān, north of Iran).

We should keep in mind that Muḥyī al-Dīn Al-Maghribī (d. June 1283 AD), was the illustrious astronomer of the first period of the Maragha Observatory. He wrote at least 26 astronomical treatises of which 10 treatises are on astrology. For further reading see Sarton (1953, Vol. 2, part 2: 1015-16), Sezgin (1978, Vol. 6, 292), Suter (1902, 154-155), Tekeli (1980), Dorcé (2002-3) and Mozaffari (2007).

Khāndmīr says about Muḥyī al-Dīn that he 'had preference over the modern and ancient astronomers in the field of astrology. Some works of his are famous among astronomers and some of his own astrological commandments, *aḥkām*, which in reality have happened are rumoured' (Khāndmīr 1954, 3, 108). Of those, he points to the military expedition of Tamerlane (1336-1415 AD) to *Rūm* in Asia Minor, which was then under the reign of Īldrum Bāyazīd (known as Bayezid I, 1354-1403 AD), the first Ottoman supporter.

This happened in 804 AH / 1402 AD, when Tamerlane was feeling danger from *Rūm* and thus consulted with his own courtier's astronomers and astrologers about making a war against Ottomans. They severely prohibited him to do so. Only Shams al-Dīn al-Mālighī encouraged him to do so, and after confronting his question 'how do you explain the astronomers' remark?', Shams al-Dīn asked for help from Shahāb al-Dīn ᶜAbd-Allāh al-Lisān who, Khāndmīr says, was foremost among the learned wise men in his own era. This time was contemporary with appearance of the comet 1402 D1, which was observed for the first time on 8 February 1402. Khāndmīr says it 'appeared from the west at the end of the day when setting of twilight, and after some days rose from the east when morning.' In mid-March, it entered solar conjunction and there were reports that it was a daylight object for eight days. In total, the comet was visible for 70 days. It reached to its own maximum brightness on 12 March (Yeomans, 2007; see also Seargent 2009, 99-102.) Shahāb al-Dīn ᶜAbd-Allāh came to Tamerlane's court with one of the Muḥyī al-Dīn's works, on which he based his announcement that 'the appearance of comet in the [sign] Aries is a symbol meaning that an army will come from the east and will dominate on *Rūm* and will imprison the *Rūm*'s patron.' Based on this astrological prognostication, Tamerlane determined to continue and on 12 March, when the comet had reached its maximum brightness, decided to start a fight that ended in his ultimate victory (Khāndmīr 1954, 3, 505).

In contrast, Sayyid Muḥammad the Astrologer says: 'any great man who has seen them [the comets] has said that they are the symbol of the inauspiciousness [...]. At 803 AH [1401 AD] I have seen a comet which was appeared on the zenith of *Rūm*. Emir Tiymūr [Tamerlane] made a military expedition with that clan [Ottoman Turks] and imprisoned Īldrum Bāyazīd and hold him in disrespect' (Sayyid Muḥammad, 214). Indeed, the date is one year out.

Tamerlane held the wretched Īldrum Bāyazīd in a cage of iron until he died there on 8 March 1403 AD. This war is referred to as the most important political event in the Middle East of the 15th century, especially for its negative effect on the establishment of the Ottoman Empire. However, what has been neglected is that some preliminary astrological considerations created an appropriate atmosphere and provided the needed social energy for doing so.

CASE 2: MARS AS THE SYMBOL OF WAR

Ghāzān Khān, the seventh patron of the Ilkhanid dynasty of Iran (28 October 1295–25 May 1304), campaigned two times to Asia Minor and Egypt, which was under the reign of the Mamlūks. It is noteworthy that he was an expert in various sciences. He invented twelve 'new' observational instruments in the second period of the Maragha Observatory; see Mozaffari and Zotti (2009) and Zotti and Mozaffari (2010). As his vizier Rashīd al-Dīn (d. 1318 AD) says, during the first campaign he moved with his army from Tabriz on Friday, 19 Muḥarram 699 AH / 16 October 1299 AD (Rashīd al-Dīn 1994, Vol. 2, 1290). One century later, Mīrkhānd (1434-1497 AD) in his own *History* dated the time of departure as Sunday, 26 Ṣafar 699 AH / 22 November 1299 AD (Mīrkhānd 2002, 5,

4235). The war happened in Ḥimṣ (now in Syria) on Wednesday, 27 First Rabīʿ / 22 December (Rashīd al-Dīn 1994, Vol. 2, 1292). Mīrkhānd says nothing about its date, but he noticed the conjunction of Mars with the Sun (*Iḥtirāq*, lit., 'ignition', 'combustion') at the time of the beginning of the war (Mīrkhānd 2002, 5, 4237). We find it on 21 December when Mars was in conjunction with the Sun in the sign Capricorn. In other words, the war was started within a day after the occurrence of the conjunction. Unsurprisingly, the solar conjunction of Mars, metaphorically as the god who ignites the embers of war, was a potentially causative factor for the Mongols to engage in the war on that specific date. The Mongols were regulating their important activities related to celestial incidents. This was their main motivation to establish the Maragha Observatory (Sayili 1960, 202). Moreover, Mīrkhānd's clear reference to the solar conjunction of Mars, as well as using it for dating the event, may ensure the direct interdependence between the phenomenon and the Mongols' decision to start the war at the time of its occurrence. In any case, it seems that Mars (so-called the powerful god of war) helped Ghāzān to reach his short-lived victory over the Mamlūk patron, because less than three years later they reversed his victory.

	Date	UT	Sign	Ecliptic Longitude
1	1297 Mar 26	19:21	Twins	~ 88.5
2	1299 Apr 18	12:47	Cancer	~116.3
3	1301 May 08	09:04	Leo	~143.2
4	1302 Dec 23	13:51	Virgo	~175.5
5	1303 Mar 06	03:33	Virgo	~172.1
6	1303 May 20	20:48	Virgo	~168.8
7	1304 Dec 31	07:11	Libra	~199.8

Figure 1.

CASE 3: THE INAUSPICIOUS CONJUNCTIONS

Our third case is concerned with Ghāzān's government during the time when the amount of water in the great rivers of Middle East such as the Nile, Amudarya and Tigris was decreased. As a result, diseases like cholera and typhoid became epidemic in Iran's prominent metropolitan areas; for instance, in Shiraz alone 50,000 people died (Mīrkhānd 2002, 5, 4231-2; Khāndmīr 1954, 3, 153). All of the historians attributed this tragedy to the sequence of conjunctions between two inauspicious planets, Saturn and Mars, especially the one occurring on 18 April 1299 AD / 15 Rajab 698 AH / 5 Urdībihisht 678 Iranian calendar. The conjunctions between Saturn and Mars usually occur biennially. The numerous conjunctions of 1302-1303 AD, as indicated in the table in Figure 1 and illustrated in Figure 2, were the result of Saturn's retrograde motion. It is worth mentioning that, in the conjunction of 1301 AD, the angular separation of the two planets was so small that the one completely overshadowed the other. This failed to happen again until two centuries later when seven conjunctions occurred within only five years, 1502-1506 AD. A reference to the various conjunctions of the planets during 909 AH / 26 June 1503 AD - 13 June 1504 AD, including three conjunctions of Saturn with Mars, can be found in Ya¬yā al-Dīn, fol. 115v.

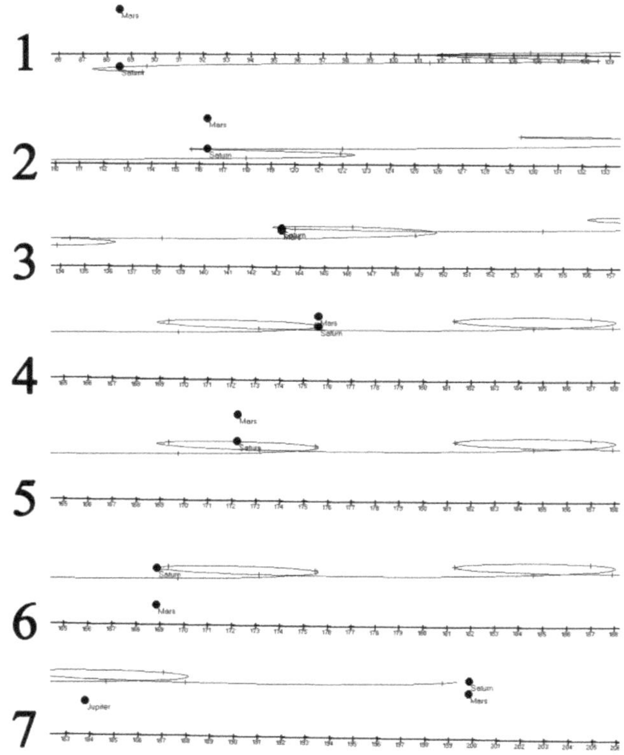

Figure 2. The thick lines show the ecliptic. The spiral lines are the path of Saturn and Jupiter.

Conclusions

In spite of the religious controversies, it was shown that the simple celestial events like planetary conjunctions or coming comets into visible can be a determinable effect on the course of history, with remarkable political and social results. Seen from such a perspective, political and sociological events in the Medieval and early Renaissance periods should be regarded to some extent as a matter of astrological causation rather than human value-based intentional efforts and activities or other peripheral circumstances.

A specific phenomenon could sometimes be presumed to be 'the cause of a historical or political event' (our two first cases) or sometimes was employed for 'justifying a social or natural happening' (our third case). In the first group, the astrological ideas appear to be determining a way through which the river of history must flow. Although the physical cause-effect relationship between the celestial phenomena and the particular historical-social events makes no sense, no one can neglect the psychological effect caused by such a belief on the human mind. This is the reason why astrology could change the natural course of the events, and hence the whole of history. As we saw in Case 1, a simple prognostication of an Arietis comet based upon the astrological doctrine of al-Maghribī changed Tamerlane's mind, and, consequently, led to a weighty political event in spite of its contradiction with the general idea of the astrologers of his court.

It should be emphasized that nothing aside from adopting a psychological point of view seems to be practical in analyzing the effect of astrology on history. 'In psychology the identity and form of the percept that actually enters into a stream of consciousness depends strongly on the intention of the probing mind: a person tends to experience what he or she is looking for, provided the potentiality for that experience is present' (Stapp 2007, 8). Broadly speaking, the impact of astrology upon history reflects the relation of the human mind with the physical world and thus constitutes a striking example to it. As mentioned earlier, the egocentric view of the human mind was a main cause of the formation of the everlasting dogmas in humanity's cosmic viewpoints. These were not only astronomical (e.g., geocentricism) or astrological, but the Aristotelian natural philosophy itself reflects an egocentric mind as well. The belief in the interaction of celestial events on terrestrial ones (likely having emerged from experiments similar to the three historical cases described before) formed the initial astral omens in the first days of history, which were then developed as astrological doctrines. They in a reverse direction conquered a part of the consciousness of the human mind. As a result, since the events have two aspects, physical and psychological, the astonishing effective action of them on the mind streams created, in a psychologically perspective, external outcomes in the physical world.

The role they played in this way is revealed in the social matrix in which they had been embedded. They penetrated into the heart of social events, and maybe nothing like them became troublesome for societies for hundreds of years. Therefore, it should now be taken for granted that astrology to a great extent coordinated the psychological aspects of man's intuitions and creeds in the Medieval period.

The interdependence of history and astrology also had more complex bases in both social and intellectual contexts than in individual and implicitly indicated ones. The astrologically important celestial phenomena were the ones that occurred rarely. It is widely believed that the conjunction between Jupiter and Saturn (Great conjunction) was the most important, indicating major political or religious events. From this perspective, an 'astrological theory of history' in Sasanian Persia (AD 224-652 AD) had already evolved (Kennedy and Pingree 1971, vi) where 'astrologers were famous as practitioners of the art of using the celestial science to reconstruct the political and religious history of the world' (Pingree 1997, 44). The theory was distributed widely and was accepted by the astrologers of other faiths or nations. Then, the important events in the history of humankind (e.g., the Deluge, the birth of prophets, the rise of dynasties or nations, etc.) were attributed to the great conjunction at the beginning of Aries (the so-called *conjunctio maxima* (Kepler 1621, VI, 5) or 'Prime Conjunction' (Abū Mu¬āmid Ghaznawī , fol. 204v)). As a result, astrologers, each one dependent on their own historical, religious or social context, engaged in preparing the long-term prognostications for the forthcoming great conjunctions (e.g., see Pingree 1985, 24-3; Goldstein, 2004; Goldstein, 2007). Therefore, these astrological doctrines also had an established social base to affect directly human consciousness and so human mind streams. However, they were opposed to religious or humanistic norms such as free will. They thus were potential factors in changing man's manner of selection of the possibilities he encountered. (Our Cases 2 and 3 illustrate this well.) They could, therefore, steer history in a specific direction. The psychological force of it is certain, and the three cases introduced earlier illustrate to what extent the story of nations and their vital and decisive choices were subject to them.

Since our examples belong to Islamic culture, other specific results may be determined. Astrology never acted against the establishment of a system of moral and intellectual values in Islam despite its everlasting survival in that domain. Also, not only it did not form a barrier versus astronomical affairs, but also inversely was a main motive for astronomical observational activities. For example, the great importance of the great conjunction made the observational astronomers (who were interested in astrology) to expect precise agreements between the calculated data and the observed ones. For this reason, al-Wābkanawī (Maragha Observatory, 1254 – c. 1320 AD) observed all four great conjunctions that occurred in his 40-year career: 1286 and the triple conjunctions of 1305-6 AD (Mozaffari 2009, 146). Above all, astrology by her doctrines seems to have offered her best service to astronomy as a science. This, as expressed in the quote from H. Poincaré at the beginning of paper, must be known as another aspect of her service to society, too.

References

Abū Mu¬āmid Ghaznawī. Kifāya al-Taᶜlīm fī Ṣināᶜa al-Tanjīm (Sufficient Knowledge for the Art of Astrology). MS. University of Tehran, No. 1914

Aristotle. *Physics*, trans. and com. Hippocrates G. Apostle (1969) Bloomington, Indiana University Press.

Avicenna/Ibn Sīnā. 1986. Fī Aqsām ᶜUlūm al-ᶜAqlīya (On the classification of the rational sciences). In ©asan ᶜÂ½ī (ed.), Ibn Sīnā's Tisᶜa Rasā'il fi al-©ikma wa al-Æabīᶜīyāt (The Seven Treatises by Avicenna on the Wisdom and the Natural Sciences). Beirut, Dār Qābis. [H 1406 H]

Dorcé, C. 2002–3. The Tāj Al-Azyāj of Muhyiddin Al-Maghribī: Methods of Computation. Suhayl 3, 193-212.

Goldstein, B. 2004. A prognostication based on the conjunction of Saturn and Jupiter in 1166 [561 AH]. In C. Burnett et al. (eds.), Studies in the history of the exact sciences in honour of David Pingree, 735–757. Leiden, Brill.

Goldstein, B. 2007. Levi ben Gerson's prognostication for the conjunction of 1345. New York, American Philosophical Society.

Gutas, D. 2003. Medical Theory and Scientific Method in the Age of Avicenna. In D. C. Reisman and A. H. Al-Rahim (eds.), Before and After Avicenna, 145–162. Leiden, Brill.

Kennedy, E. S. and Pingree, D. 1971. The astrological history of Māshā'allāh. Cambridge, MA, Harvard University Press.

Kepler, J. 1621. Epitome astronomiæ Copernicanæ. Libri V, VI and VII. Frankfurt.

Khāndmīr. 1954. ©abīb al-Sīyar. Tehran. 3 Vols.

Mīrkhānd. 2002. Tārīkh-i RawÞa al-¼afā. J. Kiyanfar (ed.). Tehran, Asātir. Vol. 5.

Mozaffari, S. M. 2007. Mathematical Basis and Functions of Astrolabe with Focus on Old Texts: "Projection of Astrolabe" by Mu¬yiddin Al-Maghribī. Unpublished M. Sc. Thesis, University of Tehran.

Mozaffari, S. M. 2009. Wābkanawī and the first scientific observation of an annular eclipse. The Observatory 129, 144–6.

Mozaffari, S. M. and Zotti, G. 2009. Ghāzān Khān's innovations of the astronomical instruments in the second period of scientific life at Maragha Observatory. Book of Abstracts of XXIII International Congress of History of Science and Technology, ICHST 2009, 413. Budapest, International Union of History and Philosophy of Science.

Neugebauer, O. 1975. A History of Ancient Mathematical Astronomy. Berlin–Heidelberg–New York, Springer–Verlag.

Pingree, D. 1985. Astronomical Works of Gregory Chioniades. Vol. 1: Zīj al-ᶜAlā'ī. Amsterdam, Gieben.

Pingree, D. 1997. From Astral Omens to Astrology; From Babylon to Bikanar. Roma, Instituto Italiano per l'Africa e l'Oriente.

Poincaré, H. 1913. The Foundations of Science. George Bruce Halsted (trans.). New York and Garrison, The Science Press.

Rashīd al-Dīn FaÅl-Allāh al-Hamidānī. 1994. Jamiᶜ al-Tawārikh. Mu¬ammad Rushan and Mustafā Mūsawī (eds.). Tehran, Alborz. 4 Vols.

Richardson, M. J. and Marsh K. L. and Schmidt, R. C. 2010. Challenging the Egocentric View of Coordinated Perceiving, Acting, and Knowing. The Mind in Context, 307–333. New York and London, Guilford Press.

Rosenthal, F. 2007. Knowledge Triumphant; The Concept of Knowledge in Medieval Islam. Leiden, Brill.

Saliba, G. 1995. History of Arabic Astronomy, New York, New York University Press.

Sarton, G. 1953. Introduction to the History of Science, Baltimore, Carnegie Institute. Vol. 2, Part 2.

Sayili, A. 1960. The Observatory in Islam. Ankara.

Seargent, D. 2009. The Great Comets in History. New York, Springer.

Sezgin, F. 1978. Geschichte Des Arabischen Scrifttums. Leiden, Brill. Vol. 6.

Shīrāzī, QuÐb al-Dīn 1. Ikhtīyārāt-i MuÞaffarī. MS. Iran, National Library, No. 3074f. (copied in 13-14th century AD, during the life of the author.)

Shīrāzī, QuÐb al-Dīn 2. Tu¬fa al-Shāhiyya. MS. Iran, Parliament Library, No. 6130. (copied at H 730 AH / 1329–30 AD.)

Sayyid Mu¬ammad the Astronomer. Latā'if al-Kalām fī A¬kām al-'Aᶜwām. MS. Tehran, Parliament Library, No. 6347, pp. 168–361.

Stapp, H. P. 2007. Mindful universe, Berlin and Heidelberg, Springer.

Suter, H. 1902. Die Mathematiker und Astronomen der Araber und Ihre Werke. Amsterdam, Leiden.

Tekeli, S. 1980. Mu¬yī al-Dīn Al-Maghribī. In C. C. Gillipsie (ed.), Dictionary of Scientific Biography, Vol. 9, 554–556. New York.

Tricot, J. 1971. De la génération et de la corruption. Paris, Librairie Philosophique J. Vrin.

Mu¬ammad al-Wābkanawī. Zīj-i Mu¬aqqaq-i SulÐānī (Testified Zīj for the Patron). MS. University of Tehran, Microfilm No. 2546.

Westerink L. 1980. La Profession de foi de Grégoire Chioniadès. Revue des études Byzantines 38, 233-245.

Ya¬yā al-Dīn b. Badr al-Dīn Anārī. Jāmiᶜ al-Qirānāt. MS. University of Tehran, No. 4883.

Yeomans, D. K. 2007. Great Comets in History. California Institute of Technology, Jet Propulsion Laboratory. Available at http:// ssd.jpl.nasa.gov/?great_ comets (accessed April 2007).

Zotti, G. and Mozaffari, S. M. 2010. Ghāzān Khān's astronomical instruments at Maragha Observatory. In L. Pigatto and V. Zanini, Astronomy and its Instruments Before and After Galileo, 157–168. Padova, Cooperativa Libraria Editrice Università di Padova (CLEUP).

THE *DUSTŪR AL-MUNAJJIMĪN* OR DOES A SOVEREIGN NEED ASTRONOMY TO STRUCTURE HIS REIGN?

PETRA G. SCHMIDL

Abstract: The *Dustūr al-munajjimīn* is an Arabic text probably written in an Eastern Ismāʿīlī context about 1100 AD. It provides astronomical, astrological and historical information. Preserved in a single copy in Paris, the history of the manuscript itself and the text of the *Dustūr al-munajjimīn* present several points of interest concerning astronomy and power. This article will introduce this unique manuscript and its history. It investigates its text, in particular its astronomical part, and shows possible reasons for its composition that might have related to the political program of the Ismāʿiliyya. This heterodox Islamic group became known to Europe mainly by two legends woven around it, the rumours and stories on the Assassins and on the Old Man of the Mountain

Keywords: Islam, astronomy, astrology, history, Ismāʿīlīs, zījes, astrological world histories

Introduction

Dustūr al-munajjimīn literally means 'instructions' or 'rules of the astrologers' or 'astronomers'; the line between these two disciplines is difficult to draw (see Ullmann 1972, 271-2; and Sezgin 1979, 11-2). It is a treatise written in Arabic, probably in an Eastern Ismāʿīlī context, about 1100 AD (for details see below). It provides astronomical, astrological and historical information–a rather uncommon combination. One conceivable reason to gather these topics in a single treatise points towards a link between astronomy and power. A closer look on the manuscript of the *Dustūr al-munajjimīn* and its history, its text and its contents will reveal more about this possible link and will help to answer the question raised in the heading: 'Does a sovereign need astronomy to structure his reign?'.

The manuscript of the *Dustūr al-munajjimīn*

PRESENT

The *Dustūr al-munajjimīn* is preserved in a single Arabic manuscript preserved in Paris, Bibliothèque Nationale de France, no. 5968 arabe (Blochet 1925, 151; abbreviated as 'P' in this article.). This voluminous copy with 346 folios is defective, incomplete and in disorder. The beginning and the end of the manuscript are missing; the title page is a later addition. These parts of a manuscript usually give information on potential readership, writing motives and supporting patrons. The copyist of the manuscript, the place and the time of its preparation are normally also to be found in these parts. Therefore other evidence is necessary to learn more about the manuscript of the *Dustūr al-munajjimīn*.

PAST

The beginning

Since the unique manuscript of the *Dustūr al-munajjimīn* is neither signed nor dated, different theories developed concerning the copyist and his time and place (Casanova, 1922; Blochet, 1923; Zimmermann, 1978). One of these theories is closely related to an account by 'Alā' al-Dīn 'Aṭā Malik Juwaynī (Iṣfahān, 1226-1283 AD), a 13th century Persian historian and author of the *Ta'rīkh-i jahān-gushāy* ('History of the world conqueror'). Juwaynī writes:

> Now when I was at the foot of Lammasar, being desirous of inspecting the library, the fame of which had spread throughout the world, I suggested to the King that the valuable books in Alamūt ought not to be destroyed. He approved my words and gave the necessary orders; and I went to examine the library, from which I extracted whatever I found in the way of copies of the Koran and [other] choice books after the manner of '*He brought forth the living from the dead*' (Qur'ān, *sūra* (*al-rūm*) 30,18(19)). I likewise picked out the astronomical instruments such as *kursīs*, armillary spheres, complete and partial astrolabes and other ... that were there. As for the remaining books, which related to their heresy and error and were neither founded on tradition nor supported by reason, I burnt them all. ('Alā' al-Dīn 'Aṭā Malik Juwaynī, ed. Boyle 1958, II, 719; Persian text in Juwaynī, ed. Qazvīnī 1937, III, 269-70).
>
> Juwaynī gives a second account of the inspection of the Alamūt library earlier in his book, this time without mentioning astronomical instruments ('Alā' al-Dīn 'Aṭā Malik Juwaynī, ed. Boyle 1958, II, 666; Persian text in Juwaynī, ed. Qazvīnī 1937, III, 186).

With regard to the *Dustūr al-munajjimīn* it will be useful to learn more on several topics Juwaynī mentioned in his account.

1. The Mongols

First, 'the King' refers to the Mongol ruler Hülegü (Mongolia, Iran, born c. 1217 AD). He was one of the grandsons of Genghis Khān (Mongolia, Iran, Central Asia, b. 1167 AD), who began, together with his Mongolian horsemen, to invade parts of China, Central Asia and Persian at the end of the 12th / beginning of the 13th century. Around 50 years later Hülegü started a new campaign, and conquered in less than 15 years the main parts of Iran, Iraq and Syria, establishing the basis of the later Īlkhānī state (see Krämer 2005, 171-9). In 1256 AD, Juwaynī accompanied Hülegü and the Mongols on their campaign against the Alamūt (see Figure 1).

2. The Alamūt and the Nizārīs

The Alamūt.

Secondly, Juwaynī mentions in his account 'the Alamut' , a mountain fortress in northern Iran near the Caspian Sea, nowadays in ruins. 'Lammasar' (or 'Lanbasar') is another stronghold situated in that region, nowadays in ruins, too. During Hülegü's and Juwaynī's time the Alamūt was the headquarters and final retreat of the Nizārīs, a heterodox Islamic group with a specific interest in learning, knowledge and science based on their religious doctrines (see below). They are Shīʿī Muslims of the Ismāʿīlī branch (see Nanji, 1995; and Krämer 2005, 114).

The Nizārīs.
In the Middle Ages, the Nizārīs became well-known–also to Europe–mainly by two legends woven around them: the rumours on the Assassins and on the Old Man of the Mountain (*shaykh al-jabal*) sitting in his mountain fortress and sending out his men to assassinate their religious and political opponents. Although these stories were originally related to the Syrian branch of the Nizārīs and were propaganda created by their opponents, they contain a kernel of truth (see Daftary 1995, especially 92-95; Daftary 1998, 13-4; and Krämer 2005, 126-7).

At the end of the 11th century, and mainly by seizing strongholds and building mountain fortresses, killing adversarial leaders and converting local inhabitants, the Nizārīs established a confederation of principalities in northern Persia and parts of Syria.

Figure 1. The siege of the Alamut by the Mongols, depicted in a 15th century manuscript of Juwaynī's historiography. Paris, Bibliothèque Nationale de France, Suppl. pers. 206, fol. 149; photo courtesy of BNF.

During these times, Persia had been controlled by the Saljūqs, a Turk dynasty. Weakened by disputes of inheritance and succession, the local population suffered from Saljūq misrule (see Hillenbrand 1996, 206). In their sphere of influence, Ḥasan-i Ṣabbaḥ (Qumm, Ray, Egypt, Alamūt, d. 1124), an Ismāʿīlī missionary (*dāʿī*), travelled widely doing missionary work and carrying out propaganda by order of the Ismāʿīlī imām and Fāṭimid caliph al-Muṣtansir bi-Allāh (Cairo, 1029-1094 AD). In 1090 AD, the takeover of the mountain fortress of Alamūt under Ḥasan-i Ṣabbaḥ's leadership was the starting signal for open opposition against the Saljūqs, and laid the foundations of an independent Nizārī state (see Daftary 1990, 339-40). Four years later, when al-Muṣtansir died, discrepancies occurred on the rightful successor to religious and political leadership. Instead of Abū Manṣūr Nizār (Cairo, b. 1045), al-Muṣtansir's eldest surviving son, his younger half-brother Abū al-Qāsim Aḥmad (Cairo, 1074-1101) had himself installed as caliph and adopted the name al-Mustaʿlī bi-Allāh. Nizār, and those who gave him their allegiance, the Nizārīs, rebelled and were finally defeated by the Mustaʿlīs, the followers of al-Mustaʿlī. While the Ismāʿīlīs of Egypt, the Yemen, and western India and parts of Syria stood by the Mustaʿlīs, the Persian communities decided in favour of the Nizārīs (see Daftary 1990, 348). The Alamūt became their major centre–not only in politics and warfare, but also in teaching and learning. There Ḥasan-i Ṣabbaḥ founded a library that would gain 'the fame of which had spread throughout the world', as Juwaynī says. More fortresses follow that example. The library of the Alamūt persisted until the Mongol conquest in 1256 AD (see Daftary 1996a, 185; and Daftary 1998, 120-4).

Nizārī interest in knowledge and science.
The Nizārīs' interest in knowledge and science had several aspects. Mainly after the separation from the Mustaʿlīs of Fāṭimid Egypt, they developed a new theological and philosophical (missionary) doctrine (*al-daʿwa al-jadīdā*) based on earlier Ismāʿīlī concepts (Daftary 1998, 131-2). Juwaynī probably includes it in his account amongst 'error and heresy'. Concerning the *Dustūr al-munajjimīn* and the relation of astronomy and power, two points are of major interest:

First, in Ismāʿīlī cosmology, man is far away from his Creator. Only knowledge may bridge this gap. Prophets, messengers and insiders sent by God communicate this knowledge to man (Daftary 1998, 57; Halm 1996, 81-2). Therefore, the Ismāʿīlīs highly respect all knowledge, from religious matters to philosophy and science, etc., including astrology (see Sezgin 1979, 11; Daftary 1990, 382; and Halm 1997, 64), and of course for its practical uses, as well. A *dāʿī* sent to foreign countries to do his missionary work has to be morally steady and intellectually sound (Daftary 1998, 92).

Secondly, Ismāʿīlīs thought relied on a semi-cyclic interpretation of time and (religious) history. This model resembles a spring with starting point and end; with each completed whorl, a new level is reached (see Daftary 1998, 53).

A specific rôle befits astrology. Its basic assumption relates all changes in the sublunar world to the movements of the celestial bodies (see Nallino 1987, 495a; von Stuckrad 2003, 15-6). They are tokens for world history and individual destiny (von Stuckrad 2003, 162). Of main importance in mundane astrology are the recurring conjunctions of the two upper planets, Jupiter and Saturn. They indicate the starting point of a new cycle, or whorl, in history. This formative concept arranges time in ever repeating cycles–or whorls–to explain–and predict–changes of religions and dynasties (see Pingree 1963, 245; and Orthmann 2006, 132-6).

Apart from interests in science and knowledge, there also might have been another, political reason. Establishing themselves as a new power the Nizārīs would have been highly interested in writing their own historiography. Being a threat to as well as threatened by religious and political opponents, they were also highly interested in political propaganda (Karīmī 2010a, 227). In both cases, they also made use of mundane astrology.

3. Interest in astronomy and astrology

Thirdly, Juwaynī specifies in his account '*kursīs*, armillary spheres, complete and partial astrolabes and other ...'. Therefore, he portrayed himself–and as a consequence Hülegü, too–as being interested in the astronomical instruments preserved in the Alamūt. Further Juwaynī mentions 'valuable books', amongst those might have been books dealing with astronomy and astrology. With his account, Juwaynī also gives evidence of Nizārī interest in these topics.

During the time of the Mongol takeover, Naṣīr al-Dīn al-Ṭūsī (Ṭūs, Baghdād, 1201-1274 AD), one of the most famous Islamic astronomers and most prolific Ismāʿīlī scholars, stayed at the Alamūt. Later al-Ṭūsī witnessed the Mongol conquest of Baghdād in 1258 AD and became the administrator of the religious foundations. As court astronomer and astrologer of Hülegü, he initiated, established, organized, and directed the observatory in Marāgha and dedicated his *Zīj-i Īlkhānī* to the Mongol ruler–giving evidence that the Mongols start to adopt the local culture and that their sovereign was interested in astronomy (Daiber and Ragep, 2002; Sayili 1960, 189-222).

One theory concerning the date and place of its copying claims that the *Dustūr al-munajjimīn* might have been one of the manuscripts rescued by Juwaynī from the Alamūt library after the Mongol campaign against the Nizārīs and the book burning that followed (see Blochet 1923, 385). Because of the initial books dealing with astronomy and astrology, the *Dustūr al-munajjimīn* is unsuspicious of being 'related to their heresy and error' Juwaynī mentions in his account. He even might have been interested in using the historical part (Casanova 1922, 135; Blochet 1923, 385; Otsuka 2010, 7-10) as he did with *Sar Gudhasht-i Sayyidnā* ('The Past of our Lord'), which contains information on the life and work of Ḥasan-i Ṣabbāḥ ('Alā' al-Dīn 'Aṭā Malik Juwaynī, ed. Boyle 1958, II, 666; Persian text in Juwaynī, ed. Qazvīnī 1937, III, 187).

The text of the *Dustūr al-munajjimīn* provides indications of its writing around 1100 AD in an Eastern Ismāʿīlī context (see below). During a close inspection of the manuscript in Paris in May/June 2010, several marginal notes in Persian were discovered. Therefore, it is still conceivable that a copy of the *Dustūr al-munajjimīn* was part of the Alamūt library (contradicting Blochet 1925, 151, 'Neskhi du XIVe siècle', and Zimmermann 1978, 185: '[...] the palaeographical evidence would suggest that the Paris copy of the *Dustūr al-munajjimīn* was produced around 1400 AD [...].').

Concerning a link between astronomy and power, it does not matter if the *Dustūr al-munajjimīn* was part of the Alamūt library or not. Juwaynī's account first most clearly portrays himself as being interested in astronomy and astrology, and the Mongols in 'the valuable books in Alamūt ought not to be destroyed'. Second, it also reflects the Nizārīs' interest, in whose central library Juwaynī found astronomical instruments. Even if the sovereign himself does not need astronomy to structure his reign, he had an interest in scholars practicing astronomy and astrology.

The end

Because the early history of the manuscript remains uncertain, it is appropriate to investigate how the copy of the *Dustūr al-munajjimīn* came to Paris. This tells a second story concerning astronomy and power.

At the end of the 19th century, the Bibliothèque Nationale de France in Paris purchased the manuscript of the *Dustūr al-munajjimīn* together with nearly 800 Arabic, Persian, and Turkish manuscripts (Blochet and Schefer 1900, ii) collected by Charles Henri Auguste Schefer (Paris, 1820-1898), a French orientalist, during his lifetime. Spending several years in the Orient, he became *dragoman*–official translator–of the French embassy at the Ottoman court in Istanbul in 1847. Ten years later, Schefer went back to Paris to take the chair of Persian language at the *École des langues orientales vivantes* in Paris (see Bouché-Leclercq 1899, 633-44). Most probably, before he resettled to France, the Ottoman sultan ʿAbd al-Mājid I (1823-1861) offered Schefer a lavish present: to select several manuscripts of the Serail library for his own collection (Blochet and Schefer 1900, ii).

Some of the manuscripts from Schefer's collection were in fact originally part of the Serail library (see manuscript no. 6080 arabe [Blochet 1925, 181], a copy of al-Bīrūnī's *India* that Eduard Sachau widely used for his edition and translation (ed. Sachau 1887, esp. viii-ix; trans. Sachau 1910). For other copies and artefacts, Schefer relied on professional tradesmen (Schefer 1895, 24). The title page of the *Dustūr al-munajjimīn* shows two seals of former owners appearing to be Ottoman (see Figure 2), but the names mentioned are illegible and not identified yet.

Figure 2. The title page of the Dustūr al-munajjimīn. Paris, Bibliothèque Nationale de France, 5968 arabe, fol. 1a; photo courtesy of BNF.

To summarize, both stories describe a link between astronomy and power. First, Juwaynī's account gives evidence of the Nizārīs' and Mongols' interest in astronomy and astrology. Hülegü ordered him 'to examine the library', where he 'picked out the astronomical instruments' and 'the valuable books'. Second, the manuscript of the *Dustūr al-munajjimīn* might have been part of the Ottoman sultan's library. Both findings point towards a storing place of the *Dustūr al-munajjimīn* in a sovereign's library–and many other manuscripts are able to tell similar stories, too. Although these results do not answer the question raised in the heading, they give evidence of sovereigns' interest in astronomy and astrology. Their motives for doing so remain unclear.

Text and contents of the *Dustūr al-munajjimīn*

TEXT

Apart from the information collected on the manuscript, the text of the *Dustūr al-munajjimīn* provides further evidence concerning a link between astronomy and power. The treatise is anonymous. Information on the author and his possible motives has to be taken from the text itself. In all likelihood, the *Dustūr al-munajjimīn* was written between 1095 and 1124 AD (Zimmermann 1978, 185-6). Apparently its author belonged to an eastern Ismā'īlī context, either in Iran or Syria–if he was not working at the Alamūt itself. Several clues support this theory. In general, book X reveals the Nizārī background of the author (see Casanova 1922, 129-33). In books I to VIII the geographical latitude of $\varphi = 36°$ plays a prominent rôle, a value that also might denote the midpoint of the fourth climate (King 1999, 32-3). Two of the main sources of the *Dustūr al-munajjimīn*, *al-Zīj al- Ṣābi'* of al-Battānī (Raqqa, d. 929 AD) and, less distinct, *al-Zīj al-Jāmi'* of Kūshyār b. Labbān (Gīlān, fl. around 1000 AD), both have calculated tables for this latitude in their astronomical handbooks (Kennedy 1989, 33a-34a). In detail, book II includes an oblique ascension table for $\varphi = 36.21°$, 'that is the latitude of *Baldat al-Iqbāl*' (P, 53a; i.e., the Alamūt, cf. 'Alā' al-Dīn 'Aṭā Malik Juwaynī, ed. Boyle 1958, II, 674; and a Persian text in Juwaynī, ed. Qazvīnī 1937, III, 200). Other medieval sources likewise use $\varphi = 36.21°$ for the Alamūt (Kennedy and Kennedy 1987, 15). This strong argument is weakened by the geographical table in book III. There, the Alamūt–or *Baldat al-Iqbāl*–is not included (Zimmermann 1978, 187). At the end of the geographical table one finds, though, one of the marginal notes in Persian discovered in Paris in May/June 2010 (see above). It reads, most probably in the handwriting of the copyist: 'Here nine folios are needed (to insert the tables (?)).' (P, 67b). This might give evidence that the geographical table in the *Dustūr al-munajjimīn* is incomplete–and it proves that the treatise was copied by a Persian. Because the text itself is written in Arabic, this might point towards an authorship or a readership not in Persia but in Syria (Daftary 1990, 325, 331-2; Daftary 1998, 121). Therefore, it is still uncertain when and where the *Dustūr al-munajjimīn* was written exactly, but the Alamūt during Ḥasan-i Ṣabbaḥ's lifetime is one reasonable possibility. One might even speculate if this text was written by or for the Nizārī leader himself and therefore displays another link between astronomy and power.

CONTENTS

To learn more about the possible motives of the anonymous author who wrote the *Dustūr al-munajjimīn*, a closer look at its contents appears to be appropriate.

The *Dustūr al-munajjimīn* comprises 10 books, each of them organized in several chapters (see Appendix). Book I to VIII contain information usually found in a *zīj*, i.e., an astronomical handbook with tables. *Zīj*es belong to a very important genre of scientific literature in pre-modern Islamic societies; 250 such texts are known. They combine tables with explanatory texts related to astronomy, trigonometry, geography and astrology. First evidence of these astronomical handbooks in Arabic texts goes back to 8th century Baghdād. In those initial years, the authors of *zij*es made use of Indian, Persian and Greek astronomical traditions (see Kennedy 1989, especially 17-23, 49 and 51; and King and Samsó 2001, especially 19-30). In a *zīj* an astronomer will find all he needs to reconstruct the sky above him at any given place and time (King and Samsó 2001, 15). Book IX of the *Dustūr al-munajjimīn* provides mundane astrology (Kennedy and van der Waerden, 1963). It uses concepts of periods and cycles of time, e.g., the recurring celestial event of the conjunctions of Saturn and

Jupiter or the revolutions of the planets (Orthmann 2006, 133-4). One of the most famous books that deals with this kind of astronomy was written in 9th century Baghdād by Abū Maʿshar (Balkh, Baghdad, Wāsiṭ, flourished around 800). It bears the title *Kitāb al-milal wa-l-duwal* ('Book of religions and dynasties – On the great conjunctions'; Burnett and Yamamoto, 2000) and brings across an idea of the astrological topics dealt with in book IX of the *Dustūr al-munajjimīn*. Book X includes historical and chronological information of the Nizārīs including pre-Islamic topics. It starts with chronological and calendrical information that is mainly taken from the *al-Qānūn al-Mas'ūdī* ('Masʿūdic Canon') and the *Kitāb al-Tafhīm li-awā'il ṣinā'at al-tanjīm* ('The book of instructions in the art of astrology'), both written after the year 1000 AD by al-Bīrūnī (Ghazna, b. 973), one of the most prolific and versatile Islamic authors. Together with the *zīj*es by al-Battānī and Kushyār b. Labbān (see above), al-Bīrūnī is also the main source in the astronomical parts of the *Dustūr al-munajjimīn* in books I to VIII. In book X the author apparently wants to form a strong basis for the following historical data. He wrote an Ismāʿīlī world history with the impetus to legitimate the Nizarī state and to establish their identity (Karīmī 2010b, 9-10).

To summarize, on the one hand, the contents of the *Dustūr al-munajjimīn* provide information of a *zīj*, although the astrological and historical data clearly outnumber what has usually been found in texts of this genre. On the other hand, the *Dustūr al-munajjimīn* resembles astrological world histories, although now in particular the astronomical data exceeds what is found usually in these texts.

Calling to mind why in particular the Nizārīs were interested in science and knowledge (see above), the contents of the *Dustūr al-munajjimīn* apparently covers most of their philosophical, theological, and political interests. Books I to VIII comprise the preliminaries that help in preparing astrological calculations, especially books I and III, which deal, respectively, with how to adopt them to a specific time and a specific place (see also the Appendix). Additionally, other subjects relate to practical purposes, mainly timekeeping and calendars. Other data is useful in performing religious duties, especially in book III, which shows how to determine the *qibla*, the sacred direction of Islam towards Mecca, and in book VII, which has calculations of the duration of twilight–useful in determining three of the five daily ritual Muslim prayers (King and Wiedemann, 1997). Book IX deals with astrological history, and might have been used for propagandistic means. Book X contains data for a Nizārī historiography.

In particular, the uncommon combination of astronomy, astrology and historiography in the *Dustūr al-munajjimīn* points towards a link between astronomy and power. This treatise covers several aspects a Nizārī sovereign–if not Ḥasan-i Ṣabbaḥ himself–might be interested in: a historiography of his people, astrological propaganda to support the own doctrine, and the astronomical preliminaries to calculate them.

Conclusion

All these findings reveal possible reasons that link the *Dustūr al-munajjimīn* with 'a sovereign who needs astronomy to structure his reign'.

First, apparently the manuscript of the *Dustūr al-munajjimīn* was regarded worthy of inclusion in a sovereign's library, probably in the Nizārīs' library of the Alamūt in the mid-13th century and possibly in the Ottoman sultan's library in Istanbul in the mid-19th century. Why these two sovereigns were interested in the *Dustūr al-munajjimīn* is a matter for speculation.

Secondly, the *Dustūr al-munajjimīn* deals with several topics closely related to Nizārī interests in science, in politics and in religion. In general, a Nizārī author provides other Nizārī scholars with knowledge in astronomy, astrology and history–and makes clear in passing that he is familiar with the scientific sources and scholarly traditions. This treatise endows in particular the Nizārīs with a proper historiography: a new or endangered power has to assure itself of its history, to harmonize, standardize and adjust it to their purposes and for their legitimization. The *Dustūr al-munajjimīn* deals with cyclic time and world-years, a major concern in Nizārī propaganda, and provides the basic skills in astronomy necessary for these calculations. The *Dustūr al-munajjimīn* shows that there is a link between astronomy and power.

Appendix: Overview of the contents of the *Dustūr*

Book I: Chronology (P,3a-25b)

chapters of book I		contents
1-5.1st half of	6	(missing)
2nd half of	6	Persian and Malikī calendars
	7	Indian calendar
	8	Jewish calendar
	9	Jewish and Christian feast and fast days
	10	Muslim feast days
	11	Persian and Zoroastrian calendars, feast and fast days
	12	Syrian feast and fast days

Book II: Trigonometry (P,26a-57b)

chapters of book II	contents
1	chord, sine
2	chord, sine, versed sine
3	interpolation of sine
4	first and second declination
5	altitude, tangent, cotangent
6	right ascension
7	oblique ascension

Book III: Mathematical geography (P,58a-67b)

chapters of book III	contents
1	cupola of the earth
2	circumference of the earth
3	geographical latitude
4	geographical longitude, distance between two places
5	geographical table
6	(Missing: meridian)
7	(Missing: *qibla*)

Book IV: Planetary positions (P,68a-165b)

chapters of book IV		contents
1–4.1st half of	5	(missing)
2nd half of	5	rising and setting, visibility and invisibility, ascendant and descendant, equation of time
	6	solar position
	7	lunar position
	8	planetary positions
	9	interpolation of planetary motions
	10	lunar latitude
	11	planetary latitudes
	12	retrograde and direct motion
	13	conjunction and opposition
	14	diameters of sun and moon
	15	distance between earth and moon
	16	ecliptic pole (?)
	17	lunar parallax
	18	solar and lunar eclipse
	19	first visibility of the lunar crescent
	20	shape of the moon (?)
	21	shift, world-year
	22	timekeeping, equation of daylight
	23	lunar transition (?)
	24	occultation and conjunction

Book V: Spherical astronomy (P,166a-182b)

chapters of book V	contents
1	rising and setting amplitude
2	declination
3	equation and arc of daylight
4	solar and stellar maximum altitude
5	culmination, transit
6	ecliptical degree rising or setting together with a star
7	timekeeping during the day
8	timekeeping during the night
9	ascendant by lunar altitude
10	ascendant by *finjān*
11	ascendant by shadow (?)
12	equalisation of the twelve houses

Book VI: Astrology, prayer times (P,183a-188b)

chapters of book VI	contents
1	ascendant of birth, *namūdhār* of conception
2	*namūdhār* according to Ptolemy
3	*namūdhār* according to the Indians
4	*namūdhār* according to Valens
5	*namūdhār* of direction (*tasyīr*)
6	twilight
7	afternoon prayer

Book VII: Mathematical astrology (P,189a-214b)

chapters of book VII	contents
1	shift of the year
2	projections of the ray
3	division (*qisma*), direction (*tasyīr*)
4	*fardārāt*
5	duration of the stages of life
6	twelfths of the zodiacal signs
7	lots

Book VIII: Lunar mansions, fixed stars, *anwā'*, and winds (P,215a-230a)

chapters of book VIII	contents
1	description of fixed stars
2	lunar mansions
3	star table
4	*anwā'*, winds

Book IX: Planetary conjunctions, world eras

chapters of book IX	contents
1	conjunction of the superior planets
2	thousands
3	periods (*adwār*)
4	trepidation

Book X: Chronology and history

chapters of book X	contents
1	varieties of days and hours
2	natural and artificial years and months
3	years and months of the peoples
4	calendars
5	leap years
6	confusions concerning calendars
7	other calendars
(?)	history of the prophets, *imāms*, kings, and the rulers of the Arabs, Indians, Persians, and Romans

Acknowledgements:

This paper was written as part of a research project on the *Dustūr al-munajjimīn* supported by the Deutsche Forschungsgemeinschaft (DFG) and located at Bonn University. I wish to thank all my colleagues who helped me in preparing this paper, in particular Benno van Dalen, Malihe Karbassian, Moḥammad Karīmī Zanjānī Aṣl, David A. King, Eva Orthmann, Osamu Otsuka, my family to back me up, and the organizing team in Gilching led by Barbara and Michael Rappenglück for the productive SEAC 2010 conference. Although carefully proofread, all remaining mistakes are the author's responsibility.

References

Abū Rayḥān al-Bīrūnī 1954-6. al-Qānūn al-Mas'ūdī, 3 Vols. Hyderabad, Osmania Oriental Publications Bureau.

'Alā' al-Dīn 'Aṭā Malik Juvaynī. 1937 (1912ff). Tārīkh-i Jahān-gushā, 3 Vols., ed. M. M. Qazvīnī. E. J. W. Gibb Memorial series, Vol. 16. Leyden, Brill; London, Luzac & Co.

'Alā' al-Dīn 'Aṭā Malik Juvaynī. 1958. The history of the world-conqueror, 2 Vols, ed. J. A. Boyle. Manchester, Manchester University Press.

Blochet, E., and Schefer, Ch. 1900. Catalogue de la collection de manuscrits orienteaux, arabes, persans, et turcs: formée par M. Charles Schefer et acquise par l'état. Paris, Leroux.

Blochet, E., and Schefer, Ch. 1923. Notes additionelles. Notices et extraits des manuscrits de la Bibliothèque Nationale 41, 383-389.

Blochet, E., and Schefer, Ch. 1925. Catalogue des manuscrits arabes des nouvelles acquisitions (1884-1924). Paris, Leroux.

Bouché-Leclercq, A. 1899. Notice sur la vie et les travaux de M. Charles Schéfer. Comptes rendus des séances de l'Académie des inscriptions et belles-lettres 27, 627-668. Paris, Diffusion de Boccard.

Burnett, Ch., and Yamamoto, K. (eds) 2000. Abu Ma'shar on Historical Astrology, The Book of Religions and Dynasties (On the Great Conjunctions), 2 Vols. Islamic Philosophy, Theology and Science: texts and studies 33 and 34. Leiden, Brill.

Casanova, P. 1922. Un nouveau manuscrit de la secte des Assassins. Journal asiatique, 11ième série 19, 126-135.

Daftary, F. 1990. The Ismā'īlīs: their history and doctrines. Cambridge et al., Cambridge University Press.

Daftary, F. 1995. The Assassin legends: myths of the Ismā'īlīs. London and New York, I. B. Tauris & Co.

Daftary, F. 1996a. Ḥasan-i Ṣabbāḥ and the origins of the Nizārī Isma'ili movement. In F. Daftary (ed.), Mediaeval Isma'ili History and Thought. Cambridge et al., Cambridge University Press, 181-204.

Daftary, F. (ed.) 1996b. Mediaeval Isma'ili History and Thought. Cambridge et al., Cambridge University Press.

Daftary, F. 1998. A short history of the Ismailis: traditions of a Muslim community. Edinburgh, Edinburgh University Press.

Daiber, H., and Ragep, F. J. 2002. art. 'al-Ṭūsī'. In The encyclopaedia of Islam (new edition) 1960-2009, 12 Vols. Leiden, E. J. Brill.

Halm, H. 1996. The cosmology of the pre-Fatimid Ismā'iliyya. In F. Daftary (ed.), Mediaeval Isma'ili History and Thought. Cambridge et al., Cambridge University Press, 75-83.

Halm, H. 1997. The Fatimids and their Traditions of Learning. London and New York, I. B. Tauris & Co. Ltd in association with the Institute of Ismaili Studies.

Hillenbrand, C. 1996. The Power Struggle between the Saljuqs and the Isma'ilis of Alamut, 487-518/1094-1124: The Saljuq Perspective. In F. Daftary (ed.), Medieval Isma'ili History & Thought, 205-220. Cambridge et al., Cambridge University Press.

Hodgson, M. G. S. 1955. The order of Assassins: the struggle of the early Nizārī Ismā'īlīs against the Islamic world. 's-Gravenhage, Mouton & Co.

Karīmī Zanjānī Aṣl, M. 2010a. From the *Dustūr al-Munajjimīn* to Ṭūsī's works: some points on the prevalent sciences among the Iranian Nizārīs through the fall of Alamūt (1256 CE/654 AH). In D. Cevenini and S. D'Onofrio (eds), Islām: collected essays (Uyūn al-akhbār – Studi sul mondo islamico 4), 223-238. Bologna, Casa editrice Emil di Odoya srl.

Karīmī Zanjānī Aṣl, M. 2010b. The combination of historical sources in the *Dustūr al-munajjimīn* (a preliminary study). Unpublished paper kindly provided by the author in 2010.

Kennedy, E. S. 1989. A survey of Islamic astronomical tables Transactions of the American Philosophical Society, new series 46:2. Philadelphia, American Philosophical Society. Reproduced c. 1989. Philadelphia, American Philosophical Society (the page numbers in the text are to the reprint).

Kennedy, E. S., and Kennedy, M. H. 1987. Geographical Coordinates of Localities from Islamic Sources. Frankfurt, Institute for the History of Arabic-Islamic Science.

Kennedy, E. S., and Van der Waerden, B. L. 1963. The World-Year of the Persians. Journal of the American Oriental Society 83, 315-327. Reprinted in Kennedy, E. S. et al. 1983. Studies in the Islamic Exact Sciences, M. H. Kennedy and D. A. King, (eds), 338-350. Beirut, American University of Beirut Press.

Kind, D. A., Samsó, J., and Goldstein, B. R. 2001. Astronomical handbooks and tables from the Islamic world (750-1900): an interim report. Suhayl 2, 9-105. Also available at: http://www.-ub.edu/arab/suhayl/volums/volum2/paper%201.pdf.

King, D. A. 1999. Bringing astronomical instruments back to earth: the geographical data on Medieval astrolabes (to ca. 1100). In L. Nauta and A. Vanderjagt (eds), Between demonstration and imagination: essays in the history of science and philosophy presented to John D. North. 3-53. Leiden et al., Brill. Reprinted and updated in King, D. A. 2005. In synchrony with the heavens: studies in astronomical timekeeping and instrumentation in medieval Islamic civilization, Vol. 2: Instruments of mass calculation (studies X-XVIII). Islamic Philosophy, Theology and Science: texts and studies 55. Leiden, E. J. Brill, XVI, 915-962.

King, D. A., and Wiedemann, E. 1997, art. 'shafak' (twilight)'. In The encyclopaedia of Islam (new edition) 1960-2009, 12 Vols. Leiden, E. J. Brill.

Krämer, G. 2005. Geschichte des Islam. München, C. H. Beck.

Nallino, C. A. 1987. art. 'Astrology'. In The encyclopaedia of Islam (1st edition) 1913-1936, 4 Vols. Leiden and London, E. J. Brill. Photomechanical Reprint in 9 vols. Leiden, E. J. Brill.

Nanji, A. 1995, art. 'Nizāriyya'. In The encyclopaedia of Islam (new edition) 1960-2009, 12 Vols. Leiden, E. J. Brill.

Orthmann, E. 2006. Astrologie und Propaganda: Iranische Weltzyklusmodelle im Dienste der Fāṭimiden. Die Welt des Orients 36, 131-142.

Otsuka, O. 2010. The *Dustūr al-munajjimīn* as a source of early Ismā'īlī history. Unpublished paper delivered at the World Congress of Middle East Studies in Barcelona July 2010 and kindly provided by the author.

Pingree, D. 1963. Astronomy and Astrology in India and Iran. Isis 54, 220-246.

Qazvīnī, M. M. 1345 AH Sh. / 1966 AD. Dustūr al-munajjimīn. In I. Afshār (ed.), Yāddāsht-hāye Qazvīnī, Vol. 8, 110-143. Tehrān, Dāneshgāh-e Tehrān.

Sachau, C. E. 1887. Alberuni's India: an account of the religion, philosophy, literature, geography, chronology, astronomy, customs, laws and astrology of India about A.D. 1030, Arabic original edition. London, Trübner & Co. Reprinted in 1993 in the series Islamic Geography 105. Frankfurt, Institute for the History of Arabic-Islamic Science.

Sachau, C. E. 1910: Alberuni's India: an account of the religion, philosophy, literature, geography, chronology, astronomy, customs, laws and astrology of India about A.D. 1030, an English edition with notes and indices, 2 Vols. 2nd edition. London, Kegan Paul, Trench, Trübner & Co. Reprinted in 1993 in the series Islamic Geography 106 and 107. Frankfurt, Institute for the History of Arabic-Islamic Science.

Sayili, A. 1960. The Observatory in Islam and its Place in the General History of the Observatory. Publications of the Turkish Historical Society, series VII 38. Ankara, Türk Tarih Kurumu Basimevi. Reprinted 1998 in the series Islamic Mathematics and Astronomy 97. Frankfurt, Institute for the History of Arabic-Islamic Science.

Schefer, Ch. 1895. Notice sur les relations des peuples musulmans avec les chinois, depuis l'extension de l'islamisme jusqu' à la fin du XVe siècle. In Centenaire de l'école des langues orientales vivantes, 1795-1895, recueil de mémoires publié par les professeurs de l'école, 1-43. Paris, Imprimerie Nationale.

Sezgin, F. 1979. Geschichte des arabischen Schrifttums, Band VII: Astrologie – Meteorologie und Verwandtes. Leiden, E. J. Brill.

Ullmann, M. 1972. Die Natur- und Geheimwissenschaften im Islam. Handbuch der Orientalistik 1. Abteilung, Band VI:2. Leiden, E. J. Brill.

Von Stuckrad, K. 2003. Geschichte der Astrologie: Von den Anfängen bis zur Gegenwart. München, C. H. Beck.

Wright, R. R. 1934. The book of instruction in the elements of the art of astrology by Abū Rayḥān Muḥammad b. Aḥmad al-Bīrūnī. London, Luzac & Co. Reprinted 1998 in the series Islamic Mathematics and Astronomy, Vol. 29. Frankfurt, Institute for the History of Arabic-Islamic Science.

Zimmermann, F. W. 1978. The Dustur al-munajjimin of ms. Paris, BN ar. no. 5968. In A. Y. al-Hassan, Gh. Karmi, and N. Namnum (eds), Proceedings of the First International Symposium for the History of Arabic Sciences, Vol. 2: papers in European languages, 184-192. Aleppo, University of Aleppo, Institute for the History of Arabic Science.

THE ORIENTATION OF PRE-ROMANESQUE CHURCHES IN SPAIN: ASTURIAS, A CASE OF POWER RE-AFFIRMATION

A. CÉSAR GONZÁLEZ GARCÍA, JUAN ANTONIO BELMONTE AND LOURDES COSTA FERRER

Abstract: Pre-Romanesque churches in the Iberian Peninsula include a number of constructions from the 4th-5th to the 11th centuries when the first Romanesque churches appear in the north of Spain. We report on preliminary results from an on-going project to measure a comprehensive set of churches along the whole Iberian Peninsula. The aim is to investigate the impact of the introduction of Christianity, the possible persistence of pre-Christian customs in the orientation of churches, and their use by the elites to confirm their power.

Keywords: Church orientations. Medieval astronomy. Medieval Church. Asturian Kingdom.

Previous studies on the orientation of churches in Europe highlight the importance of solar orientations, in particular towards the east (see McCluskey 1998, 2007).

In early Christian times, Origen, Clement of Alexandria and Tertullianus recognized the relevance of turning towards the east during prayer. The Council of Nicaea (325 AD) decreed that priests should turn eastward during prayer in church, while the Apostolic Constitutions (3rd to 5th century) prescribed that churches should be constructed looking east (Const. Apost., II, 7). This was later confirmed by Pope Vigilius in the first half of the 6th century. Isidore of Seville (7th century), in his *Etymologiae* (XV, 4) (see e.g., McCluskey, 1998) prescribes that the churches must be oriented towards the equinox. The eastern orientation had a highly symbolic meaning. East was the place where the sun rises and thus, as Christ was the Sun of Righteousness, it was where Christ would rise on the eve of time. Besides it was the area towards which the Christians in the west would be facing Jerusalem (McCluskey 1998, 2007).

Orientations of churches have been explored in a number of European countries to understand the process of Christianisation: see e.g., Hoare and Sweet (2000) and McCluskey (2007) for England; Firneis and Köberl (1989) for Austria; Ministr (1997) for Bohemia; Incerti (2001) for Italy; Liritzis and Vassiliou (2006) and Pantazis and Lambrou (2006) for Greece; Lauzikas (2008) for Lithuania; and Caval (2009) for Slovenia. Such study is missing in Spain, albeit references to the 'irregular' orientation of churches is perhaps due to an erroneous interpretation of the data (Fernández, 2004) or to peculiar illumination events of certain images in Romanesque churches reported in local media.

The pre-Romanesque period in the Iberian Peninsula includes churches built from the 4th to the 11th centuries, including the late Roman period, the Visigoth Kingdom, the original Christian kingdoms formed in the North of Spain after the Muslim conquest and their further expansion towards south prior to the construction of the first Romanesque churches in the 11th and 12th centuries (Cobreros, 2006).

Palaeo-Christian and Visigoth churches

There are still a number of basilicas from the period of the late Roman Empire in the Iberian Peninsula; some of them appear today in the basements of later re-built churches or at the sites of older Roman or even pre-Roman monuments. A few dozens of churches remain from the later Germanic period all over the Peninsula (Cobreros, 2006). The state of preservation is highly variable.

During the 5th and 6th centuries a number of Germanic tribes invaded the Iberian Peninsula, notably the Goths and Suebi who settled down and created lasting kingdoms (see e.g., Collins, 2005). The Suebi settled in the northwest of the Peninsula, mostly in the area occupied by the present day region of Galicia and the north of Portugal. The Visigoth kingdom suffered many changes during the few centuries of its existence, but for most of the 6th and 7th centuries it was the ruling power in much of the Peninsula. The southern coastal areas and the Balearic Islands were part of the Byzantine Empire for almost two centuries. After a long struggle, the Visigoth kingdom finally incorporated the whole Peninsula in the mid 7th century AD. It must be noted that, although the ruling class was of Germanic origin, most of the population was of Hispano-Roman ancestry.

A peculiarity of these kingdoms was that the Suebi were Arians when they settled in Hispania, but they were early converts to Catholicism while the Visigoths followed the Arian heresy until the 7th century. At the same time, the majority of the Hispano-Roman population was Catholic. During the struggles of the 7th century to achieve effective control of the Peninsula, the Visigoth kings Leovigildo and Recaredo realized that it was easier to obtain such control under the spell of a common faith, and finally Recaredo converted to Catholicism. At this time, Toledo as the capital of the kingdom was the metropolitan seat.

The power of the Church in this period was large and it was manifested in the periodic celebration of the 'Concilios de Toledo' (the Councils of Toledo). The bishops and high-ranking church officials met together with the Monarch and discussed a number of issues not only related to religious matters but also of the kingdom in general. A total of 18 of these councils were celebrated and the records of a number of them are still extant. In some of these, the councillors gave a number of prescriptions related to the rites to be celebrated in the churches and how the churches and monasteries should be built and provided for (Bango Torviso, 1995).

Figure 1. Visigoth church of San Juan de Baños. An inscription inside informs visitors that it was built by royal decree in the year 661. Note the horseshoe arch of the entrance.

Visigoth churches display different ground plans (Cobreros, 2006). Early churches and basilicas had a double-apsidal basilica design, which later on evolved towards a Latin/Greek cross plan with several apses and with different areas on the sides. The most characteristic aspect of these churches today is the use of the horseshoe arch, similar to that found later in Muslim mosques (see Figure 1). Decorative arts include geometric and figurative bass-reliefs together with naturalistic depictions of biblical passages and allegoric images often including the Sun and the Moon (see Figure 2).

Figure 2. Left: chancel from a church in Merida with a lunar motif. Right top: astronomical capital inside the Visigoth church of Santa María de las Viñas. Right bottom: horologium from San Pedro de la Nave, supposedly to mark the time of prayer.

Several ecclesiastic authorities of the time (Leandro, Isidore of Seville, Fructuoso) issued a number of rules for the monasteries that provided them with a peculiar personality. Under these rules, the Visigoth church developed a particular rite, which involved a characteristic inner division and use of the ritual space. At the innermost part of the church, where the consecration took place, only the priest was allowed to enter. A chancel separated this part from the second area where the rest of the monastic community was celebrating, while the last part of the church was the only area into which commoners were allowed. Specific areas for sinners and other chambers caused the inner space to be divided into several chapels and rooms (Bango Torviso, 1995; Cobreros, 2006).

Apart from prescriptions on the orientation and inner division of churches and monasteries, both in the Visigoth Concilia and in Saint Isidore's writings, there are clear prescriptions about the timings of prayer and the rules of the monastic life which demanded that the course of time be followed either by sun watching during the daytime (the horologium inside San Pedro de la Nave is a nice example of this, see Figure 2) or by inspecting the stars during the night (Saint Isidore mentions that the sacristan should perform this duty in his *Regula Monachorum*; Bango Torviso, 1995; McCluskey, 1998).

The Visigoth or Hispanic rite was different from the Roman rite and it was in use until the introduction of Romanesque churches, despite political changes.

The Asturian kingdom

In 711 AD Muslims from the north of Africa invaded the Kingdom and by 713 most of the Peninsula was under the rule of the Umayyad caliphs from Damascus. The invasion meant the imposition of the new faith in most of the territory although a number of Christian communities lived under the Muslim-controlled areas. A few isolated regions in the mountainous north kept a moderate independence and particularly after a few decades they could claim a political independence from the South. Still, the Catholic metropolitan seat was in Toledo, in the Muslim-controlled areas.

Figure 3. San Juan de Valdedios, one of the best-preserved Asturian churches. Note the entrance arch, the double-arched window above it that illuminates the Royal elevated area and the play of different heights of the naves.

The tiny but historically important Kingdom of Asturias, in the mountains of the north of the Peninsula, appears as the first area to oppose the Muslim power. A way to re-affirm the Christian nature of the kingdom against the Muslim-dominated south and to proclaim themselves as heirs of the Visigoth kingdom was to build a fairly large number of conspicuous royal churches (Arias, 1999). Thirteen of these churches are today found in present day Asturias (García de Castro Valdés, 2007).

Notably, the rites and internal structure of these churches very much resemble the Visigoth tradition. A peculiarity was the elevated area for the King placed above the entrance in the space reserved for the commoners. Although the architecture of the churches is closer to the posterior Romanesque style, these churches still include to a lesser scale the use of the horseshoe arch (Figure 3). Stucco engravings and plaster frescoes compose the internal decoration, which also includes elements with clear Muslim influence.

A notable figure in the Asturian kingdom is that of Beatus (c. 780 AD), a monk at the monastery in Liebana. Beatus wrote several books, among which *Commentaries to St. John's Apocalypse* became especially famous later on. This book mentions Saint James as the patron Saint of Spain for one of the first times (Arias, 1999).

During the reign of Alfonso II, the metropolitan bishop Elipandus from Toledo and Felix, bishop of Urgel, perhaps due to Muslim influence, defended the idea that Jesus was an adoptive Son of God, an idea later to be named Adoptionism. Beatus wrote a fierce script against this idea which was highly criticised by Elipandus. The controversy reached to the court of Charlemagne and Adoptionism was proclaimed as a heresy at a Concilium in Frankfurt in 796 AD (Arias, 1999).

This was a major strike for the prestige of the metropolitan seat in Toledo, and in a few decades the Asturian kingdom broke through with the metropolitan seat. A few years later, still under Alfonso II (810 AD), the alleged tomb of Saint James was re-discovered and a basilica was built at that place, which was later replaced by a Mozarabic one and still later by the present Romanesque Cathedral (Guerra Campos, 1982).

Mozarabic period

By the beginning of the 10th century the Asturian kingdom had expanded throughout most of the northwest of the Peninsula and the capital was changed to León. A number of Christian monks from the Muslim-controlled south, who were not allowed to build new churches or monasteries, started to move towards the north, building new churches in a new style later called Mozarabic.

The new style incorporated much influence from the Muslim south, although the rite was still the Visigoth one (later wrongly termed Mozarabic). In the Scriptorium of these monasteries, the monks produced an important number of beautifully illuminated copies of the *Commentaries to the Apocalypse* by Beatus, due to the importance of the Apocalypse by the end of the millennium.

The rite and style was common to all the Christian kingdoms in the Peninsula and thus Mozarabic churches are (mainly) found in the northern half of the Peninsula. By the mid 11th century the Pope Gregory VII asked the Christian kings to follow the Roman rite (Bango Torviso, 1995). The last kingdom to introduce the Roman rite was Castile under Alphonse VI at the time of the conquest of Toledo in 1086. The change of rite coincided with the introduction of the Romanesque style and new monastic rules mainly following the Road to Santiago.

Data

Up to the moment of writing the present paper, we have been able to study, measuring their orientation and analysing the surrounding landscape, 35 pre-Romanesque churches so far, including all the extant churches from the Asturian period (13 churches), as well as 22 churches from the previous Visigoth and later Mozarabic periods with which to compare them. Figure 4 presents the orientation diagrams for all our measurements.

All churches comply with the canonical orientation: i.e., the apse of the church or the altar is facing inside the solar arc towards the eastern part of the horizon. One exception is San Miguel de Lillo, an Asturian period royal church near the town of Oviedo, which will be commented later on.

Although our sample is still incomplete with respect to the total of Visigoth churches (we estimate that we have measured about 50% of all remaining monuments) and Mozarabic ones (about 40% measured), we consider that the trends observed might be robust, and we definitely have measured all present day existent Asturian churches. As a preliminary approach, it would then be interesting to see the orientation patterns of the three groups.

Group differences

The orientation pattern of palaeo-Christian and Visigoth churches is given in Figure 4 (b). Most of the churches concentrate towards orientations close to due east with a dispersion of a few degrees compatible with the different definitions of equinoctial or eastern orientations proposed by McCluskey (2007). A few churches are outside this range, both towards north and south, perhaps indicating the relevance of some important festivities. This point could be confirmed or rejected when the full data sample is collected.

Figure 4 (c) presents the data for all the Asturian kingdom churches still preserved. We see that all churches but one (San Miguel de Lillo, see below) are oriented within the solar arc. It is also remarkable that all churches seem to avoid the southern part of the solar arc—no Asturian church presents a negative declination. A concentration towards an azimuth of 70° must be noted. Most of these churches were built under king Alfonso II (c. 760-842).

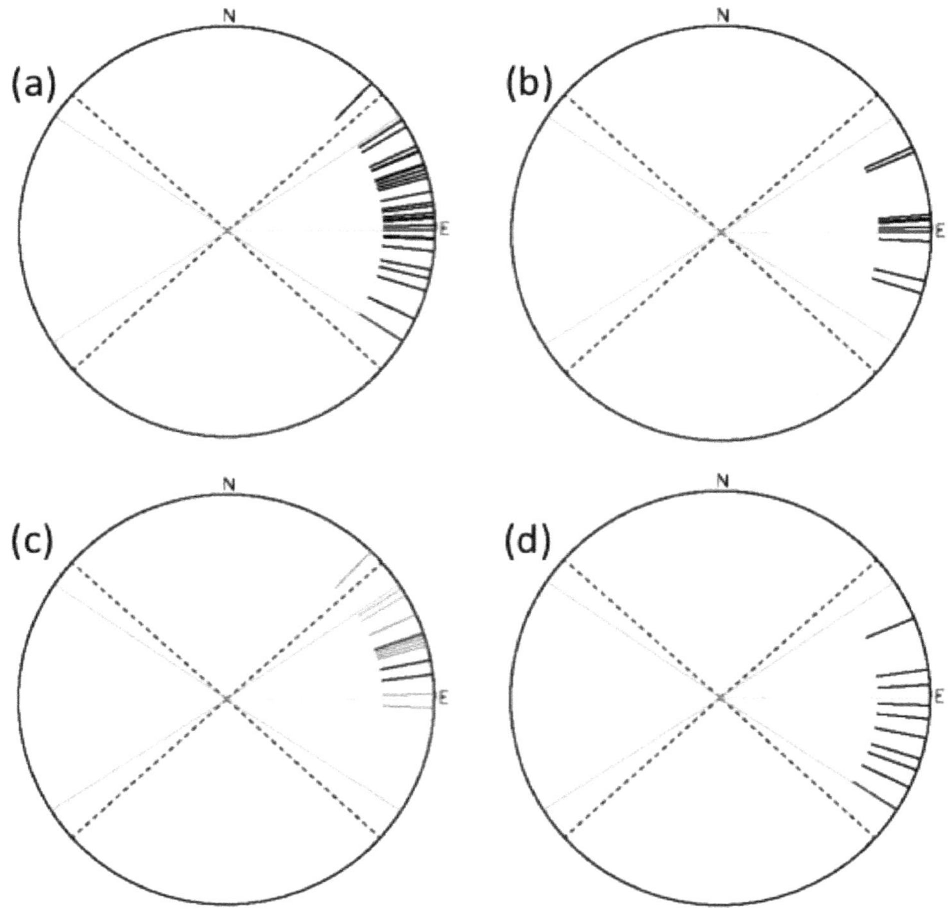

Figure 4. (a) Orientation of 35 pre-Romanesque churches in Spain. All churches are oriented according to canonical custom. Solid long lines are the solar extremes while the dashed lines give lunar ones. (b) Orientation of a representative set of palaeo-Christian and Visigoth churches. (c) Orientation of all existent Asturian churches. (d) Orientation of the Mozarabic churches measured so far. Notice the different orientation pattern relative to that of the Asturian churches.

San Miguel the Lillo was built on the slopes of Mount Naranco in the palatine complex of king Ramiro I (c.790-850). This church collapsed in early times and the remains of the church include the entrance and part of the main body of the church. Although the apsidal part was destroyed, the orientation could be measured and it is such that it is clearly north of the summer solstice rising sun. However, a possible explanation for such an anomaly is that this is a royal church and, as stated above, the king was attending the rites in an elevated area above the entrance in the commoners' area; given the elevation of the horizon in the western part, such a spot would have been illuminated by the setting sun close to the winter solstice.

The orientation of the later Mozarabic churches is given in Figure 4 (d). The orientation there is shown to be compatible with the canonical custom and there are churches with negative declination.

Discussion and preliminary conclusions

It is interesting to notice that the churches built in the period of the Asturian kingdom present a definitive orientation pattern, in principle within the canonical orientation of Christian churches (McCluskey, 1998) but certainly different from the previous and later periods.

As explained above, the Asturian kingdom had to fight for political as well as for religious independence. This fight was reflected in the Adoptionist controversy between Beatus and the Metropolitan bishop from Toledo, Elipandus.

In this sense it is interesting to compare the orientation of the Asturian churches with the mosques built in the south of the Peninsula (see Figure 5). It is worth noting that given the latitude and longitude of Al-Andalus, Muslim mosques should have been built with orientations compatible with a solar orientation (qibla around 115°) with negative declinations; however they seem to have been oriented like the Qa'aba instead of toward the Qa'aba. Both groups of monuments seem to comply with the rules for orientation of each of them but at the same time they seem to avoid each other in a remarkable way, re-affirming religious and political power by highlighting the differences in orientation.

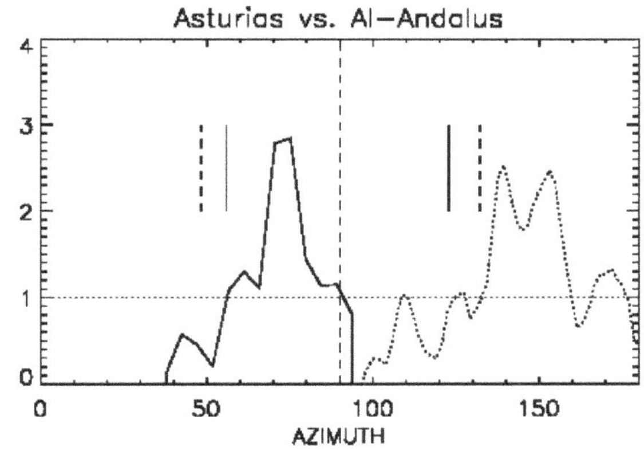

Figure 5. Histogram showing the orientations of Asturian churches (solid line) and the mosques from Al-Andalus (dashed). Vertical solid lines indicate the solar rising extremes

The Asturian kingdom fighting the Muslim powers in the south seems to have used the churches to re-affirm their sovereignty and independence by making, among many other aspects, a clear distinction with the Muslim south in the orientation of the churches, something that in previous and later times was not necessary.

Acknowledgements:

This work is partially financed under the framework of the projects P310793 'Arqueoastronomía' of the IAC, and AYA2007-60213 'Orientatio ad Sidera II' of the Spanish MICINN. A. César González García is a Ramón y Cajal fellow.

References

Arias Páramo, L. 1999. Prerrománico Asturiano: el arte de la monarquía asturiana. Oviedo, Editorial Trea.

Bango Torviso, I. 1995. Edificios e Imágenes Medievales. Historia y Significado de las Formas. Madrid, Historia 16 Temas de Hoy.

Caval, S. 2009. Astronomical Orientations of Sacred Architecture during the Medieval Period in Slovenia. In J.A. Rubiño-Martín, J.A. Belmonte, F. Prada and A. Alberdi (eds), Cosmology Across Cultures, Astronomical Society of the Pacific Conference Series (ASPCS), vol. 409, 209-219. San Francisco, ASPCS.

Cobreros, J. 2006. El Prerromanico. Madrid, Anaya.

Collins, R. 2005. La España Visigoda. Madrid, Crítica.

Fernández, C. Godoy. 2004. A los pies del templo. Espacios litúrgicos en contraposición al altar: una revision. Antigüedad y cristianismo: Monografías históricas sobre la Antigüedad tardía 21, 473-489.

Firneis, M.G. and Köberl, C. 1989. Further studies on the Astronomical Orientation of Medieval Churches in Austria. In A. F. Aveni (ed.), World Archaeoastronomy, 430-435. Cambridge, Cambridge University Press.

García de Castro Valdés, C. 2007. Arte prerrománico en Asturias. Oviedo, Ediciones Nobel.

Guerra Campos, J. 1982. Exploraciones arqueológicas en torno al sepulcro del Apostol Santiago. Santiago de Compostela, Ediciones del Cabildo de Santiago.

Hoare, P.G. and Sweet, C.S. 2000. The Orientation of Early Medieval Churches in England. Journal of Historical Geography 26, 2. 162-173.

Incerti, M. 2001. Solar Geometry in Italian Cistercian Architecture. Archaeoastronomy 16, 3-23.

Lauzikas, R. 2008. Some Cosmological Aspects of Catholic Churches in Lithuania. Archaeologia Baltica 10, 200-206.

Liritzis, I. and Vassiliou, H. 2006. Further Solar Alignments of Greek Byzantine Churches. Mediterranean Archaeology and Archaeometry, Special 6, 3, 7-26.

McCluskey, S.C. 1998. Astronomies and Cultures in Early Medieval Europe. Cambridge, Cambridge University Press.

McCluskey, S.C. 2007. Calendrical Cycles, the Eighth Day of the World, and the Orientation of English Churches. In C. Ruggles and G. Urton (eds) Skywatching in the Ancient World. New Perspectives in Cultural Astronomy, 331-353. Boulder, University Press of Colorado.

Ministr, Z. 1997. Christian Rotundas in Bohemia with Pagan Orientation to the Sun. In C. Jaschek and F. Atrio-Barandela (eds) Actas del IV Congreso de la SEAC 'Astronomía en la Cultura', 277-280. Salamanca, Universidad de Salamanca.

Pantazis, G. and Lambrou, E. 2006. On the Date of Early Christian Basilicas (Central Greece). Mediterranean Archaeology and Archaeometry, Special 6, 3, 53-60.

ASTRONOMICAL HERITAGE IN BOSNIA AND HERZEGOVINA: LATE MEDIEVAL TOMBSTONES AND ASTRAL MOTIFS

ZALKIDA HADŽIBEGOVIĆ

Abstract: Late medieval tombstones known by the name *stećak* are specific facts of astronomical heritage in Bosnia and Herzegovina. There are thousands of medieval megalithic tombstones showcasing the similarities in medieval Bosnian, Croatian, Serbian and Montenegrin cultural heritage. The first *stećci* (plural; *stećak*, singular) representing architectural-cultural means were created in 12th century and kept their primary function of tombstones until the 16th century. Over 50,000 tombstones have survived, mainly grouped by specific locations and size variations. The basic shapes of tombstones are slab, slab with upright cross, chest, tall chest, sarcophagus and, last of all, the Islamic type with the common name *nišan*. Many experts in the medieval history of the Balkans accept that *stećci* have a special cultural and artistic value. These tombstones have been under scientific investigation in a wider interdisciplinary context aimed at discovering their significance as a key for understanding the culture of *stećci* creators. A large number of tombstones were decorated with different designs from every day life (floral, zoomorphic, weapon representations) and many astral motifs. The significance of astronomy related motifs found on the tombstones could be associated with certain level of astronomical knowledge gained in that time.

Keywords: astral motifs, astronomical heritage, Bosnia and Herzegovina, medieval tombstone, stećak, stećci

Old stone tombstones found in the Balkans, locally known by their name *stećak* (plural, *stećci*), hold an important cultural, artistic and scientific dimension. Over the last five centuries they have been present in the territory of four Balkan countries: Bosnia and Herzegovina, Croatia, Serbia and Montenegro. With their genuineness and authenticity, beauty of decoration, philosophy and literary value of written messages, they deserve to soon become a treasure listed on the world heritage list. Although they principally represent tombstones, the decorations of *stećci* are of social, religious, artistic, architectonic and scientific importance, and are sublime expressions of all of the above. One of the most prominent explorers of *stećci* in Bosnia and Herzegovina, Šefik Bešlagić (2004, 16) has always accented the diversity of scientific and philosophic dimensions of the *stećci*, and the constant need for continuous research in various fields like archaeology, history, ethnology, art, literature and science in order to discover their real values. In the Balkan region, Bešlagić (2004, 13) found a total of 3162 localities and evidence of 69,356 tombstones, but 86% of them are in Bosnia and Herzegovina.

These unique cultural monuments date back to the 12th century and were built until the arrival of the Ottoman Empire in 1563 AD (Bešlagić 2004, 14). The cultural origin of the Bosnian tombstones is still debated. Some scientists associate these monuments to the Old Bosnian Church, and connect it with the Bogomils. Marian Wenzel spent several decades researching Bosnian tombstones and metalworking, and she developed a theory of a specific style, named the Bosnian Style, as a multi-ethnic phenomenon related to Bosnian monolithic tombstones and metalworking with a mixture of Gothic, Byzantine and Mediterranean-Islamic styles used by three religious groups in Medieval Bosnia, the Old Bosnian Church, Orthodox Church and Catholic Church (Wenzel, Imamović, Bartulica and Lovrenović 1999, 160). However, there are researchers that deny the sacral value of the tombstones, and emphasize only their artistic role.

The main aim of my paper is to highlight the role that Bosnian tombstones have in the astronomical heritage of medieval Bosnia and Herzegovina, which has not been previously explored. Secondly, if a hypothesis that their strong connection with astronomical knowledge is confirmed, that would bring to light some new aspects of Bosnian tombstones. Sources for research of the connection of *stećci* with astronomy are *stećci* inscriptions that have primarily a burial and committal function (Lovrenović 2009, 92). Epitaphs could contain data that linked some medieval Bosnian scholars with astronomy, or perhaps show that they recorded their knowledge in astronomy or astronomical events on *stećci*. If these monuments with their astral motifs were made for scientific purposes and built according to general scientific or astronomic postulates, one could conclude that these tombstones were not simply interment decorations, as it has been generally considered.

Classification, time and localities of the Bosnian tombstones

In the territory of Bosnia and Herzegovina tombstones are located mainly near riverbeds, plateaus and important historic settlements. Generally, Dimitrije Sergejevski classified them into two groups in 1952: horizontal and vertical stone blocks (Lovrenović 2009, 57). *Stećci* are monumental pieces of interment architecture, and because of their mass, which reached up to 30 tons, they could be considered as megalithic monuments. Around 6000 specimens were hand made in the relief form and decorated with various motifs. Bešlagić (2004, 15) found 80 main types of relief decorations. The greatest number of tombstones were made as horizontal stone blocks (around 95%) in several forms: slab, chest, tall chest, a roofed sarcophagus (*sljemenjak*) and slab with upright cross, whereas vertical blocks (stele) in the form of a pillar, cross or characteristic Islamic tombstone (*nišan*) were infrequent. The introduction of sarcophagus-shaped *stećci* is interesting because of their functionality. Sarcophagi appeared in Bosnia in the 15th century and represented tombstones, as opposed to Roman architecture where they represented tombs (Lovrenović 2009, 59). The most renowned sarcophagus-shaped *stećak* is a house-shaped *stećak* near Sarajevo in Hresa that resembles a gothic sarcophagus (Palameta 2004, 132).

Stećci have an immeasurable cultural value for the history of Bosnia and Herzegovina, whose territory was once part of the medieval Bosnian state (Ančić 2005, 43). Some *stećci* localities are exceptional, like Stolac in Herzegovina with its several necropoli among which Radimlja is worthy of mention.

Figure 1. Sarcophagus-stećak in the museum courtyard in Sarajevo, Bosnia and Herzegovina. Photo by Zalkida Hadžibegović, 2008.

Radimlja is a necropolis containing *stećci* well decorated with heraldic signs and carved texts in the languages of different ethnic groups that used similar ways of respecting the deceased, as well as to worship the power and wealth of distinct Bosnia and Herzegovina feudal families like Miloradović-Stjepanović. The Radimlja necropolis was proclaimed a Bosnian National Monumental complex in 2002.

Among various localities with Bosnian tombstones there are several localities with predominant astral representations: Biskup near Konjic in the central Bosnia with 117 monuments, the area near Blidinje in the highlands of central Bosnia with 464 monuments representing mainly symbols of the Moon and stars, and Vares near Sarajevo with 37 items representing motifs of the Sun and the Moon. In the Blidinje area there are 32 remarkably beautiful tombstones decorated entirely with Moon representations. Other localities important in the search of the association of tombstones and astronomy are an area near Stolac in Herzegovina together with a large area near Tomislavgrad with 378 tombstones decorated mainly with various forms of stars, crescents, and crosses. In the area of Buncic in southern Herzegovina, as well near Tomislavgrad, there are some tombstones ornamented with swastikas—a prehistoric symbol of the Sun and fire. Otherwise, swastikas are found on about 40 tombstones in several different localities in Bosnia and Herzegovina.

Stećci necropoli were mainly situated by churches, villages and roads in order to remind people of ever-threatening death. Preliminary analysis of the orientation of necropoli and tombstones showed that the predominant orientation was east to west, whereas north to south orientation was less represented. On some tombstone planes there were symmetrical crosses that from the astronomical perspective could represent cardinal directions of orientation.

Astral motifs on Bosnia and Herzegovina tombstones

Astral motifs predominantly found on Bosnia and Herzegovina tombstones were symbols of the Sun, Moon and the stars shown in the shape of a circle, star and rosette, semicircle and circular garland (ring). Solar presentations are infrequently found together, whereas lunar motifs were commonly grouped with the depiction of crosses and stars. Lunar symbolism was almost always represented as an early crescent, with different orientation of narrowed parts, or sometimes as a half-disk. Symbols of a full Moon could not be found, although some depictions of a plane with centred point could be interpreted as a representation of the Sun as well as the Moon. Bešlagić (2004, 77) found around 1400 specimens of astral motifs grouped in five types: crescent (most frequent), star (rosette), circle, circular garland (ring), and knobs (halfapple). Few researchers associated astral motifs on tombstones with Christian symbolism of the Sun, Moon and the stars, and most of them considered these motifs purely as decorations with no clear links with astronomy. Symbolism of astral motifs was often represented together with depiction of humans, thus associating the symbolic connection of man, Sun and life.

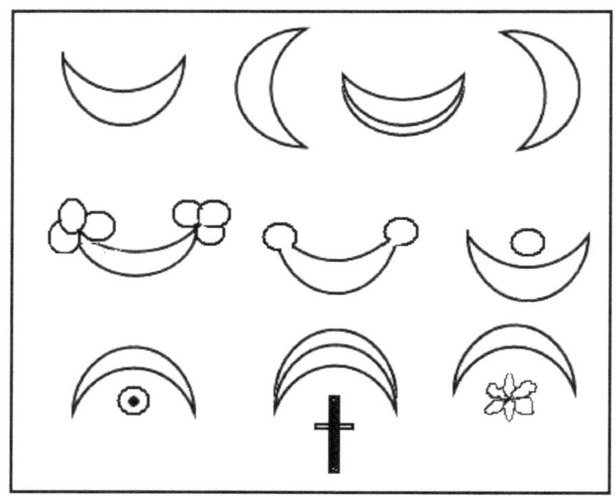

Figure 2. Most frequent crescent astral motifs on the stećci.

CRESCENT

The basic astral motif is a crescent. It occurs in few variations, unaccompanied or with star or cross combinations. Its basic form is an arch with horns downwards, upwards or sideward. According to some researchers the crescent symbolizes Bosnian knights or plays the role of a heraldic element. It has been seen with ends continuing in small circles and rosettes or as a handle of a staff, and in a hand-crescent composition. In some cases the crescent appears in a pair. If the crescent represents the Moon, its left- or right-orientated horns are possibly related to the cycle of lunar phases. The most frequent is its horizontal position on the tombstones, which shows its decorative function rather than any connection to astronomy (Figure 2).

STAR-ROSETTE

The stars as the second frequent astral motif are found on 450 tombstones (Bešlagić 2004, 112), and it is difficult to separate a star from rosette. The stars have several variants, and appear with four, five, six or eight arms, free or closed in a circle. If stars were used as an ornament rather than a symbol, they were stylized by adding circle lines, angels, or arches. Stars are very often in a combination with a crescent, Sun and cross.

Figure 3. Some of star-rosette motifs on the stećci.

Presence of the stars on the *stećci* is not investigated enough, but the star was mainly realized plastically or engraved. The stars were correlated with both Slavic cults and Christian symbolism or used as a typical heraldic element on behalf of higher feudal class members. The rosette is a round stylized flower design as a relief tombstone motif, and was mostly used in the Herzegovina region and Northeast Bosnia with four, five or eight petals.

CIRCLE, CIRCULAR GARLAND AND KNOBS

The circle is one of the relief motifs represented mainly on 175 *stećci* in central and western Bosnia (Bešlagić, 2004). It appears mostly unaccompanied, with or without a central spot. Less frequently it has rays, or it is combined with crescent and star. The circle is a universal symbol of the Sun as a source of life in many civilizations, as well as old Slavs. Circular garlands appear on 175 *stećci*, and most commonly circle stars and crosses.

As a part of anthropomorphic structures, it appears as a circular garland above human heads or in a raised hand, and very commonly in a four-fold composition with a star, crescent and cross. It is insufficiently explored as an astral motif, and it most frequently symbolizes the Sun, or a general symbol (crown) of life and a sign of salvation for Bogomils. Most scientists consider this motif as a purely decorative one, most probably because of insufficient facts on their other meanings beside artistic or architectural.

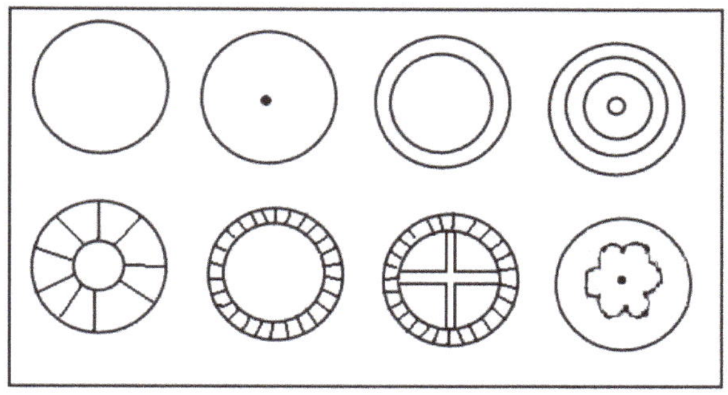

Figure 4. Circle, circular garlands (rings) and knob drawings.

KNOB (HALFAPPLE)

The knob appears as a relief motif in many variations on around 100 *stećci* in Bosnia and Herzegovina, and is not to be found in Serbia. It appears mostly as an unaccompanied circle, circular garland or a halfapple on the sarcophagus ridge. Halfapples with dentations from centre to edge may represent stars (Bešlagić 2004, 97). Knobs appear unaccompanied or in groups of two, three, or four in a specific composition. Knobs also appear on the tops of pointed stele or on sides of a prismatic stele crest (Wenzel 1965, 133). Marian Wenzel considers halfapples a result of Turkish influence (Wenzel 1965, 133), especially around Sarajevo, Travnik and Kupres.

Decorative role of spirals and crosses

Spirals are considered an artistic and floristic stylization based on religious symbolism with a predominantly ornamental role. Spirals were not considered a part of astral symbolism by most researchers. They are mainly ornaments on *stećci* in eastern Bosnia (Olovo, Kladanj, Vlasenica) and the river Drina valley (Zvornik), and appear mostly in second half of the 14th century and the first half of the 15th century.

Figure 5. Spirals on the stećak in the Botanical Garden of The National Museum of Bosnia and Herzegovina. Photo by Zalkida Hadžibegović, 2009.

Stećci spirals resemble most spirals on Greek grave steles (Wenzel 1965, 179). Bešlagić (2004, 105) described five types of spiral: simple S spiral, double spiral, spiral with cross, and spiral in composition with a cluster of grapes. Wenzel linked the appearance of spirals on *stećci* with its appearance in renaissance Italy. Most researchers find spirals as possible symbols of death and resurrection, especially when combined. The appearance of double spirals, a typical symbol of equinox (spring and autumn) since the Stone Age, on tombstones in the Balkan region could also have astronomical connotations.

Figure 6. Some of crosses which appear on the Bosnian tombstones.

In the Balkan region there are around 900 tombstones where crosses have a major decorative role (Bešlagić 2004, 86). The crosses are either incised or in relief, and vary in shapes from the simplest crosses with equal arms to those with some extensions or developments on one or more arms. There are several forms of double crosses, triple crosses and repeated crosses that are most frequent north of the river Neretva (Wenzel 1965, 92). Crosses appear in several variants based on their sepulchral functions as a cross of the Maltese type, Latin cross, Saint Andrew's cross, Greek cross, Floriated cross, Lorraine cross and Tau cross which is a late feature of *stećci* decoration. The position of the cross when represented on the horizontal surface of tombstone is always directed towards the west end. It is the position of the deceased under the tombstone in a way that the cross is projected above face of the deceased. The Egyptian hieroglyphic, a cross with ring developments on its arms, appears on the *stećci* as a cross with a single ring. If its role is put in astronomic context it could be considered as the sign of the planet Venus or simply as a symbol of life in general. Cross-shaped stones, also numerous, are not a theme in this study.

Bosnian two steles and Astronomy

In the Botanical Garden of the Bosnian National Museum in Sarajevo there are two tombstones in stele form. They are interesting from an astronomical point of connotation and sense. These two samples decorated or arranged only on astronomical purposes are linked to the Islamic style of funerary monuments that appeared in the second half of the 15th century. Both steles have been relocated from their prior locality in the Botanical Garden to a new site, without saving their original orientations. The first stele is dedicated to Mahmut Brankovic, and dates between 1471 and 1485 (Wenzel 2004, 295). Its original locality was in Petrovo polje near Rogatica in eastern Bosnia where Islamic and Christian cultures coexisted. The stele of Brankovic, who was a Turkish military authority, stands 244cm tall and around 45cm wide. The form of this stele is a stone block finishing in pyramidal top with a knob on the top and in the middle of each of its sides. On the west side of the stele is a carved inscription, with a lion and a broad sword typical in Ottoman military tradition. The east side of this stele is a double pennant lance, and on the north side is a staff with a knob on the top. The side with possible astronomical symbols is the south side of the stele with a horizontal double crescent and four knobs around the crescent and three concentrated circles above the crescent (Figure 7).

Figure 7. The stele dedicated to Mahmut Brankovic in the Botanical garden of The National Museum of Bosnia and Herzegovina in Sarajevo. Photo by Zalkida Hadžibegović, 2009.

The second stele, situated closely to the first one in the Botanical garden, belonged to Radivoj Oprašić, a Christian in the service of an Ottoman ruler who did not convert to

Islam. It is considered that it dates from around 1476 (Wenzel 2004, 297).

This stele was also moved from an Orthodox cemetery in the small village of Oprasic near Rogatica. The stele is 280cm tall and around 45cm wide and very similar to the first one in shape. It shares a similar composition of ornaments on the vertical sides. The south side of this stele is decorated with a single pennant lance, and on the east side is a decoration of a staff with a knob on the its top, and under the staff is an inscription of several words about Radivoje Oprašić as a respected knight. On the west side of the stele is a sword and on the top is a lion.

The possible astronomical symbols are on its north side with a horizontal crescent in the middle, and only one knob above but with two knobs under this crescent. The four concentric circles are above the crescent-knobs composition.

Figure 8. The stele dedicated to Radivoje Oprašić in the Botanical garden of The National Museum of Bosnia and Herzegovina in Sarajevo. Photo by Zalkida Hadžibegović, 2009.

If one puts the south side and north side of the steles in an astronomic context several hypothesis are plausible. The first one is a simple astronomical image of an arrangement of stellar bodies in a certain moment, where the form of concentric circles represents the Sun, surrounded by three planets, or by the Moon with a planet under it. The other hypothesis would set knobs as stellar bodies that were seen during a Sun eclipse. This could be supported by the horizontal position of the crescent that could never represent the Moon in that case, because it is impossible to ever see the Moon in that position in the sky. The third option would be a Moon eclipse, when three or four planets could be theoretically seen around it.

The answer as to which of these hypotheses is scientifically approvable requires detailed research in a dimension that is totally new to *stećci* exploration. It is important to mention that the arrival of the Ottoman administration brought several Ottoman scholars who understood astronomy in order to establish workshops for the astronomical time-keeping essential for determination of the timing of the five Muslim daily prayers. (Hadžibegović, 2004).

Vision of future investigation of astral characteristics of Bosnian tombstones

Astral motifs on Bosnian tombstones have not yet been systematically explored with regard to their astronomical aspects. It would be valuable to investigate astronomical aspects of these tombstones within an international project that could involve known world experts for astral symbols along with researchers from Bosnia and Herzegovina, especially Bosnian students of astronomy and archaeology. This topic should be of a greater interest for the scientific community in Bosnia and Herzegovina, especially because of the initiative of placing tombstones belonging to all four countries on the world heritage list.

References

Ančić, M. 2005. Kasnosrednjovjekovni Stolac. Povijesni prilozi 29, 41-58.

Bešlagić, S. 1967. Stećci centralne Bosne. Srednjovjekovni nadgrobni spomenici Bosne i Hercegovine IX. Sarajevo, Zavod za izdavanje udžbenika.

Bešlagić, S. 1971. Stećci i njihova umjetnost. Sarajevo, Zavod za izdavanje udžbenika.

Bešlagić, S. 1982. Stećci-kultura i umjetnost. Sarajevo, Veselin Masleša.

Bešlagić, S. 2004. Leksikon stećaka. Sarajevo, Svjetlost.

Gavazzi, M. 1965. Grobovi sa krugom od kamena. Godišnjak, Centar za balkanoloska ispitivanja 3, 188-193.

Hadžibegović, Z. 2004. Ideas and Problems Contained in the Astronomical Manuscripts Kept in Bosnia. Unpublished PhD thesis, University of Sarajevo.

Lovrenović, D. 2009. Stećci. Sarajevo, Rabic.

Palameta, M. 2004. Odjeci gotike na stećcima. STATUS Magazin za političku kulturu i društvena pitanja, (02), 130-145.

Wenzel, M. 1965. Ukrasni motivi na stećcima-Ornamental Motifs on Tombstones from Medieval Bosnia and Surrounding Regions. Sarajevo, Veselin Masleša.

Wenzel, M., Imamović, M., Bartulica, A. and Lovrenović, M. 1999. Bosanski stil na stećcima i metalu: Bosnian style on tombstones and metal. Sarajevo-Publishing.

THE STAR OF MAGI: TRANSIENT ASTRONOMICAL EVENTS AS SOURCES OF INSPIRATION IN LATE MEDIEVAL ART

M. INCERTI, F. BÒNOLI AND V.F. POLCARO

Abstract: It has long been recognized that a number of exceptional and highly impressive astronomical events are represented in Medieval artwork, such as the Bayeux Tapestry and Giotto's 'Adoration of Magi' in the Scrovegni Chapel in Padua, representing the P/Halley comet transits of 1067 and 1301, respectively, while it has been suggested that the 'Apparition of Star to Magi' fresco in the San Pietro in Valle Abbey in Ferentillo (1182) represents the 1181 Supernova. To date, however, no systematic survey of figurative Medieval and Renaissance art has been carried out in order to analyse the role of transient astronomical events as sources of inspiration in artworks in these epochs. In this work, we analyse more than a hundred artworks—dating between the 12th and 14th centuries—representing figurative elements connected with astronomy. Our statistical analysis reveals that the artists may have been inspired and even influenced by coeval astronomical events, today well known, but extraordinary and impressive to their eyes.

Keywords: Iconography Star of Magi, transient astronomical events, Medieval art

Introduction: the ACHe database

Our aim is to investigate—using statistical methods—if observed and well-documented relevant astronomical phenomena could have been a source of inspiration for medieval artistic productions.

In order to generate a homogeneous, statistically significant sample, we choose a single artistic subject. We thus built a database including all the representations of the Magi and Nativity that we were able to find, whether the Star is present or not.

The structure of the database includes information regarding the regional (region and province) and liturgical (building and patron saint) context, physical information (typology of artwork and material), time information (century, dating of the architectural or decorative element, conventional date for statistical analysis and related uncertainty), information related to the Star (shape of the star, usability of data, shape of the rays, number of rays, presence of a tail, size of the star), plus the references to the artwork and eventual comments.

Concerning the classification of the material/architectural and decorative element, we considered the following: paint, sculpture, jewellery, ivory, manuscript, metal and wood.

The conventional date for statistical analysis is the actual date of the artwork, when it is known, and the middle of the uncertainty interval if it is not. For the type of action represented we considered the following: Adoration of the Magi, Nativity, the Magi's dream, the Magi's journey, Joseph's dream, Magi as astrologers and others. Referring to the specific literature for the symbolic thematic and symbolisms, the following typologies have been considered for the shape of Star: non present, star, star with external circle, oriental star, comet, angel, rosette, flower with or without circle, sun and others. The usability of the data takes into account that some of them need to be integrated, some are out of the time window, and some are out of the representation scheme. Concerning the number of rays, we used the following classification: 0, 3, 4, 5, 6, 7, 8 and 12, not measurable and not recognizable. Concerning the size, we considered the cases of no star and stars greater than human heads, the same as heads and smaller than heads.

The research was focused on late medieval Italian artworks; however, the database contains some testimonies of the previous and following centuries. The time distribution of the artwork analysed to date is mainly concentrated on the 12th, 13th and 14th centuries (about 74% of the total). To date, the analysed artworks are 192 in total. In some cases, more narrative episodes are present in the artworks for a total of 213 episodes.

The Magi: a brief outline of the sources

The oldest-known image of the Magi dates back to the 2nd century. The three Wise Men appear in a fresco in the Catacombs of Priscilla in Rome. The Star is not visible, due to the bad repair of the superior part of the fresco. However, the Star is visible in the same Catacombs in the late 2nd-century image of Mary nursing the Infant Jesus. To the left of the Virgin, a man—who may be the diviner Balaam—is pointing to a Star apparently showing 8 rays.

The symbolism of the Star appearing in the skies to mark the birth of Jesus has been linked to some events in the Old Testament, as Balaam's oracle (Book of Numbers 24, 15-17):

> 15. Then Balaam gave voice to his oracle: the utterance of Balaam, son of Beor, the utterance of the man whose eye is true;
> 17. [...] A star shall advance from Jacob, and a staff shall rise from Israel [...];

and the prophecy by Isaiah (62, 1):

> 1. For Zion's sake I will not be silent, for Jerusalem's sake I will not be quiet, until her vindication shines forth like the dawn and her victory like a burning torch.

Matthew's spare account on the Magi (Gospel of Matthew 2, 1-12) is one of the written sources which probably inspired the representations of the Star in the medieval art.

The Nativity—and as a consequence, the Star—is not frequently represented in the first centuries: Christmas does not appear in the first lists of Christian holy days by Saint Irenaeus and Tertullianus (2nd-3rd century), whereas the Epiphany—i.e., the visit of the Magi to Infant Jesus—was already celebrated. Christmas as the celebration of the birth of Jesus of Nazareth was introduced in Rome in the 4th century and in Milan at the end of that century.

Starting in the 8th century, these themes have been diffused in western Christianity through the mediation of Armenian, Syriac, Greek and Coptic manuscripts, but most of all through the tradition of Apocryphal Gospels, literature and theological writings.

Figure 1. Abbey of Monreale (near Palermo), the Star of Magi, mosaic (1178-1192).

The success of the representation of the Epiphany in Western countries is signified by a very high number of testimonies in Palaeo-Christian art and in Christian monuments, such as sarcophagi, frescoes, mosaics, ivory tablets, devotion medals, Gospel-folders, pottery, jewels, relic cases, chairs and other objects.

Among the frescos of the Roman catacombs, there is one representation from the 2nd century, three from the 3rd century and seven from the 7th century (Cabrol et al., 1907; Vezín, 1950).

After the 8th century, the iconography of the Star in representations of Magi and Nativity is very common and widely diffuse in all the times and in all the figurative artistic finds (see e.g., Figures 1 and 2) (Chiappori, 1997; Felix, 2000; Cardini, 2000; Olson, 1984).

Figure 2. Castelseprio (near Varese), Santa Maria foris porta*, the comet of the Nativity (7th-8th century).*

PLACE	REGION	BUILDING	PATRON SAINT	TYPE OF ARTWORK	CENTURY	DATE OF THE ARTWORK	DATE UNCERTAINTY (±Y)	COMET NAME AND YEAR OF TRANSIT
Bologna	Emilia	San Petronio	San Petronio	fresco	15th	1410	2	C/1402?
Bergamo	Lombardia	Bishop's palace	-	fresco	13th	1211-1309	50	?
Castelseprio	Lombardia	Santa Maria foris porta	The Virgin Mary	fresco	7th-8th	7th-8th century	100	?
Monreale	Sicily	abbey	Our Lady of Assumption	mosaic	12th	1178-1192	7	P/Swift-Temple 1175 (+ C/1185 +C/1189?)
Assisi	Umbria	Lower Basilica	San Francesco	fresco	14th	1306-1311	2	P/Halley 1301?
Cividale del Friuli	Veneto	Tempietto Longobardo	-	fresco	14th	1300	20	P/Halley 1301?
Padua	Veneto	Cappella degli Scrovegni	-	fresco	14th	1306	1	P/Halley 1301

Figure 3. Representations of comets in late medieval artworks from the ACHe database.

ACHe database univariate statistics: action represented, star shape, number of rays and size of the star

The large majority of the artworks included in our sample represent the Adoration of Magi (45%) and the Nativity (about 32%).

The most common typologies are the Star as an 'Angel', the generic 'star' and the Star with external circles.

However, it is worth noting that in a few cases (7 in total: see Figure 3) a comet is clearly represented, though in most cases the date cannot be determined with sufficient approximation to identify the comet from Eastern sources (e.g., Ho Peng Yoke, 1962; Xu, Pankenier and Jiang, 2000).

Concerning the number of rays, when they are visible, the most common cases are 8-ray (36%) and 6-ray stars (21%). Compared to the head of the characters, only in 18 cases of artworks is the size of the star significantly large and standing over the context.

Time series statistical analysis

As said before, a large part of our database (120 out of 193 valid cases) concerns artworks made between the end of the 11th and the end of the 14th century.

This is due on one hand to the fact that strong changes in the Roman Catholic Church at the end of 11th century (see, e.g., Le Goff 2006, 104) brought a substantial renovation of many sacred buildings, provoking the destruction of the majority of the old artworks. On the other hand, the number of surviving artworks for the epochs following the beginning of the 14th century is so high that it has been so far impossible for us to perform a detailed analysis of a significant percentage of them. However, the size of the 12th, 13th and 14th-centuries sub-sample is large enough to try to perform a statistical analysis concerning these three centuries. In Figure 4 the counting of artworks representing the Star from 1101 to 1400 is presented. Data has been binned on 5-years bins. The straight red line is the average over the period (2.4 artworks per 5 years).

Vertical error bars represent the standard deviation of the distribution (corresponding to 1.6 artwork per 5 years). The RMS average uncertainty of the date (2.5 years) is given by the symbol size.

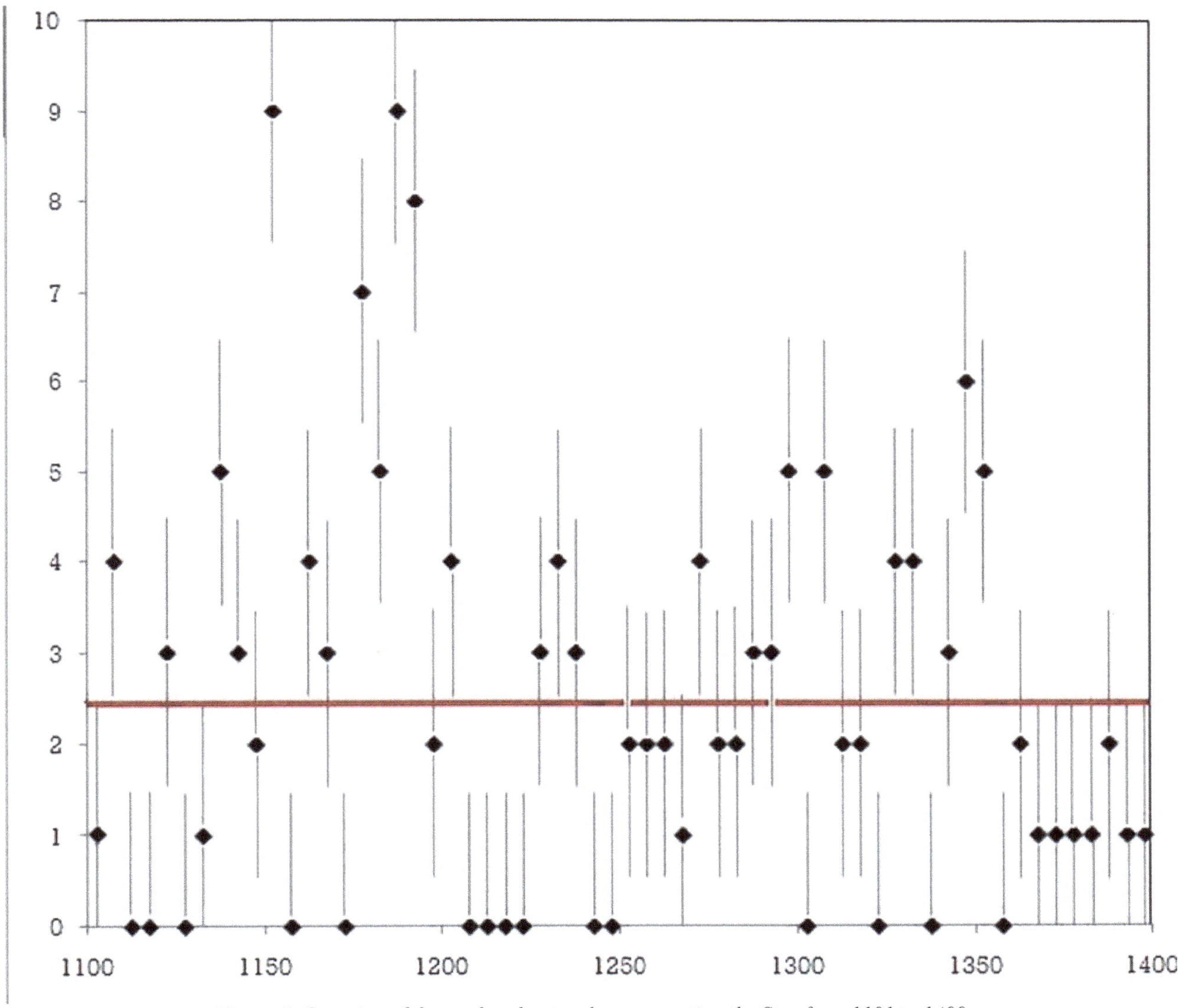

Figure 4. Counting of the analysed artworks representing the Star from 1101 to 1400.

As can be seen, most of the representative points fall at less than 2σ from the average, showing that the variations in counting are due to statistics. Only a few points stand more than 2σ over the average, suggesting that they are due to non-random reasons. These points correspond to the 1150-1154 bin and to the three bins covering the period 1175-1194.

In a previous work (Incerti, Bònoli and Polcaro, 2009) we presented the counting of the artwork representing the Star in the same time interval (1101-1400), but based on a smaller sample, obtaining two significant statistical bins over 2σ: 1141-1150 and 1181-1190. In that work we noticed that the intervals corresponded to the two most outstanding astronomical phenomena of the considered time window: the transit of comet P/Halley in 1145, followed, as reported by Oriental chronicles (Ho Peng Yoke, 1962; Xu, Pankenier and Jiang, 2006), by the transit of two bright comets in the following four years and the Supernova 1181, whose representation in the San Pietro in Valle Abbey (Ferentillo, Italy) has been previously suggested by Polcaro (2006).

It is worth noting that a study of the astronomical records in the annals of the Hawaiian 'king-priests' (Masse, 2010) gives maxima in the very same time interval, thus proving that the excesses in the artwork counting that we detected were not due to chance.

The narrower time resolution due to the higher statistics we use in the work allows us to confirm that the first excess is actually nearly contemporary to the transit of the P/Halley in 1145. As far as the second excess is concerned, in light of latest analysis it seems to be due not to the Supernova 1181 alone, but to four astonishing astronomical events which occurred between 1175 and 1190: the transit of the bright P/Swift-Temple in 1175; the Supernova 1181 and the transit of two bright comets, reported by the Oriental chronicles in 1185 and 1189 (Xu et al., 2006). Supposing that the number of analysed artworks and their dating are affected by random errors and that the binned counts can thus be represented by Gaussians centred on the known most impressive astronomical events of the considered time interval, we can build a mathematical model, where the height of the peaks is a free parameter, that fits the experimental data quite well (see Figure 5).

Conclusions

We can thus suggest that, as a minimum, the most impressive astronomical events could actually have been the source of inspiration for medieval artists, or of their commissioners, who increased the number of artwork where a star is represented when these astonishing events happened and in the years immediately following.

However, the increase in the number of artworks follows the event by 1-2 years (as it is natural, taking into account the time needed to produce the artwork).

Furthermore, the effect, apart from a few cases, is small and becomes noticeable only when more than one event happens in a few years.

Lastly, only a few artists (such as the Master of Ferentillo and Giotto) represented the event in a realistic way, while usually the Star is represented following a 'standard' iconography (see Figure 6).

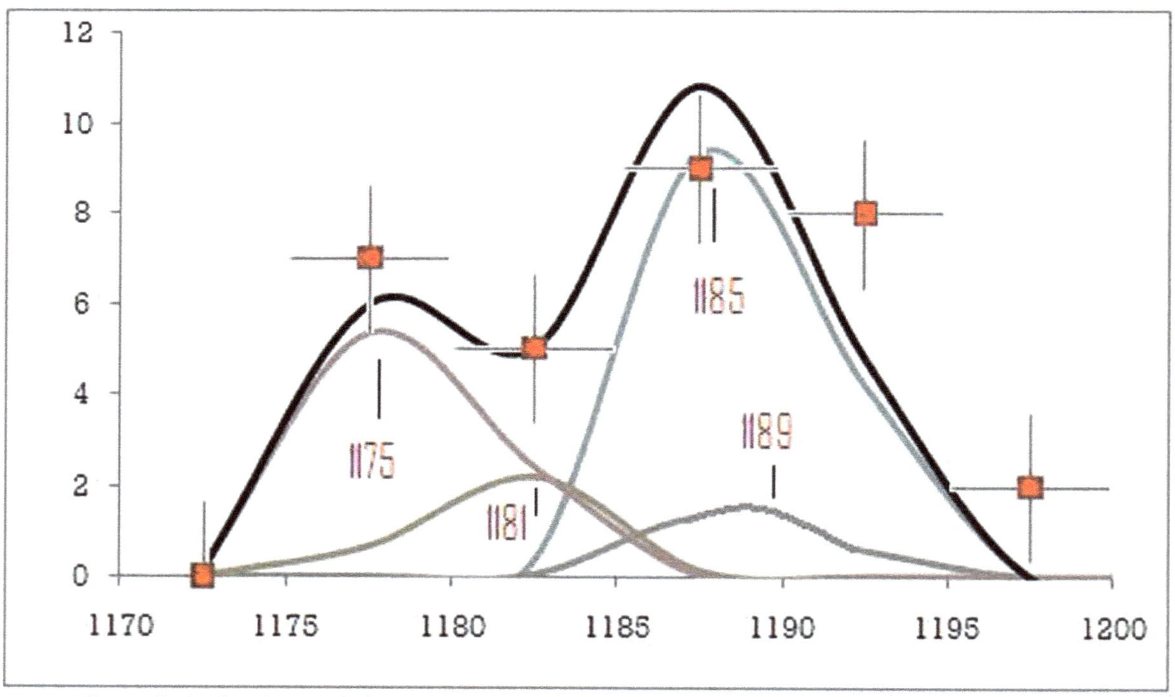

Figure 5. Analysed artworks between 1170 and 1200 and their interpretative model (see text for details).

Figure 6. Cathedral of Altamura (near Bari), Magi in the portal (14th century).

References

Cabrol, F and Leclercq, H. (eds) 1907. V. Mages. In Dictionnaire d'archéologie chrétienne et de liturgie, Vol. 1-2. Paris, Librairie Letouzey et Ané.

Cardini, F. 2000. I re magi: storia e leggende. Venezia, Marsilio (in Italian).

Chiappori M. G. 1997. Magi. In Enciclopedia dell'arte medievale, Vol. VII, 1228-1230 (in Italian). Rome, Istituto dell'Enciclopedia Italiana.

Felix M. and Boni, D. 2000. I Re Magi. Milano, Jaca Book (in Italian).

Ho Peng Yoke. 1962. Ancient and Medieval Observations of Comets and Novae in Chinese Sources. Vistas in Astronomy 5, 127-225.

Incerti, M., Bònoli F. and Polcaro V.F. 2011. Transient astronomical events as inspiration sources of Medieval and Renaissance art. In The Inspiration of Astronomical Phenomena VI (Venice, Oct. 2009) Vol. 441, 139. .

Le Goff, J. 2006. Il cielo sceso in terra. Bari Laterza (Italian translation).

Masse, W. B. 2010. The celestial engine at the hearth of traditional Hawaiian culture. In M. Rappenglück (ed.), Astronomy and Power: How worlds are structured. Oxford, British Archaeological Reports (this volume).

Olson R. J. M. 1984. ... And They Saw Stars: Renaissance Representations of Comets and Pretelescopic Astronomy. In Art Journal 44, no. 3, Art and Science: Part II, Physical Sciences, 216-224.

Polcaro V. F. 2006. A possible European Witness of Supernova 1181 in Todd W. Bostwick and Bryan Bates (eds), Viewing the Sky through Past and Present Cultures. Oxford VII International Conferences on Archaeoastronomy, Pueblo Grande Museum Anthropological Papers no. 15, Phoenix AZ, 399-426.

Vezin G. 1950. L'adoration et le cycle des Mages dans l'art chrétien primitif: étude des influences orientales et grecques sur l'art chrétien. Paris, Presses Universitaires de France.

Xu, Z., Pankenier, D. W. and Jiang, Y. T. 2000. East-Asian Archaeoastronomy: Historical Records of Astronomical Observations of China, Japan and Korea. Amsterdam, Gordon and Breach Science.

A SOCIAL HISTORY OF MEDIEVAL ASTRONOMY

A. MARTOCCHIA AND V.F. POLCARO

Abstract: This work is a summary of the first chapters of the book *Storia sociale dell'Astronomia* (*A social history of astronomy*), currently in preparation. We analyze the main factors that played a role in suppressing or re-awakening interest in astronomical observations and events throughout the centuries. Among such elements we include: the loss of Greek-language knowledge as a vector of scientific knowledge; Christian and Islamic conceptions of astrology; religious practices connected to sky observations; the birth of universities; and the evolution of the social figure of the scientist in the West–from monks to aristocrats. In this contribution we focus attention especially on the role of the Christian and Islamic religions and on the complex relationships between Europe and Islam in the evolution of astronomy in the Middle Ages.

Keywords: History of Astronomy, Medieval Astronomy

The collapse of the Western Roman Empire and the loss of the Greek language

The Roman Empire culture was basically bilingual, with the Greek language being known by literate people as well as Latin. The collapse of the Western Roman Empire determined the end of this culture. In Western Europe, when the Roman schools disappeared, the Greek language was lost.

Unfortunately, most of the classic scientific literature was written in that language and typically used for science, as English is used today.

Thus the astronomical knowledge, born in Egypt and Mesopotamia, transmitted to the Greek civilization and then to the Roman Empire mainly by the Phoenician and Alexandrinian Kingdoms and culminating with Ptolemy's work, also was lost.

Only a minor number of texts written in Latin (today we would call these works 'popular science')–and in particular the ones of Plinius, Macrobius and Martianus Capella–remained available (see, e.g., Le Goff, 2004). These works reported the results of ancient science but not the methods used to reach these results: this circumstance made it impossible to further develop previous knowledge.

Furthermore, with the collapse of the social organization and depopulation of the cities during the 5th century, the only schools that remained operative were those of the monasteries: and these schools were not interested in gaining further knowledge of nature, since they were based on the principle that the whole truth was written in the Bible and that the knowledge of natural phenomena was only useful to better understand the Holy Books.

The role of Christian religion

However, there are at least two reasons why the centuries following the collapse of the Western Roman Empire to the 12th century must be considered important for the development of the modern scientific thought and of modern astronomy in particular.

First, the diffusion into the whole West of the Christian religion, based on an unique God who is creator of the universe, cancelled the crowd of gods, semi-gods and spirits that the ancient world associated with natural phenomena, smoothing the way for their subsequent scientific explanation. In the 7th Century, Isidorus, bishop of Seville, demonstrated in his work *Ethimologiarum libri XX* that all natural phenomena and human events can be described in a way independent of any mythical explanation. Furthermore, in this widely diffused work, Isidorus introduced an important distinction, ignored by classical science, between astronomy and astrology: the first studied the motion of the skies, the Sun, the Moon and planets; the second, being in contradiction with the basic Christian principle of the 'free will', had to be condemned as a superstition.

Figure 1. Christ Pantocrator (mosaic, Monreale Cathedral, c. 1190).

Second, Christian liturgy demanded the solution to a number of problems connected with time reckoning. In those days, the Julian calendar was still used as a civil calendar and there was no problem about using it also for the 'fixed feasts' like Christmas, Epiphany and the Annunciation. However, the day of the most important Christian feast, Easter, and thus all the others that derive from the date of Easter (the beginning of Lent, the Ascension and Pentecost) are not fixed but determined on the basis of rules coming from the Hebrew lunar calendar. The first Council of Nicaea (325 AD) actually stated that Easter has to be celebrated on the Sunday following the first full Moon after the spring equinox. A number of non-trivial computations, implying a deep knowledge of the Sun and Moon's apparent motions, are thus needed to determine the date of Easter ('Epact computation') in the Julian calendar, which is a solar one. This problem initiated many studies, and a final solution came in the work by the British Benedictine monk Beda Venerabilis,

De temporum ratione, written on 725, a masterpiece of mathematical astronomy.

Figure 2. Portrait of Beda Venerabilis from a manuscript of the 12th Century (from www.hs-augsburg.de).

A further problem for the monks was the computation of the 'canonical hours', i.e., the proper time to perform liturgical functions. Liturgical time has its own sacrality, which requires that it be measured through the most reliable methods: the observation of the Sun in the daytime and of the stars in the night.

Monastic astronomy as observational astronomy

Monks thus became assiduous and careful observers of the sky. Of course, they were not experts in astronomy like their contemporary Eastern astronomers. It is indeed true that writing chronicles had a quite different aim in medieval Europe in comparison with the Eastern countries. But Western chronicles cannot be defined as 'rough', as is too often stated by scholars of historical astronomy (i.e. the subfield of cultural astronomy devoted to the recovery of astrophysical data from pre-telescopic observation).

When monks observed a relatively 'common phenomenon', such as an eclipse or a comet or a meteoric shower, they were precise in their descriptions. However, when they had to deal with an 'unusual phenomenon'–one outside of their expertise, such as a 'new star' (i.e. nova or supernova)–it was more difficult for them to describe it in their 'Latin' (as standard and conventional written language), so they did so with common words or tropes related to religious or popular beliefs that were able to suggest an explanation of the phenomenon. For the latter cases we need a much more critical effort to interpret their descriptions (see Ghignoli and Polcaro, 2007).

A significant role in what could be defined as 'monastic observational astronomy' was played by Irish monks, who transferred the traditional care for sky observation of the Celtic people into a Christian environment, and disseminated this knowledge throughout Western Europe where they established many new monasteries. St. Patrick himself is the author of astronomical books (Gaspani, 1999).

Especially with the so-called 'Carolingian Renaissance', the careful observation of the sky became a habit also in the 'Schools of the Episcopates', where laymen were admitted, too (see e.g. Lindberg, 1992). In these schools, the 'secular' part of the education included the 'Seven liberal arts', one of which was astronomy. Of course, their teaching was based on the Latin texts cited before, but it was integrated with experimental practice. Furthermore, the increased availability of economical resources allowed the 'Schools of the Episcopates'–as incubators of the universities that were established in the following centuries–to introduce the first astronomical instruments used in Europe since the end of the Roman Empire. These were just tubes, used to isolate the star in the viewing field and to measure its alt-azimuthal coordinates; nevertheless, their use is a proof of the increasing interest in the precision of observations.

At the same time, Arabic astronomical knowledge and related observational techniques started to reach Europe. In 1050, the Benedictine Hermann of Reichenau wrote a book on the *astrolabe*, thus helping to introduce in Europe a powerful and versatile instrument allowing the precise measurement of stars' positions and complex astronomical computations (Poulle, 1954).

The Islamic astronomy

It has been stated that Islam's interest in astronomy derived from religious needs: the prayers to be performed at fixed times and in the direction of Mecca (the *Qibla*), as well as the time of the beginning of Ramadan, determined by astronomical conditions (see, e.g., Bausani, 1977; Ruis, 2008). It has also been stated that the origin of Islamic astronomy is related to the gift, in 772 AD, of an astronomical book by some Hindi scholars to the Caliph of Baghdad al Mansur, who immediately ordered Ibn Ibrahim al Fazari to translate this book (named in Arabic *Zij al Sindhind*, i.e., Hindi astronomical tables) with the help of the same Hindi scholars. Other authors suggested that the interest of Arabic people in science and technology derived from the very nature of Islamic civilization itself: a tolerant, multi-racial and highly educated society which was ready to assimilate the best of the other peoples' knowledge.

All these statements are at least partially true, but they are not enough to explain the role of astronomy in Islam. Most probably, the reason why astronomy became the most important science in the Islamic civilization has to be related to its origins in the Arabic nomadic people, travelling through the sand desert of Arabia, where no geographical reference points are visible and the stars are the only reliable reference. Also the role of the Zoroastrian religion–spread in the entire area from Arabia to Central Asia in pre-Islamic times–and the Sabian religion has to be taken into account. These religions had a significant influence on the development of astronomy in Islamic world. For sure, Al-Sabi Thabit ibn Qurra al-Harrani (826-901), one of the first Islamic mathematicians, translator of Euclid's *Elements* and a renowned astronomer, was a Sabian.

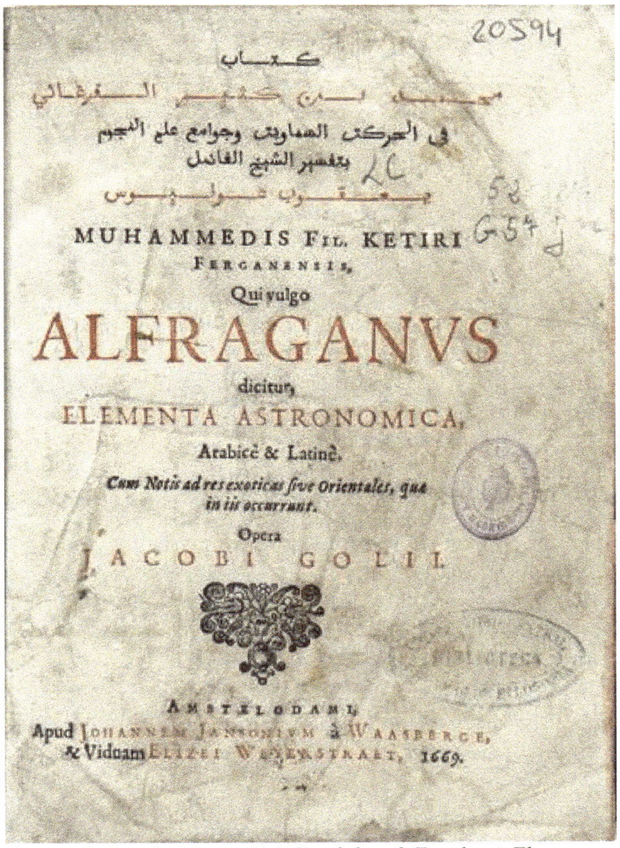

Figure 3. Dutch edition (1669) of the al-Farghani Elements of Astronomy, in Arabic with Latin translation.

On the other hand, the contact with these cultural heritages, as well as the preservation of the classical astronomy, also brought a stronger interconnection between astronomy and astrology than in the classical world. Actually, Islam does not accept any form of divination, including astrology, and in some cases holds the latter as 'magic' or even 'idolatry', both capital offences. However, Islamic scholars were socially highly considered and thus enjoyed a generous tolerance. Furthermore, the Abbasid caliphs were nearly obsessed with the recovery of the Greek and Roman cultures, considering them a powerful tool in the struggle (which was not only economic and military, but also cultural) with the Byzantine Empire for the control of Near East. This is why they were very inclined to make some exceptions to the doctrine in order to compete with the Byzantines on the cultural ground. It is a matter of fact that all Islamic astronomers were also (and in some cases mainly, as was Masha'allah, Baghdad, c. 800) astrologers. It would be too long here to name all the great Islamic astronomers, from Avicenna to Ulugh Beg, and to list all the paramount developments of astronomy in the Islamic world between the 7th and 14th centuries (see, e.g., Bausani, 1977; King, 2004-2005). Here we stress just a few of these contributions to the rediscovery, and then overcoming, of Ptolemaic astronomy. The major astronomical work of Ptolemy, e.g., is known everywhere as the 'Almagest' (from the Arabic *al-Magisti*, an adaptation of the Greek 'Megalè syntaxis', i.e., 'Great Treatise'), whose first Arabic translation was done by al Haggiag ibn Yussuf ibn Matar in the 9th century by order of the Abbasid caliph Harum al Rashid. The recovery of classical astronomy was immediately followed by the resumption of the experimental activities. For instance, the caliph al-Mamun and his team of astronomers led by the famous Ahmad ibn Muhammad ibn Kathir al-Farghani (Alfraganus) are responsible for the building of two astronomical observatories and two extraordinary astronomical exploits: the extremely precise measurement of the obliquity of the ecliptic as 23° 33' in 830 AD and that of the meridian arc length in 828 AD. Worthy of special mention is the work of Nasir al-Din al-Tusi (1201-1274) and his team (including even a Chinese astronomer!) at Maragha Observatory in Persia, where a careful process of analysis and revision of the Ptolemaic model was performed. This work, often properly named the 'Maragha Revolution', demonstrated the incompatibility of the hypothesis of non-uniform motion and the introduction of the 'equant' with the Aristotelian principles that were the basis of the Ptolemaic model. Though there is no documented proof on this point, it is very probable that Copernicus' work actually started from al-Tusi's, given the analogy of the demonstrations used. For a deeper discussion on the influence of Maragha astronomy on Western astronomy see e.g. Saliba (2009) and Dobrzycki and Kremer (1996), who also consider the influence on Western astronomy by the later Syrian astronomer Ibn ash-Shatir.

The 'Scientific revolution' of the 12th Century

Islamic influence on Western astronomy (and on science as a whole) became of paramount importance during the 12th century: this fact demonstrates that, despite the Crusades, between Western Europe and Islam there was not only a clash, but an encounter too, though complex and harassed (see, e.g., Le Goff, 2004) .

In this framework, the role of Emperor Frederick II (1194-1250) as a highly innovative ruler and intellectual is well known, for him being *trait d'union* between Christian and Islamic civilizations, as well as a promoter of the dawn of 'secular science' (see, e.g., Bressler, 2010). His masterpiece as a scholar is the book *De arte venandi cum avibus*–not only a 'hunt handbook' but a deep treatise on zoology, where various birds are described in detail from an experimental point of view; the author does not hesitate to criticize Aristotle himself, when his views are in contradiction with experimental evidence.

Frederick II did not leave us any text on astronomy, but his point of view on this field, fully inspired by Islamic Ptolemaic astronomy and astrology, is shown by the architectural design of his castles. For sure, Frederick II was the sponsor of Leonardo Fibonacci, the mathematician who started modern mathematics by introducing the

Islamic innovations in this science (see, e.g., Sigler, 2002), and Michael Scotus, the scholar who reintroduced astrology in European astronomy. Most probably the inclination of Frederick II towards astrology was due to his admiration for Islamic culture and his opposition to Papacy.

Figure 4. Frederic II in the Manfred edition of his De Arte Venandi cum Avibus (Book on Falconry, 1239). Rome, Biblioteca Apostolica Vaticana, Pal. lat. 1071, fol. 1 v°. Image from Wikipedia.

However, the reintroduction of astrology must not only be considered as a cultural involution (what is of course true) but also as one of the reasons for the subsequent impetuous development of astronomy. Actually, during the Middle Ages, European astronomers were monks and clerics, performing a technical service for their communities, and only a few people dedicated their life to such a task like astronomy. When astrology was reintroduced in Europe, astronomy became a lucrative and respected profession: the number of astronomers, and thus the number of astronomical studies and results, started to increase, also supported by the need for astronomical knowledge brought by the development of long range navigation in the new economical and social framework from the 13th century onwards.

References

Bausani, A. 1977. Appunti di astronomia e astrologia arabo-islamiche. Venezia, Cafoscarina.

Bressler, R. 2010. Frederick II: The Wonder of the World. Chicago, Westholme Publishing.

Dobrzycki, J., and Kremer, R. L. 1996. Peurbach and Maragha Astronomy? The Ephemerides of Johannes Angelus and their Implications. Journal for the History of Astronomy 27, 187-237.

Gaspani, A. 1999. L'astronomia dei primi monaci irlandesi. L'Astronomia 199, 30-39.

Ghignoli, A., and Polcaro, V. F. 2007. Eleventh Century Astronomical Events as Recorded in Contemporary European Sources. Mediterranean Archaeology and Archaeometry 6, 61-66.

King, D. A 2004-5. In Synchrony with the Heavens. Studies in Astronomical Timekeeping and Instrumentation in Islamic Civilization, 2 Vols. Leiden, Brill.

Le Goff, J. 2004. Il cielo sceso in terra. Bari, Laterza.

Lindberg, D. C. 1992. The Beginnings of Western Science. Chicago, University of Chicago Press.

Poulle, E. 1954. L'astrolabe médiéval d'après les manuscrits de la Bibliothèque nationale. Bibliothèque de l'école des chartes 112, no. 1, 81-103.

Ruis, M. 2008. Finding the Sacred Direction: Medieval Books on the Qibla. In J. A. Rubiño-Martín, J. A. Belmonte, F. Prada and A. Alberdi (eds), Cosmology Across Cultures, Astronomical Society of the Pacific Conference Series 409, 177-182.

Saliba, G. 2009. Tusi, Nasir-al-Din as Mathematician and Astronomer. In Encyclopædia Iranica, available at: http://www.iranica.com/articles/tusi-nasir-al-din (accessed 25 July 2011).

Sigler, L. 2002. Fibonacci's Liber Abaci. New York, Springer.

ASTRONOMY AND POLITICS IN THE MODERN AGE

ASTRONOMY AND POLITICIANS

MAGDA STAVINSCHI AND CĂTĂLIN MOSOIA

Abstract: There are many issues that can be of interest when looking at the relationship between astronomy and power. If we just consider the influence over the centuries on the progress of human civilization or the spatial techniques developed by few countries and the astronomical results obtained, there are many subjects to deal with on special symposia for each particular case. We are proposing a close perspective on astronomy and high political representatives.
Keywords: Romania, Oradea , IYA2009, IAU, Belgium

Introduction

Man on earth once noticed the sky and was fascinated by its beauty and hidden mysteries. He then tried to understand it and then use it for personal and community needs–starting from better knowing himself and the place where he was born, to knowing people around him, and also the planet, which offered suitable conditions for life: better knowing the people and the planet for the next generations to come.

In this way, astronomy became, along with the evolution of humanity, a social science, in the sense that this term is accepted in the contemporary world: namely in relation to the redistributive policies of governments which aim to apply resources in the public interest–social security, for example. The protection of society and especially the protection of the planet itself are in the hands of governments and politicians.

Each society has various levels of interest regarding space, ranging from theoretical to practical aspects, such as making calendars, determining favorable moments for having a successful fight or war, to know and conquer the cosmos by launching space missions for very pragmatic purposes, including military, or just for competition.

In this paper, we will try to give a brief description of the above statements.

Figure 1. Hrisant Notara.

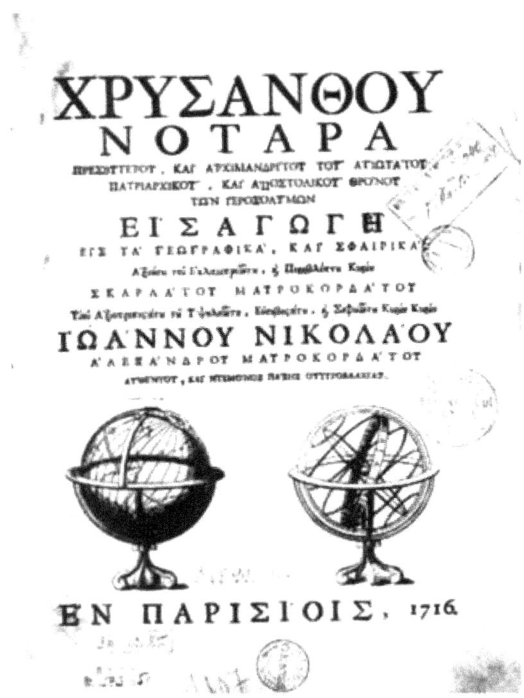

Figure 2. Introductio ad geographiam et sphaeram *(Paris, 1716).*

HRISANT NOTARA

History has in a way a repetitive pace. Like today, people in the past who wanted an elevated education for their children guided them to the most important universities of the time. Thus, Constantin Brancoveanu, prince of Wallachia, wanted his sons to be highly educated so they would be able to solve the political challenges of late 17th century, as well as being familiar with western European culture. He sent Hrisant Notara, one of his sons' teachers, to train at the best universities of the time, including the Paris Observatory, which was founded in 1667 by Giovanni-Domenico Cassini (1625-1712). Notara became a close collaborator of Cassini and published the book *Introductio ad geographiam et sphaeram* (Paris, 1716), which proved Notara's remarkable scientific knowledge. This was just one dimension of his personality as he also played an important political role in the eastern part of Europe. Furthermore, Notara reached the top of the church hierarchy as he became the patriarch of Jerusalem.

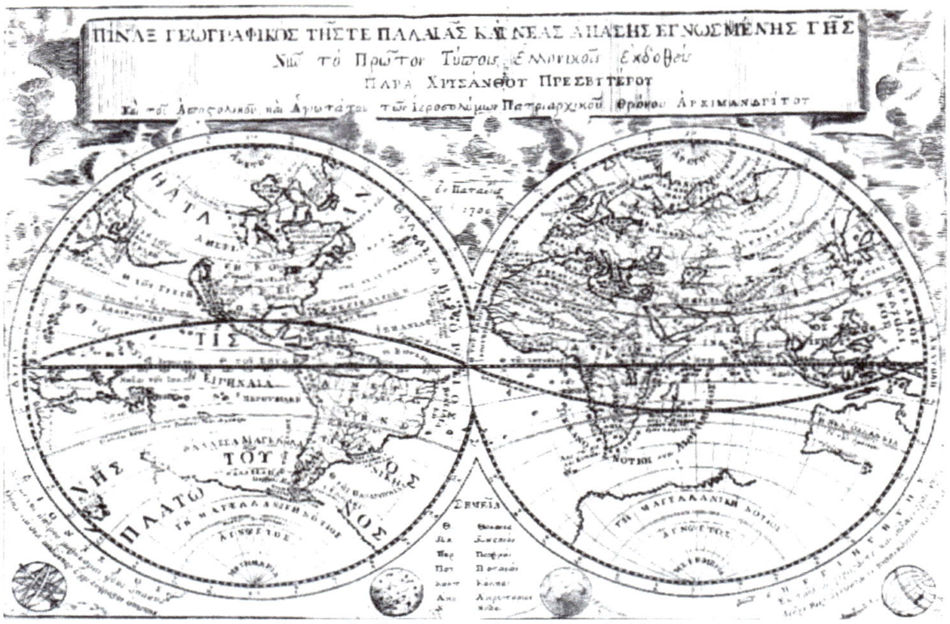

Figure 3. Constantin Cantacuzino's Map.

CONSTANTIN CANTACUZINO

It is also worth mentioning the first scientific astronomical observations performed by the erudite Constantin Cantacuzino (1650-1716), high chancellor and uncle of Brancoveanu, by means of his own telescope.

Figure 4. Ioan Vitez.

IOAN VITEZ

History tells us that there were many church representatives who had an interest in astronomy (e.g., Jan Vitéz). The word 'vitez' literally means knight in Bosnian, Croatian, Slovenian, Hungarian and Serbian. Vitéz was teacher of Prince John Hunyadi's (1406-1456) children. The Observatory in Oradea, an important centre of Renaissance culture in that period, was created by Ioan Vitéz in 1445, more than a century before the one set up at Uraniborg by Tycho Brahe (c. 1576-1580). So that the goal of the Observatory should be fulfilled at the highest level, he invited the Viennese physicist Georg Peuerbach to Oradea.

Figure 5. Oradea's Observatory (1445).

The latter also brought there his best student, Johannes Muller, also called Regiomontanus. Regiomontanus' observations, as well as the ones made by a pupil of Ioan Vitéz, became the basis of the famous *Tabele oradene (Tabula Varadiensis)*, used by Columbus to discover Cuba and later by Tycho Brahe and Kepler.

ALBA IULIA'S OBSERVATORY

The second observatory was set up in 1795 at Alba Iulia by Bishop Ignatius Batthyany, who founded the famous library that was to be named after him. The observatory ended its activity in 1860.

BUCHAREST'S OBSERVATORY

An example of a special relationship between astronomy and politics can be found at the Astronomical Observatory of Bucharest. The current library of the institute housed the headquarters of the first Romanian government in 1859 as a consequence of the Union of the Principalities. Constantin Bozianu, ex-counselor of the first Romanian Prince Alexandru Ioan Cuza and for a short time Prime Minister, had a beautiful house in the hills south of the capital where important personalities of the time met; it was also the meeting place of students at the Law School where he was dean. Tradition says that in this house the Union was prepared in secret. After his death the house remained the property of the Romanian state, and soon after became part of the Bucharest Observatory heritage.

Figure 6. Constantin Bosianu.

SPIRU HARET

Later, Spiru Haret (1851-1912), Minister of Cults and Public Instruction, was the first Romanian to defend a PhD dissertation abroad after the Unification of 1859. He dedicated his doctoral thesis (1878) to the study of the invariability of the major axes of planetary orbits. He also signed the founding act of the Bucharest Astronomical Observatory on 1 April 1908.

Figure 7. The Bosianu building at Bucharest Observatory. Credit: Magda Stavinschi.

We have given just a few examples from Romanian culture but there are more in the western world and beyond.

Figure 8. Spiru Haret.

THOMAS JEFFERSON

Thomas Jefferson fell in love with the science of stars and planets when he, himself, observed the annular solar eclipse of 17 September 1811. It is true; by that time, he was no longer the US President, having served from 1801-1809. His biographer James Parton said he could 'calculate an eclipse, survey an estate, tie an artery, plan an edifice, try a cause, break a horse, dance a minuet, and play the violin'.

ALBERT EINSTEIN

One more example: in 1905 Albert Einstein elaborated three theories, each worthy of a Nobel Prize: one that explained the Brownian motion, another the photoelectric effect, and the last one, the theory of relativity. The third one became the most celebrated. Completed between 1913 and 1916, it needed confirmation. A chance for it was immediate: the total solar eclipse of 21 August 1914. Who had to perform the observations? A German team on the Russian territory! The result? The expected one: not even a photo. But the Englishman Arthur Eddington, who had knowledge of Einstein's epochal results, initiated on his own account an expedition intended to observe another eclipse, that of 29 May 1919. A British test to corroborate a German theory, here is a perfect proof of scientific internationalism. This proof was consequently used to hinder any boycott against German scientists after the war.

IAU

An indubitable example of the relationship between astronomy and politics is the foundation of the International Astronomical Union in 1919. It is the first international scientific community which understands that, through discussions at the same table, people can surpass political borders; and history teaches us these were hard and tragic times, as World War I had just ended in 1918.

BELGIUM

Closer to today, we would give the example of King Baudouin of Belgium, who did astronomy research at high level. We would add more names, maybe not so famous, but all of them prove that astronomy did influence great politicians or Church representatives from all over the world.

IYA2009

During the International Year of Astronomy 2009, some politicians understood and agreed to lead science committees of this unique world event. The final report shows that at least 815 million people in 148 countries participated in the world's largest science event in decades. We have to quote from this document:

> Political interest in IYA2009 was also high, which in itself is an achievement for any popularisation initiative. In the United States of America, the House of Representatives passed a resolution supporting IYA2009. The Spanish Congress of Deputies also passed a law supporting astronomy in the framework of IYA2009. Heads of State were keen to express their support for the Year. The President of the Portuguese Republic, Prof. Dr. Aníbal Cavaco Silva, personally presided over the Portuguese IYA2009 Honour Committee. The President of the Republic of Slovenia, Dr. Danilo Turk, became the patron of IYA2009 in Slovenia. Lech Kaczynski presided over the Polish IYA2009 Honour Committee, while Prince Felipe of Spain (Prince of Asturias) did the same for the Spanish IYA2009 Honour Committee. The former Belgium Prime Minister and current President of the European Council, Herman Van Rompuy, voiced support of astronomy during an IYA2009 event in Belgium held in April 2009. The former European Commissioner for Science and Research and present European Commissioner for Environment, Janez Potočnik, expressed his support of astronomy during the European opening of IYA2009 in Prague, Czech Republic. In the US, the event celebrating IYA2009 at the White House with President Obama and the First Family on 5 October 2009 made headlines. In Nepal the total solar eclipse observation event on 22 July 2009 was attended by the Prime Minister of Nepal, Madhav Kumar together with thousands of members of the public. Iran's President Mahmoud Ahmadinejad pointed that IYA2009 provided a chance for young scientists to develop a more vivid vision of man's future during his inaugural speech of the 3rd International Astronomy and Astrophysics Olympiad in Tehran, Iran. Pope Benedict XVI gave an eloquent speech in which he said "The International Year of Astronomy is meant not least to recapture for people throughout our world the extraordinary wonder and amazement which characterised the great age of discovery in the sixteenth century.

The list of examples could continue but we want also to note the build up of astronomical societies as a consequence of the political situation. This is the case of the South-East Branch of the European Astronomical Society. Countries from this part of Europe faced a special history during the five decades of communism. Isolation from the western world and common destitutions led them to build a special society; although it was part of the big European society and the IAU, it was more suitable for better solving their common problems, such as the brain drain of young researchers or recovering the lost years.

Finally, we cannot end this very short review of the astronomy-politics relationship without mentioning the very up to date subject of taking care of our planet. Anyone might observe that there is an increasing attention from astronomers as well as the whole society to this subject. More precisely, there is a global concern for preserving the planet and decreasing light pollution that can lead us with today's fast pace to atmospherically conditions similar to Venus, which is bad.

COMETS IN POLITICAL CARICATURES: EXAMPLES FROM THE 18TH TO 21ST CENTURY

BARBARA RAPPENGLÜCK

Abstract: Comets have been interpreted as portents of political fortune since ancient times. The use of comets in the iconography of political caricatures from the 18th to the 21st century substantiates that comets have always been and still are a suitable metaphor for expressing political statements. But in the caricatures the comet as an icon does not serve the interests of powerful people any more, but is used by a critical public to ridicule and challenge their power. The use and the design of the comet as an icon thereby reflects the knowledge and theories of comets, which were relevant at the time in question: the character of comets as celestial objects, comets being solid objects, the question of habitable comets, the calculability of cometary paths etc. This aspect comes along with a tendency to trivialize comets. On the other hand, the rationalisation of comets has contributed to keep old fears alive in a modified form. A subliminal fascination of comets as scary objects has always lived on, and the caricatures show that allusions to the traditional interpretation of comets as portents of a divine tribunal can still be activated to intensify political messages.
Keywords: caricature, cartoon, comets, comet research, comet fear, doomsday

Introduction

Comets have been interpreted as portents of political fortune since ancient times. According to the course, size, shape, colour and other attributes of a comet, luck and calamity of rulers and of peoples have been predicted (Schechner 1997, 51): the impending death of a ruler, the outcome of a war, the fortune of a political important wedding, social upheaval and natural catastrophes. Such interpretations were closely related to traditional knowledge of comets. Until the modern era the most influential theory was the Aristotelian one. In his cosmological concept the Earth was encircled by different spheres, the sublunary sphere being closest to the Earth. Comets were interpreted as exhalations of earthly origin which in the sublunary sphere are ignited by the motion of the celestial sphere (Schechner 1997, 92). On this basis relations could be assumed between comets and events on Earth. Comets were, for example, understood as symptomatic indicators of meteorological conditions which could also result in storms and earthquakes. This means that comets did not cause these phenomena but signalled a predisposition for their occurrence and hence served as foretelling omens. In the interpretations of antiquity up to the early modern era, this signalling character of comets was not limited to natural disaster, but included social and political calamities as well as the divine tribunal of the Last Judgement. At the beginning of the modern era these interpretations of comets were even fuelled again by Martin Luther and Philip Melanchthon, 'who claimed that the comet was not merely a portent natural phenomenon, but that it was created by God to instill horror, and it was a sign of the last days'. (Heidarzadeh 2008, 43)

On the other hand, since the early modern epoch essential new knowledge has been gained concerning the characteristics of comets (for this paragraph see Briese 1998, 182-187). At the end of the 16th century Tycho Brahe (1546-1601) realised that comets are not of earthly origin (i.e., earthly exhalations) but extraterrestrial objects, far beyond the Moon. Isaac Newton (1643-1727) noticed that comets are solid bodies and that they may have a regular orbit. Edmond Halley (1656-1742) succeeded in computing cometary orbits and their periodicity, successfully predicting a comet's return in 1705. Such insights provided the basis for new theories and interpretations. Comets were no longer only seen as foretelling signs, but as directly threatening objects. The awareness that comets are solid bodies and the calculability of cometary orbits gave a reasonable basis to the fear that a comet might hit the Earth, or that life on Earth might be harmed by the passage of the Earth through a comet's tail. The theologian and physicist William Whiston (1667-1752) claimed in his very influential book, *A New Theory of the Earth* (1696), that the Biblical deluge had been caused by the Earth's passage through the tail of a comet, and the newly gained knowledge of cometary orbits fuelled the fear that the 'Flood Comet' might return and again cause a disaster. Other tendencies tried to downplay and ridicule such threats, and yet others raised arguments against the probability of such threats becoming true. These attitudes changed several times during modern times, depending on political and economic conditions, ruling philosophical and theological trends, and the actual approach of remarkable comets. In this mixture of fears, minimization and rational approaches, comets became a common metaphor in literature and anew served to express and illustrate political ideas, e.g., in political caricatures.

A caricature depicts a person or a circumstance by exaggerating its typical characteristics. The purposes of a caricature are mockery, unmasking and criticism. Political caricatures which satirize politicians have been known since the time of classical antiquity. But only since the early modern era have caricatures satirizing social circumstances appeared; and it is only since the 18th century that political events in general have become a subject of caricatures.

This article will show some examples that reveal how comets were and are used in the metaphoric iconography of political caricatures. In each case it will give a short insight into the political background of the caricature and will demonstrate how the use of the comet metaphor reflects contemporary astronomical knowledge and theories as well as continuing traditional interpretations.

'The Gallic Magi led by the Imperial Comet'

Figure 1. 'The Gallic Magi led by the Imperial Comet'; © The Trustees of the British Museum.

This caricature shows Emperor Napoleon Bonaparte with his wife Marie Louise seated on a throne and encased by drapery with imperial attributes. A nursemaid holds the baby Napoleon François Joseph. Napoleon as well as his baby son are looking through telescopes at a comet. The nucleus of the comet is formed by a naked child. By its attributes–the plumed bicorn hat which is worn by Napoleon as well as his son, and the cock, a traditional symbol of France, on the posterior of the baby–the child is depicted to be the image of Napoleon Francois Joseph. A massive exhalation from the baby's posterior forms the tail of the comet. Four elderly noblemen, who unroll star maps which show the comet, are subserviently prostrating in front of the royal family. The caricature has the title: 'The Gallic Magi led by the Imperial Comet'.

The caricature was created by the British printmaker William Elmes and published as a broadside on 24 September 1811 (British Museum, a). In the year 1811 Napoleon Bonaparte had extended his influence to many countries of Europe. The Continental System that had been established to defeat Britain fuelled anti-Napoleonic feelings. In this situation anti-Napoleonic caricatures flourished in Britain. This specific caricature relates to the birth of Napoleon's most eagerly awaited heir, Napoleon François Joseph, the later Napoleon II who also held the title 'King of Rome'. The birth of this son gave Napoleon the chance to establish an imperial dynasty. This is the political background of the caricature, but there also exists an astronomical one. The Great Comet of 1811 could be seen by the naked eye for nearly nine months (Kronk 2003, 19). Its appearance was very spectacular. Napoleon, who 'was assuredly aware of the tradition wherein comets were associated with kings and great rulers' and 'adopted various comets as his protecting genii', 'enthusiastically' greeted the Great Comet of 1811 'as his guiding star and the controller of his destiny'. (Olson et al. 1999, 144) Caricaturists often referred to his obsession with comets.

This caricature is in some aspects linked to the traditional iconography of comets. First there is the identification of a comet and a ruler. But the identification exceeds the traditional pattern: the connection between the comet and the ruler is not only a spiritual one, with the comet betokening the rulers fate, but here the comet has an identity with the person, in this case the baby which forms the nucleus of the comet. This depiction corresponds with the widespread tendency in the literature of the 18th and 19th century to personalise comets. Research into the use of such personalisation of comets has shown that it had the more or less unconsciousness function of downplaying the threat of comets (Briese 1998, 189f.). At the same time the caricature plays with the traditional relation of a comet and a ruler and satirises it by two facts: 1), it is a baby, not a mighty emperor who forms the nucleus of the comet; and 2), the tail of the comet is nothing else than an exhalation of the baby. Johannes Kepler (1571-1630), among others, promoted the theory that the tail of a comet might have a poisonous influence on the air (Heidarzadeh 2008, 66), an idea that had its followers even at the start of the 20th century (see Figure 3 below, 'Roerens Bitte an den Kometen Halley'). The allusion to another old tradition is used to illustrate Napoleon's arrogance: the arrangement of the imperial family, the adoring noblemen, the comet and the title 'The Gallic Magi led by the Imperial Comet' is reminiscent of the scene of Bethlehem, with the Holy Family, the Three Kings and the Star of Bethlehem. With this exaggeration Napoleon's son is satirically identified with Christ as the redeemer of the world. At the same time the caricature creates a drastic discrepancy between the comet as the 'personification' of the 'redeemer' and the allusion to the baneful effects of its tail.

Finally this caricature reflects the eagerness to observe comets, which rapidly spread in the late 18th century. Scientists and amateurs founded astronomical societies for sky-watching in general and searching for comets in particular. The diaries of Caroline Herschel (1750-1848) testify the enormous interest of high-ranking persons from all over Europe in watching the sky. Hence, this caricature showing the Emperor Napoleon, who watches a comet through a telescope, refers on one hand to Napoleon's personal enthusiasm for comets but on the other hand also to a common activity at this time. That way the Emperor is placed on the same level with ordinary people. Maybe the caricature in this way alludes to the discrepancy between Napoleon's imperial aspirations and his origin from lower nobility.

'The Terrifying Comet'

The personalisation of comets, mentioned in context of the previous caricature, can further be demonstrated by a look at another British satirical print, 'The Terrifying Comet' by William Dent, from 1792. It shows a comet whose nucleus is formed by a somewhat grotesque head which is labelled as the British king 'George III'. Another, smaller head spits out the comet's tail, which is formed by soldiers and military camps. Concerning the political background (British Museum, b), here it is sufficient to know that the caricature satirizes a political debate concerning the introduction of a new system of military manoeuvres. The illustration also addresses the serious attacks of madness which George III suffered from severely during his reign.

Figure 2. 'The Terrifying Comet'; © The Trustees of the British Museum.

This caricature revisits the traditional relation between a comet and a ruler by identifying the nucleus of the comet with the king's head. The madness of the king is stressed by the grotesque appearance of the head. Again, the comet is personalised, as well as the stars which surround it, which are formed from the heads of important politicians of this era. An additional aspect is given by the comet's tail made up of soldiers and their camps. Living beings as dwellers of comets required that the comets be solid bodies, as had been recognized by Tycho Brahe (Briese 1998, 236). This aspect is closely connected to the theory of multiple inhabited worlds, which, based on ancient and medieval theories, was actualised and popularised by Bernard le Bovier de Fontenelle (1657-1757) in his work *Entretiens sur la pluralité des mondes* (1686). In the 18th century it was a popular theory that celestial bodies in general, and comets as well, are populated (Briese 1998, 236-253). The depiction of the tail of the comet with the soldiers and their camps is inspired by such ideas. In a rather strange reasoning the conviction that comets are populated was understood as some kind of guarantee that they would never hit and destroy the Earth. Even famous astronomers like Johann Elert Bode (1747-1826) spread this notion (Briese 1998, 242f.). In the personalisation of comets, the idea of inhabited comets again had the–probably mostly unconsciousness–function to calm the fears of comets (Briese 1998, 243).

'Roerens Bitte an den Kometen Halley' ('Roerens appeal to the comet Halley')

This caricature by Gustav Brandt (Brandt, 1910), published in the German satirical magazine *Kladderadatsch* in April 1910, consists of two images and a poem. The first image shows rocking buildings, the white tail of a comet and the explosion cloud of the comet's impact. The buildings can easily be identified as three prominent landmarks in Berlin: from left to right the Siegessäule (victory column), the Brandenburger Tor falling into ruins, and the cathedral of Berlin. The poem identifies Berlin with the Biblical town of Sodom, which had, due to the sins of its citizens, been wiped out by a divine tribunal. The comet Halley is addressed like a redeemer and is asked to destroy the sinful Berlin by knocking it to pieces and by covering it with poison, sulphur and fire, while the rest of the country shall be unharmed. The poem formulates the vision of a young shepherd finding the fragment of the statue of a naked woman and expressing his concern about the immorality of people who had been in favour of such statues. The second image illustrates this scene. The whole arrangement has the title 'Roerens Bitte an den Halleyschen Kometen' ('Roeren's appeal to the comet Halley').

Figure 3. 'Roerens Bitte an den Kometen Halley' ('Roerens appeal to the comet Halley'); © Ruprecht-Karls-Universität Heidelberg.

The caricature refers to a scandal that rocked Germany's capital Berlin in 1908-9. From 1908 Germany's first nude dancer, Olga Desmond, was very successfully performing in Berlin. Furthermore, she entered the vaudeville stage as the living (and nude) imitation of famous statues. The conservative public, media and politicians were upset about her 'immoral' performances. In 1909 the conservative parliamentarian Hermann Roeren, an outspoken advocate of moral behaviour, even made the scandal a subject of parliamentarian debates and accused the dancer of being responsible for the decay of morals (Runge 2009, 51-60; 174f.). Desmond was forbidden to perform without a costume.

The caricature links to the old tradition of interpreting comets as portents of a divine tribunal. But instead, here the comet is a tool of the punishing God, and the comet itself is addressed as a personality with its own will and morals. Here again we see an example of the personalisation of comets. At the same time the 'divine tribunal' has lost much of its former apocalyptic character, because the petitioner imagines it to be limited to Berlin and harming nothing else. Likewise the caricature plays with the fear of a catastrophic impact of a comet on the Earth, something that had become a concrete threat since the orbits of comets could be calculated with a good degree of exactness.

The text of the poem wishes that poison from the comet will cover Berlin. This notice very probably refers to the fear that the tail of the comet might harm people by poisonous gas. In 1864 Giovanni Battista Donati had 'made the first visual spectroscopic observation of a comet' (Heidarzadeh 2008, 231). In 1908 spectroscopy enabled astronomers to get an initial idea of the composition of the tail of a comet. They discovered the very toxic gas cyanogen in the tail of comet Morehouse (Ridpath 1985, 35). When Halley returned in 1910, cyanogen was detected in its tail, too (*New York Times*, 1910). Since calculations ascertained that the Earth would pass through the tail of comet Halley, people were afraid of the poisonous gas, although scientists tried to calm fears by stressing its very minimal concentration. The poem of the caricature hence plays with a traditional fear which had been revitalized by very actual scientific results.

'Kennedy's Comet'

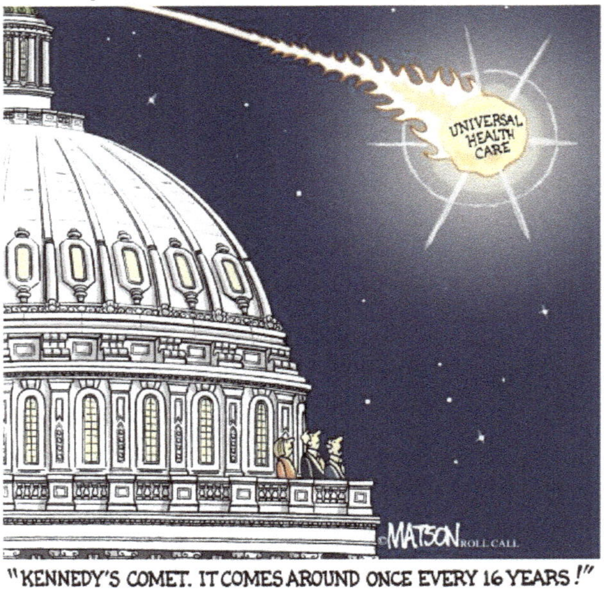

Figure 4. 'Kennedy's Comet'; © Robert Matson.

This US-American caricature shows the cupola of the United States Capitol. A few people are standing on the gallery, gazing at a comet which has the inscription 'Universal Health Care'. The caption below reproduces the exclamation of one of the observers: 'Kennedy's comet. It comes around once every 16 years!'

On 22 July 2009, this caricature was published in the newspaper *Roll Call*. On the same day President Barack Obama gave a speech to promote the project of health care for almost every citizen of the United States. The project had been very controversially discussed among the public as well as in Congress. Sceptical observers remembered the repeated vain efforts to establish a universal system of health care in the United States (Matson, 2010). In the last four decades health care had, in more or less regular intervals, been a major topic in politics, Senator Edward Kennedy being one of its most vehement advocates.

The caption 'Kennedy's comet. It comes around once every 16 years!' refers to these repeated efforts and compares them with the regular appearance of a periodic comet. In the caricature the comet no longer has the character of anything punishing, frightening or surprising. It is observed as a well-known occurrence. It is a remote, celestial event and as always it will pass by without any effect. The manner the comet in this caricature is used reflects an attitude to comets that is widespread nowadays in science: the comet has become rationalised, it is calculable and successfully examinable as never before. Comets are visited by probes like 'Stardust', 'Rosetta' or 'Deep Impact', which take close-up photographs, gather samples, and carry out experiments.

But apart from this rationalized perception of comets, comets are still associated with fascination and a shudder, too. This is illustrated by the last caricature:

'The deficit'

Figure 5. 'The deficit'; © Geoff Hook.

This Australian caricature was published in April 1986. It shows a comet with the inscription 'The deficit', blazing worryingly low above the heads of two men. One of them comments on the appearance of the comet, uttering: 'I thought it was supposed to be getting smaller!' In April 1986 the Australian Prime Minister and the Treasurer were facing a growing deficit in the national economy (The Bob Hawke Prime Ministerial Library). At the same time comet Halley had its brightest appearance, visible from the southern hemisphere. Once more we see an allusion to the rational perception of comets: the caption refers to the 'calculability of comets'. On the other hand the cartoon resumes the old idea of comets as threatening objects and as portents of bad political circumstances.

Conclusions

The use of comets in political caricatures from the 18th to the 21st centuries shows multiple aspects of ambivalent character: the caricatures reflect the knowledge and theories of comets, which were relevant at the time in question: the character of comets as celestial objects, comets being solid objects, the habitability of comets, the calculability of cometary paths etc. At the same time these caricatures illustrate that a comet had been and is still a suitable metaphor for expressing a close relation between the celestial object and the mighty ones, be it an emperor, a capital, or a senator. But in the case of the caricatures the association of a comet and a mighty person or institution is no longer used by the powerful ones themselves to illustrate their power, but is used by a critical public to ridicule and challenge their power. This aspect comes along with a tendency to trivialize comets, as becomes evident, e.g., in the personifications of comets. On the other hand, the rationalisation of comets has contributed to keep old fears alive in a modified form. A subliminal fascination of comets as scary objects has always lived on and allusions to the traditional interpretation of comets as portents of a divine tribunal can still be activated to intensify political messages.

Acknowledgements:

I thank Robert Matson (www.rjmatson.com/index_js.htm) for permitting the reproduction of his cartoon "Kennedy's Comet" and for so kindly providing me with information concerning its background. I am solely responsible for any misinterpretation. I thank Geoff Hook (www.geoffhook.com) for permitting the reproduction of his cartoon "The deficit".

References

The Bob Hawke Prime Ministerial Library. Available at: http://www.library.unisa.edu.au/bhpml/anniversary/1986.asp (Accessed 6 November 2012).

Brandt, G. 1910. Roerens Bitte an den Kometen Halley. Kladderadatsch 63, no. 14 (3.4.1910), 8. Beibl.

Briese, O. 1998. Die Macht der Metaphern. Blitz, Erdbeben und Kometen im Gefüge der Aufklärung. Stuttgart, Metzler.

British Museum, a. Available at: http://www.britishmuseum.org/research/search_the_collection_database/search_object_details.aspx?objectid=1479206&partid=1&searchText=comet+george&fromADBC=ad&toADBC=ad&numpages=10&images=on&orig=%2Fresearch%2Fsearch_the_collection_database.aspx&currentPage=1. (Accessed 6 November 2012).

British Museum, b. Available at: http://www.britishmuseum.org/research/search_the_collection_database/search_object_details.aspx?objectid=1634889&partid=1&output=People%2F!!%2FOR%2F!!%2F4336%2F!%2F4336-1-9%2F!%2FAssociated+with+Right+Hon+Michael+Angelo+Taylor%2F!%2F%2F!!%2F%2F!!!%2F&orig=%2Fresearch%2Fsearch_the_collection_database%2Fadvanced_search.aspx&currentPage=5&numpages=10 (Accessed 6 November 2012).

Heidarzadeh, T. 2008. A history of physical theories of comets, From Aristotle to Whipple. New York, Springer.

Kronk, G. W. 2003. Cometography. A catalog of comets. Vol. 2, 1800-1899. Cambridge, Cambridge University Press.

Matson, R. 2010. Personal communication concerning the political background of the caricature 'Kennedy's Comet'.

New York Times. 1910. Available at: <http://query.nytimes.com/gst/abstract.html?res=9407E4DF1430E233A2575BC0A9649C946196D6CF>. (Accessed 6 November 2012).

Olson, R. J. M., Pasachoff, J. M., and Pillinger, C. 1999. Fire in the Sky: Comets and Meteors, the Decisive Centuries, in British Art and Science. New York, Cambridge University Press.

Ridpath, I. 1985. A comet called Halley. Cambridge, Cambridge University Press.

Runge, J. 2009. Olga Desmond: Preußens nackte Venus. Friedland/Meckl., Steffen Verlag.

Schechner, S. 1997. Comets, popular culture, and the birth of modern cosmology. Princeton, Princeton University Press.

www.ingramcontent.com/pod-product-compliance
Ingram Content Group UK Ltd.
Pitfield, Milton Keynes, MK11 3LW, UK
UKHW061845041026
15591UKWH00015B/924

9 781407 314419